MATHEMATICS
TEACHER RESOURCE HANDBOOK

A Practical Guide for
K-12 Mathematics Curriculum

CORWIN PRESS, INC.
A Sage Publications Company
Thousand Oaks, California

Consulting Editor:

Dan Dolan
Deputy Director of the Project to Increase
Mastery of Mathematics and Science
Wesleyan University

Editorial and Production Staff:

L. Meredith Phillips, *Managing Editor*
Joy E. Runyon, *Senior Editor*
Liza Pleva, *Production Editor*
Lois Hilchey, *Editorial Assistant*

Cover design:
Sonja Originals; adapted by Pat Tanner

Cover illustrations:
Rick Powell

First Printing 1993
Printed in the United States of America

Library of Congress Cataloging-in-Publication Data
Mathematics teacher resource handbook; a practical guide
    for K–12 mathematics curriculum
            p.   cm.
        1.Mathematics—Study and teaching (Elementary)—
United States—Handbooks, manuals, etc. 2. Mathematics—
Study and teaching (Secondary)—United States—Handbooks,
manuals, etc. 3. Curriculum development—United States—
Handbooks, manuals, etc.
QA13.M349   1993   93-16939
510'.71'273—dc20

CONTENTS

PART III: TEXTBOOKS, CLASSROOM MATERIALS, AND OTHER RESOURCES

PUBLISHER'S FOREWORD

T HE *Mathematics Curriculum Resource Handbook* is one of a new series of practical references for curriculum developers, education faculty, veteran teachers, and student teachers. The handbook is designed to provide basic information on the background of mathematics curriculum, as well as current information on publications, standards, and special materials for K-12 mathematics. Think of this handbook as the first place to look when you are revising or developing your mathematics curriculum—or if you need basic resource information on mathematics any time of the year.

This handbook does not seek to prescribe any particular form of curriculum, nor does it follow any set of standards or guidelines. Instead, the book provides a general grounding in the mathematics curriculum, so that you can use this information and then proceed in the direction best suited for your budget, your school, and your district. What this handbook gives you is a sense of the numerous *options* that are available—it is up to you to use the information to develop the appropriate curriculum or program for your situation.

How To Use This Handbook

There are various ways to use this resource handbook. If you are revising or creating a mathematicss curriculum, you should read the Introduction (for an overall sense of the different philosophies of curriculum and how this will affect the program you develop), chapter 1 (for basic background on the trends and research in K-12 mathematics), and chapter 2 (for a how-to guide to developing curriculum materials). With this background, you can go through the other

chapters for the specific information you need— ranging from topics to be covered at various grade levels (chapter 4) to state requirements (chapter 5) to publishers and producers (chapter 12).

If you know what type of information you need, then check the Table of Contents for the most appropriate chapter, or check the Index to see where this material is covered. For instance:

1. If you are looking for ideas on developing special projects, turn to chapter 10.
2. If you are looking for a new textbook or supplementary materials (book, video, or software), turn to chapter 12.
3. If you need to contact state departments of education for mathematics curriculum documents, check the list provided in chapter 6.

What's in the Handbook

The *Introduction* provides an overview of the ideologies and philosophies that have affected American curriculum through the years. This section will acquaint you with the various ideologies, so that you can determine whether your school is following one such philosophy (or a combination), and how this might influence the development of your curriculum. The Introduction is generic by design, since these ideologies pertain to all subject areas.

Chapter 1 provides an overview of *Trends and Issues in Mathematics Curriculum*. This chapter discusses the development of present-day curriculum and looks at the directions the curriculum is taking. The major research works are cited so that you can get more detailed information on particular topics.

Chapter 2 is a step-by-step description of

Curriculum Process and Design. It is meant to be a practical guide to creating or revising mathematics curriculum guides. This chapter is also somewhat generic, but includes examples specific to mathematics.

Chapter 3, *Funding Curriculum Projects,* lists funding for programs that are studying or developing curriculum. Along with addresses and phone numbers, the names of contact persons are provided (wherever possible) to expedite your gathering of information.

Chapter 4 outlines *Topics in the Mathematics Curriculum, Grades K-12.* This is not meant to be a pattern to follow, but instead is a reflection of what most schools cover and what current research recommends.

Chapter 5, *State-Level Curriculum Guidelines: An Analysis,* describes the statewide frameworks and discusses the various emphases, philosophies, and coverage among the state materials.

Chapter 6, *State-Level Curriculum Guidelines: A Listing,* supplements the previous chapter by listing addresses of state departments of education and their publication titles.

Chapter 7, *Development of a Mathematics Assessment Program,* discusses the considerations and methods involved in developing a program to assess the effectiveness of your mathematics curriculum.

Chapter 8 is a selection of *Curriculum Guides* for mathematics.

Chapter 9 reprints a mathematics curriculum guide to use as an example in creating your own curriculum materials.

Chapter 10 discusses *Ideas for Special Projects in Mathematics.* The chapter defines a special mathematics project, the goals, and the methods used to create it. Also included is information on national mathematics contests.

Chapter 11 gives information on *Children's Trade Books* that can be used as supplementary texts in mathematics classrooms. This chapter discusses the bibliographic tools to use in finding these trade books; it also cites the various published lists of children's books for mathematics.

Chapter 12 is an annotated list of *Curriculum Material Producers* of textbooks, videos, software, and other materials for use in K-12 mathematics.

Chapter 13, *Statewide Text Adoption,* lists the mathematics textbooks adopted by each state.

Chapter 14 is an *Index to Reviews* of mathematics textbooks and supplementary materials. Since these items are reviewed in a wide variety of publications, we have assembled the citations of appropriate reviews in index form (cited by title, author, publisher/distributor, subject, and grade level).

Chapter 15 provides a list of *Kraus Curriculum Development Library* (KCDL) subscribers; KCDL is a good source for models of curriculum guides in all K-12 subject areas.

Acknowledgements

The content of this handbook is based on numerous meetings and discussions with educators and curriculum specialists across the country. Our thanks go to the curriculum supervisors in schools across the United States; the faculty at education departments in the colleges and universities we visited; and curriculum librarians. Special thanks go to the members of the Curriculum Materials Committee (CMC) and the Problems of Access and Control of Education Materials (PACEM) committee of the Association of College and Research Libraries' Education and Behavioral Sciences Section (ACRL/EBSS). Our meetings with the committees during American Library Association Conferences continue to provide Kraus with valuable ideas for the handbooks and for future curriculum projects.

We also acknowledge with thanks the assistance of Mary Kirsch, whose mathematical expertise contributed greatly to the accuracy of this publication.

Your Feedback

We have a final request of our readers. At the back of this handbook is a user survey that asks your opinions about the book, its coverage, and its contents. Once you have used this book, please fill out the questionnaire—it should only take a minute or so—and mail it back to us. If the form has already been removed, please just send us a letter with your opinions. We want to keep improving this new series of handbooks, and we can do this only with your help! Please send questionnaires or other responses to:

Kraus International Publications
358 Saw Mill River Road
Millwood, NY 10546-1035
(914) 762-2200 / (800) 223-8323
Fax: (914) 762-1195

SERIES INTRODUCTION

P. Bruce Uhrmacher

Assistant Professor of Education
School of Education, University of Denver, Denver, Colorado

HEN I travel by airplane and desire conversation, I inform the person sitting next to me that I'm in education. Everyone has an opinion about education. I hear stories about teachers (both good and bad), subject matter ("The problem with the new math is . . ."), and tests ("I should have gotten an A on that exam in seventh grade"). Many people want to tell me about the problems with education today ("Schools aren't what they used to be"). Few people are apathetic about schooling. When I do not wish to be disturbed in flight, however, I avoid admitting I'm in education. "So, what do you do?" someone trying to draw me out asks. I reply matter-of-factly, "I'm a curriculum theorist." Unless they persist, my retort usually signals the end of the dialogue. Unlike the job titles *farmer, stockbroker,* or even *computer analyst,* for many people *curriculum theorist* conjures few images.

What do curriculum theorists do? The answer to this question depends in part on the way curriculum theorists conceive of curriculum and theory. The term *curriculum* has over 150 definitions. With so many different ways of thinking about it, no wonder many curriculum theorists see their task differently. In this introduction, I point out that curriculum theorists have a useful function to serve, despite the fact that we can't agree on what to do. In short, like economists who analyze trends and make recommendations about the economy (and, incidentally, who also

agree on very little), curriculum theorists generate a constructive dialogue about curriculum decisions and practices. Although curricularists originally fought over the word *curriculum,* trying to achieve conceptual clarity in order to eliminate the various differences, in time educators recognized that the fight over the term was unproductive (Zais 1976, 93). However, the problem was not simply an academic disagreement. Instead, curricularists focused on different aspects of the educational enterprise. At stake in the definition of curriculum was a conceptual framework that included the nature of the role of the curricularist and the relationships among students, teachers, subject matter, and educational environments. Today, most curricularists place adjectives before the term to specify what type of curriculum they're discussing. Thus, one often reads about the intended, the operational, the hidden, the explicit, the implicit, the enacted, the delivered, the experienced, the received, and the null curriculum (see glossary at the end of this chapter). Distinctions also can be made with regard to curricularist, curriculum planner, curriculum worker, and curriculum specialist. I use the terms *curricularist* and *curriculum theorist* to refer to individuals, usually at the college level, who worry about issues regarding curriculum theory. I use the other terms to refer to people who actually take part in the planning or the implementation of curriculum in schools.

In order to trace the development that has

brought the field of curriculum to its present state, I will begin with a brief overview of the progression of curriculum development in the United States. First, I examine issues facing the Committee of Ten, a group of educators who convened in 1892 to draft a major document regarding what schools should teach. Next, I focus on the perennial question of who should decide what schools teach. Curriculum was not a field of study until the 1920s. How were curriculum decisions made before there were curriculum specialists? How did curriculum become a field of study? We learn that the profession began, in part, as a scientific endeavor; whether the field should still be seen as a scientific one is a question of debate. Finally, I provide a conceptual framework that examines six curriculum "ideologies" (Eisner 1992). By understanding these ideologies, educators will discern the assumptions underlying various conceptions of curriculum. Then they should be able to decide which ideology they wish to pursue and to recognize its educational implications.

What Should Schools Teach?

In the nineteenth century, curriculum usually meant "the course of study," and what many educators worried about was what schools should teach. Under the theoretical influence of "mental discipline" (derived from the ideas of faculty psychologists), many educators believed that certain subjects strengthened the brain, much like certain exercises strengthened body muscles. Greek, Latin, and mathematics were important because they were difficult subjects and thus, presumably, exercised the brain. By the 1890s, however, with the great influx of Italian, Irish, Jewish, and Russian immigrants, and with the steady increase of students attending secondary schools, a concern grew over the relevance and value of such subjects as Greek and Latin. Why should German or French be any less worthy than Greek or Latin? In addition, students and parents raised further questions regarding the merits of vocational education. They wanted curricula that met their more practical needs.

While parents pressed for their concerns, secondary school principals worried about preparing students for college, since colleges had different entrance requirements. In 1892 the National Education Association (NEA) appointed

the Committee of Ten to remedy this problem. Headed by Charles W. Eliot, president of Harvard University, the committee debated and evaluated the extent to which a single curriculum could work for a large number of students who came from many different backgrounds with many different needs. In its final report, the committee suggested that colleges consider of equal value and accept students who attended not only the classical curriculum program, but also the Latin scientific, the modern language, and the English programs.

By eliminating the requirement of Greek for two of the programs and by reducing the number of required Latin courses, the committee broke with the traditional nineteenth-century curriculum to some degree. Yet, they were alert to the possibility that different kinds of curriculum programs taught in different ways could lead to a stratified society. Eliot had argued that the European system of classifying children into "future peasants, mechanics, trades-people, merchants, and professional people" was unacceptable in a democratic society (Tanner and Tanner 1975, 186). The committee believed all should have the opportunity for further studies under a "rational humanist" orientation to curriculum, a viewpoint that prizes the power of reason and the relevance and importance of learning about the best that Western culture has to offer.

The committee's report met with mixed reviews when it came out. One of its foremost opponents was G. Stanley Hall, a "developmentalist," who argued that the "natural order of development in the child was the most significant and scientifically defensible basis for determining what should be taught" (Kliebard 1986, 13). According to Hall, who had scientifically observed children's behavior at various stages of development, the committee did not take into account children's wide-ranging capabilities, and it promulgated a college-bound curriculum for everyone, even though many high school students would not go to college. Rather than approaching curriculum as the pursuit of a standard academic experience for all students, Hall and other developmentalists believed that knowledge of human development could contribute to creating a curriculum in harmony with the child's stage of interest and needs.

Thus far I have indicated two orientations to curriculum: the rational humanist and the developmentalist. We should understand, however, that at any given time a number of interest groups struggle for power over the curriculum. Historian Herbert Kliebard observes:

> We do not find a monolithic supremacy exercised by one interest group; rather we find different interest groups competing for dominance over the curriculum and, at different times, achieving some measure of control depending on local as well as general social conditions. Each of these interest groups, then, represents a force for a different selection of knowledge and values from the culture and hence a kind of lobby for a different curriculum. (Kliebard 1986, 8)

Who Should Decide What Schools Teach?

Thinking about curriculum dates back in Western culture to at least the ancient Greeks. Plato and Aristotle, as well as Cicero, Plutarch, and Rousseau, all thought about curriculum matters in that they debated the questions of what should be taught to whom, in what way, and for what purposes. But it wasn't until 1918 that curriculum work was placed in the professional domain with the publication of *The Curriculum* by Franklin Bobbitt, a professor at the University of Chicago. Although supervisors and administrators had written courses of study on a piecemeal basis, "Professor Bobbitt took the major step of dealing with the curriculum in all subjects and grades on a unified and comprehensive basis" (Gress 1978, 27). The term *curriculum theory* came into use in the 1920s, and the first department of curriculum was founded at Teachers College, Columbia University, in 1937. Of course, the question arises: If curricularists (a.k.a. curriculum specialists, curriculum theorists, and curriculum workers) were not making decisions about what should be taught in schools prior to the 1920s, then who was?

As we have seen, national commissions made some of the curricular decisions. The NEA appointed the Committee of Ten to address college–high school articulation in 1892 and the Committee of Fifteen to address elementary school curriculum in 1895. In the early 1900s the NEA appointed another committee to develop fundamental principles for the reorganization of secondary education. Thus, university professors, school superintendents, and teachers made some curricular decisions as they acted in the role of acknowledged authorities on national commissions.

Along with commissions, forces such as tradition have shaped the curriculum. One long-time student of curriculum, Philip Jackson, observes:

> One reason why certain subjects remain in the curriculum is simply that they have been there for such a long time. Indeed, some portions of the curriculum have been in place for so long that the question of how they got there or who decided to put them there in the first place has no answer, or at least not one that anyone except a historian would be able to give. As far as most people are concerned, they have just "always" been there, or so it seems. (Jackson 1992, 22)

Jackson also notes here that subjects such as the three R's are so "obviously useful that they need no further justification"—or, at least, so it seems.

Texts and published materials have also been factors in shaping the curriculum. Whether it was the old *McGuffey Readers* or the modern textbooks found in almost any classroom in the United States, these books have influenced the curriculum by virtue of their content and their widespread use. According to some estimates, text materials dominate 75 percent of the time elementary and secondary students are in classrooms and 90 percent of their time on homework (Apple 1986, 85). Textbook writers are de facto curriculum specialists.

National Commission committees, tradition, textbooks, instructional materials, and the influence from numerous philosophers (e.g., Herbart and Dewey) were focal in deciding what schools should teach. Of course, parents, state boards of education, and teachers had their own convictions as to what should be in the curriculum. However, as the United States moved toward urbanization (30 percent of 63 million lived in cities in 1890; over 50 percent of 106 million lived in cities in 1920 [Cremin 1977, 93]), new factors influenced schooling. In particular, the industrial and scientific revolutions commingled in the minds of some to produce new ways of thinking about work. Franklin Bobbitt applied these new ideas to education. Influenced by Frederick Winslow Taylor, the father of the scientific management movement, Bobbitt assumed that the kinds of accomplishments that had been made in business and industry could be made in education. What was needed was the application of scientific principles to curriculum.

Briefly, Bobbitt believed that "educational engineers" or "curriculum-discoverers," as he called them, could make curriculum by surveying the array of life's endeavors and by grouping this broad range of human experience into major fields. Bobbitt wrote:

> The curriculum-discoverer will first be an analyst of human nature and of human affairs. . . . His first task . . . is to discover the total range of habits, skills, abilities, forms of thought . . . etc., that its members need for the effective performance of their vocational labors; likewise the total range needed for their civic activities; their health activities; their recreations, their language; their parental, religious, and general social activities. The program of analysis will be no narrow one. It will be as wide as life itself. (Bobbitt 1918, 43)

Thus, according to Bobbitt, curriculum workers would articulate educational goals by examining the array of life's activities. Next, in the same way one can analyze the tasks involved in making a tangible object and eliminate waste in producing it, Bobbitt believed one could streamline education by task analysis, by forming objectives for each task, and by teaching skills as discrete units.

Bobbitt's push for the professionalization of curriculum did not replace other factors so much as it added a new dimension. By arguing that schools needed stated objectives and that curricularists should be chosen for the task since they were trained in the science of curriculum, Bobbitt opened up a new line of work. He and his students would be of direct help to practitioners because they would know how to proceed scientifically (analyze the range of human experience, divide it into activities, create objectives) in the making of curriculum, and this knowledge gave curricularists authority and power. The world was rapidly changing in communications, in agriculture, in industry, and most of all in medicine. Who could argue with the benefits of science?

If Franklin Bobbitt created the field of professional curriculum activities, Ralph Tyler defined it. In his short monograph, *Basic Principles of Curriculum and Instruction* (1949), Tyler offered a way of viewing educational institutions. He began his book by asking four fundamental questions that he believed must be answered in developing curriculum:

1. What educational purposes should the school seek to attain?
2. What educational experiences can be provided that are likely to attain those purposes?
3. How can these educational experiences be

effectively organized?
4. How can we determine whether these purposes are being attained? (Tyler 1949, 1)

Tyler devoted one chapter to each question. Unlike some curricularists, Tyler did not say what purposes a school should seek to attain. He recognized that a school in rural Idaho has different needs from an urban one in Boston. Rather, Tyler suggested that schools themselves determine their own purposes from three sources: studies of the learners themselves, studies of contemporary life, and studies from subject matter specialists.

Tyler, like Bobbitt before him, wished to bring order to the complex field of education. Although there are differences between the two men, both believed there was work to be done by professional curricularists. Both men trained students in the field of curriculum, and both believed in the liberal ideals of rationality and progress. Curricularist Decker Walker summarizes the tradition that Bobbitt and Tyler started as follows:

> Since Bobbitt's day, planning by objectives (PBO) had developed into a family of widely used approaches to curriculum improvement. As a method of curriculum materials design, PBO focuses early attention on developing precise statements of the objectives to be sought. If the process is to be fully scientific, the selection of objectives must be rationally justifiable and not arbitrary. (Walker 1990, 469)

While Bobbitt and Tyler taught students how to become professional curricularists and encouraged them to conduct research, to write, and to attain university positions, differences of opinion on what curricularists should be doing soon mounted. At issue was not only the utility of scientific curriculum making, but also the specific endeavors many curricularists pursued.

A Framework for Thinking about Curriculum

Tyler produced a seminal work that provided curriculum workers with a way of thinking about curriculum. While some elaborated on his ideas (Taba 1962), others wondered whether indeed Tyler provided the best questions for curricularists to think about. During the 1970s, numerous educators began to seek other ways of thinking about curriculum work. William Pinar, for example, asked, "Are Tyler's questions . . . no longer pertinent or possible? Are they simply cul-de-sacs?" (Pinar 1975, 397). Reconceptualizing the term *curriculum* (race course) from the verb of the Latin root, *currere* (to run a race), Pinar goes on to argue:

> The questions of *currere* are not Tyler's; they are ones like these: Why do I identify with Mrs. Dalloway and not with Mrs. Brown? What psychic dark spots does the one light, and what is the nature of "dark spots," and "light spots"? Why do I read Lessing and not Murdoch? Why do I read such works at all? Why not biology or ecology? Why are some drawn to the study of literature, some to physics, and some to law? (402)

More will be said about Pinar's work later. My point here is that curriculum theorists do not necessarily agree on how one should approach thinking about curriculum. By trying to redefine curriculum entirely, Pinar drew attention to different aspects of the educational process.

Out of this continuing discussion among curricularists, various ideologies—beliefs about what schools should teach, for what ends, and for what reasons—have developed (Eisner 1992). In this section, I present six prominent curriculum ideologies that should prove useful in thinking about developing, adapting, or implementing curriculum. While these ideologies are important, they are not the only ones. Elliot Eisner writes of religious orthodoxy and progressivism and excludes multiculturalism and developmentalism. Some authors may include constructivism rather than developmentalism.

I remind the reader that few people actually wear the labels I describe. These conceptualizations are useful in helping one better articulate a set of assumptions and core values. They help us see the implications of a particular viewpoint. They also help us understand issues and concerns that may otherwise be neglected. Sometimes ideologies are specified in mission statements or some other kind of manifesto; at other times, ideologies are embedded in educational practice but are not made explicit. Rarely does a school adhere to one curriculum ideology—though some do. More often, because public schools are made up of people who have different ideas about what schools should teach, a given school is more likely to embrace an array of curricular ideas. While some readers may resonate strongly with a particular ideology because it expresses their inclinations, some readers may appreciate particular ideas from various ideolo-

gies. In either case, it may be a good idea to examine the strengths and weaknesses of each one. Later in this chapter I argue that one does not need to be ideologically pure in order to do good curriculum work.

Rational Humanism

We have already seen, in the historical example of Charles Eliot and the Committee of Ten, an early exemplar of rational humanism. During Eliot's day, rational humanists embraced the theory of mental discipline, which provided a handy rationale for traditional studies. Why study Greek and Latin? Because these subjects exercised the mind in ways that other subjects did not. While mental discipline fell by the wayside, rational humanism did not. From the 1930s through the 1950s, Robert Maynard Hutchins and Mortimer Adler championed the rational humanistic tradition, in part by editing *The Great Books of the Western World.* Hutchins argued that the "great books" offer the best that human beings have thought and written. Thus, rather than reading textbooks on democracy, science, and math, one ought to read Jefferson, Newton, and Euclid.

Today, one may find the rational humanist ideology in some private schools and in those public schools that have adopted Adler's ideas as represented in the *Paideia Proposal* (Eisner 1992, 310). In short, the Paideia plan provides a common curriculum for all students. Except for the choosing of a foreign language, there are no electives. All students learn language, literature, fine arts, mathematics, natural science, history, geography, and social studies.

While Adler endorses lecturing and coaching as two important teaching methods, the aspect of teaching Adler found most engaging was maieutic or Socratic questioning and active participation. In essence, maieutic teaching consists of a seminar situation in which learners converse in a group. The teacher serves as a facilitator who moves the conversation along, asks leading questions, and helps students develop, examine, or refine their thinking as they espouse particular viewpoints. This process, according to Adler, "teaches participants how to analyze their own minds as well as the thought of others, which is to say it engages students in disciplined conversation about ideas and values" (Adler 1982, 30).

Another important educational feature of these seminars is that one discusses books and art but not textbooks. In a follow-up book to *The Paideia Proposal,* Adler (1984) provides a K–12 recommended reading list in which he recommends for kindergarten to fourth grade Aesop, William Blake, Shel Silverstein, Alice Walker, Jose Marie Sanchez-Silva, Langston Hughes, and Dr. Seuss, among other authors. I indicate these authors in particular because the charge that Adler's program embraces only the Western European heritage is not entirely accurate. While Adler would argue that some books are better than others, and that, in school, students should be reading the better ones, one can see that Adler includes authors who are not elitist and who are from culturally diverse backgrounds.

Developmentalism

Another approach to curriculum theory, which was discussed briefly in the historical section of this chapter, is developmentalism. Although a range of scholars falls under this heading, the basic point is that, rather than fitting the child to the curriculum, students would be better served if the curriculum were fitted to the child's stage of development. Why? One argument is that doing otherwise is inefficient or even detrimental to the child's development. It would be ridiculous to try to teach the Pythagorean theorem to a first grader, and it could be harmful (to use a fairly noncontroversial example) to teach a fourth grader to master throwing a curve ball. By understanding the range of abilities children have at various ages, one can provide a curriculum that meets the needs and interests of students. Of course, while the stage concept cannot pinpoint the development of a particular child at a given age, it serves as a general guide.

One might also pay attention to the idea of development when creating or adapting curriculum because of the issue of "readiness for learning." There are two ways of thinking about readiness. Some educators, in their interest to hurry development, believe that encouraging learners to perform approximations of desired behaviors can hasten academic skills. In this case, one tries to intervene in apparently natural development by manipulating the child's readiness at younger and younger ages. The research findings on whether one can greatly enhance one's learning processes are somewhat mixed, but, in my opinion, they favor the side that says "speed learning" is inefficient (Duckworth 1987, Good and Brophy 1986, Tietze 1987). I also think

the more important question, as Piaget noted, is "not how fast we can help intelligence grow, but how far we can help it grow" (Duckworth 1987, 38).

A different way of thinking about readiness for learning concerns not how to speed it up, but how to work with it effectively. Eleanor Duckworth, who studied with Piaget, believes the idea of readiness means placing children in developmentally appropriate problem situations where students are allowed to have their own "wonderful ideas." She believes that asking "the right question at the right time can move children to peaks in their thinking that result in significant steps forward and real intellectual excitement" (Duckworth 1987, 5). The challenges for teachers are to provide environmentally rich classrooms where students have the opportunity to "mess about" with things, and to try to understand children's thought processes. Students should have the opportunity to experiment with materials likely to afford intellectual growth, and teachers should learn how their students think. In this approach to curriculum, mistakes are not problems; they are opportunities for growth.

The developmental approach to curriculum teaches us to pay attention to the ways humans grow and learn. One basic idea underlying the various theories of human development in regard to curriculum is that the curriculum planner ought to understand children's abilities and capabilities because such knowledge enables one to provide worthwhile educational activities for students.

Reconceptualism

As noted earlier with Pinar's use of the term *currere,* during the 1970s numerous individuals criticized the technical aspects and linear progression of steps of the Tyler rationale. Loosely labelled as reconceptualists, some educators felt the following:

> What is missing from American schools . . . is a deep respect for personal purpose, lived experience, the life of imagination, and those forms of understanding that resist dissection and measurement. What is wrong with schools, among other things, is their industrialized format, their mechanistic attitudes toward students, their indifference to personal experience, and their emphasis on the instrumental and the out of reach. (Pinar 1975, 316)

Reconceptualists have focused on Dewey's

observation that one learns through experience. Given this assertion, some important questions arise. For example, how can teachers, teacher educators, or educational researchers better understand the kinds of experiences individual students are having? To answer this question, reconceptualists employ ideas, concepts, and theories from psychoanalysis, philosophy, and literature.

Another question that arises when one reflects on understanding experience is, How can teachers provide worthwhile conditions for students to undergo educational experiences? Maxine Greene divides educational experiences into two types: "an education for having" and "an education for being." Education for having is utilitarian—for example, one may learn to read in order to get a job. Some students need this kind of experience. Education for being is soulful— one may learn to read for the sensual qualities it can provide. All students, she says, need the latter kind of experience. One problem is that the latter has often been neglected or, if not, often provided for the talented or gifted at the expense of others (Green, 1988a).

In their effort to reperceive education, reconceptualists such as Maxine Greene, Madeleine Grumet, and William Pinar do not usually offer specific educational ideas that are easily implemented. In part, this is because the kind of education with which they are concerned is not easy to quantify or measure. In general, reconceptualists do not believe their theories and ideas need quick utilization in schools in order to validate their worth. If in reading their writings you think more deeply about educational issues, then I think they would be satisfied.

Nevertheless, I can think of two practical challenges for education that stem from their writing. First, how could you write a rigorous and tough-minded lesson plan without using objectives? What would such a lesson plan look like? Second, if you wanted to teach students a concept such as citizenship, how would you do it? Rational humanists would have students read Thomas Jefferson or Martin Luther King, Jr. Reconceptualists, however, would wonder how teachers can place students in problematic situations (i.e., in the classroom or on the playground) where students would grapple with real issues concerning citizenship.

Critical Theory

The idea of critical theory originated at the Institute for Social Research in Frankfurt ("the Frankfurt school") in the 1920s. Today, scholars who continue to recognize the value and importance of Marxist critiques of society and culture draw from and build on ideas from critical theory. In education they reveal, among other things, that schooling comprises a value-laden enterprise where issues of power are always at play.

For instance, while many people perceive schools as neutral institutions, places that will help any hard-working student to get ahead in life, critical theorists suggest that, on the contrary, schools do not operate that way. Michael Apple points out, "Just as our dominant economic institutions are structured so that those who inherit or already have economic capital do better, so too does cultural capital act in the same way" (Apple 1986, 33). According to Apple, schools reflect the general inequities in the larger society. Rather than changing society through cultural transformation (teaching students to question or to be independent thinkers), schools actually maintain the status quo through cultural reproduction.

Unlike some curricularists who try to appear neutral in exercising judgments about curriculum matters, Apple's values are well known. He believes in John Rawls's insight that "for a society to be truly just, it must maximize the advantage of the least advantaged" (Apple 1979, 32). Apple encourages curricularists to take advocacy positions within and outside of education. While critical theory makes for a powerful theoretical tool, one question frequently asked of critical theorists is how this information can be used in the classroom. Teachers point out that they may not be able to change the school structure, the kinds of material they must cover, or the kinds of tests that must be given. Although admittedly application is difficult, one high school English educator in Boston who employs the ideas of critical theory is Ira Shor.

In an activity called "prereading," for example, Shor tells students the theme of a book they are about to read and has them generate hypothetical questions the book may answer. At first students are reluctant to respond, but after a while they do. Shor believes this kind of exercise has numerous functions. First, it provides a bridge for students to decelerate from the "rush of mass culture" into the slow medium of the printed word. Habituated to rock music and MTV, students need a slow-down time. Also, after creating a list of questions, students are curious how many will actually be addressed. Students may still reject the text, says Shor, but now it won't be a result of alienation. Perhaps most importantly, prereading demystifies the power of the written word. Rather than approaching the text as some kind of untouchable authority, "students' own thoughts and words on the reading topic are the starting points for the coordinated material. The text will be absorbed into the field of their language rather than they being ruled by it" (Shor 1987, 117).

Critical theory offers a radical way of thinking about schooling. Particularly concerned with students who are disenfranchised and who, without the critical theorists, would have no voice to speak for them, critical theory provides incisive analyses of educational problems.

Multiculturalism

In some ways, multiculturalists have an affinity with the critical theorists. Though critical theory traditionally is more concerned with class, most critical theorists have included race and gender in their analyses and discussions. Multiculturalism, however, deserves its own category as a curriculum ideology because it is rooted in the ethnic revival movements of the 1960s. Whether the purpose is to correct racist and bigoted views of the larger community, to raise children's self-esteem, to help children see themselves from other viewpoints, or to reach the child's psychological world, the multicultural ideology reminds educators that ethnicity must be dealt with by educators.

One major approach to multicultural education has been termed "multiethnic ideology" by James Banks (1988). According to Banks, Americans participate in several cultures—the mainstream along with various ethnic subcultures. Therefore, students ought to have cross-cultural competency. In addition to being able to participate in various cultures, Banks also suggests that when one learns about various cultures, one begins to see oneself from other viewpoints. The multiethnic ideology provides greater self-understanding.

When teaching from a multiethnic perspective, Banks advises that an issue not be taught from a dominant mainstream perspective with other points of view added on. This kind of

teaching still suggests that one perspective is the "right one," though others also have their own points of view. Rather, one should approach the concept or theme from various viewpoints. In this case, the mainstream perspective becomes one of several ways of approaching the topic; it is not superior or inferior to other ethnic perspectives. In addition to what takes place in the classroom, Banks also argues that a successful multiethnic school must have system-wide reform. School staff, school policy, the counseling program, assessment, and testing are all affected by the multiethnic ideology.

Cognitive Pluralism

According to Eisner, the idea of cognitive pluralism goes back at least to Aristotle; however, only in the last several decades has a conception of the plurality of knowledge and intelligence been advanced in the field of curriculum (Eisner 1992, 317). In short, cognitive pluralists expand our traditional notions of knowledge and intelligence. Whereas some scientists and educators believe that people possess a single intelligence (often called a "g factor") or that all knowledge can ultimately be written in propositional language, cognitive pluralists believe that people possess numerous intelligences and that knowledge exists in many forms of representation.

As a conception of knowledge, cognitive pluralists argue that symbol systems provide a way to encode visual, auditory, kinesthetic, olfactory, gustatory, and tactile experiences. If, for example, one wants to teach students about the Civil War, cognitive pluralists would want students not only to have knowledge about factual material (names, dates, and battles), but also to have knowledge about how people felt during the war. To know that slavery means by definition the owning of another person appears quite shallow to knowing how it feels to be powerless. Cognitive pluralists suggest students should be able to learn through a variety of forms of representation (e.g., narratives, poetry, film, pictures) and be able to express themselves through a variety of forms as well. The latter point about expression means that most tests, which rely on propositional language, are too limiting. Some students may better express themselves through painting or poetry.

One may also think about cognitive pluralism from the point of view of intelligence. As I mentioned, some scholars suggest that intelligence may be better thought of as multiple rather than singular. Howard Gardner, a leading advocate of this position (1983), argues that, according to his own research and to reviews of a wide array of studies, a theory of multiple intelligences is more viable than a theory about a "g factor." He defines intelligence as follows:

> To my mind, a human intellectual compe-
> tence must entail a set of skills of problem-
> solving—enabling the individual to resolve
> genuine problems or difficulties that he or
> she encounters and, when appropriate, to
> create an effective product—and must also
> entail the potential for finding or creating
> problems—thereby laying the groundwork for
> the acquisition of new knowledge. (Gardner
> 1983, 60–61)

Gardner argues that there are at least seven distinct kinds of human intelligence: linguistic, musical, logical–mathematical, spatial, bodily–kinesthetic, interpersonal, and intrapersonal. If schools aim to enhance cognitive development, then they ought to teach students to be knowledgeable of, and to practice being fluent in, numerous kinds of intelligences. To limit the kinds of knowledge or intelligences students experience indicates an institutional deficiency.

Applying Curriculum Ideologies

While some teachers or schools draw heavily on one particular curriculum ideology (e.g., Ira Shor's use of critical theory in his classroom or Mortimer Adler's ideas in Paideia schools), more often than not, a mixture of various ideologies pervade educational settings. I don't believe this is a problem. What Joseph Schwab said in the late 1960s about theory also applies to ideologies. He argued that theories are partial and incomplete, and that, as something rooted in one's mind rather than in the state of affairs, theories cannot provide a complete guide for classroom practice (1970). In other words, a theory about child development may tell you something about ten-year-olds in general, but not about a particular ten-year-old standing in front of you. Child developmentalists cannot tell you, for example, whether or how to reprimand a given child for failing to do his homework. Schwab suggested one become eclectic and deliberative when working in the practical world. In simpler terms, one should know about various theories and use them when applicable. One does not need to be ideologically pure. One should also reflect upon

one's decisions and talk about them with other people. Through deliberation one makes new decisions which lead to new actions which then cycle around again to reflection, decision, and action.

To understand this eclectic approach to using curriculum ideologies, let's take as an example the use of computers in the classroom. Imagine you are about to be given several computers for your class. How could knowledge of the various curriculum ideologies inform your use of them?

Given this particular challenge, some ideologies will prove to be more useful than others. For example, the rational humanists would probably have little to contribute to this discussion because, with their interests in the cultivation of reason and the seminar process of teaching, computers are not central (though one of my students noted, that, perhaps in time, rational humanists will want to create a "great software" program).

Some developmentalists would consider the issue of when it would be most appropriate to introduce computers to students. Waldorf educators, who base their developmental ideas on the writings of philosopher Rudolf Steiner (1861–1925), do not believe one should teach students about computers at an early age. They would not only take into account students' cognitive development (at what age could students understand computers?), but they would also consider students' social, physical, and emotional development. At what age are students really excited about computers? When are their fingers large enough to work the keyboard? What skills and habits might children lose if they learned computers at too early an age? Is there an optimum age at which one ought to learn computers? Waldorf educators would ask these kinds of developmental questions.

Developmentalists following the ideas of Eleanor Duckworth may also ask the above questions, but whatever the age of the student they are working with, these educators would try to teach the computer to children through engaging interactive activities. Rather than telling students about the computer, teachers would set up activities where students can interact with them. In this orientation, teachers would continue to set up challenges for students to push their thinking. Sustaining students' sense of wonder and curiosity is equally important. In addition to setting new challenges for students, teachers would also monitor student growth by trying to understand student thought processes. In short, rather than fitting the child to the curriculum, the curriculum is fitted to the child.

Reconceptualists' first impulse would be to consider the educational, social, or cultural meaning of computers before worrying about their utility. Of course, one should remember that there isn't one party line for any given ideological perspective. Some reconceptualists may be optimistic about computers and some may not. Although I don't know William Pinar's or Madeleine Grumet's thoughts on computers, I imagine they would reflect on the way computers bring information to people. Pinar observes that place plays a role in the way one sees the world (Pinar 1991). The same machine with the same software can be placed in every school room, but even if students learn the same information, their relationship to this new knowledge will vary. Thus, to understand the impact of computers one needs to know a great deal about the people who will learn from and use them. Having students write autobiographies provides one way to attain this understanding. Students could write about or dramatize their encounters with technology. After such an understanding, teachers can tailor lessons to meet student needs.

Critical theorist Michael Apple has examined the issue of computers in schools. Though he points out that many teachers are delighted with the new technology, he worries about an uncritical acceptance of it. Many teachers, he notes, do not receive substantial information about computers before they are implemented. Consequently, they must rely on a few experts or pre-packaged sets of material. The effects of this situation are serious. With their reliance on purchased material combined with the lack of time to properly review and evaluate it, teachers lose control over the curriculum development process. They become implementers of someone else's plans and procedures and become deskilled and disempowered because of that (Apple 1986, 163).

Apple also worries about the kind of thinking students learn from computers. While students concentrate on manipulating machines, they are concerned with issues of "how" more than "why." Consequently, Apple argues, computers enhance technical but not substantive thinking. Crucial political and ethical understanding atrophies while students are engaged in computer proficiency. Apple does not suggest one avoid computers because of these problems. Rather, he wants teachers and students to engage in social, political, and ethical discussions while they use the new technology.

Multiculturalists would be concerned that all students have equal access to computers. Early research on computer implementation revealed that many minority students did not have the opportunity to use computers, and when they did, their interaction with computers often consisted of computer-assisted instruction programs that exercised low-level skills (Anderson, Welch, Harris 1984). In addition to raising the issue that all students should have equal access to computers, multiculturalists would also investigate whether software programs were sending biased or racist viewpoints.

Finally, cognitive pluralists, such as Elliot Eisner, would probably focus on the kinds of knowledge made available by computers. If computers were used too narrowly so that students had the opportunity to interact only with words and numbers, Eisner would be concerned. He would point out, I believe, that students could be learning that "real" knowledge exists in two forms. If, however, computers enhance cognitive understanding by providing multiple forms of representation, then I think Eisner would approve of the use of this new technology in the classroom. For example, in the latest videodisc technology, when students look up the definition of a word, they find a written statement as well as a picture. How much more meaningful a picture of a castle is to a young child than the comment, "a fortified residence as of a noble in feudal times" (*Random House Dictionary* 1980, 142).

In addition to learning through a variety of sensory forms, Eisner would also want students to have the opportunity to express themselves in a variety of ways. Computers could be useful in allowing students to reveal their knowledge in visual and musical as well as narrative forms. Students should not be limited in the ways they can express what they know.

Each curriculum ideology offers a unique perspective by virtue of the kinds of values and theories embedded within it. By reflecting on some of the ideas from the various curriculum ideologies and applying them to an educational issue, I believe educators can have a more informed, constructive, and creative dialogue. Moreover, as I said earlier, I do not think one needs to remain ideologically pure. Teachers and curricularists would do well to borrow ideas from the various perspectives as long as they make sure

CURRICULUM IDEOLOGIES

Ideology	Major Proponent	Major Writings	Educational Priorities	Philosophical Beliefs	Teachers, Curriculum, or Schools Expressing Curriculum Ideology	Suggestions for Curriculum Development
Rational Humanism	R. M. Hutchins M. Adler	The Paideia Proposal (Adler 1982) Paideia Problems and Possibilities (Adler 1983) The Paideia Program (Adler 1984)	Teaching through Socratic method. The use of primary texts. No electives.	The best education for the best is the best education for all. Since time in school is short, expose students to the best of Western culture.	Paideia Schools. See Adler (1983) for a list of schools.	Teach students how to facilitate good seminars. Use secondary texts sparingly.
Developmentalism	E. Duckworth R. Steiner	Young Children Reinvent Arithmetic (Kamii 1985) "The Having of Wonderful Ideas" and Other Essays (Duckworth 1987) Rudolf Steiner Education and the Developing Child (Aeppli 1986)	Fit curriculum to child's needs and interests. Inquiry-oriented teaching.	Cognitive structures develop as naturally as walking. If the setting is right, students will raise questions to push their own thinking.	Pat Carini's Prospect School in Burlington, VT.	Allow teachers the opportunity to be surprised. Rather than writing a curriculum manual, prepare a curriculum guide.
Reconceptualism	W. Pinar M. Grumet	Bitter Milk (Grumet 1988) Curriculum Theorizing (Pinar 1975) Curriculum and Instruction (Giroux, Penna, Pinar 1981)	Use philosophy, psychology, and literature to understand the human experience. Provide an "education for having" and an "education for being."	One learns through experience. We can learn to understand experience through phenomenology, psychoanalysis, and literature.	See Oliver (1990) for a curriculum in accordance with reconceptualist thinking.	Write lesson plans without the use of objectives. Curriculum writers ought to reveal their individual subjectivities.
Critical Theory	M. Apple I. Shor P. Freire	Ideology and Curriculum (Apple 1979) Teachers and Texts (Apple 1986) Pedagogy of the Oppressed (Freire 1970) Freire for the Classroom (Shor 1987)	Equal opportunities for all students. Teaching should entail critical reflection.	A just society maximizes the advantage for the least advantages. Schools are part of the larger community and must be analyzed as such.	See Shor's edited text (1987) for a number of ideas on implementing critical theory.	Curriculum writers ought to examine their own working assumptions critically and ought to respect the integrity of teachers and students.
Multiculturalism	J. Banks E. King	Multiethnic Education (Banks 1988) Multicultural Education (Banks and Banks 1989)	Students should learn to participate in various cultures. Approach concept or theme from various viewpoints.	Students need to feel good about their ethnic identities. All people participate in various cultures and subcultures.	See King (1990) for a workbook of activities teaching ethnic and gender awareness.	Make sure that text and pictures represent a variety of cultures.
Cognitive Pluralism	E. Eisner H. Gardner	"Curriculum Ideologies" (Eisner 1992) The Educational Imagination (Eisner 1985) Frames of Mind (Gardner 1983)	Teach, and allow students to express themselves, through a variety of forms of representation. Allow students to develop numerous intelligences.	Our senses cue into and pick up different aspects of the world. Combined with our individual history and general schemata, our senses allow us to construct meaning.	The Key School in Indianapolis.	Curriculum lesson plans and units ought to be aesthetically pleasing in appearance. Curriculum ought to represent a variety of ways of knowing

they are not proposing contradictory ideas.

The chart on page 11 summarizes the major proponents, major writings, educational priorities, and philosophical beliefs of each curriculum ideology covered in this chapter. (Of course, this chart is not comprehensive. I encourage the reader to examine the recommended reading list for further works in each of these areas.) In the fifth column, "Teachers, Curriculum, or Schools Expressing Curriculum Ideology," I indicate places or texts where readers may learn more. One could visit a Paideia school, Carini's Prospect School, or the Key School in Indianapolis. One may read about reconceptualism, critical theory, and multiculturalism in the listed texts. Finally, in the sixth column, "Suggestions for Curriculum Development," I also include interesting points found in the literature but not necessarily contained in this chapter.

Recommended Reading

The following is a concise list of recommended reading in many of the areas discussed in this chapter. Full bibliographic citations are provided under *References*.

Some general **curriculum textbooks** that are invaluable are John D. McNeil's *Curriculum: A Comprehensive Introduction* (1990); William H. Schubert's *Curriculum: Perspective, Paradigm, and Possibility* (1986); Decker Walker's *Fundamentals of Curriculum* (1990); and Robert S. Zais's *Curriculum: Principles and Foundations* (1976). These books provide wonderful introductions to the field.

The recently published *Handbook of Research on Curriculum* (Jackson 1992) includes thirty-four articles by leading curricularists. This book is a must for anyone interested in research in curriculum.

For a discussion of **objectives** in education, Tyler (1949) is seminal. Also see Kapfer (1972) and Mager (1962). Bloom refines educational objectives into a taxonomy (1956). Eisner's (1985) critique of educational objectives and his notion of expressive outcomes will be welcomed by those who are skeptical of the objectives movement.

Good books on the **history of curriculum** include Kliebard (1986), Schubert (1980), and Tanner and Tanner (1975). Seguel (1966), who discusses the McMurry brothers, Dewey, Bobbitt, and Rugg, among others, is also very good.

Some excellent books on the **history of education** include the following: Lawrence Cremin's definitive book on progressive education, *The Transformation of the School: Progressivism in American Education, 1876–1957* (1961). David Tyack's *The One Best System* (1974) portrays the evolution of schools into their modern formation; and Larry Cuban's *How Teachers Taught: Constancy and Change in American Classrooms, 1890–1980* (1984) examines what actually happened in classrooms during a century of reform efforts. Philip Jackson's "Conceptions of Curriculum and Curriculum Specialists" (1992) provides an excellent summary of the evolution of curriculum thought from Bobbitt and Tyler to Schwab.

For works in each of the ideologies I recommend the following:

To help one understand the **rational humanist** approach, there are Mortimer Adler's three books on the **Paideia school**: *The Paideia Proposal: An Educational Manifesto* (1982), *Paideia Problems and Possibilities* (1983), and *The Paideia Program: An Educational Syllabus* (1984). Seven critical reviews of the Paideia proposal comprise "The Paideia Proposal: A Symposium" (1983).

For works in **developmentalism** based on Piaget's ideas see Duckworth (1987, 1991) and Kamii (1985). Among Piaget's many works you may want to read *The Origins of Intelligence* (1966). If you are interested in Waldorf education see Robert McDermott's *The Essential Steiner* (1984) and P. Bruce Uhrmacher's "Waldorf Schools Marching Quietly Unheard" (1991). Willi Aeppli's *Rudolf Steiner Education and the Developing Child* (1986), Francis Edmunds's *Rudolf Steiner Education* (1982), and Marjorie Spock's *Teaching as a Lively Art* (1985) are also quite good.

A general overview of the developmental approach to curriculum can be found on pages 49–52 of Linda Darling-Hammond and Jon Snyder's "Curriculum Studies and the Traditions of Inquiry: The Scientific Tradition" (1922).

Two books are essential for examining **reconceptualist** writings: William Pinar's *Curriculum Theorizing: The Reconceptualists* (1975) and Henry Giroux, Anthony N. Penna, and William F. Pinar's *Curriculum and Instruction* (1981). Recent books in reconceptualism include William Pinar and William Reynolds's *Understanding Curriculum as Phenomenological and Deconstructed Text* (1992), and William Pinar and Joe L. Kincheloe's *Curriculum as Social Psychoanalysis: The Significance of Place* (1991).

Some excellent works in **critical theory** include Paulo Freire's *Pedagogy of the Oppressed* (1970) and *The Politics of Education* (1985). Apple's works are also excellent; see *Ideology and Curriculum* (1979) and *Teachers and Texts* (1986). For an overview of the Frankfurt School and the application of Jürgen Habermas's ideas, see Robert Young's *A Critical Theory of Education: Habermas and Our Children's Future* (1990).

For an application of critical theory to classrooms see the Ira Shor–edited book, *Freire for the Classroom* (1987) with an afterword by Paulo Freire.

In **multicultural education** I recommend James Banks's *Multiethnic Education: Theory and Practice* (1988) and Banks and Banks's *Multicultural Education: Issues and Perspectives* (1989). Also see Gibson (1984) for an account of five different approaches to multicultural education. Nicholas Appleton (1983), Saracho and Spodek (1983), and Simonson and Walker (1988) are also important. Edith King's *Teaching Ethnic and Gender Awareness: Methods and Materials for the Elementary School* (1990) provides useful ideas about multicultural education that could be used in the classroom. John Ogbu's work (1987) on comparing immigrant populations to involuntary minorities is also an important work with serious educational implications.

Important works in the field of **cognitive pluralism** include Elliot Eisner (1982, 1985, 1992) and Howard Gardner (1983, 1991). Some philosophical texts that influenced both of these men include Dewey (1934), Goodman (1978), and Langer (1976).

For $20.00, the Key School Option Program will send you an interdisciplinary theme-based curriculum report. For more information write Indianapolis Public Schools, 1401 East Tenth Street, Indianapolis, Indiana 46201.

Glossary of Some Common Usages of Curriculum

delivered curriculum: what teachers deliver in the classroom. This is opposed to Intended curriculum. Same as operational curriculum.

enacted curriculum: actual class offerings by a school, as opposed to courses listed in books or guides. *See* official curriculum.

experienced curriculum: what students actually learn. Same as received curriculum.

explicit curriculum: stated aims and goals of a classroom or school.

hidden curriculum: unintended, unwritten, tacit, or latent aspects of messages given to students by teachers, school structures, textbooks, and other school resources. For example, while students learn writing or math, they may also learn about punctuality, neatness, competition, and conformity. Concealed messages may be intended or unintended by the school or teacher.

implicit curriculum: similar to the hidden curriculum in the sense that something is implied rather than expressly stated. Whereas the hidden curriculum usually refers to something unfavorable, negative, or sinister, the implicit curriculum also takes into account unstated qualities that are positive.

intended curriculum: that which is planned by the teacher or school.

null curriculum: that which does not take place in the school or classroom. What is not offered cannot be learned. Curricular exclusion tells a great deal about a school's values.

official curriculum: courses listed in the school catalogue or course bulletin. Although these classes are listed, they may not be taught. *See* enacted curriculum.

operational curriculum: events that take place in the classroom. Same as delivered curriculum.

received curriculum: what students acquire as a result of classroom activity. Same as experienced curriculum.

References

Adler, Mortimer J. 1982. *The Paideia Proposal: An Educational Manifesto*. New York: Collier Books.

———. 1983. *Paideia Problems and Possibilities*. New York: Collier Books.

———. 1984. *The Paideia Program: An Educational Syllabus*. New York: Collier Books.

Aeppli, Willi. 1986. *Rudolf Steiner Education and the Developing Child*. Hudson, NY: Anthroposophic Press.

Anderson, Ronald E., Wayne W. Welch, and Linda J. Harris. 1984. "Inequities in Opportunities for Computer Literacy." *The Computing Teacher: The Journal of the International Council for Computers in Education* 11(8):10–12.

Apple, Michael W. 1979. *Ideology and Curriculum*. Boston: Routledge and Kegan Paul.

———. 1986. *Teachers and Texts: A Political Economy of Class and Gender Relations in Education*. New York: Routledge and Kegan Paul.

Appleton, Nicholas. 1983. *Cultural Pluralism in Education*. White Plains, NY: Longman.

Banks, James A. 1988. *Multiethnic Education: Theory and Practice*. 2d ed. Boston: Allyn and Bacon.

Banks, James A., and Cherry A. McGee Banks, eds. 1989. *Multicultural Education: Issues and Perspectives*. Boston: Allyn and Bacon.

Bloom, Benjamin S., ed. 1956. *Taxonomy of Educational Objectives: The Classification of Educational Goals, Handbook 1: Cognitive Domain*. New York: McKay.

Bobbitt, Franklin. 1918. *The Curriculum*. Boston: Houghton Mifflin.

Cremin, Lawrence A. 1961. *The Transformation of the School: Progressivism in American Education, 1876–1957*. New York: Vintage Books.

———. 1977. *Traditions of American Education*. New York: Basic Books.

Cuban, Larry. 1984. *How Teachers Taught: Constancy and Change in American Classrooms 1890–1980*. White Plains, NY: Longman.

Darling-Hammond, Linda, and Jon Snyder. 1992. "Curriculum Studies and the Traditions of Inquiry: The Scientific Tradition." In *Handbook of Research on Curriculum: A Project of the American Educational Research Association*, ed. Philip W. Jackson, 41–78. New York: Macmillan.

Dewey, John. 1934. *Art as Experience*. New York: Minton, Balch.

Duckworth, Eleanor. 1987. *"The Having of Wonderful Ideas" and Other Essays on Teaching and Learning*. New York: Teachers College Press.

———. 1991. "Twenty-four, Forty-two, and I Love You: Keeping It Complex. *Harvard Educational Review* 61(1): 1–24.

Edmunds, L. Francis. 1982. *Rudolf Steiner Education*. 2d ed. London: Rudolf Steiner Press.

Eisner, Elliot W. 1982. *Cognition and Curriculum: A Basis for Deciding What to Teach*. White Plains, NY: Longman.

———. 1985. *The Educational Imagination*. 2d ed. New York: Macmillan.

———. 1992. "Curriculum Ideologies." In *Handbook of Research on Curriculum: A Project of the American Educational Research Association*, ed. Philip W. Jackson, 302–26. New York: Macmillan.

Freire, Paulo. 1970. *Pedagogy of the Oppressed*. Trans. Myra Bergman Ramos. New York: Seabury Press.

———. 1985. *The Politics of Education*. Trans. Donaldo Macedo. South Hadley, MA: Bergin and Garvey.

Gardner, Howard. 1983. *Frames of Mind*. New York: Basic Books.

———. 1991. *The Unschooled Mind: How Children Think and How Schools Should Teach*. New York: Basic Books.

Gibson, Margaret Alison. 1984. "Approaches to Multicultural Education in the United States: Some Concepts and Assumptions." *Anthropology and Education Quarterly* 15: 94–119.

Giroux, Henry, Anthony N. Penna, and William F. Pinar. 1981. *Curriculum and Instruction: Alternatives in Education*. Berkeley: McCutchan.

Good, Thomas S., and Jere E. Brophy. 1986. *Educational Psychology*. 3d ed. White Plains, NY: Longman.

Goodman, Nelson. 1978. *Ways of Worldmaking*. Indianapolis: Hackett.

Greene, Maxine. 1988a. "Vocation and Care: Obsessions about Teacher Education." Panel discussion at the Annual Meeting of the American Educational Research Association, 5–9 April, New Orleans.

———. 1988b. *The Dialectic of Freedom*. New York: Teachers College Press.

Gress, James R. 1978. *Curriculum: An Introduction to the Field*. Berkeley: McCutchan.

Grumet, Madeleine R. 1988. *Bitter Milk: Women and Teaching*. Amherst: Univ. of Massachusetts Press.

Jackson, Philip W. 1992. "Conceptions of Curriculum and Curriculum Specialists." In *Handbook of Research on Curriculum: A Project of the American Educational Research Association*, ed. Philip W. Jackson, 3–40. New York: Macmillan.

Kamii, Constance Kazuko, with Georgia DeClark. 1985. *Young Children Reinvent*

Arithmetic: Implications of Piaget's Theory. New York: Teachers College Press.

Kapfer, Miriam B. 1972. *Behavioral Objectives in Curriculum Development: Selected Readings and Bibliography.* Englewood Cliffs, NJ: Educational Technology.

King, Edith W. 1990. *Teaching Ethnic and Gender Awareness: Methods and Materials for the Elementary School.* Dubuque, IA: Kendall/Hunt.

Kliebard, Herbert M. 1986. *The Struggle for the American Curriculum, 1893–1958.* Boston: Routledge and Kegan Paul.

Langer, Susanne. 1976. *Problems of Art.* New York: Scribners.

McDermott, Robert A., ed. 1984. *The Essential Steiner.* San Francisco: Harper & Row.

McLaren, Peter. 1986. *Schooling as a Ritual Performance: Towards a Political Economy of Educational Symbols and Gestures.* London: Routledge and Kegan Paul.

McNeil, John D. 1990. *Curriculum: A Comprehensive Introduction.* 4th ed. Glenview, IL: Scott, Foresman/Little, Brown Higher Education.

Mager, Robert. 1962. *Preparing Instructional Objectives.* Palo Alto, CA: Fearon.

Ogbu, John. 1987. "Variability in Minority School Performance: A Problem in Search of an Explanation." *Anthropology and Education Quarterly* 18(4): 312–34.

Oliver, Donald W. 1990. "Grounded Knowing: A Postmodern Perspective on Teaching and Learning." *Educational Leadership* 48(1): 64-69.

"The Paideia Proposal: A Symposium." 1983. *Harvard Educational Review* 53 (4): 377–411.

Piaget, Jean. 1962. *Play, Dreams and Imitation in Childhood.* New York: Norton.

————. 1966. *Origins of Intelligence.* New York: Norton.

Pinar, William F., ed. 1975. *Curriculum Theorizing: The Reconceptualists.* Berkeley: McCutchan.

Pinar, William F., and Joe L. Kincheloe, eds. 1991. *Curriculum as Social Psychoanalysis: The Significance of Place.* Albany: State Univ. of New York Press.

Pinar, William F., and William M. Reynolds, eds. 1992. *Understanding Curriculum as Phenomenological and Deconstructed Text.* New York: Teachers College Press.

The Random House Dictionary. 1980. New York: Ballantine.

Saracho, Olivia N., and Bernard Spodek. 1983. *Understanding the Multicultural Experience in Early Childhood Education.* Washington, DC: National Association for the Education of Young Children.

Schubert, William H. 1980. *Curriculum Books: The First Eight Years.* Lanham, MD: Univ. Press of America.

————. 1986. *Curriculum: Perspective, Paradigm, and Possibility.* New York: Macmillan.

Schwab, Joseph J. 1970. *The Practical: A Language for Curriculum.* Washington, DC: National Education Association.

Seguel, M. L. 1966. *The Curriculum Field: Its Formative Years.* New York: Teachers College Press.

Shor, Ira, ed. 1987. *Freire for the Classroom: A Sourcebook for Liberatory Teaching.* Portsmouth, NH: Heinemann.

Simonson, Rick, and Scott Walker, eds. 1988. *The Graywolf Annual Five: Multi-Cultural Literacy.* St. Paul, MN: Graywolf Press.

Spock, Marjorie. 1985. *Teaching as a Lively Art.* Hudson, NY: Anthroposophic Press.

Taba, Hilda. 1962. *Curriculum Development: Theory and Practice.* New York: Harcourt Brace Jovanovich.

Tanner, Daniel, and Laurel N. Tanner. 1975. *Curriculum Development: Theory into Practice.* New York: Macmillan.

Tietze, Wolfgang. 1987. "A Structural Model for the Evaluation of Preschool Effects." *Early Childhood Research Quarterly* 2(2): 133–59.

Tyack, David B. 1974. *The One Best System: A History of American Urban Education.* Cambridge: Harvard Univ. Press.

Tyler, Ralph W. 1949. *Basic Principles of Curriculum and Instruction.* Chicago: Univ. of Chicago Press.

Uhrmacher, P. Bruce. 1991. "Waldorf Schools Marching Quietly Unheard." Ph.D. diss., Stanford University.

Walker, Decker. 1990. *Fundamentals of Curriculum.* New York: Harcourt Brace Jovanovich.

Young, Robert. 1990. *A Critical Theory of Education: Habermas and Our Children's Future.* New York: Teachers College Press.

Zais, Robert S. 1976. *Curriculum: Principles and Foundations.* New York: Thomas Y. Crowell.

TRENDS IN SCHOOL MATHEMATICS: CURRICULUM AND TEACHING STANDARDS OVERVIEW

by Jay Stepelman
Chairperson, Math Department (Retired)
George Washington High School, New York, New York

T HE Jeffersonian ideal of universal education for the purpose of furnishing our nation an educated citizenry has a long history in the United States.

The earliest European settlers brought with them strong convictions about the importance of education. While their first concern was, of course, survival in a foreign environment, they did not neglect the teaching of reading, writing, and computational skills to their children. These were taught at home, mostly by mothers, while fathers were at work farming or building. In the early 1800s, public schools were still not available for all children. But eventually, state by state, laws were passed mandating the establishment of such schools.

The era of the industrial revolution produced large urban centers that existed alongside agricultural communities. Meanwhile, waves of immigrants continued to arrive. The cities, and consequently their schools, began to feel the effects of overcrowding and mixed populations. The early grades focused on reading and arithmetic while later on there was preparation for college, which usually meant preparation for work in the ministry. However, many college students had little interest in pursuing church positions. For those who were not inclined to attend college, factory owners offered employment to youths who could work well with their hands and who knew how to handle machines. With an ever-increasing size and diversity of population, the formal education usually found in public school buildings came to seem irrelevant.

Students in a typical school were grouped heterogeneously and all studied the same basic subjects together. Most remained in school only as long as they were required by law.

The uniquely American concept of a comprehensive school was refined a bit with the introduction of elective courses such as science, history, and foreign language. When the Great Depression of the 1930s arrived, the schools received much criticism. Something, after all, must be terribly wrong with the curriculum if so many people from every level of education could be unemployed! Most elective courses were condemned and the study of foreign languages deemed a waste of time.

A transformation occurred in the school curriculum during and after World War II.

Mathematics, science, history, and languages became popular once again. After all, nuclear energy, the incipient stages of rocketry, computers, and the unfolding of events after World War II convinced even the most hardened skeptic of the need to continue studying those subjects.

The following years saw changes in the mathematics curriculum that attempted to make it even more meaningful. The emphasis was on understanding concepts that would give meaning to the underlying mathematical processes taught in school. In addition to a focus on underlying meaning, the so-called new math attempted to introduce more rigorous language, as well as set theory, in the hopes that more talented students would be better prepared for careers as mathematicians, scientists, or engineers. The new math also downplayed practice in computational skills.

In reaction to the perceived failure of the new math, the 1970s marked an era of "back to basics" for American youth. Many parents, politicians, business leaders, and the press insisted that rote drill of number facts and memorization were the answers to poor performance on competitive exams within our own schools as well as in comparison with those abroad.

A New Direction

The National Council of Teachers of Mathematics (NCTM), a society of professional educators first created in 1920, determined that it had a special obligation to try to resolve conflicting opinions and present a responsible and knowledgeable viewpoint of the directions mathematics education ought to be taking. They presented a program known as *An Agenda for Action*, which resulted in recommendations called *School Mathematics of the 1980s*. The proposed curriculum for school mathematics of the 1980s stated that:

1. Problem solving should be the focus of school mathematics.
2. Basic skills should mean more than mere computational facility.
3. Calculators and computers should be incorporated into the curriculum at all levels.

4. Student learning should be evaluated by both traditional and other-than-traditional means.
5. More study of mathematics should be required of all students.
6. A greater range of elective courses should be offered.

Vision of the Future

During the late 1980s, NCTM's Commission on Standards for School Mathematics directed four working groups to draft a set of standards for mathematics curricula which would essentially reform every aspect of school mathematics in American schools (K-12). The working groups consisted of classroom teachers, supervisors, educational researchers, teacher educators, and university mathematicians. The four areas of focus were: K-4, 5-8, 9-12, and evaluation. The two documents which were issued as a result of this collaboration, *Curriculum and Evaluation Standards for School Mathematics* (1989) and *Professional Standards for Teaching Mathematics* (1991) now form the cornerstone of reform in school mathematics for the 1990s.

Precisely what ought to be changed, and what retained? What should the mathematics curriculum include in terms of content and emphasis? These documents offer recommendations for all aspects of school mathematics— curriculum, teaching strategies, evaluation techniques, professional development of teachers, and even the public's responsibilities.

The proposed curriculum calls for an integration of several different content areas such as mathematics, art, history, geography, music, current events, and sports. It encourages critical thinking, develops creative solutions to realistic problems, supports a team effort to resolve mathematical challenges and recommends the use of calculators and computer technology.

Identical goals were envisioned for all students for all levels, namely that they:
1. learn to think mathematically
2. become confident in their ability to do mathematics
3. become mathematical problem solvers
4. learn to communicate mathematically

5. learn to reason mathematically (NCTM 1989, 5)

The document divides the proposed curriculum into three groupings: the early grades K-4, middle school grades 5-8, and high school grades 9-12.

The K–4 Standards Curriculum

Attention is focused on providing a learning environment derived from the child's everyday experiences. Children should no longer be subjected to rote memory learning. Education for the early years will now include opportunities for students to explore mathematical problem solving as a way of interpreting their world. Children will be encouraged to explore, develop, test, discuss, and apply ideas. Manipulatives and hand calculators will be available for experimentation and mathematical challenges will be open-ended, so youngsters may propose and discuss ideas.

Assessment strategies will include a variety of new tools, such as open-ended problems, portfolios of student work, and observation of students' approaches to problems and their ability to communicate ideas.

The proposed curriculum is aimed at increasing skills in these areas:

1. Problem solving: There will be a strong emphasis on mathematical concepts that support the development of problem-solving strategies.
2. Communication: Learning the mathematics of the *Standards* will result in an appropriate, dynamic environment where students will be encouraged to discuss and describe their results, orally and in both graphic and written forms.
3. Reasoning: The teacher will create an environment where learning is an outcome of students' exploring, justifying, representing, and predicting.
4. Measurement and estimation: Manipulatives will provide support as students work with counters, interlocking cubes, geometric models, rulers, spinners, die, playing cards, poker chips, dominoes, geoboards, and graph paper in addition to creatively constructed homemade models.
5. Whole number operations, fractions, and decimals: The curriculum will include a broad range of mathematical content. Applications of fractions and decimals will help children better understand their world of giving and sharing.
6. Spatial relationships: To become mathematically literate, one must know more than just arithmetic. A foundation for further study of mathematics, art, and architecture is acquired by developing skill in three-dimensional drawings and visualizations.
7. Statistics and probability: Simple picture graphs and probability games will introduce students to important concepts that appear often in everyday life. These topics are easily integrated with history, geography, sports, and current events.
8. Patterns and relations: The recognition of pattern and the ability to communicate that recognition are important skills in the ability to apply mathematics to the everyday world. Students should be given experience in recognizing and in creating patterns.

Developing a new curriculum that meets the standards is a formidable and important task. It should be a goal of every thoughtful mathematics educator to create, develop, and implement exemplary and challenging programs where students are not bored with endless repetition and drill.

Most youngsters begin to develop what will be their lifelong attitudes toward mathematics in the early elementary grades. Their initial enthusiasm and positive attitudes often disappear in the grades that follow.

Typical Activities for the K–4 Standards

Activity 1: Each child is given six two-sided counters. The child must make as many different combinations of counters as he or she can. For example, if the counters are red on one side and yellow on the other, some combinations are: one yellow and five red, two yellow and four red, etc.

Activity 2: Find the favorite ice cream flavor of each student. Display the information in several different ways.

Activity 3: A figure is displayed using an overhead projector for a few seconds. The students are asked to draw what they remember. Then the figure is displayed again and the students discuss what they saw (see figure 1).

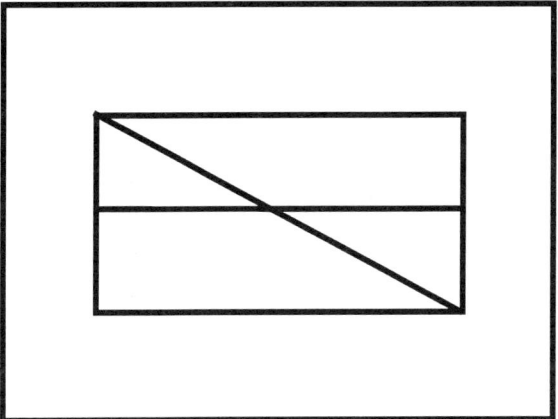

The 5–8 Standards Curriculum

Middle grade students are a more restless, energetic group, easily subject to peer influence and extremely self-conscious about both their appearance and their awareness of new emotional and sexual stimulants. During this period, they seem to be especially responsive to hands-on and visual activities. While attempts to impart an appreciation for the beauty and power of mathematics for these youngsters may not be a totally lost cause, it is surely a challenge for teachers.

Besides universal agreement on the continued importance and value of basic skills related to reading, writing, and mathematics in grades 5–8, the recommendation of the *Standards* for this age level is to increase attention to the areas of:

1. Problem solving, by challenging students with open ended problems, projects, and puzzles
2. Communication, by increasing skills through writing, reading, listening, and discussing
3. Reasoning, by inductive and deductive analyses of television commercials, newspaper ads, political pronouncements, spatial relationships, graphs, and tables
4. Connections, by connecting mathematics with the worlds of art, science, business, the supermarket, music, and sports
5. Number and number relationships, by encountering concrete materials, number lines, area models, graphs, calculators, and computers
6. Number systems and number theory, by recognizing the need for numbers beyond whole numbers and by exposure to the underlying structure of mathematics
7. Computation and estimation, by using estimates to check the reasonableness of an answer, by continued exposure to integers, decimals, fractions, ratios, proportion, and percent
8. Patterns and functions, by investigating, observing, and describing patterns in the world around them and by using algebraic expressions to describe functions and solve problems
9. Algebra, by using an intuitive approach to solving equations and by informal investigation of direct and inverse variations, of inequalities, and nonlinear equations
10. Statistics, by collecting, organizing, and describing data, by constructing, reading, and interpreting tables, charts, and graphs
11. Probability, by modeling situations, carrying out experiments, constructing sample spaces, by making predictions based on experimental or theoretical probabilities
12. Geometry, by exposure to a variety of two- and three-dimensional geometric figures, by constructing regular polyhedra and prisms, by analyzing the properties of geometric figures, by investigating patterns, rotations, reflections, translations, and dilations
13. Measurement, by estimating, making, and using measurements to describe and compare, by selecting appropriate units and tools to measure to a reasonable degree of accuracy, by extending understanding of concepts or perimeter, area, volume, angle measure, capacity, weight, and mass

In this new framework, the teacher must assume the role of facilitator and thus ensure equitable opportunity for each student to be a full participant in the learning process. Calculators, computers, and manipulative aids are most appropriate for fulfilling the mandate of the *Standards*. Assessment strategies include the traditional multiple-choice types, enhanced multiple-choice tests that allow for an answer with an explanation, and short-answer and fill-in types, plus the more lengthy multistep exercises, all in addition to the teacher's observations and evaluations of the student, in conjunction with the student's portfolio. The use of open-ended questions and projects will continue at this level.

Although the teacher has a multitude of roles to play in the classroom—facilitator, questioner, encourager—parental assistance with student attitudes toward mathematics is not only welcome, it is a necessity. Parents are essential partners in the process to counter negative attitudes toward mathematics that are so often displayed by students, adults, and even teachers. Children can learn to enjoy mathematics if parents display a positive attitude.

All children will acquire a more positive attitude when they observe that a partnership exists between teacher and parent. Additional ways in which parents can help include the following:

- Challenge their own myths about school mathematics.
- Visit mathematics classrooms and labs.
- Encourage innovation in schools through P.T.A. meetings.
- Encourage the media to present positive images of mathematics.

Typical Activities for the 5–8 Standards

Activity 1: Find the number of diagonals in a quadrilateral, pentagon, hexagon, and so on. Write a general rule for a polygon with n sides.
Activity 2: Given a certain perimeter, find the greatest possible area for a rectangle with the given perimeter.
Activity 3: Given data for Judy's last 50 times at bat, determine the probability of her getting a hit, a walk, a home run, or a strike out.

The Standards and the 9–12 Curriculum

By the time students complete high school, they must be prepared for daily living as a worker-citizen or for entry into a postsecondary education program. The basic-core curriculum suggested by the *Standards* reflects the goal

that all students become productive, literate citizens, even though their performance levels and interests may vary greatly. Mathematics will play an important part in achieving that goal.

Both teacher and student roles have become fulcrums for change in a curriculum that will have its priorities reordered. Familiar teaching and testing techniques have been expanded to incorporate instructional methods that develop strategies aimed at increasing a student's analytic power. This new student will also possess a substantial degree of skill in the latest technologies involving calculators and computers.

New emphases on instructional practices will also take effect upon implementation of the mathematics curriculum of the *Standards*. The *Standards* maintains that:

- Effective questioning is but one key to effective teaching. Another key is the environment in which a lesson unfolds. Still another is the classroom discourse that evolves. These are among the new emphases that will emerge in classrooms of the nineties. The teacher will select from a variety of strategies those that might be most useful for different types of lessons (i.e., whole class, small group, peer instruction, individual exploration, project work).
- Student communication develops through preparation of daily written logs, journals, and comments both written and oral.
- Scientific calculators and graphics utilities will be used to do mathematics.
- Daily, as well as other regular, assessments of the lesson will be implemented.

Typical Activities for the 9–12 Standards

Activity 1: Graph $y = x^2 - 4$, $y = x^2$ and $y = x^2 + 4$ with a graphing utility. Discuss the difference and similarities between these graphs. Repeat the activity for $y = 2x^2$, $y = -x^2$, and $y = 0.5 x^2$.
Activity 2: Use an interactive computer software package to explore properties of rhombi. Formulate conjectures about the observed properties and prove or disprove the conjectures.
Activity 3: Use a random-number table or a computer program to stimulate a probability experiment. Repeat the experiment and record results so that a connection can be made

between a large number of trials and mathematical probability.

The Core Curriculum

In terms of the *Standards*, the core curriculum is "the three years of required study of mathematics in high school that centers around a core of subject matter and methodology that is differentiated by variations in treatment of topics as well as the nature of the applications themselves" (NCTM 1989, 129).

That statement presupposes that all students, regardless of ability level, will experience the entire scope of topics in the curriculum. Thus, students who might have been selected previously for a general or business track, with its own specialized curriculum, will now be exposed to the same subject matter as the college-bound, even though they are at different levels of preparedness. In the *Standards* (NCTM 1989, 132–36), two examples are given that illustrate how the same content may be presented at different levels even through teaching strategies will vary in accordance with levels of interest, skills, and future goals. What follows is a brief summary of those examples.

The first begins with the problem of finding the amount of money that will be in a savings account at the end of ten years given the amount of the original deposit ($100), and an interest rate compounded annually (6%).

At **Level 1,** students use calculators to solve the problem by determining the amount after each successive year. They might also use a computer spreadsheet. Students are encouraged to find an underlying pattern, for example:

Amount at end of 1 year = $100(1.06)^1$
Amount at end of 10 years = $100(1.06)^{10}$

At **Level 2,** students generalize the problem in stages, finally arriving at this formula: $A_n = A_0(1 + r)^n$ where A_n is the amount after n years, A_0 is the amount of the original deposit, and r is the annual interest rate.

At **Level 3,** students further generalize the formula so that they can explore problems in which the rate is compounded semiannually, quarterly, monthly, or daily.

At **Level 4,** students should be able to solve the formula given in Level 2 for any of the variables. Possible extensions of the topics are:
- the derivation of the rule of 72
- solving problems in which compound growth in biology provides an appropriate application
- proving the results using mathematical induction
- making a connection between this topic and the development of the irrational number e.

The second example is a problem in which students must find the dimensions of a cylinder that will give a certain volume with a minimum surface area.

At **Level 1,** students must develop the formulas for volume and surface area of a cylinder. Then, students use algebra to rewrite the volume formula for the height, h, given the required volume. Students use a calculator to create a table that shows, for different values of the radius, r, the corresponding values of h and the surface area. By analyzing their tables, students find an approximate solution to the problem.

At **Level 2,** students develop algebraic expressions as in Level 1 and design a computer algorithm to print out a table of values. The students analyze the output and modify the program in order to get a better approximation of the solution.

At **Level 3,** students use the given volume and the formulas to express the surface area as a function of the radius of the cylinder. A graphic utility is used to plot the function, and the graph is analyzed to find the value of r that corresponds to the minimum value of the surface area function. That value of r is used to find the corresponding value of h.

At **Level 4,** students explore different volume values to discover a general relationship between the radius and the height of a cylinder that minimizes its surface area for a given volume.

At **Level 5,** students would prove formally that the surface is minimal when $h = 2r$.

The following areas are recommended by the *Standards* for inclusion in the core curriculum for grades 9–12.

1. coordinate and transformational geometry

2. development of mini-postulational geometry
3. three-dimensional geometry
4. structure of number systems
5. matrix algebra with applications
6. use of technology to explore/solve problems: scientific calculators, computers, graphic utilities
7. real-world problems across the board— in the areas of geometry, algebra, trigonometry, and functions
8. statistics
9. probability
10. discrete mathematics
11. conceptual underpinnings of calculus.

The major advantage of a core curriculum is that all students can keep their options open. Students will not be "tracked out" of the mathematics they need should they decide to go on to college or vocational training. The core curriculum will better prepare the non-college-bound student for the world of work. At the same time, all students will encounter more interesting and important mathematics and experience the power of mathematics to solve real problems.

Reaching for the Standards

Major textbook publishers are shifting their attention to reflect the *Standards* curriculum for elementary, middle and junior high school grades, as well as for an integrated (or unified) *Standards* mathematics program for the secondary level. New directions in mathematics assessment that are being developed by the New Standards Project and the Mathematics Science Education Board also bear watching for their impact on trends. The NCTM, in Reston, Virginia, can provide the latest information on these and other projects.

If there were ever any contemplation of a national education curriculum for mathematics in the United States, the *Standards* would surely fit the bill. However, the *Standards* would have to be embraced fifty times, so to speak, before it became truly national. Several states have already embarked on innovative programs that may eventually lead to universal

acceptance of the *Standards* curriculum, with variations that recognize regional nuances.

Innovative programs that are being developed by some states already reflect at least a partial acceptance of the *Standards*. It should be made clear, however, that there is a long way to go before the universal vision of the *Standards* is achieved.

The Standards Classroom

What the *Standards* is attempting to create is something quite different from what we experienced as students and even quite different from what we may be doing now as classroom teachers. No document alone will change what goes on in the classroom. It is up to teachers to implement that. However, questions remain that must be answered: What will classrooms be like in which students encounter genuine uses of calculators, computers, concrete and pictorial models? What exactly is meant by "engaging students in mathematical reasoning"? How will students make conjectures, present arguments, and construct proofs at the various grade levels?

One thing is certain: The *Standards* cannot be used to standardize teaching. There are, after all, real-world reasons for learning mathematics. While art is seldom seen as a consideration in problem solving except as an aid for enhanced visualization, fractals have brought a new dimension to mathematics that has been transformed into art. There are examples of computer animation, science, and language arts being used at every grade level.

Students in every generation represent the hope of the future. The purpose of the *Standards* is to guide mathematics teachers closer to ideal practices and persuade them to abandon traditional methods that have not proven themselves particularly effective. The vision for teaching the generation of the nineties will include movement toward greater independence from a teacher and subsequently more self-reliance; fewer memorization procedures and more opportunities for conjecturing, inventing, and formulating and problem solving; and connecting mathematics with other disciplines.

What Should a Teacher Know?

The document *Professional Standards for Teaching Mathematics,* which NCTM published in 1991, focused on the professional development of teachers. The writers recommend that "teachers at all grade levels have not only a thorough understanding of the mathematics they are teaching but also a vision of where that mathematics is leading" (135).

The report suggest that the mathematical content teachers should know for grades K-4 is at the core of a series of concentric rings, with the second ring representing mathematics for teachers of grades 5-8 and the third ring representing mathematics for teachers of grades 9-12 (NCTM 1991, 136-39).

For teachers of K-4, the NCTM recommends a well-developed number sense and a knowledge of number systems through real numbers. Teachers should understand how geometry helps describe our world and solves real problems. They should be able to analyze two- and three-dimensional figures and be familiar with synthetic, coordinate, and transformational geometry. With regard to measurement, teachers should know its historical development, be able to work with standard and nonstandard units of measure, and have experience with formulas for perimeter, area, and volume and with indirect measurement. Teachers should also have extensive experience with elementary statistics and probability and an understanding of the concepts of functions (NCTM 1991, 136-37).

Teachers of grades 5-8 need further work with number systems to include the complex numbers. They should also study algebraic structures such as matrices. In geometry, the emphasis should be on solving problems as well as making and testing conjectures. In their study of statistics, teachers should encounter concepts of descriptive statistics as well as dispersion and central tendency. Misuse of statistics and misconceptions about probability should be discussed. Function, graphs, and the concept of limits should be explored. Teachers should understand how the concepts of calculus are used to solve real problems in science and mathematics (NCTM 1991, 137-38).

The teachers of grades 9-12 should investigate algebraic structures such as groups, rings, integral domains, and fields. They should

explore new topics in number theory such as coding theory. Non-Euclidean geometries, geometric transformations, and the interplay of algebra and geometry should be explored. Teachers should investigate both descriptive and inferential statistics and both discrete and continuous probability. In addition to calculus and analysis, teachers need familiarity with new topics in discrete mathematics such as symbolic logic, induction and recursion, relations, and sequences and series (NCTM 1991, 138-39).

In summary, the council is calling for a more extensive background in mathematics for all teachers. It therefore recommends that undergraduate preparation include nine semester hours of coursework in content mathematics for grades K-4 teachers, fifteen hours for grades 5-8 teachers and the equivalent of a major in mathematics for grades 9-12 teachers (NCTM 1991, 139). The report also recommends that teachers be given the kinds of mathematical experiences—problem solving, communicating, analyzing, and so forth—that we want them to be able to use with their own students.

References

Abbott, E. A. 1884. *Flatland—A Romance of Many Dimensions.* Reprint. New York: Dover.

Bakst, A. 1952. *Mathematics—Its Magic and Mystery.* New York: Van Nostrand.

Balka, D. 1985. *Attribute Logic Block Activities.* Oak Lawn, IL: Ideal School Supply Company.

Baur, G. R., and L. O. George. 1976. *Helping Children Learn Mathematics.* Menlo Park, CA: Benjamin Cummings.

Bayha, B., and B. Katherine. 1985. *Pattern Block Activities.* Palo Alto, CA: Dale Seymour.

Bell, E. T. 1965. *Men of Mathematics.* New York: Simon & Schuster.

Brown, S. I., T. J. Cooney, and D. Jones. 1985. *Research in Mathematics Teacher Education.* New York: Macmillan.

Buckeye, D. A., W. A. Ewbank, and J. L. Ginther. 1989. *Cloudburst of Creative Mathematics Activities.* Vols. 1 and 2. Pacific Grove, CA: Midwest.

Burton, L. 1986. *Girls into Math Can Go.* New York: Holt, Rinehart and Winston.

California Department of Education. 1992. *Mathematics Framework for California Public Schools, Kindergarten through Grade 12.* Sacramento.

CEEB. 1992. *Advanced Placement Mathematics Course Description.* New York: The College Board.

Charles, L. H., and M. Randolof-Brummett. 1989. *Connections: Linking Manipulatives to Mathematics, Grades 1 through 6.* Sunnyvale, CA: Creative Publications.

Cheney, W., and D. Kincaid. 1985. *Numerical Mathematics and Computing.* Monterey, CA: Brooks/Cole.

Courant, R., and H. Robbins. 1958. *What Is Mathematics?* New York: Oxford University Press.

D'Abey, D. A. 1985. *The Place Value Connection.* Palo Alto, CA: Dale Seymour.

Dantzig, T. 1954. *Number: The Language of Science.* Garden City, NY: Doubleday.

Davidson, N., ed. 1990. *Cooperative Learning in Mathematics.* Menlo Park, CA: Addison-Wesley.

Davidson, P., and R. Willcut. 1981. *From Here to There with Cuisenaire Rods—Area, Perimeter, and Volume.* New Rochelle, NY: Cuisenaire.

———. 1983. *Spatial Problem Solving with Cuisenaire Rods.* New Rochelle, NY: Cuisenaire.

Dunham, W. 1990. *Journey through Genius.* New York: John Wiley.

Edmiston, M. C. 1991. *Merlin Book of Logic Puzzles.* New York: Sterling.

Freeman, M. 1986. *Creative Graphing.* New Rochelle, NY: Cuisenaire.

Gamow, G. 1954. *One, Two, Three ... Infinity.* New York: Viking Press.

Gnadesikan, M., R. L. Schaeffer, and J. Swift. 1987. *The Art and Techniques of Simulation.* Palo Alto, CA.: Dale Seymour.

Guenther, J. E., and M. K. Corbitt. 1985. *Using the Newspaper in Secondary Mathematics, a Guide for Teachers.* Washington, DC: American Newspaper Publishers Association Foundation.

Hardy, G. H. 1992. *A Mathematician's Apology.* Reprint. New York: Cambridge University Press.

Hirsch, C. R., ed. 1985. *The Secondary School Mathematics Curriculum. 1985 Yearbook of The National Council of Teachers of Mathematics.* Reston, VA.:

Jensen, R., and D. Spector. 1984. *Teaching Mathematics to Young Children.* Palo Alto, CA: Dale Seymour.

Kasner, E., and J. Newman. 1940. *Mathematics and the Imagination.* New York: Simon & Schuster.

Keenan, E. P., and E. Dressler. 1990. *Integrated Mathematics Course II.* New York: Amsco School.

Kelly, B. 1992. *Using the TI 81 Graphics Calculator to Explore Functions.* Evanston, IL: McDougal, Littell.

Kennedy, M., and S. Bezuska. 1987. *Tesselations Using Logo.* Palo Alto, CA: Dale Seymour.

Kline, M. 1964. *Mathematics in Western Culture.* New York: Oxford University Press.

Munroe, M. E. 1969. *Historical Topics for the Mathematics Classroom. NCTM 31st Yearbook.* Washington, DC: NCTM.

———. 1980. *Problem Solving in School Mathematics. NCTM 1980 Yearbook.* Reston, VA: NCTM.

National Council of Teachers of Mathematics. 1989. *Curriculum and Evaluation Standards.* Reston, VA: NCTM.

———. 1991. *Professional Standards for Teaching Mathematics.* Reston, VA: NCTM.

———. 1992. *Catalog of Educational Materials.* Reston, VA: NCTM.

National Research Council. 1989. *Everybody Counts: A Report to the Nation on the Future of Mathematics Education.* Washington, DC: National Academy Press.

———. 1990. *Reshaping School Mathematics: A Framework for Curriculum.* Washington, DC: National Academy Press.

Oakes, J. 1985. *Keeping Track: How Schools Structure Inequality.* New Haven, CT: Yale University Press.

Payne, J. N. 1975. *Mathematics Learning in Early Childhood.* Reston, VA: National Council of Teachers of Mathematics.

Peitgen, H. O., E. Maletsky, H. Jurgens, T. Perciante, D. Saupe, and L. Yunker. 1991. *Fractals for the Classroom: Strategic Activities,* Vols. 1 and 2. New York: Springer Verlag/NCTM.

Polya, G. 1945. *How To Solve It.* Princeton, NJ: Princeton University Press.

Posamentier, A. S., and J. Stepelman. 1992. *Teaching Secondary School Mathematics.* Columbus, OH: Macmillan.

Resnick, L. B. 1987. *Education and Learning To Think*. Washington, DC: National Academy Press.

Romberg, T. A. 1984. *School Mathematics: Options for the 1990s*. Washington, DC: U.S. Government Printing Office.

Roper, A., and L. Harvey. 1980. *The Pattern Factory*. Palo Alto, CA: Creative.

Ropskopf, M. F. 1970. *Mathematics Education: Historical Perspectives. 1970 Yearbook of the National Council of Teachers of Mathematics*. Reston, VA: NCTM.

Rosen, K. H. 1988. *Discrete Mathematics and Its Applications*. New York: Random House.

Sawyer, W. W. 1955. *Prelude to Mathematics*. Baltimore: Penguin.

Seymour, D., and E. Beardsley. 1988. *Critical Thinking Activities in Patterns, Imagery and Logic: Grades K–3; 4–6*. Palo Alto, CA: Dale Seymour.

Steen, L. A. 1989. "Teaching Mathematics for Tomorrow's World. " *Educational Leadership* 4: 18-23.

Stenmark, J. K. 1989. *Assessment Alternatives in Mathematics*. Berkely, CA: EQUALS, University of California.

Stenmark, J. K., V. Thompson, and R. Cossey. 1986. *Family Math*. Berkely, CA: Lawrence Hall of Science.

Swan, M. 1987. *The Language of Functions and Graphs*. Nottingham, UK: Shell Centre for Mathematics Education.

Townsend, C. B. 1991. *World's Most Baffling Puzzles*. New York: Sterling.

Walter, M. 1985. *The Mirror Puzzle Book*. New York: Parkwest.

Weiss, I. 1989. *Science and Mathematics Education Briefing Book*. Chapel Hill, NC: Horizon Research.

Wiggington, E. 1986. *Sometimes a Shining Moment*. Garden City, NY: Anchor Press.

Willoughby, S. S. 1990. *Mathematics Education for a Changing World*. Alexandria, VA: Association for Curriculum Supervision and Development.

CURRICULUM GUIDES: PROCESS AND DESIGN

by Jurg Jenzer
Principal
Fairfield Center School, Fairfield, Vermont

DESIGNING a quality mathematics curriculum is a complex process. District size, geographic location, funding capability, philosophy, state statutes, and demographic characteristics are some of the many factors that must be considered. Because so many students, teachers, and community members will be affected by the curriculum, many people should participate in its development.

Currently, with the dramatic mathematics education reform movement spearheaded by the National Council of Teachers of Mathematics (NCTM 1989), a nationwide effort to boost mathematics achievement through curriculum change is underway.

A curriculum states how mathematics will be developed in classrooms and schools. Since classroom teachers are the primary users of mathematics curriculum guides, their participation is essential. In addition to giving teachers the opportunity to share their expertise, the development process helps them understand the organization of the curriculum across grade levels and both the usefulness and limitations of curriculum guides.

Highly specific curriculum guides may impress on appearance, but they leave little room for teacher choice and flexibility. Brief and highly generic guides, on the other hand, do give teachers flexibility, but may do little in terms of implementing management objectives or state standards.

Textbooks still drive the curriculum in many districts. Mathematics textbooks offer a scope of mathematics in terms of topics, as well as the sequence in which they are to be taught and learned. As a result, textbooks exert a powerful influence on the curriculum materials developed by states or single school districts. It is not surprising to see textbook language and design reflected in many locally developed mathematics curricula.

Designing curriculum materials at the district level challenges both the textbook industry and the dependency teachers have on textbooks. A curriculum development process, moreover, allows teachers to discover some surprising facts. Alternatives to textbook-style scope and sequence programs in mathematics are not only possible, but can offer more interesting instructional methods as well.

Veteran teachers are already curriculum designers in practice. They tend to use textbooks eclectically, identifying strengths and weaknesses in each book, and picking and choosing from other available resources.

Curriculum is not so much a document as it is professional development. At its best, it is a learning process for all those involved with its development and implementation. The actual curriculum in mathematics, or in any other subject, is the one being taught. The taught curriculum may or may not resemble the planned curriculum (textbook or curriculum guide). This point is crucial. A curriculum process that is organized as a challenging learning opportunity will affect instruction far more than a document alone. A local curriculum process challenges the separation of planning and implementation, an issue all too often obscured by arguments for state or national curriculum initiatives. Unless teachers understand the processes and decisions that generate a curriculum guide, implementation may be in jeopardy.

A curriculum development process with and for teachers provides insight into the decision-making process that generates both textbooks and curriculum guides. It is the process, rather than the document, which provides the time to examine familiar assumptions and practices, and to discuss and test new ones. It also offers a reasonable incentive to get involved.

Finally, curriculum development should not be a one-time event. Schools and districts that have an ongoing curriculum revision process can best react to new standards and methods in teaching mathematics, and therefore can best serve their students and prepare them for the world beyond graduation.

This chapter outlines the numerous steps curriculum designers and committees may take when developing or revising elementary, middle, and/or high school mathematics. For some of these steps, key factors that are considered in order to make informed and responsible decisions are examined.

- Performing a needs assessment
- Defining the mission
- Forming a curriculum committee
- Public input
- Scheduling the project
- Budgeting
- Looking at standards, key topics, design options
- Population analysis (target students)
- Field testing
- Editing, ratification, production, and dissemination
- Adoption process
- Staff development and support
- Monitoring and supervision
- Evaluation and revision

Note that the steps described here are only *typical* of the process involved. The actual steps may differ in your district—there may be more steps or fewer ones, and some steps may occur in a different order. But you will most likely encounter many of these steps at some stage in the process, and this chapter is meant to acquaint you with these steps and many of the related decisions.

Performing a Needs Assessment

The needs assessment is an important part of curriculum planning in any subject area, in that it will provide a direction for the curriculum. The assessment defines the priorities of the curriculum (under local and state standards, and in view of recommendations from national organizations), the goals of curriculum development in mathematics, and the gaps that exist in the current curriculum.

In order to get a clearer picture of the school's mathematics needs, curriculum planners may wish to compare their current and planned curriculum program with those of other states and other districts.

Defining the Mission

Closely related to the needs assessment is the definition of the mission of the school or district. While curriculum implementation may be the province of teachers and school administrators, the mission statement should be developed with members of the school board, teachers, parents, students, state education officials, the private sector, and others within the community. When the mission has already been defined, it is important to relate the curriculum reform effort to this broader perspective. A widely shared understanding of the mission of the school district in curricular and programmatic terms greatly enhances the odds for successful implementation. Formulating a new curriculum is expected to do the following:

- establish a relationship between district goals and instructional programs and methods
- establish a relationship among local programs, state and national standards, laws, and policies
- link curriculum and educational programs with important policy and budgetary decisions
- inform communities about the schools' direction and programs
- ensure a planned and coordinated educational program

Forming a Curriculum Committee

Curriculum designers, whether they are administrators, classroom teachers, or education professors, must be able to work effectively with others. In addition, curriculum designers must have an excellent grasp of the subject, have classroom experience, and communicate effectively in order to be accepted and respected by teachers. At least some of the people on the committee should have previous experience in curriculum development.

In forming a curriculum committee, questions such as these arise about its composition:

- Who wants to participate? Who should be recruited?
- Have prospective participants worked with districtwide curricula before?
- Which teachers can play leadership roles? Who will chair the committee?
- What are the advantages of small versus large committees?
- What types of incentives are available to recruit quality committees?
- How committed are the school board and the administration to the committee's work?
- Are department heads, administrators, program specialists, guidance counselors, par-

ents, students, board members, and business and community representatives on the committee?

In addition, the following questions should be addressed:

- Should all affected schools and grade levels be represented?
- Should elementary committees work separately from middle level or secondary level committees? How will they coordinate transitions between levels?
- How will strand continuity be ensured across grade levels?
- How will a cohesive body of curriculum within a grade level be ensured?

Public Input

Schools—and curriculum development—can benefit a great deal from collaborating with parents, community-based organizations and agencies, the private sector, and institutions of higher education. The question for the curriculum designer is: how can this be organized and what do we do with the information?

The public should certainly be involved in the mission-defining statement. It is also appropriate to include some members of the community on the curriculum committee. However, it cannot be assumed that members of the public know curriculum the way that educators do.

Scheduling the Project

In order to develop a reasonable timetable for a curriculum development process, numerous factors must be considered. These include:

- the scope of the project (e.g., K–4, 5–8, 9–12, K–12)
- the extent of the project (complete revision, partial revision, etc.)
- the number of employees and students affected by the curriculum
- number of participants in the project and their respective work schedules; meeting times
- staff development time that will be required to implement the new curriculum

- available resources (also see below, under "Budgeting")
- existing deadlines (state, local education agency, federal agency mandates)

Since teachers have relatively inflexible classroom schedules, a method must be chosen that will facilitate their participation. Four of the more common methods used are *pullout projects, after-school projects, summer projects,* and *course projects.* Each method offers significant fiscal and procedural advantages.

Pulling teachers out of the classrooms for full-day work sessions allows them the opportunity to concentrate and focus their energies on the project. Typically, this reduces the number of meetings necessary to complete the project. On the other hand, this method requires the services of substitute teachers, which cost money and also requires considerable effort by the teachers, who must prepare entire days of lessons for the substitute. In addition, the frequency with which teachers are pulled out of the classroom must be carefully managed to avoid any potentially negative impact on students.

After-school projects avoid most of the negative implications of the pullout project. Some costs remain if contracts require compensation to personnel for extra duty. The most significant problem with this method, however, is that teachers are often tired after spending a day in the classroom; this can minimize their energy level and can affect the quality of their work. This method also increases the number of meetings needed, since after school meeting time is limited.

Summer projects allow participants to concentrate on the task without the competing demands of the school year. This method incurs personnel costs.

Course projects are usually planned in collaboration with institutions of higher education. The strength of this method is its flexibility of scheduling. Course projects can be organized as evening courses or as intensive two- or three-week work sessions during school vacations. Graduate credits issued by the cooperating college or university can be an attractive incentive for teachers who apply credits toward graduate degrees and/or salary schedules. One well-known problem with this method lies in potential conflicts between

school district personnel and higher education faculty over controlling the project, the mission, and the curriculum itself.

Budgeting

The costs of the project can include:

- compensation for participants (may include hiring substitute teachers)
- consulting costs
- computer costs
- secretarial costs
- production costs (layout, paper, printing, copying, distribution, binders, graphics, etc.)
- administrative costs
- legal costs (reviews for compliance with state and federal laws)

Curriculum developers might want to explore the possibility of outside funding to help support the extra costs involved in the project. Some national foundations provide grants for development of particular curriculum; these organizations often have regional restrictions or will fund only for certain types of curriculum. In addition, some corporations fund educational projects in their state or region. Chapter 3 provides a listing of foundations that offer grants for education projects.

Looking at Standards

It is essential to consult standards—and to keep track of emerging standards—when designing or adapting curriculum materials. As arguments over national curriculum and testing systems fly back and forth, curriculum designers are first and foremost concerned with understanding and evaluating current trends in mathematics instruction. Undoubtedly, national standards will make a difference, but major curriculum decisions remain at the local level.

Every school district has standards which emerge from: *(a)* a multitude of community values; *(b)* the successes and failures of local reform and restructuring efforts in response to pressures from state and federal agencies, and *(c)* the curriculum in use. Curriculum designers must decide how current standards can best be adapted to their local needs.

Curriculum documents, no matter how well designed, will become outdated as new standards and trends emerge across the nation. Teachers in school districts with an ongoing curriculum development process will be better equipped to adapt to new standards from any source.

State and Local Standards

Copies of all statutes, regulations, and policies regulating a curriculum development or revision project should be made available to all who are participating in or are affected by the curriculum process.

Under the U.S. Constitution, the state has the ultimate responsibility for education, and state education agencies define standards and conditions under which schools operate. But the degree to which states regulate curriculum process and development varies widely. In some cases, the state sets standards for mathematics and defines acceptable instructional resources for implementing that curriculum. In other cases, the curriculum process is largely controlled by local educational agencies (LEAs). Curriculum designers need to ascertain the nature and scope of the set standards, and the degree of flexibility LEAs have in interpreting and implementing them (for further details regarding particular states, see chapter 5).

If state regulations on mathematics are specific, diversity among local school districts is restricted. On the other hand, if state regulations are generic, they allow for significant local variation. This opens the door to local emphases on issues such as computers and calculators in elementary mathematics, or the use of manipulatives in teaching fractions; it is possible for some of these emphases to find their way into curriculum guides and school board policies.

In addition to state regulation, there are other factors which may have resulted in de facto standards in local school districts. The district may have adopted a specific instructional methodology that unifies instruction in some way (i.e., common textbook series, adoption of a specific K–8 problem solving strategy, etc.).

Teachers, too, can establish standards when they "import" new ideas, materials, or methods into the school and act as role models for other teachers. In fact, this is the most common way in which mathematics is transformed at the classroom level, and is a good indicator of what can be termed "local standards."

The curriculum document should state clearly that state and local regulations have been addressed.

National Standards

In 1989, the National Research Council published "Everybody Counts: A Report to the Nation on the Future of Mathematics Education." It notes that enrollment in mathematics studies has decreased while its importance for the average citizen has increased. The report states that "Quantitative literacy provides the foundation of technological expertise in the workplace" (National Research Council 1989, 12). The lack of sufficient mathematical training is cited as particularly acute among women and minorities. The report further states that the "transformation of mathematics from a core of abstract studies to a powerful family of mathematical sciences is reflected poorly, often not at all, by the traditional mathematics curriculum" (National Research Council 1989, 43).

The standards issued in 1989 by the National Council of Teachers of Mathematics (NCTM) call for sweeping changes in the way we teach K–12 mathematics. School districts across the country are being asked to review and revise their current curricula in order to incorporate these changes. The general goals of the NCTM standards are that students learn to value mathematics, become confident in their ability to do mathematics, become mathematical problem-solvers, learn to communicate mathematically, and learn to reason mathematically.

In addition to the standards, the NCTM is publishing a series of Addenda which give specific suggestions or amplifications either by grade level or by topic.

Based on the assumption that students are attracted to applications which reflect their world and experiences, the NCTM standards will be most effective when they are translated by curriculum designers and teachers into the local context.

The topics and themes shown in table 1 are based on NCTM recommendations (1989) and should be compared to other sources of topics,

Table 1. Overview of K–12 Mathematics, in Accordance with NCTM Standards (1989)		
Grades K–4	**Grades 5–8**	**Grades 9–12**
Mathematics as problem solving	Mathematics as problem solving	Mathematics as problem solving
Mathematics as communication	Mathematics as communication	Mathematics as communication
Mathematics as reasoning	Mathematics as reasoning	Mathematics as reasoning
Mathematical connections	Mathematical connections	Mathematical connections
Estimation	Number and number relationships	Algebra
Number sense and numeration	Number systems and number theory	Functions
Concepts of whole number operations	Computation and estimation	Geometry from a synthetic perspective
Whole number computation	Patterns and functions	Geometry from an algebraic perspective
Geometry and spatial sense	Algebra	Trigonometry
Measurement	Statistics	Statistics
Statistics and probability	Probability	Probability
Fractions and decimals	Geometry	Discrete mathematics
Patterns and relationships	Measurement	Conceptual underpinnings of calculus
		Mathematical structure

issues, themes, and strands in K–12 mathematics.

Examining Key Topics in Mathematics

While the research base and literature regarding elementary, middle, and high school mathematics is rich and at times overwhelming, curriculum designers cannot afford to ignore it (for a basic introduction to the literature, see chapter 1). The most critical decision to be made is how to deal with it at all. Should teachers be involved in reviewing and discussing this literature, or should the facilitator conduct a review and then brief the participants? This review allows participants to gain a critical perspective on current controversies in mathe-

matics, as well as their own assumptions on what a mathematics curriculum is and will be in the future.

One approach is to review other school districts' curricula or current textbooks. On the one hand, this review yields a great deal of practical comparisons between districts as well as a sense of security (commonality) in the decisionmaking process. On the other hand, it may promote the continued use of potentially obsolete curriculum design features and instructional topics.

Reviewing three sources of information—the publishing sector (textbooks), the school sector (current curricula), and the academic sector (research)—will ensure a balanced overview regarding the issues of mathematics education. The following is a selection of topics under discussion at the time of publication (for a more detailed rendition of current controversies, turn to chapters 1 and 4):

- translating (NCTM) standards into curriculum
- influence of standardized tests on curriculum
- alternative assessments
- selecting appropriate textbooks and software, and the criteria for selection
- developing communication skills in mathematics
- developing spatial thinking
- mathematics phobia—reality or myth?
- gender/race/ethnic group differences in mathematics achievement
- presenting real-world applications of mathematics
- teaching mental arithmetic and estimation skills
- developing number sense in elementary school children
- the role and effectiveness of manipulatives
- students' use of technology in the classroom
- appropriate grade, age, or developmental levels for algebra
- eliminating tracking in mathematics 9–12
- increasing the mathematics requirement for high school graduation
- updating secondary mathematics to reflect the growth of topics and applications in contemporary mathematics

Curriculum designers are advised to develop their own checklists. This allows participants to share their questions and personal philosophies on mathematics instruction, and on topics and controversies in which they are interested. The use of checklists also supports the idea that all opinions are important, and that some compromise is a natural part of the curriculum development process.

The next step could be the most difficult in your curriculum development and design procedures. The curriculum committee must make choices regarding the organization of the guide, including the identification of strands or domains (i.e., numeration, operations, geometry, measurement, statistics), the relationship between strands (i.e., should "problem solving" be separated from or integrated into other

strands?), and the specific topics to be covered at each grade level, grade cluster, or developmental level.

Choosing Curriculum Features and Design Options

When considering the design of the curriculum document, developers must analyze the teachers who are going to use it as an audience, and decide what type of curriculum guide they would find most useful. The curriculum designers' own classroom experience and interaction with teachers will greatly facilitate the assessment of what teachers need and/or want.

- Which topics should be taught at certain grade levels in order to secure a high degree of achievement for most learners?
- Should a curriculum be much like a textbook and describe all aspects of instructional work; should it specify the local or state goals and objectives?
- Should a curriculum be brief and merely outline the instructional scope and sequence "blueprint fashion," and thus leave most curriculum decision making to teachers?
- Should curriculum guides contain one feature or multiple features or types of information?
- Does the actual curriculum content or design make any difference to teachers?
- Can curriculum guides be engaging to teachers, and thus shift the incentive for implementation from top-down mandates to the inherent benefits of the document itself?
- If teachers have greatly varied professional needs and practices, should the curriculum in fact consist of multiple documents with varied content?

Mathematics curriculum guides may include the following features:

- list of objectives
- suggestions for instructional strategies and/or student activities
- list of available resources
- evaluation/testing

The design of the guide would be influenced by the characteristics of the school district, the mission of the curriculum project, administrative agendas to empower teachers or increase centralized control, and other decisive factors. There is little evidence, if any, suggesting that one format works better than another.

The search for alternatives to traditional curriculum designs and features has been an ongoing process. Some school districts have experimented with integrated or interdisciplinary designs, challenging many fundamental assumptions not only about the nature of curriculum but the nature of schooling as well. For an historical perspective on the interdisciplinary curriculum, see Vars 1991.

Alternative designs, for the most part, have been developed at the local level, often in conjunction with more traditional subject-based guides. Two possible designs, the thematic curriculum and the interdisciplinary curriculum, have emerged.

In the thematic model, a theme—chosen by teachers, administrators, students, parents, or combinations of these groups—serves as a focus for study and investigation for a given period of time (e.g., day, week, month). The theme is then examined from a variety of perspectives or traditional disciplines such as mathematics, art, science, home economics, and physical education.

For a comprehensive review of interdisciplinary curriculum models and procedures, see Palmer 1991; Jacobs 1989; Miller et al. 1990; Drake 1991; and Pappas et al. 1990.

Population Analysis (Target Students)

While teachers are the audience for whom curricula are written, the real beneficiaries of high quality guides are the students. The demographic characteristics of the student population is a key factor in defining curriculum content, and in choosing features, scope, resource specifications, etc. For example, a curriculum developed in an affluent suburb may work well there, but fail completely when implemented in an inner-city school. This consideration is particularly important for districts with diverse student populations, and for schools in districts with diverse cultural and economic settings. In such instances, a districtwide curriculum must be carefully monitored.

The characteristics of a given student population determine the teacher's ability to implement a curriculum guide. While such guides are usually formatted to outline performance expectations or specific topics for, say, "all" ninth graders or "all" fifth graders, such "generic" design assumptions about student populations may simply be ignoring realities teachers must deal with on a daily basis. In fact, any given grade level classroom will include students above and below grade level in mathematics ability. Some classes may be comprised of up to 25 percent of students who have been identified as needing some special education, requiring the teacher to make drastic adaptations to curriculum expectations. In this light, the blame for "failing to implement" may rest not with the teacher but with the guide itself. It also explains the surging interest in curriculum guides designed not for grade levels but for developmental levels.

Field Testing

There is no need to separate the drafting process from the field testing of a new guide during curriculum development; indeed, it may be counterproductive. The best field testing involves a limited number of teachers who are either participants on the curriculum committee or who have been experimenting with changing a current mathematics program (curriculum-in-use). Tests should be conducted in cooperation with the appropriate administrators and faculty.

Encouraging cooperative field testing during the drafting stage has numerous advantages. It allows teachers an early look at upcoming changes in the mathematics program. If the new program, for example, builds heavily upon the NCTM standards, the transition from a traditional curriculum to an NCTM-type curriculum will require time for extensive staff development activities. Building ownership for a new mathematics program throughout the design process will greatly facilitate successful implementation.

The final stage of field testing should be conducted with a completed curriculum at a building or district level. One or two years of a district level field test may be required for a gradual transition; teachers should be asked to provide feedback to the curriculum committee in order to contribute practical information for final editing work. This should include, if possible, one-on-one interviews with each teacher who participated in the field test preferably conducted by curriculum committee participants. Consider the following questions:

- Have you found the curriculum useful? How has it been useful?
- Do you have any reservations or concerns you would like to share?
- Are suggested activities useful for implementation?
- What do you need to implement this curriculum?
- Is there anything else you would like the curriculum committee to know?

Editing, Ratification, Production, and Dissemination

The tasks of editing, producing, and disseminating the guide are closely tied to the available resources, budget, and timetable of the committee.

Curriculum designers need to discuss ratification procedures with the administration in order to comply with state statutes, regulations, and local school board policies. Such procedures often involve local school board action.

Adoption Process

The adoption of a curriculum guide at the building level is by far the most challenging phase of the curriculum process. This is the crucial point at which curriculum is actually translated into instruction. The complexity of the curriculum, the degree to which assumptions about mathematics are spelled out or hidden, the volume of the curriculum, and the teachers' disposition toward the curriculum guide are all factors that affect adoption.

Although a curriculum may be mandated, it is not necessarily implemented. Teachers must first decide whether or not they will work with that curriculum. Curriculum adoption at the classroom level is, at least in part, a personal decision, and curriculum designers must take

several factors into consideration which influence that decision:

- degree to which a teacher "owns" the curriculum
- level of experience working with district curriculum
- degree to which the teacher depends on the textbook
- the administration's flexibility and willingness to support teachers
- availability of necessary resources and materials
- level of support from community, school board, state officials, etc.
- availability and quality of staff development opportunities

Staff Development and Support

There is little disagreement among authors that staff development can be an effective implementation strategy (Fullan 1990; Joyce et al. 1988; Goodlad 1990; Holmes 1990; Loucks-Horsely et al. 1987; Schon 1987; Hirsh 1989; Sergiovanni 1982). Teachers charged with implementing the curriculum may require specific training. Curriculum designers often see multiple training needs. To assess those needs as accurately as possible and develop an effective staff development program, administrators need to examine several factors:

- Who are the users of the new curriculum? This information will determine the scale of training which will have to be provided to the district (i.e., how many classroom teachers per grade level, how many specialists, etc).
- Which local, state, or national goals and standards are being adopted? This information must be included in order to effectively train and inform staff about such standards or goals.
- Has a timetable been adopted? Ideally, staff development targeted to facilitate curriculum implementation should be planned over several years and linked with a clear message to teachers that adoption is a longitudinal learning process.
- What types of resources are needed/ available? Available resources such as trainers and staff development consultants

must be identified and included in the district's staff development budget. In the case of mathematics, federal assistance (Eisenhower Mathematics and Science Grants, Chapter II funds, etc.) may be available.

• Who has served on the curriculum committee? Educators who have participated in the curriculum design process can lead discussions or model implementation in their own classrooms.

• What are current topics in K-12 mathematics? Current topics, controversies, or problems in K-12 mathematics should have been identified in the review of the literature; this information can be helpful in defining topics for staff development workshops. In addition, authors and researchers may be available to serve as trainers or lecturers.

• Which design format has been adopted? If the chosen format differs significantly from that of previously used district curricula, some staff development may be required to introduce teachers to the new format.

• What are the results of the field test? Teachers will often reveal staff development needs and topics when interviewed during the field testing process.

Monitoring and Supervision

Supervisors who observe teachers are in a prime position to help with the implementation process. These supervisors provide opportunities for teachers to share and discuss problems and uncertainties with regard to their implementation efforts. In order for supervisors to be effective in this role, they should keep the following guidelines in mind:

• Prove to teachers that supervision aims to support the teachers' many tasks. In fact, it may be necessary to separate the supervisor's two roles, and to emphasize the *supervision* role over the *evaluation* role.

• Be familiar with the curriculum and the teachers' professional development: focus on issues, students, lesson plans, and instructional techniques that relate specifically to that teacher's classroom.

• Provide feedback to the curriculum designer or committee that handles staff development or curriculum revision.

Evaluation and Revision

Evaluation

After all the dust has settled, all the decisions have been made, and, at long last, a new mathematics guide has been adopted for implementation, all participants surely deserve a rest. For the curriculum designer, however, challenges still lie ahead.

Curriculum documents require a thorough evaluation. Changes in the field of mathematics, new research information, new standards, and better instructional methods appear quickly, making curricular adjustments necessary on an ongoing basis. School districts must maintain a curriculum process beyond the mere document. Curriculum designers should communicate this issue to all educators, parents, and policymakers.

Traditionally, achievement scores are held to be indicators of the quality of a mathematics curriculum. Achievement scores, however, are only one source of curriculum evaluation data, and they are useful only insofar as the tested curriculum correlates with both the planned and the taught curriculum (English 1980).

Again, curriculum designers face several critical decisions:

• What is the purpose of evaluating the curriculum? Is it a tool to evaluate teachers, students, or schools?

• Who will participate in evaluating the curriculum? Should the same curriculum committee be used, or would a different perspective—and therefore a fresh committee—serve better? Another option is to hire an independent evaluator of the process and the document.

The curriculum evaluation committee should be empowered to investigate the strengths and weaknesses of an implemented curriculum in full pursuit of the truth.

Curriculum designers should clarify the evaluation process in advance. Included in this process is a clarification of the following:

• who the evaluation committee *participants* will be

- to what degree curriculum process *goals* have been achieved
- whether the overall *mission* of the curriculum project has been reached
- whether curriculum *content* is appropriate in light of district characteristics and mission objectives
- whether *instruction* is based on the curriculum
- whether the *assessment* process measured the taught curriculum against the planned curriculum

The evaluation, then, will yield a *needs assessment* to clarify what types of resources, time demands, training and workshops, and supervision strategies must be in place for successful implementation to continue. In addition, a *revision plan* should be issued for the curriculum guide itself, giving specifics for additions or deletions.

Table 2 provides a sampling of the goals and questions that could be addressed during evaluation of a curriculum.

Revision

Planning for the next generation mathematics program begins now. The first step consists of educating the school board, the staff, and the community, that implementation will yield a variety of positive experiences as well as numerous problems. Information of this sort must be collected and organized in view of future revisions of the curriculum. Any curriculum has room for improvement and must be dynamic enough to incorporate upcoming changes in the field of mathematics.

Curriculum designers should therefore present a *curriculum development process,* if not at the outset, then certainly at the time of adoption and implementation. This will clarify the ground rules for all concerned. A development process at the district level should incorporate all curriculum areas. However, in order to remain manageable, the district should avoid revising all curricula during the same year.

Revising a mathematics curriculum can be as complex as designing it in the first place. Numerous decisions must be made and numerous sources of information considered:

- Is a revision necessary in view of available information, or should the revision cycle be changed (to revise sooner or later than planned)?

- Should a committee be established to carry out the revision?
- Have conditions changed since the mathematics curriculum was first implemented (new standards, new testing systems, new staffing patterns, significant changes in enrollment, budget crises, etc.)?

Curriculum designers often battle the assumption that completed curriculum guides finalize the curriculum process—in reality, a curriculum guide marks the *beginning* of a curriculum. A curriculum guide that looks the same five years after it was written will most likely be outdated. Teachers tend to leave them on the shelf, and for good reason.

Each school or district must have a process in place for curriculum development and revision. The schools and districts that have an ongoing curriculum revision process can best react to new standards and methods in teaching mathematics, and in all other K–12 subjects—and these are the schools that can best serve their students and prepare them for the world beyond graduation.

References

Anderson, S. A., et al. 1987. *Curriculum Process: Yale Public Schools.* Yale, MI: Yale Public Schools.

Apple, M. W. 1979. *Ideology and Curriculum.* Boston: Routledge.

Argyris, C. 1982. *Reading, Learning, and Action: Individual and Organizational.* San Francisco: Jossey-Bass.

Aronowitz, S., and H. A. Giroux. 1985. *Education under Siege.* South Hadley, MA: Bergin & Garvey.

Bradley, L. H. 1985. *Curriculum Leadership and Development Handbook.* Englewood Cliffs, NJ: Prentice-Hall.

Caine, R. N., and G. Caine. 1991. *Making Connections—Teaching and the Human Brain.* Alexandria, VA: Association for Supervision and Curriculum Development.

California State Department of Education. 1982. *Handbook for Planning an Effective Mathematics Program K–12.* Sacramento, CA.

Campbell, M., et al. 1989. "Board Members Needn't Be Experts To Play a Vital Role in

Table 2. Curriculum Evaluation: K–5 Mathematics		
Rationale The evaluation will identify strengths and weaknesses in the K-5 mathematics program.	**Participants** Curriculum committee members; supervisory personnel; consultant	**Goals** Quality of document (design, format, volume, type of information); effectiveness of staff development; effectiveness of supervision; quality of achievement
Mission To what degree has the curriculum been implemented? Have adopted standards and goals been achieved? Has mathematics achievement been raised? Have teachers adopted the curriculum?	**Revision** K section on problem solving; revise 4th-grade program; redesign 4-6 staff development program; incorporate NCTM standards for 3rd grade	**Content** Is the information in the document accurate, verifiable, measurable, "teachable," developmentally appropriate?
Needs Assessment K teachers must be involved in redesigning the problem-solving strand. The grade 4 curriculum is too demanding, geared to advanced students. Staff development efforts for 4-6 teachers ineffective. Grade 3 goals do not meet intent of NCTM standards.	**Assessment** Does the testing program cover this curriculum? Have teachers changed their assessment tools and strategies effectively to accommodate the curriculum? Have achievement scores changed with implementation of this curriculum?	**Instruction** Do teachers have sufficient time to teach the curriculum? Have teachers changed instructional methods? Have teachers used the curriculum in planning instruction? Are instructional resources (textbooks, manipulatives, software, etc.) available?

Curriculum." *American School Board Journal* 176 (April): 30–32.

Carr, J. F., and D. E. Harris. 1992. *Getting It Together: A Process Workbook for Curriculum Development, Implementation, and Assessment.* Boston: Allyn & Bacon.

Connelly, F. M., and D. J. Clandinin. 1988. *Teachers as Curriculum Planners: Narratives of Experience.* New York: Teachers College Press.

Doll, R. C. 1989. *Curriculum Improvement: Decision Making and Process.* 7th ed. Boston: Allyn & Bacon.

Drake, S. M. 1991. "How Our Team Dissolved the Boundaries." *Educational Leadership* 49 (October): 20–22.

Driscoll, M., and J. Confrey. 1986. *Teaching Mathematics: Strategies That Work.* Portsmouth, NH: Heinemann.

English, F. W. 1980. "Improving Curriculum Management in the Schools." Occasional Paper 30. Washington, DC: Council for Basic Education.

———. 1988. *Curriculum Auditing.* Lancaster, PA: Technomic.

Frey, K., et al. 1989. "Do Curriculum Development Models Really Influence the Curriculum?" *Journal of Curriculum Studies* 21 (November-December): 553–59.

Fullan, M. G. 1990. "Staff Development, Innovation, and Institutional Development." In *1990 ASCD Yearbook*, 3–25. Alexandria,

VA: Association for Supervision and Curriculum Development.

Giroux, H. A. 1983. *Theory and Resistance in Education.* South Hadley, MA: Bergin & Garvey.

Glassberg, S. *Developmental Models of Teacher Development.* ERIC: ED 171 658.

Glatthorn, A. A. 1987. *Curriculum Leadership.* Glenview, IL: Scott, Foresman.

Glickman, C. 1990. *Supervision of Instruction: A Developmental Approach.* 2d ed. Boston: Allyn & Bacon.

Goodlad, J. I. 1990. *Teachers for Our Nation's Schools.* San Francisco: Jossey-Bass.

Harris, D. E., and J. Jenzer. 1990. *The Search for Quality Curriculum Design: Four Models for School Districts.* Paper presented at National ASCD Conference, San Antonio, TX.

Harste, J. 1989. *New Policy Guidelines for Reading: Connecting Research and Practice.* Urbana, IL: National Council of Teachers.

Hirsh, S. 1989. "Long-Range Planning for Staff Development." In *The Developer.* Oxford, OH: National Staff Development Council.

Holmes Group. 1990. *Tomorrow's Schools: Principles for the Design of Professional Development Schools.* East Lansing, MI.

Jacobs, H. H. 1989. *Interdisciplinary Curriculum—Design and Implementation.* Alexandria, VA: Association for Supervision and Curriculum Development.

Joyce, B., and B. Showers. 1988. *Student Achievement through Staff Development.* White Plains, NY: Longman.

Kanpol, B., and E. Weisz. 1990. "The Effective Principal and the Curriculum—A Focus on Leadership." *NASSP Bulletin* 74 (April): 15–18.

Kaplan, R. G., T. Yamamoto, and H. P. Ginsberg. 1989. "Teaching Mathematics Concepts." In *The 1989 ASCD Yearbook: Toward the Thinking Curriculum: Current Cognitive Research.* Alexandria, VA: Association for Supervision and Curriculum Development.

Kelman, P., et al. 1983. *Computers in Teaching Mathematics.* Reading, MA: Addison-Wesley.

Loucks-Horsely, S., et al. 1987. *Continuing To Learn: A Guidebook for Teacher Development.* Andover, MA: Regional Laboratory for Educational Improvement of the Northeast and the Islands.

McNeil, J. D. 1985. *Curriculum—A Comprehensive Introduction.* 3d ed. Boston: Little, Brown.

McNergney, R., and C. Carrier. 1981. *Teacher Development.* New York: Macmillan.

Miller, J., B. Cassie, and S. M. Drake. 1990. *Holistic Learning: A Teacher's Guide to Integrated Studies.* Toronto: Ontario Institute for Studies in Education.

Montana State Office of the Superintendent of Public Instruction. 1990. *The Curriculum Process Guide: Developing Curriculum in the 1990s.* Helena, MT.

National Council of Teachers of Mathematics. 1982. *How To Evaluate Your Mathematics Curriculum.* Reston, VA.

———. 1989. *Curriculum and Evaluation Standards for School Mathematics.* Reston, VA.

National Research Council. 1989. *Everybody Counts: A Report to the Nation on the Future of Mathematics Education.* Washington, DC.

Oja, S. N., and M. Ham. 1987. *A Collaborative Approach to Leadership in Supervision.* Project funded by U.S. Department of Education (OERI), ED no. 400-85-1056. Washington, DC.

Palmer, J. M. 1991. "Planning Wheels Turn Curriculum Around." *Educational Leadership* 42 (October): 57–60.

Pappas, C. C., B. Z. Kiefer, and L. S. Levstik. 1990. *An Integrated Language Perspective in the Elementary School.* White Plains, NY: Longman.

Schon, D. 1987. *Educating the Reflective Practitioner.* New York: Basic Books.

Sergiovanni, T. 1982. "Toward a Theory of Supervisory Practice: Integrating Scientific, Clinical, and Artistic Views." In *Supervision of Teaching.* Alexandria, VA: Association for Supervision and Curriculum Development.

Sprinthall, N. A., and L. Thies-Sprinthall. 1983. "The Teacher as an Adult Learner: A Cognitive-Developmental View." In *Eighty-second Yearbook of the National Society for the Study of Education*, 13–35. Chicago, IL: National Society for the Study of Education.

Thies-Sprinthall, L. 1986. "A Collaborative Approach to Mentor Training: A Working Model." *Journal of Teacher Education* 19 (November/December): 13–20.

United States Department of Education. 1983. *A Nation At Risk: The Imperative for Educational Reform.* Washington, DC.

Vars, G. F. 1991. "Integrated Curriculum in Historical Perspective." *Educational Leadership* 49 (October): 14–15.

Wood, R. C., ed. 1986. *Principles of School Business Management.* Reston, VA: Association of School Business Officials International.

Young, H. J. 1990."Curriculum Implementation: An Organizational Perspective." *Journal of Curriculum and Supervision* 5 (Winter): 132–49.

FUNDING CURRICULUM PROJECTS

THE greatest challenge curriculum developers often face is locating money to finance their projects. We hear that money is available for such projects, but are at a loss to find how it can be accessed. Frequently, it requires as much creativity to locate financing as to generate the curriculum. This chapter includes information on three types of funding available for education projects:

1. Federal programs that provide money for special school projects
2. Foundations that have recently endowed education projects, programs for those with a primary language other than English, or services directed at increasing literacy
3. Foundations that identify education, including special projects, as a mission.

Although not all of the projects reviewed are specifically for a school setting, their mission is similar: to increase math proficiency in school-age children through twelfth grade. It is important to keep your goal in mind when seeking funding sources. Think of other areas that are similar to yours. If you are seeking money for a math enrichment program for disadvantaged children, don't overlook foundations that have supported other projects for the economically disadvantaged. Also consider foundations that underwrite undertakings for children, youth, or other minority groups.

When seeking a potential funding source for the project, first review any information that is available about the foundation. Look specifically at the areas of:

- Purpose: Is a mission of the foundation to provide money for education?
- Limitations: Are there specific geographic requirements? Are there some areas that are disqualified?
- Supported areas: Does the foundation provide funding for special projects?
- Grants: After reviewing the education projects that have been funded, does it appear that the organizations and projects are similar to yours?

Your search will be most useful if you also keep these questions in mind:

- Has the foundation funded projects in your subject area?
- Does your location meet the geographic requirements of the foundation?
- Is the amount of money you are requesting within the grant's range?
- Are there foundation policies that prohibit grants for the type of support you are requesting?
- Will the foundation make grants to cover the full cost of a project? Does it require costs of a project to be shared with other foundations or funding sources?
- What types of organizations have been supported? Are they similar to yours?
- Are there specific application deadlines and procedures, or are proposals accepted continuously?

This information can be found in the annual report of the foundation or in *Source Book Profiles*. Many of the larger public libraries maintain current foundation directories. If yours does not, there are Foundation Center Libraries located at:

79 Fifth Avenue
New York, NY 10003-3050
(212) 620-4230

312 Sutter Street
San Francisco, CA 94180
(415) 397-0902

1001 Connecticut Avenue, NW
Suite 938
Washington, DC 20036
(202) 331-1400

1442 Hanna Building
1442 Euclid Avenue
Cleveland, OH 44115
(216) 861-1934

Identifying appropriate foundations is the first step in your quest for money. The next step is initiating contact with the foundation, either by telephone or letter. It is a good idea to direct your inquiry to the person in charge of giving; otherwise, your letter could easily go astray. A phone call to the foundation will provide you with the necessary information.

Federal Programs That Provide Money for Special School Projects

Jacob B. Javits Gifted and Talented Students
Research Applications Division
Programs for the Improvement of Practice
Department of Education
555 New Jersey Avenue, NW
Washington, DC 20202-5643
(202) 219-2187
Provides grants for establishing and operating model projects to identify and educate gifted and talented students.

Eisenhower Mathematics and Science
 Education—State Grants
Division of School Effectiveness
School Improvement Programs
Office of Elementary and Secondary Education
Department of Education
400 Maryland Avenue, SW—Room 2040
Washington, DC 20202
(202) 732-4062
Supports programs to improve teacher skills and instruction in mathematics and science in elementary and secondary schools and institutions of higher learning.

Teacher Preparation and Enhancement
Division of Teacher Preparation and
 Enhancement
National Science Foundation
1800 G Street, NW
Washington, DC 20550
(202) 357-7073
Provides funding for projects that support networking activities to encourage talented men and women to remain in K-12 science- and math-teaching careers. Restricted to areas of astronomy, atmospheric sciences, biological sciences, chemistry, computer science, earth sciences, engineering, information science, mathematical sciences, oceanography, physics, social sciences, and technology.

Materials Development, Research, and Informal
 Science Education
National Science Foundation
1800 G Street, NW
Washington, DC 20550
(202) 357-7452
Provides funds to support the development of new and improved research models and materials resources for the precollege educational system in mathematics, science, and technology.

Minority Math Science Leadership Develop-
 ment Recognition
Department of Energy
MI, Room 5B-110
1000 Independence Avenue, SW
Washington, DC 20585
(202) 586-1593
Promotes education of minorities in the areas of science and math.

National Program for Mathematics and Science
    Education
FIRST
Office of Educational Research and
    Improvement
Department of Education
Washington, DC 20208-5524
(202) 357-6496
Provides funds for projects of national signifi-
cance in mathematics and science instruction in
public and private elementary and secondary
schools.

The Secretary's Fund for Innovation in
    Education
Department of Education
FIRST
Office of Educational Research and
    Improvement
Washington, DC 20208-5524
(202) 219-1496
Funding for educational programs and projects
that identify innovative educational approaches.

Foundations That Have Recently Endowed Education Projects Related to Mathematics

The Abell Foundation, Inc.
116 Fidelity Building
210 North Charles Street
Baltimore, MD 21201-4013
(301) 547-1300
Contact: Robert C. Embry Jr., President
  · $16,000 to Rowland Park Country School
    for teacher training, focusing on use of
    geometry in teaching mathematics.
Giving limited to Maryland, with a focus on
Baltimore.

ARCO Foundation
515 South Flower Street
Los Angeles, CA 94104
(415) 421-2629
Contact: Eugene R. Wilson, President
  · $10,000 to Mathcounts, Alexandria, VA,
    for high school programs in Alaska,
    Colorado, and Texas.
Giving primarily in San Francisco Bay
area, CA.

The Bush Foundation
East 900 First National Bank Building
332 Minnesota Street
St. Paul, MN 55101
(612) 227-0891
Contact: Humphrey Doermann, President
  · $455,000 to American Indian Science
    Engineering Society, Boulder, CO, toward
    planning grant to design math/science
    program for Indian students in Minnesota,
    North Dakota, and South Dakota.
Giving primarily in Minnesota, North Dakota,
and South Dakota.

Carnegie Corporation of NY
437 Madison Avenue
New York, NY 10022
(212) 371-3200
Contact: Dorothy W. Knapp, Secretary
  · $25,000 to Center for Applied Linguistics,
    Washington, DC, for dissemination of
    national project to improve math and
    science education in middle schools
    through language instruction.
  · $93,750 to the University of Chicago
    toward an integrated secondary school
    mathematics curriculum.
  · $159,000 to National Academy of Sci-
    ences, Washington, DC, for production
    and disemination of Spanish-language
    version of elementary school mathematics
    kit for parents and children.

The Chatlos Foundation, Inc.
P.O. Box 915048
Longwood, FL 32791-5048
(407) 862-5077
Contact: William J. Chatlos, President
  · $13,076 to School Board of Seminole
    County, Sanford, FL, for summer camp in
    mathematics.

The Edna McConnell Clark Foundation
250 Park Avenue, Room 900
New York, NY 10017
(212) 986-7050
Contact: Peter Bell, President
  · $110,000 to Interface Institute, Oakland,
    CA, for programs helping underachievers
    excel in math and science in three junior
    high schools in Oakland.

The Cleveland Foundation
1422 Euclid Avenue, Suite 1400
Cleveland, OH 44115-2001
(216) 861-3810
Contact: Steven A. Minter, Executive Director
 • $55,000 to Cleveland Education Fund as a four-year grant for Model Mathematics Project in Glenville and John Adams high schools.
 • $15,000 to East Cleveland City Schools for development of mathematics component of East Cleveland School District's Science/Mathematics Enrichment Center.
Giving limited to greater Cleveland area. Initial approach should be through a letter.

Cray Research Foundation
1440 Northland Drive
Mendota Heights, MN 55120
(612) 683-7386
Contact: William C. Linder-Scholer, Executive Director
 • $75,000 to Mathcounts, Alexandria, VA, for national sponsorship.
Giving primarily in Minnesota and Wisconsin for science and engineering education.

The Cullen Foundation
P.O. Box 1600
Houston, TX 77251
(713) 651-8835
Contact: Joseph C. Graf, Executive Secretary
 • $40,000 to Kinkaid School, Houston, TX, for Houston Independent School Disctrict Math/Science Institute.
Limited to Texas, with emphasis on Houston.

The Aaron Diamond Foundation
1270 Avenue of the Americas, Suite 2624
New York, NY 10020
(212) 757-7680
Contact: Vincent McGee, Executive Director
 • $10,000 to Institute for Schools of the Future for early childhood mathematics program.
Giving limited to New York City.

Geraldine R. Dodge Foundation, Inc.
163 Madison Avenue
P.O. Box 1239
Morristown, NJ 07962-1239
(201) 540-8442
 • $10,000 to Newark Academy, Livingston, NJ, toward summer institute pilot program of mathematics and language arts for motivated 7th- and 8th-grade minority students, in joint venture with Newark public school system.
 • Support to various programs benefitting disadvantaged children.
Giving primarily in New Jersey, with support for local projects limited to Morristown–Madison area.

Exxon Education Foundation
225 East John W. Carpenter Freeway
Irving, TX 75062-2298
(214) 444-1104
Contact: E. F. Ahnert, Executive Director
 • $40,280 to The National Council of Teachers of Mathematics, Reston, VA, to develop strategies for including discrete mathematics topics in high school curriculum.
 • Funding to multiple school districts throughout U.S. to implement math programs in early grades.

Fireman's Fund Foundation
777 San Marin Drive
Novato, CA 94998
(415) 899-2757
Contact: Barbara B. Friede, Director
 • $10,000 to Center for Excellence in Education, McLean, VA, to improve teaching skills of teachers and provide programs for talented U.S. high school students in areas of math and science.
Giving primarily in the counties of San Francisco, Marin, and Sonoma, CA.

The Ford Foundation
320 East 43rd Street
New York, NY 10017
(212) 573-5000
Contact: Barron M. Tenny, Secretary
 • To cooperating school districts of the St. Louis suburban area, St. Louis, MO: (1) $60,000 to administer collaboratives designed to strengthen mathematics teaching in inner-city high schools and (2) $60,000 for supplement to administer urban mathematics collaborative.

Ford Motor Company Fund
The American Road
Dearborn, MI 48121
(313) 845-8712
Contact: Leo J. Brennan, Junior Executive Director

• $20,000 to Atlanta, GA, public schools for math and science project.

Giving primarily in areas of company operations nationwide, with special emphasis on Detroit and Michigan.

Gates Foundation
3200 Cherry Creek South Drive, Suite 630
Denver, CO 80209-3247
(303) 722-1881
Contact: F. Charles Froelicher, Executive Director
• $50,000 to Colorado Department of Education for testing of 24,000 students between the ages of 9 and 13, to see how they compare with their counterparts in 22 foreign countries.

Giving limited to Colorado, especially the Denver area, except for foundation-initiated grants.

General Motors Foundation, Inc.
13-145 General Motors Building
3044 West Grand Boulevard
Detroit, MI 48202-3091
(313) 556-4260
Contact: D. R. Czarnecki
• $75,000 to Mathcounts, Alexandria, VA, for general support.

Giving primarily in plant cities where company has significant operations.

The Gold Family Foundation
159 Conant Street
Hillside, NJ 07205
(908) 353-6269
Contact: Meyer Gold, Manager
• $10,000 to Mason Early Education Foundation, Princeton, NJ, to enhance the elementary school math and expository writing programs.

Support primarily for Jewish organizations.

The Greenwall Foundation
Two Park Avenue, 24th floor
New York, NY 10016
(212) 679-7266
Contact: William C. Stubing, President
• $35,000 to Playing to Win, New York City, for MathTech, a project involving teachers, parents, and children in Community School District 4.

GTE Foundation
One Stamford Forum
Stamford, CT 06904
(203) 965-3620
Contact: Maureen Gorman, Secretary and Director, Corporate Social Responsibility
• $10,000 to Mathcounts, Alexandria, VA.

Giving limited to areas of company operations.

The Hearst Foundation, Inc.
East of Mississippi River:
888 Seventh Avenue, 27th floor
New York, NY 10106-0057
(212) 586-5404
Contact: Robert M. Frehse, Junior Executive Director
West of Mississippi River:
90 New Montgomery Street, Suite 1212
San Francisco, CA 94105
(415) 543-0400
Contact: Thomas Eastham, Vice-President and Western Director
• $25,000 to Institute for the Development of Urban Educators, Project Interface, Oakland, CA, for tutoring programs in math and science for low-income minority secondary students.

Vira I. Heinz Endowment
30 GNC Tower
625 Liberty Avenue
Pittsburgh, PA 15222-3115
(412) 391-5122
Contact: Alfred W. Wishart Jr., Executive Director
• $26,631 to University of Pittsburgh for mathematics learning activities aimed at improving college preparation for 8th- and 9th-grade black students.

Giving limited primarily to Pittsburgh and western Pennsylvania; support to other areas will be considered for projects on a national or international basis.

The Hitachi Foundation
1509 22nd Street, NW
Washington, DC 20037
(202) 457-0588
Contact: Robyn L. James
• $120,000 to Maine Center for Educational Services, Auburn, ME, to help five high schools with mathematics and science curriculum.

The Martha Holden Jennings Foundation
710 Halle Building
1228 Euclid Avenue
Cleveland, OH 44115
(216) 589-5700
Contact: Dr. Richard A. Boyd, Executive
Director
- $16,550 to Cleveland Public Schools,
Cleveland, OH, for Whole School Ap-
proach to renovate math instruction.
Limited to Ohio.

W. K. Kellogg Foundation
400 North Avenue
Battle Creek, MI 49017-3398
(616) 968-1611
Contact: Nancy A. Sims, Executive Assistant,
Programming
- $2,500,000 to Battle Creek School District
for an improved math/science curriculum
for K–12.

The Medtronic Foundation
7000 Central Avenue, NE
Minneapolis, MN 55432
(612) 574-3029
Contact: Jan Schwarz, Manager
- $12,500 to Minneapolis Public Schools to
work with the community on general
curriculum development.
Giving primarily in areas of company
operations.

The Morgan Guaranty Trust Company of NY
Charitable Trust
60 Wall Street
New York, NY 10260
(212) 648-9672
Contact: Roberta Ruocco, Vice-President,
Morgan Guarantee Trust Co. of NY
- $20,000 to Phillips Academy, Andover,
MA, for math and science programs.
Giving limited to New York City, except for
selected institutions of higher education.

The David and Lucile Packard Foundation
300 Second Street, Suite 200
Los Altos, CA 94022
(415) 948-7658
Contact: Colburn S. Wilbur, Executive Director
- To East Side Union High School District,
San Jose, CA: (1) $195,000 for third year
of Valdes Mathematics Project and (2)

$171,000 for Jose Valdes Mathematics
Program at Andrew Heil High School.
- $11,500 to Sequoia Union High School
District, Redwood City, CA, for Math A
Program at Sequoia High School.

The Ralph M. Parsons Foundation
1055 Wilshire Boulevard, Suite 1701
Los Angeles, CA 90017
(213) 482-3185
Contact: Christine Sisley, Executive Director
- $15,000 to Mathcounts, Alexandria, VA,
for precollege math couching and competition program.
- $12,000 to Remedial Reading and Learn-
ing Center, Los Angeles, CA, for services
of math specialist in after-school tutoring
program.
Giving limited to Los Angeles County, with the
exception of some grants to higher education.
Does not support programs for which other
funding is readily available.

The San Francisco Foundation
685 Market Street, Suite 910
San Francisco, CA 94105-9716
(415) 495-3100
Contact: Robert M. Fisher, Director
- $75,000 to San Francisco University
Foundation for Interactive Mathematics
Project, a radically new mathematics
curriculum in Bay Area high schools.
Giving limited to Bay Area counties of
Alameda, Contra Costa, Marin, San Francisco,
and San Mateo.

Stuart Foundation
425 Market Street, Suite 2835
San Francisco, CA 94105
(415) 495-1144
Contact: Theodore E. Lobman, President
- $223,504 to Developmental Studies Cen-
ter, San Ramon, CA, for Cooperative
Mathematics Project, to develop new
curriculum for grades 2–6 emphasizing
problem solving and cooperative learning
strategies.
Giving primarily in California; applications
from Washington will be considered.

Weingart Foundation
P.O. Box 17982
Los Angeles, CA 90017-0982
(213) 482-4343
Contact: Charles W. Jacobson, President

• $75,000 to Los Angeles Educational Partnership for first of three installments toward expansion of PLUS Project for mathematics education.

Giving limited to southern California. No support to religious programs.

U.S. West Foundation
7800 East Orchard Road, Suite 300
Englewood, CO 80111
(303) 793-6661
Contact: Larry J. Nash, Director of Administration
  • $100,000 to Tacoma Youth Investment, Tacoma, WA, as second payment of a three-year grant, to support public-private partnership to increase math and reading skills of low achievers.

Limited to states served by US WEST calling areas. Address applications to local US WEST Public Relations Office or Community Relations Team.

Westinghouse Foundation
c/o Westinghouse Electric Corporation
11 Stanwix Street
Pittsburgh, PA 15222
(412) 642-3017
Contacts: G. Reynolds Clark, President; C. L. Kubelick, Manager of Contributions and Community Affairs
  • $10,000 to Mathcounts, Alexandria, VA.

Giving primarily in areas of company operations.

Foundations That Fund Education, Including Special Projects, as a Mission

Aetna Foundation, Inc.
151 Farmington Avenue
Hartford, CT 06156-3180
(203) 273-6382
Contact: Diana Kinosh, Management Information Supervisor

The Ahmanson Foundation
9215 Wilshire Boulevard
Beverly Hills, CA 90210
(213) 278-0770
Contact: Lee E. Walcott, Vice-President & Managing Director
Giving primarily in southern California.

Alcoa Foundation
1501 Alcoa Building
Pittsburgh, PA 15219-1850
(412) 553-2348
Contact: F. Worth Hobbs, President
Giving primarily in areas of company operation.

The Allstate Foundation
Allstate Plaza North
Northbrook, IL 60062
(708) 402-5502
Contacts: Alan F. Benedeck, Executive Director; Allen Goldhamer, Manager; Dawn Bougart, Administrative Assistant

American Express Minnesota Foundation
c/o IDS Financial Services
IDS Tower Ten
Minneapolis, MN 55440
(612) 372-2643
Contacts: Sue Gethin, Manager of Public Affairs, IDS; Marie Tobin, Community Relations Specialist
Giving primarily in Minnesota.

American National Bank & Trust Co. of Chicago Foundation
33 North La Salle Street
Chicago, IL 60690
(312) 661-6115
Contact: Joan M. Klaus, Director
Giving limited to six-county Chicago metropolitan area.

Anderson Foundation
c/o Anderson Corp.
Bayport, MN 55003
(612) 439-5150
Contact: Lisa Carlstrom, Assistant Secretary

The Annenberg Foundation
St. Davids Center
150 Radnor-Chester Road, Suite A-200
St. Davids, PA 19087
Contact: Donald Mullen, Treasurer

AON Foundation
123 North Wacker Drive
Chicago, IL 60606
(312) 701-3000
Contact: Wallace J. Buya, Vice-President
No support for secondary educational institutions or vocational schools.

Atherton Family Foundation
c/o Hawaiian Trust Co., Ltd.
P.O. Box 3170
Honolulu, HI 96802
(808) 537-6333; Fax (808) 521-6286
Contact: Charlie Medeiros
Limited to Hawaii.

Metropolitan Atlanta Community
    Foundation, Inc.
The Hurt Building, Suite 449
Atlanta, GA 30303
(404) 688-5525
Contact: Alicia Philipp, Executive Director
Limited to metropolitan area of Atlanta and
surrounding regions.

Ball Brothers Foundation
222 South Mulberry Street
Muncie, IN 47308
(317) 741-5500; Fax (317) 741-5518
Contact: Douglas A. Bakker, Executive
    Director
Limited to Indiana.

Baltimore Gas & Electric Foundation, Inc.
Box 1475
Baltimore, MD 21203
(301) 234-5312
Contact: Gary R. Fuhronan
Giving primarily in Maryland, with emphasis
on Baltimore.

Bell Atlantic Charitable Foundation
1310 North Courthouse Road, 10th Floor
Arlington, VA 22201
(703) 974-5440
Contact: Ruth P. Caine, Director
Giving primarily in areas of company
operations.

Benwood Foundation, Inc.
1600 American National Bank Building
736 Market Street
Chattanooga, TN 37402
(615) 267-4311
Contact: Jean R. McDaniel, Executive Director
Giving primarily in Chattanooga area.

Robert M. Beren Foundation, Inc.
970 Fourth Financial Center
Wichita, KS 67202
Giving primarily for Jewish organizations.

The Frank Stanley Beveridge Foundation, Inc.
1515 Ringling Boulevard, Suite 340
P.O. Box 4097
Sarasota, FL 34230-4097
(813) 955-7575; (800) 356-9779
Contact: Philip Coswell, President
Giving primarily to Hampden County, MA, to
organizations that are not tax-supported.

F. R. Bigelow Foundation
1120 Norwest Center
St. Paul, MN 55101
(612) 224-5463
Contact: Paul A. Verret, Secretary-Treasurer
Support includes secondary education in greater
St. Paul metropolitan area.

The Blandin Foundation
100 Pokegama Avenue, North
Grand Rapids, MN 55744
(218) 326-0523
Contact: Paul M. Olson, President
Limited to Minnesota, with an emphasis on
rural areas.

Borden Foundation, Inc.
180 East Broad Street, 34th Floor
Columbus, OH 43215
(614) 225-4340
Contact: Judy Barker, President
Emphasis on programs to benefit disadvantaged
children in areas of company operations.

The Boston Globe Foundation II, Inc.
135 Morrissey Boulevard
Boston, MA 02107
(617) 929-3194
Contact: Suzanne Watkin, Executive Director
Giving primarily in greater Boston area.

The J. S. Bridwell Foundation
500 City National Building
Wichita Falls, TX 76303
(817) 322-4436
Support includes secondary education in Texas.

The Buchanan Family Foundation
222 East Wisconsin Avenue
Lake Forest, IL 60045
Contact: Huntington Eldridge, Jr., Treasurer
Giving primarily in Chicago.

The Buhl Foundation
Four Gateway Center, Room 1522
Pittsburgh, PA 15222
(412) 566-2711
Contact: Dr. Doreen E. Boyce, Executive
    Director
Giving primarily in southwestern Pennsylvania,
particularly the Pittsburgh area.

Edyth Bush Charitable Foundation, Inc.
199 East Welbourne Avenue
P.O. Box 1967
Winter Park, FL 32790-1967
(407) 647-4322
Contact: H. Clifford Lee, President
Giving has specific geographic and facility
limitations.

California Community Foundation
606 South Olive Street, Suite 2400
Los Angeles, CA 90014
(213) 413-4042
Contact: Jack Shakley, President
Orange County:
13252 Garden Grove Boulevard, Suite 195
Garden Grove, CA 92643
(714) 750-7794
Giving limited to Los Angeles, Orange, Riverside, San Bernadino, and Ventura counties.

The Cargill Foundation
P.O. Box 9300
Minneapolis, MN 55440
(612) 475-6122
Contact: Audrey Tulberg, Program & Adminis-
    trative Director
Giving primarily in the seven-county Min-
neapolis–St. Paul metropolitan area.

H. A. & Mary K. Chapman Charitable Trust
One Warren Place, Suite 1816
6100 South Yale
Tulsa, OK 74136
(918) 496-7882
Contacts: Ralph L. Abercrombie, Trustee;
    Donne Pitman, Trustee
Giving primarily in Tulsa.

Liz Claiborne Foundation
119 West 40th Street, 4th Floor
New York, NY 10018
(212) 536-6424
Limited to Hudson County, NJ, and the
metropolitan New York area.

The Coca-Cola Foundation, Inc.
P.O. Drawer 1734
Atlanta, GA 30301
(404) 676-2568

The Columbus Foundation
1234 East Broad Street
Columbus, OH 43205
(614) 251-4000
Contact: James I. Luck, President
Giving limited to central Ohio.

Cowles Media Foundation
329 Portland Avenue
Minneapolis, MN 55415
(612) 375-7051
Contact: Janet L. Schwichtenberg
Limited to Minneapolis area.

Dade Community Foundation
200 South Biscayne Boulevard—Suite 4770
Miami, FL 33131-2343
(305) 371-2711
Contact: Ruth Shack, President
Funding limited to Dade County, FL.

Dewitt Families Conduit Foundation
8300 96th Avenue
Zelland, MI 49464
Giving for Christian organizations.

Dodge Jones Foundation
P.O. Box 176
Abilene, TX 79604
(915) 673-6429
Contact: Lawrence E. Gill, Vice-President,
    Grants Administrator
Giving primarily in Abilene.

Carrie Estelle Doheny Foundation
911 Wiltshire Boulevard, Suite 1750
Los Angeles, CA 90017
(213) 488-1122
Contact: Robert A. Smith III, President
Giving primarily in Los Angeles area for non-
tax-supported organizations.

The Educational Foundation of America
23161 Ventura Boulevard, Suite 201
Woodland Hill, CA 91364
(818) 999-0921

The Charles Engelhard Foundation
P.O. Box 427
Far Hills, NJ 07931
(201) 766-7224
Contact: Elaine Catterall, Secretary

The William Stamps Farish Fund
1100 Louisiana, Suite 1250
Houston, TX 77002
(713) 757-7313
Contact: W. S. Farish, President
Giving primarily in Texas.

Joseph & Bessie Feinberg Foundation
5245 West Lawrence Avenue
Chicago, IL 60630
(312) 777-8600
Contact: June Blossom
Giving primarily in Illinois, to Jewish
organizations.

The 1525 Foundation
1525 National City Bank Building
Cleveland, OH 44114
(216) 696-4200
Contact: Bernadette Walsh, Assistant Secretary
Primarily in Ohio, with emphasis on Cuyahoga
County.

The Flinn Foundation
3300 North Central Avenue, Suite 2300
Phoenix, AZ 85012
(602) 274-9000
Contact: John W. Murphy, Executive Director
Limited to Arizona.

The Edward E. Ford Foundation
297 Wickenden Street
Providence, RI 02903
(401) 751-2966
Contact: Philip V. Havens, Executive Director
Funding to independent secondary schools.

George F. & Sybil H. Fuller Foundation
105 Madison Street
Worcester, MA 01610
(508) 756-5111
Contact: Russell E. Fuller, Chairman
Giving primarily in Massachusetts, with emphasis in Worcester.

The B. C. Gamble & P. W. Skogmo
Foundation
500 Foshay Tower
Minneapolis, MN 55402
(612) 339-7343
Contact: Patricia A. Cummings, Manager of
Supporting Organizations
Giving primarily for disadvantaged youth,
handicapped, and secondary educational institutions in the Minneapolis–St. Paul metropolitan area.

Golden Family Foundation
40 Wall Street, Room 4201
New York, NY 10005
(212) 425-0333
Contact: William T. Golden, President
Primary interest is funding for higher education
sciences. Limited to New York.

The George Gund Foundation
1845 Guildhall Building
45 Prospect Avenue West
Cleveland, OH 44115
(216) 241-3114; Fax (216) 241-6560
Contact: David Bergholz, Executive Director
Giving primarily in northeastern Ohio.

The Haggar Foundation
6113 Lemmon Avenue
Dallas, TX 75209
(214) 956-0241
Contact: Rosemary Haggar Vaughan, Executive
Director
Limited to areas of company operations in
Dallas and south Texas.

Gladys & Roland Harriman Foundation
63 Wall Street, 23rd Floor
New York, NY 10005
(212) 493-8182
Contact: William F. Hibberd, Secretary

Hasbro Children's Foundation
32 West 23rd Street
New York, NY 10010
(212) 645-2400
Contact: Eve Weiss, Executive Director
Funding for children, under the age of 12, with
special needs.

The Humana Foundation, Inc.
The Humana Building
500 West Main Street
P.O. Box 1438
Louisville, KY 40201
(502) 580-3920
Contact: Jay L. Foley, Contribution Manager
Giving primarily in Kentucky.

International Paper Company Foundation
Two Manhattanville Road
Purchase, NY 10577
(914) 397-1581
Contact: Sandra Wilson, Vice-President
Giving primarily in communities where there
are company plants and mills.

Walter S. Johnson Foundation
525 Middlefield Road, Suite 110
Menlo Park, CA 94025
(415) 326-0485
Contact: Kimberly Ford, Program Director
Giving primarily in Alameda, Contra Costa,
San Francisco, San Mateo and Santa Clara
counties in California and in Washoe, NV.
There is no support to private schools.

W. Alton Jones Foundation, Inc.
232 East High Street
Charlottesville, VA 22901
(804) 295-2134
Contact: John Peterson Myers, Director

Donald P. & Byrd M. Kelly Foundation
701 Harger Road, #150
Oak Brook, IL 60521
Contact: Laura K. McGrath, Treasurer
Primarily in Illinois, with emphasis on Chicago.

Carl B. & Florence E. King Foundation
5956 Sherry Lane, Suite 620
Dallas, TX 75225
Contact: Carl Yeckel, Vice-President
Giving primarily in Dallas area.

Knight Foundation
One Biscayne Tower, Suite 3800
Two Biscayne Boulevard
Miami, FL 33131
(305) 539-2610
Limited to areas where Knight-Ridder newspapers are published. Initial approach should be
through letter.

Thomas & Dorothy Leavey Foundation
4680 Wiltshire Boulevard
Los Angeles, CA 90010
(213) 930-4252
Contact: J. Thomas McCarthy, Trustee
Primarily in southern California, to Catholic
organizations.

Levi Strauss Foundation
1155 Battery Street
San Francisco, CA 94111
(415) 544-2194
Contacts: Bay Area: Judy Belk, Director of
    Contributions; Mid-South Region: Myra
    Chow, Director of Contributions; Western
    Region: Mario Griffin, Director of Contribu-
    tions; Rio Grande: Elvira Chavaria, Director
    of Contributions; Eastern Region: Mary Ellen
    McLoughlin, Director of Contributions
Generally limited to areas of company
operations.

Lyndhurst Foundation
Suite 701, Tallan Building
100 West Martin Luther King Boulevard
Chattanooga, TN 37402-2561
(615) 756-0767
Contact: Jack E. Murrah, President
Limited to southeastern U.S., especially
Chattanooga.

McDonnell Douglas Foundation
c/o McDonnell Douglas Corp.
P.O. Box 516, Mail Code 1001440
St. Louis, MO 63166
(314) 232-8464
Contact: Walter E. Diggs, Jr., President
Giving primarily in Arizona, California, Flor-
ida, Missouri, Oklahoma, & Texas.

James S. McDonnell Foundation
1034 South Brentwood Boulevard, Suite 1610
St. Louis, MO 63117
(314) 721-1532

Meadows Foundation, Inc.
Wilson Historic Block
2922 Swiss Avenue
Dallas, TX 75204-5928
(214) 826-9431
Contact: Dr. Sally R. Lancaster, Executive
    Vice-President
Limited to Texas.

The Milken Family Foundation
c/o Foundation of the Milken Families
15250 Ventura Boulevard, 2d floor
Sherman Oaks, CA 91403
Contact: Dr. Jules Lesner, Executive Director
Giving limited to Los Angeles area.

The New Hampshire Charitable Fund
One South Street
P.O. Box 1335
Concord, NH 03302-1335
(603) 225-6641
Contact: Deborah Cowan, Associate Director
Limited to New Hampshire.

The New Haven Foundation
70 Audubon Street
New Haven, CT 06510
(203) 777-2386
Contact: Helmer N. Ekstrom, Director
Giving primarily in greater New Haven and the
lower Naugatuck River Valley.

Dellora A. & Lester J. Norris Foundation
P.O. Box 1081
St. Charles, IL 60174
(312) 377-4111
Contact: Eugene Butler, Treasurer
Funding includes secondary education.

The Northern Trust Company Charitable Trust
c/o The Northern Trust Company
Corporate Affairs Division
50 South LaSalle Street
Chicago, IL 60675
(312) 444-3538
Contact: Marjorie W. Lundy, Vice-President
Limited to metropolitan Chicago area.

O'Donnell Foundation
1401 Elm Street, Suite 3388
Dallas, TX 75202
(214) 698-9915
Contact: C. R. Bacan
Giving primarily in Texas for science and
engineering projects.

The Principal Financial Group Foundation, Inc.
711 High Street
Des Moines, IA 50392-0150
(515) 247-5209
Contact: Debra J. Jensen, Secretary
Primarily in Iowa, with emphasis on Des
Moines area.

Z. Smith Reynolds Foundation, Inc.
101 Reynolds Village
Winston-Salem, NC 27106-5197
(919) 725-7541; Fax (919) 725-6067
Contact: Thomas W. Lambeth, Executive
Director
Limited to North Carolina. Will provide
funding for special projects for K–12.

Sid W. Richardson Foundation
309 Main Street
Forth Worth, TX 76102
(817) 336-0497
Contact: Valleau Wilkie Jr., Executive Vice-
President
Limited to Texas.

R.J.R. Nabisco Foundation
1455 Pennsylvania Avenue, NW, Suite 525
Washington, DC 20004
(202) 626-7200
Contact: Jaynie M. Grant, Executive Director

The Winthrop Rockefeller Foundation
308 East Eighth Street
Little Rock, AR 72202
(501) 376-6854
Contact: Mahlon Martin, President
Funding primarily in Arizona, or for projects
that will benefit Arizona.

Community Foundation of Santa Clara County
960 West Hedding, Suite 220
San Jose, CA 95126-1215
(408) 241-2666
Contact: Winnie Chu, Programs Officer
Limited to Santa Clara County, CA.

John & Dorothy Shea Foundation
655 Brea Canyon Road
Walnut, CA 91789
Giving primarily in California.

Harold Simmons Foundation
Three Lincoln Center
5430 LBJ Freeway, Suite 1700
Dallas, TX 75240-2697
(214) 233-1700
Contact: Lisa K. Simmons, President
Limited to Dallas area.

Sonart Family Foundation
15 Benders Drive
Greenwich, CT 06831
(203) 531-1474
Contact: Raymond Sonart, President

The Sosland Foundation
4800 Main Street, Suite 100
Kansas City, MO 64112
(816) 765-1000; Fax (816) 756-0494
Contact: Debbie Sosland-Edelman, Ph.D
Limited to Kansas City areas of Missouri and
Kansas.

Community Foundation for Southeastern
   Michigan
333 West Fort Street, Suite 2010
Detroit, MI 48226
(313) 961-6675
Contact: C. David Campbell, Vice-President of
   Programs
Giving limited to southeastern Michigan.

Springs Foundation, Inc.
P.O. Drawer 460
Lancaster, SC 29720
(803) 286-2196
Contact: Charles A. Bundy, President
Limited to Lancaster County and/or the town-
ships of Fort Mill and Chester, SC.

Steelcase Foundation
P.O. Box 1967
Grand Rapids, MI 49507
(616) 246-4695
Contact: Kate Pew Wolters, Executive Director
Limited to areas of company operations. Initial
contact by letter.

Strauss Foundation
c/o Fidelity Bank, N.A.
Broad & Walnut Streets
Philadelphia, PA 19109
(215) 985-7717
Contact: Richard Irvin, Jr.
Giving primarily in Pennsylvania.

T.L.L. Tempee Foundation
109 Tempee Boulevard
Lufkin, TX 75901
(409) 639-5197
Contact: M. F. Buddy Zeagler, Assistant
   Executive Director & Controller
Giving primarily in counties constituting the
East Texas Pine Timber Belt.

Travelers Companies Foundation
One Tower Square
Hartford, CT 06183-1060
(203) 277-4079/4070
Funding for school programs limited to
Hartford.

Turrell Fund
111 Northfield Avenue
West Orange, NJ 07052
(201) 325-5108
Contact: E. Belvin Williams, Executive
   Director
Giving limited to New Jersey, particularly the
northern urban areas centered in Essex County.
Also giving in Vermont.

Philip L. Van Every Foundation
c/o Lance, Inc.
P.O. Box 32368
Charlotte, NC 28232
(704) 554-1421
Primarily in North Carolina and South Carolina.

Joseph B. Whitehead Foundation
1400 Peachtree Center Tower
230 Peachtree Street, NW
Atlanta, GA 30303
(404) 522-6755
Contact: Charles H. McTier, President
Giving limited to metropolitan Atlanta.

Winn-Dixie Stores Foundation
5050 Edgewood Court
Jacksonville, FL 32205
(904) 783-5000
Contact: Jack P. Jones, President
Limited to areas of company operation.

The Zellerbach Family Fund
120 Montgomery Street, Suite 2125
San Francisco, CA 94104
(415) 421-2629
Contact: Edward A. Nathan, Executive Director
Giving primarily in San Francisco Bay area.

This chapter includes a sampling of foundations that can be contacted for funding your curriculum project. By no means are these all the resources that can be tapped. Remember, think creatively! Are there any community service organizations, such as the Jaycees, Lions Club, or Rotary International, that can be contacted? Is there a local Community Fund that supports education projects? Ask friends and neighbors about the organizations they support. Ask if you can use their names as references—and be sure to get the names of the people to contact. Make many initial contacts, and don't be discouraged by rejections. The money is there for you; all you need is to be persistent!

References

The Foundation Directory. 1992. New York: Foundation Center.
Information about private and community grantmaking foundations in the US.

The Foundation Grants Index. 1992. New York: Foundation Center.
Provides funding patterns and other information about the most influential foundations in the US.

Government Assistance Almanac. 1992. Detroit, MI: Omnigraphics.
A comprehensive guide to federal programs that provide financial assistance.

Source Book Profiles. 1992. New York: Foundation Center.
Information on the one thousand largest U.S. foundations.

4

TOPICS IN THE MATHEMATICS CURRICULUM, GRADES K–12

by Charles E. Lamb
Associate Professor of Mathematics
The University of Texas at Austin, Texas
and

Barbara Montalto
Mathematics Curriculum Specialist
Texas Education Agency, Austin, Texas

MORE than a decade ago, The National Council of Teachers of Mathematics (NCTM), published *An Agenda for Action: Recommendations for School Mathematics of the 1980s.* This was followed closely by *The Agenda in Action,* an NCTM yearbook, in 1983. The *Agenda* called for changes in the content to be taught and the methods of teaching mathematics. Since then, there has been a growing interest in the restructuring of mathematics in the United States. In 1983, *A Nation At Risk* sounded the alarm to alert the nation to the declining quality of education in America. This document was followed by a report focusing on mathematics education, *Everybody Counts: A Report to the Nation on the Future of Mathematics Education* (1989). The NCTM established the Commission on Standards for School Mathematics in 1986 for the purpose of developing a set of standards for mathematics curricula. In the spring of 1989, *Curriculum and Evaluation Standards for School Mathematics* was published. A companion volume on teaching standards followed in 1991. The *Standards* represent a consensus of

what the mathematics community believes should be the fundamental content and processes for learning the mathematics curriculum and standards by which to evaluate these programs. Any attempt to list the major topics of grades K–12 without referring to the NCTM standards would be of questionable credibility. In curriculum discussions, teachers often ask these questions:

· What topics do I need to teach my students at this grade or course level?
· What topics do I need to review and extend for my students?
· What can I expect them to know from their previous instruction or experiences?

The purpose of this chapter is to provide you, the teacher, with an overview of the important topics of each grade level and course in the mathematics curriculum. Just as important as the topics to be taught is the way in which they are taught. The emphasis at all levels is on problem solving as the focus of the curriculum.

Before beginning a discussion of the topics in each grade level, it is important to understand the changing nature of mathematics.

Changes in the workplace require changes in the mathematics that children learn. The "basics" of addition, subtraction, multiplication, and division are no longer sufficient; students need other topics and the ability to use mathematics in meaningful ways. Technology has changed the ways in which mathematics is learned, and has influenced the emphasis placed upon many topics traditionally taught to students.

To help students understand the meaning and value of mathematics, a good mathematics program should use the following four components as the basis for the curriculum.

Mathematics as Problem Solving

Teachers should approach instruction through problem solving. This means using problem solving as a daily way of introducing topics so that students gain confidence in their ability to do mathematics and learn to relate mathematics to their activities. The traditional way of students doing a lot of computation problems followed by a word problem or two is backwards and nonproductive. By posing a problem and then "discovering" or exploring various options for solving the problem, the students attach meaning to their work. Students need to be involved in a variety of problem-solving activities, including nonroutine problems, multistep problems, and "messy" problems, which require students to work cooperatively for a common solution. In so doing, students will develop a repertoire of problem-solving strategies and the insight to select the appropriate ones for a particular situation. Students should be involved in creating problems and relating what they already know to new and different situations.

Mathematics as a Form of Communication

Teachers and students need to interact verbally. Asking students to verbalize their thought patterns and talk about mathematics helps them to internalize mathematical concepts and deepen their understanding. Good questions that have been carefully thought out will act as signposts or markers that aid students in discovering and constructing their own mathematical concepts. By asking probing questions which require more than a recall of information, teachers increase students' understanding of mathematics. This also allows the teacher to correct

misconceptions and faulty thinking. Students learn the language of mathematics through reading, writing, and discussing mathematical ideas and concepts. They need to be able to demonstrate their understanding of mathematical situations using concrete, pictorial, graphical, and algebraic models.

Mathematics as a Way of Reasoning

Students should be asked to justify their answers as a way to teach them to think logically and realize that mathematics makes sense. They should use models and manipulatives to demonstrate their conceptual understanding and justify their thinking. Students should begin to make the transition from concrete thinking to more formal reasoning and abstract thinking as they progress in school. Teachers need to ask questions that require students to explain their reasoning and strategies, to validate and value their ideas, and to evaluate and appreciate the power of mathematics.

Mathematical Connections

Connecting their mathematics to other subject areas and to their everyday lives helps students to recognize that mathematics is more than computation and more than something you do only at school. Students should be offered opportunities to show the connections between different topics in mathematics as well as connections to other disciplines. They should develop an understanding of the usefulness of mathematics in their lives and the important role it will play in their future.

This listing of topics is a combination of current common practices, the NCTM *Curriculum and Evaluation Standards for School Mathematics*, and emerging patterns and trends of curriculum as presented in new state-level curriculum frameworks. With the publication of NCTM's *Standards* in 1989, many states began efforts to reform their programs in mathematics education. Among them are California, Florida, Georgia, Kentucky, Maine, North Carolina, Ohio, South Carolina, South Dakota, and Texas. With regard to the following lists of topics, the reader might find it helpful to know that much of the material in the sections for grades K–4 and 5–8 was developed using curriculum documents from the states of Ohio (1988) and Texas (1991), and the material in the section on grades 9–12 draws on Texas's curriculum document (1991).

Grades K–4

The early years of mathematical study should concentrate on several areas of mathematical learning. First, teaching and learning should be conceptually based. Second, this conceptual approach should be undertaken by using instruction which makes children "doers" of mathematics, not merely spectators. In addition, mathematical reasoning should be developed and encouraged. Furthermore, a broad range of content should be introduced as real-world applications and appropriate technology are employed.

In grades K–4, there are seven broad areas of coverage:

1. Problem solving
2. Patterns, relations, and functions
3. Number and numeration concepts
4. Operations and computations
5. Measurement
6. Geometry
7. Probability, statistics, and graphing.

Kindergarten

In kindergarten, the emphasis is on developing ideas at the readiness level. It is a time for exploration, discovery, and individual growth and development. Children are to be provided with the opportunity to gain experience with the major ideas of elementary school mathematics. These opportunities are to include many activities with concrete manipulatives. Prenumber experiences include describing sets of objects to build vocabulary and gain experience in noticing their attributes, one of which is, of course, number; and matching, comparing, and ordering sets of objects to develop the concept that a number can be an attribute of more than one set. Preoperational activities include using manipulatives to model problems involving the joining and separating of objects. The study of patterns using attribute shapes or other manipulatives includes both the continuations of a given pattern and the building and describing of a pattern by the student.

Premeasurement activities focus on comparing and ordering objects by size using estimation rather than actual measurement. Geometric topics include vocabulary development of spatial relationships such as top/bottom, over/under, inside/outside; learning to recognize common geometric shapes—square, rectangle, circle, and triangle; and learning to sort, classify, and describe them. Graphing is introduced by having students build graphs using objects in the classroom. In addition, students are to be involved with the use of oral language as it relates to mathematical problem solving. Problem-solving strategies developmentally appropriate for kindergarten include making a model using manipulatives, acting out the problem, making a picture, and guessing and checking.

First Grade

At this level, there is a heavier emphasis on the topic of problem solving. This is approached via concentration on the use of language for description and discussion of relevant problem attributes. Material introduced in kindergarten is expanded and extended. Number concepts include place value for two- and three-digit numbers; using ordinals through tenth; counting by ones, twos, fives, and tens; reading and writing numerals through the highest number conceptualized; comparing and ordering numbers using the symbols < and >; and identifying fractional parts of a whole and of a set of objects. Operations on numbers include having the students demonstrate with concrete materials the properties of addition (commutative, associative, identity), explore multiplication and division by joining and separating sets of objects, recall sums and differences for basic facts through 10 using their understanding of the inverse relationship, and recognize and construct models for addition sentences. Measurement topics include measuring objects using nonstandard units, centimeters, and inches, comparing two objects by weight, and telling time to the hour and half-hour. In geometry, students identify the interior and exterior of plane figures and identify and discuss shapes and their properties. They also explore three-dimensional objects, describing their differences and similarities using appropriate mathematical language. In first grade, students begin collecting their own data, using it to make pictographs and bar graphs, draw simple conclusions, and make informal predictions. Students are exposed to patterns, relations, and functions by learning to identify and extend patterns from concrete objects, symbols, geometric figures, and real-life situations. Problem-solving strategies to be developed include communicating an understanding of a problem orally, using models, and working with written forms.

Second Grade

In the second grade, students continue to develop basic mathematical concepts from earlier grades. There is a beginning and gradual movement to the use of symbols in mathematics. Patterns and relations concepts are extended to include the construction of a set of ordered pairs by pairing concrete objects according to a given rule and by using the commutative property to show patterns in addition and multiplication. Number concepts that are new in the second grade include having the student demonstrate the use of models for ones, tens, and hundreds to show place value and addition and subtraction of two-, three-, and four-digit numbers; demonstrate the properties of multiplication (commutative, identity, and zero facts) with concrete models; use models to show equivalent fractions and fractions with numerators greater than one; and to demonstrate and write the value of various collections of coins (the prerequisite to introducing decimals). Addition and subtraction are expanded to regrouping, including the use of appropriate technology such as calculators and problem solving, including selecting the correct operation, writing the number sentence and solving real-life problems appropriate for second grade students. Problem solving is also expanded in the curriculum by applying it to other content areas such as measurement and geometry. In addition, connections are made between mathematical topics as in the case of perimeter, which relates to both measurement and geometry. The emphasis in measurement is on learning to estimate length, width, height, and weight/mass of objects and then measuring them. Students should estimate time and tell time on traditional clocks. New topics in geometry are the investigation of congruence and symmetry using models, drawings, and computer graphics, and the exploration of the concept of perimeter using concrete models.

Probability, statistics, and graphing are extended to describing data on a graph, using information from a graph to solve application and nonroutine problems, drawing conclusions from graphed data, and making predictions based on graphed data. Probability is explored by having students discuss the likelihood of an event occurring (e.g., never, always, sometimes, how often).

Third Grade

Problem solving is further extended by having students generate their own problems. They should use the problem-solving strategies of restating the problem in their own words and identifying missing or irrelevant information. The topic of patterns is applied to the content areas of numbers (whole numbers and decimals). Students should also be taught to describe how terms of a pattern are created, predict additional terms, and extend the pattern. Numeration topics to be introduced include symbols for fractions, reading and writing the value of amounts of money in words and symbols, and developing place-value concepts of tenths and hundredths using concrete models. The topic of operations includes recall of basic multiplication facts, multiplication with a one-digit multiplier and two- and three-digit multiplicands, and using a variety of models to explore division. New measurement concepts related to temperature are introduced, and students use concrete materials to explore the finding of perimeter, area, and volume. In geometry, students are asked to identify characteristics of two and three-dimensional figures, to replicate a solid using cubes, and to develop their spatial skills by describing a three-dimensional object from different perspectives. In graphing, students are introduced to locating points on a grid. In addition, situations are presented which show the application of mathematics to the real world.

Fourth Grade

Problem solving continues to be an important focus of the curriculum. Emphasis is placed on developing an organized approach to the solution of application and nonroutine problems involving many of the topics in the curriculum. Patterns are investigated as they relate to patterns of problems with inverses such as multiplication and the multiplication table. Computers and calculators are employed whenever the opportunity presents itself for a meaningful enhancement of mathematical learning. New ideas related to number theory are introduced; they include primes, composites, and factors. Numeration topics include the use of models to represent mixed numbers, place value for decimals, comparing and ordering of decimals, equivalent fractions, and factor pairs of numbers. At this time, students make the transition from manipulatives and models to symbols in division of whole numbers and in adding and subtracting decimals. In measurement, students measure irregular areas using grids and estimation, use the concept of perimeter to solve problems, estimate and measure capacity using standard and nonstandard units, and solve problems involving measurement units. Geometry is extended to the study of angles, parallelism, and perpendicularity. Students use their spatial skills to illustrate the reflection, rotation, and translation of geometric shapes using concrete models. The study of statistics now includes hypothesis generation, prediction, and solving application and nonroutine problems based on organized data. Probability includes the listing of all possible outcomes in a given situation. Graphing involves plotting points on a coordinate plane representing a problem situation and enlarging or reducing a figure using grids.

Grades 5–8

There is currently a strong interest in the middle-school level of the American educational system. This is especially true of the mathematics curriculum. Children at this level are in transition, both mentally and physically. Efforts must be made which allow them to continue developing their reasoning skills and move from the concrete to the abstract ideas of mathematics. Instruction should be designed so as to keep the doors "open" for all. Too often, students at this level turn away from mathematics, which may hinder their ability to pursue further learning or the career of their choice. Topics must be chosen which allow extension of content from the earlier grades and builds toward the more advanced mathematical ideas of high school (algebra, geometry, etc.).

Fifth Grade

New topics are developed as an extension of previously learned material. For example, patterns are used to explore the rules of divisibility and properties of powers of 10. New number concepts and skills in the fifth grade include identifying prime factors of a number, finding common factors of a set of numbers, using factors and multiples to write equivalent fractions and common denominators of two or more fractions, and developing the concept of ratio using models. In problem solving, students are expected to select an appropriate operation/ strategy or set of operations, justify the selection, estimate the answer, write and solve the problem, and determine the reasonableness of the solution. Operations include addition and subtraction of decimals and fractions along with properties of estimation and rounding. In measurement, models are used to develop and apply the formulas for area and circumference with the major emphasis on developing the concept of volume through concrete models, estimation, nonstandard and standard units, and the relationship between volume units in the metric system. Conversion between units within the same measurement system using multiples and submultiples is a new topic in the fifth grade. In geometry, students are asked to construct examples of the rigid transformations, and constructions using a protractor and a ruler are studied. Probability and statistics concepts are extended to include measures of central tendency as a way of summarizing data, the use of averaging in solving problems, determining simple experimental and theoretical probabilities, and relating fractions to probability.

Sixth Grade

Patterns are used to explore other numeration systems as well as exponents. Students explore functions by building simple functions with concrete models and generating a corresponding rule. Number concepts are extended to the meaning of percents using models; relating fractions, decimals, and percents and identifying the appropriate use of each; comparing and ordering positive rational numbers; and developing the meaning of integers using problem-solving situations. Operations with decimals and fractions continue their development with the four basic operations. There is an emphasis at this grade level on the development of estimation and problem solving using ratios, proportions, and percents. Measurement focuses on area: developing and applying area formulas for common polygons and circles, approximating areas of irregular figures using grids, estimating answers, exploring the effect of changing one dimension of a figure on the area of the figure, and expanding to surface area. Further emphasis is placed on the development of real-world applications of geometry. Spatial skills are used to construct angles, parallel lines, and so forth, and students are expected to classify lines, angles, and polygons, by their characteristics. Computers and calculators are employed in all areas of the curriculum. Students are given experiences that allow them to compare theoretical and experimental probabilities. Graphing techniques are expanded to the circle graph. Problem-solving topics include writing expressions for word phrases, writing and solving simple linear equations from real-world situations, checking the reasonableness of an answer, and using denominate numbers in application problems.

Seventh Grade

Topics developed in patterns involve investigating patterns generated by repeating and terminating decimals. Functions and relations are expanded to include introducing the concept of using letters to represent variables, using concrete models to develop the concept of operations with variables, evaluating algebraic expressions, and investigating solutions to simple open sentences (equalities and inequalities). Number concepts focus on comparing and ordering integers, exploring the absolute value of an integer, developing an understanding of squares and square roots, expressing numbers in scientific notation, using divisibility rules, and simplifying expressions involving exponents. Operations and computation topics include using the order of operations to solve multistep problems, applying the four basic operations to rational numbers and integers, and exploring concepts using a calculator or a computer. In measurement, students develop the general formulas for the volume and surface area of prisms/cylinders and cones/pyramids. They explore the relationship between the dimensions and the volumes of similar solids by varying one of the dimensions at a time. Spatial skills

are reinforced by students constructing and copying one-, two-, and three-dimensional figures from pictorial views, a real object, or a description. Geometry also includes a conceptual development and practical use of the Pythagorean theorem. Statistics and graphing topics include the use of box and whisker graphs, stem and leaf plots, and histograms to display data, comparing different graphic representations of the same data to determine appropriateness of the graph; and drawing inferences and constructing convincing arguments based on data analysis. Probability topics focus on constructing sample spaces in multiple ways, finding the probability of simple events and using permutations and combinations in application problems. Problem-solving skills involve writing expressions and equations using variables, identifying dependent and independent variables, choosing a correct strategy, analyzing the solution for reasonableness, recognizing alternative solutions to a problem, and generating and extending a good problem.

Eighth Grade

If students are not ready for algebra and need another year of preparation, then it is very important that the content of the eighth-grade mathematics program be a challenging and rewarding experience and not a rehash of previous material. Most of the topics will be prealgebra concepts to strengthen preparation for algebra. Patterns are used to develop the concept of negative exponents and to explore other number patterns such as Pascal's triangle, the Fibonacci sequence, and triangular and other figurate numbers. Functions and relations bridge from concrete to graphic and symbolic representations. Logic topics include logical connectives (if . . . then, either . . . or, etc.) and some simple inductive and deductive reasoning. Number concepts are reinforced using a calculator, including irrational numbers and higher roots. All the basic number concepts need to be reinforced and related to algebraic applications. Computation focuses on selecting the appropriate operation and/or strategy and justifying the selection. Students solve linear equations and inequalities by trial and error, graphing, and algebraic methods with integer, rational, and decimal solutions. Measurement concepts are used in application and nonroutine problems. In geometry, topics include using ratio and proportion to find missing parts of similar figures, extending perimeter, area, surface area, and volume concepts to congruent and similar figures. New probability and statistics activities are simulations using a computer, finding the probability of compound events, identifying uses and misuses of probability and statistics, and investigating bias as it affects the validity of conclusions and predictions for a given set of data.

Grades 9–12

Courses offered at the high-school level range from basic to more advanced topics. Typical courses are as follows:

1. Fundamentals of Mathematics or Basic Mathematics
2. Consumer Mathematics
3. Prealgebra
4. Informal Geometry
5. Algebra 1
6. Algebra 2
7. Geometry
8. Trigonometry
9. Elementary Analysis
10. Analytic Geometry
11. Precalculus
12. Mathematics of Consumer Economics, Business Mathematics, or Applied Mathematics
13. Computer Mathematics 1
14. Computer Mathematics 2
15. Probability and Statistics
16. Calculus
17. Number Theory
18. Linear Algebra
19. Linear Programming
20. History of Mathematics
21. Survey of Mathematics
22. Advanced Mathematics for Business
23. Mathematics of Money

The requirement for high school graduation is two or three units of high school mathematics in most states. The courses listed above are some of the choices offered to students as they plan their education. The list is designed to allow all students to achieve the mathematical sophistication necessary for life in the twenty-first century. Whether they wish to enter college, pursue a technical career, or go directly into the work force, completion of a strong program will allow them to be successful. Topical outlines are provided for a selection of the courses.

Fundamentals of Mathematics or Basic Mathematics

This course is being phased out in many states, primarily because it no longer provides the mathematics necessary for students to function in today's society. Topics found in most textbooks and state frameworks are:

1. Concepts and skills associated with numbers and place value
   a. basic operations of addition, subtraction, multiplication on whole numbers, fractions, decimals, and integers
   b. compare and round decimals and fractions
   c. use ratio, proportion, and percent to solve problems
2. Geometry and measurement concepts and skills
   a. identify and use units of measure
   b. convert units within a measuring system
   c. use formulas to find perimeter, area, surface area, and volume
3. Personal finance concepts and skills
   a. compute wages, income, and taxes
   b. compare consumer costs and savings
   c. interpret payroll deductions
   d. work with budgets, checking accounts, and savings accounts
4. Probability, statistics, and graphing
   a. collect, organize, and graph data
   b. find the mean, median, and mode and use them appropriately
   c. make predictions from organized data
   d. calculate the probability of simple and compound events
5. Problem Solving (this should be integrated throughout the course but is frequently taught as a separate topic)

Consumer Mathematics

This is also a below-grade-level course used as an alternative course for students wishing to avoid algebra. Topics usually included in this course are:

1. Statistics and graphing
   a. gathering data and constructing charts and graphs
   b. interpreting charts, graphs, and tables
   c. mean, median, and mode
   d. solving problems using data from graphs, charts, and tables
2. Probability
   a. list all the possible outcomes for a situation and determine probabilities
   b. construct and use tree diagrams
   c. use combinations and permutations
   d. make predictions based on probabilities
3. Cost comparisons
   a. solve problems involving costs of housing (renting or owning), food, clothing, insurance (health and life), and transportation
   b. make decisions and support using cost comparisons
   c. solve problems involving sales, rebates, discounts, coupons, etc.
4. Financing procedures and cost comparisons
   a. credit cards and charge accounts
   b. financing a car
   c. home mortgages
   d. secured and unsecured personal loans
5. Banking and investments
   a. use deposit slips, checks, bank statements
   b. compare different types of investments
6. Local, state, and federal taxes
   a. use tax forms to compute taxes
   b. compare different types of taxes (hidden, excise)
7. Income and budgets
   a. calculate wages and deductions
   b. prepare budgets (personal and household)

Prealgebra

Prealgebra is a course intended to prepare students for algebra, a bridge from the concrete to the abstract. Topics usually included in a prealgebra course are:
1. Number and numeration
   a. place value
   b. factors, prime factors, and multiples using divisibility rules

   c. use scientific notation with calculator applications
   d. write symbolic expressions for word phrases
   e. least common multiple and greatest common factor applications
   f. properties and characteristics of subsets of the real numbers
2. Operations on numbers
   a. basic operations on the rationals using a variety of tools
   b. properties and order of the operations
   c. absolute value
   d. the laws of exponents including zero and negative integral exponents
   e. squares and square roots
   f. solve simple linear equations
3. Functions, relations, and their graphs
   a. simplify polynomials
   b. use direct and indirect variation
   c. graph solutions of equalities and inequalities on a line
   d. graph equations and inequalities in a coordinate plane
   e. slope and intercepts
   f. systems of equations
4. Geometry and measurement
   a. basic geometric figures and their characteristics
   b. use metric and customary units
   c. use indirect measurement with similar figures
   d. investigate right triangle properties including the Pythagorean theorem
   e. solve problems using denominate numbers
5. Probability, statistics, and graphing of data
   a. collect, organize, and graph data
   b. analyze and interpret graphs and data
   c. determine mean, median, and mode and their uses
   d. use counting procedures and various ways of determining possible outcomes
   e. find the probability of independent and dependent events
6. Problem solving
   a. how to select and use a variety of problem-solving strategies including make a model, act it out, solve a simpler problem, guess and check, draw a picture, diagram, chart, or

graph, find a pattern, work backwards, guess and check
b. appropriate methods and materials for
solving problems
c. estimating solutions and evaluating
their reasonableness
d. how to generate, extend, and generalize problems and solutions

Algebra I

The current trend is algebra for everyone.
Algebra is the gate to many job fields. It is not
only for college-bound students, but also for
those who plan to enter two-year programs,
vocational and technical schools, trade schools,
and many occupations for which there is no
formal education or training required beyond
high school. Therefore, many schools are
changing the way in which they teach algebra.
They are allowing students to spend more time
($1\frac{1}{2}$ to 2 years), and they are using different
methods to ensure that all students are successful. These methods include algebra tiles and
other manipulatives, cooperative learning
groups, more actively engaged time, graphing
calculators, and interactive computer programs.

The philosophy of algebra instruction is
reflected in the following statement from
NCTM's *Curriculum and Evaluation Standards
for School Mathematics* (1989):

> Algebra is the language through which
> most of mathematics is communicated. It
> also provides a means of operating with
> concepts at an abstract level and then
> applying them, a process that often fosters
> generalizations and insights beyond the
> original context.

Algebraic symbols may represent objects rather
than numbers, as in "p + q" representing the
sum of two polynomials. This more sophisticated understanding of algebraic representation
is a prerequisite to further formal work in
virtually all mathematical subjects, including
statistics, linear algebra, discrete mathematics,
and calculus. Moreover, the increasing use of
quantitative methods, both in the natural
sciences and in such disciplines as economics,
psychology, and sociology, have made algebraic processing an important tool for applying

mathematics. The proposed algebra curriculum
will move away from a tight focus on
manipulative facility to include a greater
emphasis on conceptual understanding, on
algebra as a means of representation, and on
algebraic methods as a problem-solving tool.
This represents a trade-off in instructional time
as well as in emphasis. Although an appropriate
level of proficiency is important, available and
projected technology forces a rethinking of the
level of skill expectations.

Changes in emphasis require more than
simple adjustments in the amount of time to be
devoted to individual topics; they also will
mean changes in emphasis within topics. For
example, although students should spend less
time simplifying radicals and manipulating
rational exponents, they should devote more
time to exploring examples of exponential
growth and decay that can be modeled using
algebra. Similarly, students should spend less
time plotting curves point by point, but more
time interpreting graphs, exploring properties of
graphs, and determining how these properties
relate to the forms of the corresponding
equations.

In the algebra classroom, students should
have access to graphing calculators at all times,
there should be a computer and overhead
projection device available for demonstration
purposes and students should have access to
computers for individual and group work as
needed.

Topics to be addressed in Algebra I
include:
1. The real number system and its
subsystems
a. properties of the real numbers and
subsystems
b. density property of real numbers
c. order of operations
d. absolute value
2. Algebraic representation, solution, and
evaluation of problem situations
a. how to write a linear expression from
a verbal expression and vice versa
b. model the problem-solving process
c. justify the solution to a problem,
orally, by modeling, and in writing
d. solve literal equations for any variable
e. solve systems of equations by various
methods

f. solve absolute value equations and inequalities

g. applications

3. Linear relations, functions, and inequalities

   a. properties of linear relations, functions, and equalities

   b. domains and ranges

   c. graphing linear equations, functions, and inequalities

   d. graphing linear equations from their characteristics

   e. writing linear equations and inequalities from their graphs or descriptions

   f. using linear equations and inequalities to model real world situations

   g. graphing linear inequalities in two variables

   h. graphing systems of linear equalities and inequalities

4. Quadratic equations

   a. graphing quadratic equations

   b. investigating the effects of parameter changes on the graphs of quadratic equations

   c. using the quadratic formula and a calculator/computer to approximate solutions to quadratic equations

5. Polynomials

   a. classifying polynomials by degree and number of terms

   b. performing the four basic operations on polynomials

   c. factoring polynomials

6. Rational expressions

   a. evaluating rational expressions

   b. operations on rational expressions

   c. solving rational equations

   d. solving problems using ratio and proportion

   e. applications

7. Roots

   a. approximating roots using a calculator

   b. simplifying radical expressions

   c. performing basic operations on algebraic and numerical radical expressions

   d. solving simple radical equations including use of the Pythagorean theorem

Algebra II

Students enrolled in Algebra II should have access to graphing calculators, a computer, and projection device for demonstration purposes, and access to computers for individual and group activities. Topics in Algebra II include:

1. Mathematical structure of the complex number system

   a. subsets and their characteristics

   b. finite and infinite subsystems

   c. complex numbers, their characteristics, and operations

2. Quadratic relations and functions including the conic sections

   a. quadratic formula

   b. other methods of solution

   c. writing the equation, given the roots

   d. graphs of quadratic relations and functions

   e. investigating the effect of parameter changes on the graphs of quadratic relations and functions

   f. writing an equation from the graph of the relation or function

   g. applications to real world problems

3. Systems of equations

   a. methods of solution of two or three variable linear systems, linear combinations, augmented matrices, calculators and/or computers when appropriate

   b. linear programming techniques to solve real world problems

   c. solving quadratic-linear and quadratic-quadratic systems

4. Numerical methods and higher degree polynomials and equations

   a. methods of solution: synthetic division, graphing (using a graphing calculator or computer), successive approximation, iterative process

   b. Fundamental Theorem of Algebra and Factor theorem

5. Exponential and logarithmic functions

   a. extending the properties of exponents to rational and algebraic exponents

   b. developing the concept of a logarithm from exponential functions and their inverses

   c. converting between exponential, logarithmic, and radical forms of an expression

   d. solving equations using the properties of logarithms

   e. applications

6. Rational algebraic functions

a. graphs of rational algebraic functions
b. intuitive concept of a limit
c. simplifying complex fractions
d. using direct and inverse variation functions
e. applications
7. Sequences and series
   a. using patterns to discover the formulas or rules for sequences and series
   b. recursive and generator formulas
   c. nth term of a geometric or arithmetic series
   d. nth partial sums of geometric and arithmetic series
   e. graphs of geometric and arithmetic sequences
   f. convergent and divergent series
   g. applications to real world problems
   h. Binomial theorem and expansion of binomial expressions
   i. enumeration problems involving permutations and combinations

Geometry

Students in geometry should have access to graphing calculators, a computer and projection device in the classroom and access to a computer for individual and group activities. Although formal proof is a component of the geometry curriculum, it is no longer the focus of the course. Rather, the emphasis is on the connections between geometry and algebra, between geometry and the real world, and between various approaches to geometry. While students may work separately in synthetic, transformational, and coordinate geometry, the focus will be on the interconnectedness of the three and the students' ability to choose which is the more appropriate for a given situation. Topics currently taught in geometry include:

1. Axiomatic systems
   a. inductive and deductive reasoning
   b. conditional statements and logical arguments
   c. the relationship between a conditional statement and its converse, inverse, and contrapositive
   d. patterns of inference and valid conclusions
   e. distinguishing between validity of an argument and truth of a statement
   f. Euclidean geometry as an axiomatic system

g. applications
2. Lines, segments, and angles
   a. synthetically: construct figures; write formal proofs related to parallel lines and the angles associated with parallel lines and transversals
   b. use a coordinate approach to: develop distance, midpoint, and betweeness properties; investigate slopes of parallel and perpendicular lines; write equations for lines given conditions such as locus
3. Triangles
   a. synthetically: construct triangle congruence postulates; write formal proofs of triangle congruences; explore triangle similarity postulates and proportionality; develop the trigonometric functions; write formal proofs of angle theorems related to triangles; investigate triangular inequalities
   b. use a coordinate approach to: investigate properties of segments associated with triangles (altitudes, angle bisectors, medians, etc.); make generalizations about these properties and make a convincing argument to support these generalizations
   c. use transformations to: explore congruence and similarity under transformations; investigate translations, rotations, reflections, and dilations of triangles
4. Polygons
   a. synthetically: write formal proofs related to quadrilaterals and other polygons; explore angle and other relationships in polygons and make generalizations inductively about these relationships
   b. use a coordinate approach to: explore properties of quadrilaterals; make generalizations and make convincing arguments to validate these generalizations
   c. use transformations to: explore symmetry of polygons; generalize concepts of congruence and similarity
5. Circles
   a. synthetically: investigate lines, segments and angles related to circles; write formal proofs related to circles;

use theorems to solve problems related to circles

b. use a coordinate approach to: discover the distance formula; develop the equations for circles; graph equations of circles; from the graphs of circles write the equations

c. applications

6. Measurement
   a. synthetically: develop the area and volume formulas; use the concept of limits to investigate the circumference and area formulas for a circle; use ratios to investigate arc length and areas of sector of a circle; develop the relationships between linear measures, areas and volumes of similar figures
   b. use a transformational approach to: explore invariants of geometric figures under translations, rotations, and reflections; explore the ratio of distances as an invariant under dilations; explore angle measure as an invariant under translations, rotations, reflections, and dilations
   c. applications

7. Three-dimensional geometry
   a. synthetically: investigate the properties of cylinders, cones, spheres and other polyhedra using concrete models and computers; build and draw three-dimensional models
   b. use transformations to explore symmetry of polyhedra and spheres in space
   c. applications

Precalculus

The topics in precalculus are the main topics from three one-semester courses: trigonometry, elementary analysis, and analytic geometry. By combining them in one course a student is exposed to all three subjects in one year, this provides a better foundation for calculus. Topics in precalculus are:

1. Properties of the real numbers
   a. structure of mathematical systems
   b. group, rings, and fields
   c. matrices
   d. ordered fields

2. Analysis of relations and functions
   a. properties and graphs
   b. inverse and composite functions
   c. increasing and decreasing functions
   d. continuous and periodic functions
   e. operations on functions
   f. discrete functions

3. Polynomial and rational functions
   a. the Division theorem and the Factor theorem
   b. synthetic division
   c. Rational Root theorem
   d. Descartes' Rule of Signs
   e. Upper and Lower Bounds theorem
   f. Fundamental Theorem of Algebra
   g. graphs of polynomial functions
   h. graphs and solutions for polynomial inequalities

4. Properties and graphs of special functions
   a. exponential functions
   b. logarithmic functions
   c. circular functions

5. Properties, graphs, and applications of trigonometric functions
   a. right triangle properties
   b. general triangle properties
   c. trigonometric equations
   d. fundamental identities
   e. proof and use of graphs

6. Complex numbers and polar coordinates
   a. basic operations on complex numbers
   b. graphic representations
   c. transform equations from polar coordinates to rectangular and vice versa
   d. trigonometric form of complex numbers

7. Quadratic relations and functions
   a. equations of the conic sections
   b. graphs
   c. translation and rotation of axes
   d. solution of quadratic equations
   e. applications

8. Sequences and series
   a. summation proofs
   b. Binomial theorem including factorials and combinations
   c. limits of sequences
   d. convergent and divergent series

9. Vectors and parametric equations
   a. vector equations for lines and planes
   b. dot and cross products of vectors
   c. derive and graph parametric equations

d. applications

Calculus

Many schools do not offer calculus at the high school level. When calculus is offered to students the topics are usually taken from the College Board curriculum for calculus, since the majority of students who are successful in high school calculus take one of the advanced placement examinations. AB Calculus is equivalent to the first semester of college calculus and BC Calculus is considered equivalent to a full year of college calculus. For further information, write to Advanced Placement Program, P.O. Box 6670, Princeton, NJ 08542-6670.

State Frameworks

As mentioned earlier, there have been many state initiatives toward mathematics education reform. Documents from some of those states were used as references for the previous topic listings and discussions. In addition, the authors of this chapter felt that it was important to provide an example of one of the state frameworks. The following is a summary of the California Framework for Mathematics, K–12.

The California Framework

In 1990, the State of California published a new framework. It was updated in 1992. Emphases are placed on hands-on, highly interactive learning experiences for children. Problem solving is a primary goal. In addition, new forms of assessment are being used. The broad goal is to produce students who are capable of being competitive in the world market. This goal suggests that there are new expectations for students as well as teachers. New goals for students are:
1. mathematical thinking
2. ideas
3. tools and techniques
4. communication

In order for students to achieve these goals, teachers will be required to do new things and try new methods. As an example, new assessment and evaluation techniques will need to be developed. In addition, the new goals for

mathematics require that both teachers and students see mathematics learning in a new light. Mathematical knowledge should be transmitted in such a way that children make sense of their learning as they construct mathematics for themselves. A key to this new type of learning will be the amount of confidence children acquire as they learn. Successful teaching and learning will be the result of implementation of this new approach.

The framework has essential characteristics as follows.
1. full student participation
2. large mathematical ideas and their interconnections
3. coherent units of multi-day tasks
4. understanding
5. assessment is integrated into instruction
6. teachers as facilitators
7. cooperative learning
8. manipulatives, computers, and calculators
9. communication about mathematics
10. worthwhile tasks

The teacher's role is important in the implementation of this program. Teachers should change the way they view themselves. Instead of being imparters of knowledge, they should become resources for students. Teachers and students should become partners in the learning and teaching process.

Grades K–5
There are several components of a strong mathematics program at the elementary school level.
1. A goal is to develop number sense (representation, operations, and interpretation).
2. A foundation for all aspects of mathematics should come out of elementary school mathematics. Topics include: geometry, measurement, data analysis, probability, and discrete mathematics.
3. Understanding should develop over time, across grade levels.
4. Procedures and algorithms
5. Basic facts and knowledge

Grades 6–8
Middle school mathematics programs have attributes of both elementary and high school programs. Key ideas for middle school are:

1. representation, summarization, and interpretation of data; sampling
2. functional relationships; graphs, tables
3. probability
4. rational numbers
5. proportional reasoning
6. similarity and scaling
7. spatial reasoning and visualization
8. accuracy and precision of measurement
9. relationships among perimeter, area, and volume
10. variable expressions to generalize patterns and relationships

These topics should be employed to emphasize the practical power of mathematics.

Grades 9–12
A primary goal of mathematics at the secondary level is to develop symbol sense (representation, operations, and interpretation). In addition, secondary mathematics should cover the entire spectrum of mathematical science. That is, topics should include:

1. algebra
2. geometry
3. data analysis
4. discrete mathematics
5. optimization

In order to achieve this, the framework suggests a common sequence for high school. After grades 7 and 8, all students would take the same three course sequence. Then, students would either follow the calculus path or take a course in probability and statistics. The framework does not suggest specific courses, but rather indicates broad content suggestions as directed by the NCTM *Standards*. More specific suggestions are available from the California *Framework*.

Summary

As one can see, there is a great deal of change occurring in the field of mathematics education. At the state and national levels, professionals are recommending that the nation's schools change the way mathematics is viewed and taught to children. This requires a commitment of all those involved: educators, parents, teachers, and students. Teachers can take guidance from publications such as the *Professional Standards for Teaching School Mathe-*

matics published by NCTM in 1991. In addition, it is no longer acceptable for people to believe that mathematics is unimportant and that it is okay for one to be mathematically ignorant. As well as the new directions being pursued by mathematics educators, it is time for an attitude change on the part of the American public (Cain, Carry, Carter, Lamb, and Madsen 1991).

References

Cain, R. W., L. R. Carry, H. L. Carter, C. E. Lamb, and A. L. Madsen. 1991. "Attitudes about Mathematics Education Must Change if Students Are To Improve." *On Campus* 30 Sep.: 2.

California State Department of Education. 1992. *Mathematics Framework for California Public Schools, K–12.*

National Commission on Excellence in Education. 1983. *A Nation At Risk: The Imperative for Educational Reform.* Washington, DC: U.S. Government Printing Office.

National Council of Teachers of Mathematics. 1980. *An Agenda for Action: Recommendations for School Mathematics of the 1980s.* Reston, VA.

———. 1983. *The Agenda in Action: 1983 Yearbook.* Reston, VA.

———. 1989. *Curriculum and Evaluation Standards for School Mathematics.* Reston, VA.

———. 1991. *Professional Standards for Teaching Mathematics.* Reston, VA.

National Research Council. 1989. *Everybody Counts: A Report to the Nation on the Future of Mathematics Education.* Washington, DC: National Academy Press.

Ohio State Department of Education. 1988. *Mathematics Monographs.* Columbus, OH.

Texas Education Agency. 1991. *State Board of Education Rules for Curriculum.* Austin.

STATE CURRICULUM GUIDES: AN ANALYSIS

Jay Stepelman
Chairperson, Math Department (Retired)
George Washington High School, New York, New York

A mathematics curriculum guide is a document that enumerates the specific mathematics content for every grade level, core understandings, and classroom activities for each topic. Rather than listing specific details about what to teach in each school, the guide offers a broad outline of curriculum topics and concepts. Each school or district can then determine the specifics of attaining these objectives.

An effective curriculum guide is one that steers state policies and programs toward:

- the adoption of appropriate instructional materials
- the creation of a vision for mathematics education
- the strategies for more productive learning and teaching
- a core of K–12 learnings
- the professional development of mathematics teachers
- a support system outside of the education bureaucracy that translates the overall ideas into classroom, school, and district practice

From school mathematics topics for the classroom to teaching strategies in the classroom, this analysis adheres closely to the vision and spirit of the National Council of Teachers of Mathematics's *Curriculum and Evaluation Standards* and *Professional Standards for Teaching Mathematics* to determine the criteria by which the state curriculum guides in this volume can be evaluated.

In the past, some guides placed an undue emphasis on the procedures of mathematics, such as speed and computational proficiency. At the time, these procedures suited our students well, but that emphasis is hardly enough for guides that reflect a vision for the last part of this decade. Estimation, data analysis, use of technology, justifying answers, and making mathematical connections with the real world are some of the new standards on which school mathematics of the future will develop.

Framework Analysis

The framework that was used to analyze each curriculum document is outlined below. The items are listed sequentially, even though their corresponding numerical designations may differ from one document to another. Unique features of each curriculum document are also noted.

A. Curriculum Content
  1. arrangements and divisions of grades and levels for K–12, followed by a general description of the curriculum used by that state
  2. core areas of broad curricular guidelines
  3. references to special needs of college-preparatory students
  4. specific tasks or activities listed in the particular guide
  5. topics related to other content areas as well as other areas of mathematics
  6. depth of commitment to using the technology of calculators and computers in the classroom

B. Teaching Standards
  1. the variety of classroom teaching strategies that are offered
  2. connections with other areas of mathematics or content areas
  3. the use of communication skills in mathematics classrooms, such as reading and writing for understanding, presenting arguments, and convincing others
  4. student involvement by engaging in tasks, solving problems, explaining to others, and convincing them
  5. elevation of student expectations to higher levels
  6. the role in the mathematics classroom of minorities and women—the underachievers of the past

C. Professional Standards
  1. membership in professional organizations
  2. anticipation of an era of cooperation with business, government, and other partners in education
  3. vision for the future of mathematics education
  4. textbook selection criteria to guide the appropriate personnel
  5. infusion of career education

The NCTM's Standards have been widely accepted as reflecting the vision that is needed at this time, and thus is the model used to determine the states' own standards of teaching and vision for the future. The variety and quality of documents that were submitted for analysis reflect the diverse purposes for which they were designed as well as the philosophy

and standards of each state department of education. Some provide extensive, detailed information for mandated courses of study, while others are meant only as suggested guidelines. (Note: No documentation was submitted for Colorado, Hawaii, Kansas, Maine, Massachusetts, Nebraska, New Jersey, Rhode Island, and Texas.)

Traditional vs. Modern (Standards) Approaches

The listing of states in this chapter is alphabetical rather than by geographic region, since the variance in orientation of the submitted documents was quite broad, with no single region displaying a preponderance of either the traditional or the more modern NCTM standards approach.

Rather than list specific topics to be taught in each grade for a particular state, the curriculum topics are described as reflecting traditional, modern/traditional, or modern influences. Unique features of each curriculum:

Traditional influence. Consists of: set concepts, whole number and fraction concepts and operations, decimals, percent, length/area volume, two- and three-dimensional geometric concepts, deductive reasoning, patterns, estimating and measuring, gathering and describing data, circle/bar/line graphs, statistics, probability, problem solving, factors and multiples, elementary algebra, congruent and similar figures, coordinate geometry, right-triangle trigonometry, functions, conic sections, Pythagorean Theorem, scientific notation, metric system, powers, roots, logarithms.

Traditional high school mathematics sequence. Consists of: general mathematics, consumer mathematics, pre-algebra, algebra I, geometry, algebra II, trigonometry, advanced mathematics (theory of equations, polar coordinates, matrices, mathematical induction, analytic geometry, precalculus, calculus).

Modern/traditional strands. Consists of: postulational geometry, random sampling, curve fitting, discrete mathematics, simulations, systems of equations, inequalities, matrices and determinants, test conjectures, technology, use of scientific and graphing calculators, computer awareness, shapes, reflections, transformations, exponential growth, decay, codes, problem-solving strategies, open sentences, identification of trends, connections, inductive reasoning, symbolic logic.

Modern strands. Consists of: ability to communicate mathematics through reading/writing/speaking/listening, relate mathematics to surrounding world, estimate and predict reasonableness of results, connect and extend patterns and relationships with other areas of mathematics, ability to appreciate mathematics.

Alabama

K–12 Alabama Course of Study—Mathematics
A. Curriculum Content
  1. Grades K–6: traditional topics; outcomes are organized under topics within individual areas
  2. Grades 7–12: outcomes are organized according to instructional objectives within individual areas
  3. Core areas for K–6: traditional sequence, with emphasis on algebra and geometry; traditional core sequences of courses for 7–12
  4. College-preparatory students: no references
  5. Tasks or activities: none given
  6. Related topics: none indicated
  7. Technology: a minimum amount of time is to be devoted to reinforcing computational skills; calculators are to be used frequently so that students may focus on concepts and new content; appropriate computer-assisted instruction is a part of every mathematics course at each grade level
B. Teaching Standards
  1. Teaching strategies: cooperative learning groups
  2. Connections: no references
  3. Communication skills: see teaching strategies above
  4. Student involvement: use of manipulatives to help students make connections between concrete experiences and abstract mathematical representations
  5. Student expectations: no references
  6. Role for minorities/women: no references

C. Professional Standards
  1. Professional organizations: names are listed in connection with supplementary reading recommendations
  2. Business/government agencies: none listed
  3. Vision: no reference
  4. Textbook selection: no reference
  5. Career infusion: no reference

Alaska

K–12 Elementary and Secondary Model Curriculum Guides
A. Curriculum Content
  1. Grade K: traditional influence
  2. Grades 1–8: traditional topics
  3. Grades 9–12: traditional courses
  4. Core framework: traditional topics
  5. College-preparatory students: no reference
  6. Tasks: several graded sample activities are listed for each topic
  7. Related topics: none indicated, except in an occasional activity such as "graph wildlife population numbers over time"
  8. Technology: reference made to using calculators and computers across the board in a meaningful way; calculator and computer units are included in general and consumer math courses
B. Teaching Standards
  1. Teaching strategies: no references
  2. Connections: no listing other than incidental references as indicated above for related topics
  3. Communication skills: no reference
  4. Student involvement: only as indicated in sample learning activities
  5. Student expectations: no references
  6. Role for minorities/women: prefatory remarks note that "districts should attend to the problem of under-representation of females and some minorities in higher level mathematics courses"
C. Professional Standards
  1. Professional organizations: none listed
  2. Business/government agencies: none listed
  3. Vision: no reference
  4. Textbook selection: no reference

5. Career infusion: a unit in career education is offered in general math

Arizona

K–12 Essential Skills for Mathematics
A. Curriculum Content
1. Grades K–3: traditional, followed by 33 grade-level appropriate outcomes
2. Grades 4–8: traditional, followed by 77 grade-level appropriate outcomes
3. Grades 9–12: traditional, followed by 38 grade-level appropriate outcomes
4. Core areas for K–3: number, measurement, geometry, patterns, data analysis, analytical reasoning; for 4–8, algebra and probability are added to K–3 list
5. College-preparatory students: traditional
6. Tasks or activities: none listed
7. Related topics: none indicated
8. Technology: strongly encourages the incorporation of computers and calculators but no direction is given
B. Teaching Standards
1. Teaching strategies: suggests using a "spiral arrangement" of the curriculum and having students work in cooperative learning groups
2. Connections: no references
3. Communications skills: no reference
4. Student involvement: advocates solving routine and nonroutine problems without specific illustrations
5. Student expectations: no reference
6. Role for minorities/women: no reference
C. Professional Standards
1. Professional organizations: none listed
2. Business/government agencies: none listed
3. Vision: no reference
4. Textbook selection: no reference
5. Career infusion: no reference

Arkansas

1–12 Mathematics
A. Curriculum Content
1. Grades 1–8: traditional topics and skills; each topic is amplified by a list of basic

skills, developmental skills, and extensions (enrichment)
2. Grades 9–12: traditional high school topics and course of study
3. Core areas: traditional secondary school topics; extensions (enrichment) are included for all areas
4. College-preparatory students: no reference
5. Activities or tasks: none listed
6. Related topics: none indicated
7. Technology: encourages extensive use of calculators and computers, but no reference to ways
B. Teaching Standards
1. Teaching strategies: none listed
2. Connections: none indicated
3. Communication skills: no references
4. Student involvement: no references
5. Student expectations: no references
6. Role for minorities/women: no references
C. Professional Standards
1. Professional organizations: no reference
2. Business/government agencies: no references
3. Vision: no reference
4. Textbook selection: no reference
5. Career infusion: no reference

California

K–12 Mathematics Framework for California Public Schools
A. Curriculum Content
The State of California recommends its own unique modern/traditional curriculum
1. Grades K–5 units: attributes and classification; numbers and numeration; arithmetic operations; process of measurement; visualizing and representing shapes; measuring geometric figures; location and mapping; dealing with data; exchange; games and rules; sharing
2. Grades 6–8 units: objects, shapes and containers; maps and scale drawings; growth; motion; expressing proportional relationships; fairness; meters, gauges, and decimal representation; connections among arithmetic operations; units of measurement; descriptive geometry;

3. Grades 9–12 units: representing and analyzing physical change over time; measuring inaccessible distances; families of functions; amortization; geometrical solids; looking back at triangles; encryption and decryption; demographics; geodesics; scheduling and distributing
4. Core areas: endorses the NCTM's Standards for a core high school curriculum and draws content from traditional, widely used subject categories of mathematics
5. College-preparatory students: mathematics is recommended every year of high school
6. Tasks: encourages the development of mathematical power at every grade level
7. Related topics: interweaving of strands including activities from statistics, problem solving, geometry, algebra, and graphs, as well as art and communication
8. Technology: use of manipulatives, such as calculators and computers, to help make sense of mathematical concepts is recommended for all students

B. Teaching Standards
1. Teaching strategies: flexible methods of instruction that include situational lessons, concrete materials, corrective instruction/remediation, and cooperative learning groups
2. Connections: analysis of interrelationship of courses such as algebra and geometry
3. Communication skills: use of written and oral communication skills in the form of discussions, presentations, portfolios, and diaries
4. Student involvement: is a necessary outcome of specific exercises and activities
5. Student expectations: contains a list of expectations for student work in class as well as for successful completion of school studies
6. Role for minorities/women: a section on student diversity appears at the end of the guide

C. Professional Standards
1. Professional organizations: none listed
2. Business/government agencies: none listed
3. Vision: a large section is devoted to "visualizing the future of mathematics education"

4. Textbook selection: no reference
5. Career infusion: no reference

Connecticut

K–12 Curriculum Development Guide
A. Curriculum Content
1. Grades K–8: traditional influence
2. Grades 9–12: course sequences—nine alternative sequences totaling 18 courses are proposed, giving pupils a wide range of course choices
3. Core areas: indicated in listing entitled "Ten Basic Skill Areas"
4. College-preparatory students: the algebra I to precollege math sequence is acceptable
5. Tasks or activities: several activities for below grade 7 are included; activities for other grades should be developed by the teacher
6. Related topics: indicates related business and consumer topics
7. Technology: shows advantages of calculator use and computer literacy courses for some pupils

B. Teaching Standards
1. Teaching strategies: analyzes instructional strategies for handicapped, gifted, and older students
2. Connections: presents several model lesson plans
3. Communication skills: notes use
4. Student involvement: appears in proposed activities
5. Student expectations: indirect references
6. Role for minorities/women: no references

C. Professional Standards
1. Professional organizations: lists statewide and regional professional associations as well as regional educational service centers and other sources of information
2. Business/government agencies: no contact names or addresses are suggested, even though interdisciplinary connections are noted in the areas of art, business, driver education, languages, health education, industrial arts, music, science, and social studies
3. Vision: no reference

4. Textbook selection: sample Mathematics Textbook Adoption and Evaluation Form is shown
5. Career infusion: none given for classroom lessons, but there is a general listing of mathematics-related careers

Delaware

1-12 Mathematics Content Standards
A. Curriculum Content
   1. Grades 1-8: traditional influence
   2. Grades 9-12 courses: traditional high school mathematics sequence
   3. Core areas: traditional influence
   4. Tasks and activities: provides lists of expected outcomes for each topic that is taught
   5. Related topics: none indicated
   6. Technology: *Computer Literacy Compliance Standards,* a supplementary guide concerning the role of computers in society, has been issued
B. Teaching Standards
   1. Teaching strategies: none listed
   2. Connections: none indicated
   3. Communications skills: none indicated
   4. Student involvement: none indicated
   5. Student expectations: no reference
   6. Role for minorities/women: no reference
C. Professional Standards
   1. Professional organizations: none listed
   2. Business/government agencies: none listed
   3. Vision: no reference
   4. Textbook selection: no reference
   5. Career infusion: no reference

Florida

6-12 Two Curriculum Framework Documents
A. Basic Programs
   1. Grades 6-8 document: contains 15 traditional middle/junior high school mathematics course frameworks; a supplementary document includes six computer education course outlines

2. Grades 9-12 document: contains 51 traditional high school and adult secondary mathematics course frameworks
3. Core areas for grades 6-8: includes traditional courses with addition of calculator and computer courses
   Core areas for grades 9-12: includes the traditional high school mathematics sequence, although linear and abstract algebra, differential equations, discrete mathematics, business mathematics, and liberal arts mathematics are added
4. College preparatory students: no reference
5. Tasks or activities: none listed
6. Technology: use of computers and calculators is advocated, although no specific direction or guidance is offered
B. Teaching Standards
   1. Teaching strategies: none suggested
   2. Connections: none listed
   3. Communication skills: no reference
   4. Student involvement: no reference
   5. Student expectations: no reference
   6. Role for minorities/women: no reference
C. Professional Standards
   1. Professional organizations: no reference
   2. Business/government agencies: no reference
   3. Vision: none indicated
   4. Textbook selection: no reference
   5. Career infusion: no reference

Georgia

K-12 Quality Core Curriculum
A. Curriculum Content
   1. Grades K-8: 312 traditional topics/ concepts
   2. Grades 9-12: 90 objectives-general/ vocational I, II, and III
      55 objectives—pre-algebra (college-prep readiness)
      41 objectives—algebra I
      46 objectives—geometry or informal Geometry
      51 objectives—algebra II
      51 objectives—advanced algebra and trigonometry
      57 objectives—senior mathematics
      18 objectives—analysis

29 objectives—statistics

30 objectives (approx.)—computer mathematics

85 objectives—discrete mathematics

33 objectives—calculus

3. Core areas: listed above for K–8 and 9–12
4. College-preparatory students: a sequence is recommended
5. Tasks or activities: none indicated
6. Technology: advocated without illustrations, examples, or models

B. Teaching Standards
   1. Teaching strategies: no references
   2. Connections: no references
   3. Communication skills: no references
   4. Tasks or activities: no references
   5. Student expectations: no references
   6. Role for minorities/women: no references

C. Professional Standards
   1. Professional associations: none listed
   2. Business/government agencies: none listed
   3. Vision: none indicated
   4. Textbook selection: no reference
   5. Career infusion: no reference

Idaho

K–8 Mathematics Course of Study Guide
A. Curriculum Content
   1. Grades K–8 critical components: traditional topics plus technology, career awareness and algebraic principles
   2. Grades 9–12 courses: traditional high school mathematics sequence
   3. Core area skills: traditional topics
   4. College-preparatory students: traditional program
   5. Tasks: none listed
   6. Related topics: none listed
   7. Technology: indicates that schools should make both the computer and calculator available, if at all possible, and allow their use in secondary mathematics classes, but no example or guidance is offered

B. Teaching Standards
   1. Teaching strategies: should be consistent with traditionally accepted teaching techniques

2. Connections: none made
3. Communication skills: no reference
4. Student involvement: no reference
5. Student expectations: none listed
6. Role for minorities/women: no reference

C. Professional Standards
   1. Professional organizations: only resources are listed, such as books, posters, and teaching aids, with addresses that must be indicated on the material
   2. Business/government agencies: no references
   3. Vision: prefatory remarks refer to "expanding our view of mathematics"
   4. Textbook selection: no reference
   5. Career infusion: suggested goal of every lesson from kindergarten through eighth grade

Illinois

Grades 3, 6, 8, 10, 12 State Goals for Learning
A. Curriculum Content
   1. Grades 3, 6, 8, 10, 12: traditional content
   2. Each goal for each grade is followed by sample learning objectives; each local school district is expected to develop its own objectives
   3. Core areas: traditional influence
   4. College-preparatory students: no reference
   5. Tasks or activities: none indicated
   6. Related topics: none indicated
   7. Technology: no references

B. Teaching Standards
   1. Teaching strategies: no reference
   2. Connections: no reference
   3. Communication skills: no reference
   4. Student involvement: no reference
   5. Student expectations: prefatory remarks indicate that State Goals for Learning represent endpoint expectations for students
   6. Role for minorities/women: no reference

C. Professional Standards
   1. Professional organizations: no reference
   2. Business/government agencies: no references
   3. Vision: no reference
   4. Textbook selection: no reference
   5. Career infusion: no reference

Indiana

K–12 Mathematics Proficiency Guide
A. Curriculum Content
1. Grades K–4: grade-appropriate modern/ traditional proficiencies are indicated
2. Grades 5–8: grade-appropriate modern/ traditional proficiencies are indicated, with these additions at appropriate points: algebra concepts and processes, making predictions, transformations, Venn diagrams
3. Grades 9–12: grade-appropriate modern/ traditional proficiencies are indicated
4. Core areas: are not specifically indicated, although the curriculum reflects a modern/ traditional approach
5. College-preparatory students: a specific curriculum is recommended
6. Activities and tasks: dozens of activities listed for pupils of every grade and for every topic
7. Related topics: offered abundantly with diagrams, graphs, and statistical data
8. Technology: calculators and computers recommended for every pupil at every grade level, together with the appropriate activities
B. Teaching Standards
1. Teaching strategies: small, cooperative-learning, as well as large-group strategies and activities are recommended for specific topics only
2. Connections: are made with many areas from a variety of disciplines such as science, geography, and industrial art
3. Communication skills: are recommended via oral presentations, journals, and newspapers
4. Student involvement: students as doers and thinkers pervades the entire curriculum guide
5. Student expectation: high because of a massive infusion of confidence-building activities
6. Role for minorities/minorities: no significant reference
C. Professional Standards
1. Professional organizations: listed

2. Business/government agencies: no listing
3. Vision: same as NCTM Standards
4. Textbook selection: no guidelines
5. Career infusion: recommendation is that infusion ought to be a significant addition to every lesson, but no specifics appear in the guide

Iowa

K–12 A Guide to Curriculum Development in Mathematics
A. Curriculum Content
1. Grades K–8: traditional strands
2. Grades 9–12: traditional and enriched courses
3. Core areas: traditional course sequences
4. College-preparatory students: provides a framework for college preparatory work
5. Tasks and activities: suggested as strategies for problem solving
6. Related topics: no references
7. Technology: encouraged, but the guide has few calculator or computer-appropriate tasks
B. Teaching Standards
1. Teaching strategies: indicated for each strand, in addition to instructional materials and manipulatives
2. Connections: none indicated
3. Communication skills: none indicated
4. Student involvement as doers: many illustrations and suggested activities
5. Student expectations: no reference
6. Role for minorities/women: no reference
C. Professional Standards
1. Professional organizations: no reference
2. Business/government agencies: no reference
3. Vision: no reference
4. Textbook selection: no reference
5. Career infusion: no reference

Kentucky

K–12 A List of Valued Outcomes
A. Curriculum Content
1. Grades K–12: a variation of the traditional curriculum that includes additional and unique features such as mathematical

procedures, space and dimensionality, mathematical structure, and data as related to certain and uncertain events
2. Core areas: six learning goals with 28 valued outcomes
3. College-preparatory students: no reference
4. Tasks or activities: none suggested
5. Related topics: none indicated
6. Technology: using technology is an outcome of one of the six goals

B. Teaching Standards
1. Teaching strategies: none recommended
2. Connections: none indicated
3. Communication: the first goal refers to communication and math skills
4. Tasks and activities: goal five involves student activities related to thinking and problem solving
5. Student expectations: no reference
6. Roles for minorities/women: no reference

C. Professional Standards
1. Professional organizations: no reference
2. Business/government agencies: no reference
3. Vision: no reference
4. Textbook selection: no reference
5. Career infusion: only general reference is made to vocational studies and career paths that include postsecondary school, jobs, and the military

Louisiana

K–8 Mathematics Curriculum Guide
Introduction to Algebra Curriculum Guide
Geometry Curriculum Guide
Computer Literacy Curriculum Guide
Computer Science Curriculum Guide
Advanced Mathematics Curriculum Guide
A. Curriculum Content
1. Grades K–8: traditional influence
2. *Introduction to Algebra Curriculum Guide*: traditional mathematics content
3. *Geometry Curriculum Guide*: traditional mathematics content plus optional topics such as volume, construction, coordinate geometry
4. *Computer Literacy Curriculum Guide*: the computer's impact on society today, the development of computers, computer hard-

ware/software/applications, elementary programming, hardware and software selection
5. *Computer Science Curriculum Guide*: history of computers, the design and functional use of computers, algorithm construction, flowcharting, writing a structured computer program, using a symbolic language, computer design and operation
6. *Advanced Mathematics Curriculum Guide*: traditional advanced algebra, or precalculus topics plus matrices
7. Core areas: indicated in the first three guides listed above
8. College-preparatory students: no reference
9. Tasks: many problems, activities, and tasks for pupils are demonstrated
10. Related topics: none indicated
11. Technology: activities that involve the use of computers and calculators are given and explained

B. Teaching Standards
1. Teaching strategies: techniques are suggested for teaching many of the topics
2. Connections: none made
3. Communications skills: some written and oral communication skills are suggested
4. Student involvement: much opportunity for student involvement
5. Student expectations: no reference
6. Role for minorities/women: no reference

C. Professional Standards
1. Professional associations: none listed
2. Business/government agencies: no reference
3. Vision: none indicated
4. Textbook selection: several textbook recommendations are made; a large number of optional and supplemental materials listed
5. Career infusion: no reference

Maryland

K–12 A Maryland Curricular Framework— Mathematics
A. Curriculum Content
1. Grades K–8: traditional topics
2. Grades 9–12: traditional courses

3. Core areas: traditional scope and sequence are outlined for each grade level in 6 goals and 31 subgoals
4. College-preparatory students: no reference
5. Tasks or activities: model activities for every grade level appear under the heading "Instructional Units"
6. Related topics: none indicated
7. Technology: uses are noted

B. Teaching Standards
1. Teaching strategies: refers to considering individual differences and similarities
2. Connections: no reference
3. Communication skills: references to developing language skills in order to communicate about mathematics
4. Tasks: sample tasks indicate the value of hands-on activities
5. Student expectations: refers only to development of "useful" skills
6. Role for minorities/women: no references

C. Professional Standards
1. Professional organizations: no references
2. Business/government agencies: no references
3. Vision: none
4. Textbook selection: no references
5. Career infusion: none
6. Special feature in appendix: a comparison between algebra and geometry

Michigan

K–12 Model Core Curriculum Outcomes

A. Curriculum Content (selected outcomes)
1. High school: modern/traditional secondary school topics
2. Middle/junior high: modern/traditional topics
3. Elementary: traditional/modern topics
4. Core areas: modern/traditional topics
5. College-preparatory students: no reference
6. Tasks or activities: none recommended
7. Related topics: none
8. Technology: the use of calculators and computers is advocated in general terms

B. Teaching Standards
1. Teaching strategies: teamwork is emphasized
2. Connections: no reference

3. Communication skills: provides a formal program of communication skills that includes reading, writing, listening, and speaking in all subject areas
4. Student involvement: no references
5. Student expectations: no references
6. Roles for minorities/women: no references

C. Professional Standards
1. Professional organizations: no reference
2. Business/government agencies: reference is made to the necessity of collaboration among educators, parents, business, industry, and community members, although no list of names and addresses is given
3. Vision: appears in the form of seven broad outcomes that describe what a Michigan student will be like
4. Textbook selection: no reference
5. Career infusion: no reference

Minnesota

K–12 Model Learner Outcomes for Mathematics Education

A. Curriculum Content
1. Grades K–4: traditional influence
2. Grades 5–8: enriched traditional influence
3. Grades 9–12: modern/traditional influence
4. Core areas: follows recommendations of NCTM Standards
5. College-preparatory students: follows recommendation of NCTM Standards
6. Tasks or activities: those that require decision making are interspersed throughout all grade levels
7. Related topics: few references
8. Technology: calculator use is noted in selected activities

B. Teaching Standards
1. Teaching strategies: calculator use is indicated for some activities
2. Connections: are made within the mathematics curriculum itself for the early grades in integration charts for K–4
3. Communication skills: the value of communication skills is noted for every grade
4. Student involvement as doers: recommends that students be involved as problem solvers

5. Student expectations: the desirability of high student and parent expectations is noted in the section "Achieving Excellence"
6. Role for minorities/women: a section entitled "Achieving Equity" expresses concerns for women and minority issues in education

C. Professional Standards
  1. Professional organizations: guide contains a list of Minnesota professional organizations
  2. Business/government agencies: none listed
  3. Vision: included in the section "Priorities for the 1990s"
  4. Textbook selection: no reference
  5. Career infusion: no reference
  6. Unique feature: reports on the value of high school calculus by the Mathematical Association of America and the NCTM appear in a special section

Mississippi

K–12 Curriculum Structure—Mathematics
A. Curriculum Content
  1. Grades K-12: presents 642 skills for traditional topics to be introduced/mastered in grades K-8, specifically 4-8, 6-8, and 7-8
  2. Grades 9-12: traditional high school course sequence
  3. Core areas: preceded by "C" in a list of 642 skills
  4. College-preparatory students: the only reference to the goal of teaching Mississippi pupils is that they be able to make a wise college or career choice
  5. Tasks or activities: no reference
  6. Related topics: no reference
  7. Technology: no reference
B. Teaching Standards
  1. Teaching strategies: no reference
  2. Connections: no reference
  3. Communication skills: refers to the goal of encouraging communication skills for the purpose of learning to reason logically and solve problems creatively
  4. Tasks or activities: none given
  5. Student expectations: no reference
  6. Role for minorities/women: no reference

C. Professional Standards
  1. Professional organizations: none listed
  2. Business/government agencies: none listed
  3. Vision: no reference
  4. Textbook selection: no criteria are listed
  5. Career infusion: no reference

Missouri

K–10 First Grade Mathematics Test and *Core Competencies and Key Skills for Missouri Schools, Grades 2–10*
A. Curriculum Content
  1. Grade 1: unique categories, although it appears to be a disguised traditional curriculum; purpose for which numbers are used, mathematical relationships, conventions of mathematics; pupils are tested at the end of the first grade to determine second grade mathematics placement
  2. Grades 2-6: Revised Core Competencies/Key Skills Guide— traditional topics
  3. Grades 7-10: the traditional middle/junior high topics in algebra, geometry, and business are added to grades 2-6 skills and topics
  4. Core areas: as indicated above
  5. College-preparatory students: no reference
  6. Tasks or activities: a large number of graded tasks/problems is presented with answers indicated for starred problems
  7. Related skills are presented, though not related topics
  8. Technology: no reference
B. Teaching Standards
  1. Teaching strategies: specific methods of teaching certain topics are proposed
  2. Connections: no references
  3. Communication skills: all types are encouraged
  4. Student involvement: the document seems to indicate a preoccupation with test taking and testing strategies
  5. Student expectations: no reference
  6. Role for minorities/women: no reference
C. Professional Standards
  1. Professional organizations: none listed
  2. Business/government agencies: none listed
  3. Vision: none given

4. Textbook selection: no reference
5. Career infusion: no reference

Montana

K–12 Montana School Accreditation—Mathematics: Model Learner Goals
A. Curriculum Content
   1. Primary: modern/traditional level-appropriate topics
   2. Intermediate: modern/traditional level-appropriate topics
   3. Upon graduation: modern/traditional topics
   4. Core areas: modern/traditional leaning toward traditional
   5. College-preparatory students: no reference
   6. Tasks or activities: none listed
   7. Related topics: no reference
   8. Technology: integrated throughout
B. Teaching Standards
   1. Teaching strategies: no reference
   2. Connections: no reference
   3. Communication skills: no reference
   4. Student involvement: no reference
   5. Student expectations: no reference
   6. Role for minorities/women: no reference
C. Professional Standards
   1. Professional organizations: no reference
   2. Business/government agencies: no reference
   3. Vision: no reference
   4. Textbook selection: no reference
   5. Career infusion: no reference

Nevada

K–8 Elementary Course of Study
A. Curriculum Content
   1. Kindergarten (11 objectives): traditional influence
   2. Grade 3 (51 objectives): traditional influence
   3. Grade 6 (84 objectives): traditional influence; four additional objectives in "career awareness" for grade 6
   4. Grade 8 (64 objectives): traditional influence; four additional objectives in "career awareness" for grade 8
   5. Core areas: traditional influence
   6. Related topics: no references
   7. Technology: the significance of computers and calculators is emphasized, although no specific strategies concerning their use are noted
B. Teaching Standards
   1. Teaching strategies: no reference
   2. Connections: no reference
   3. Communication skills: no reference
   4. Student involvement as doers: no reference
   5. Student expectations: no reference
   6. Role for minorities/women: no reference
C. Professional Standards
   1. Professional organizations: no reference
   2. Business/government agencies: no reference
   3. Vision: no reference
   4. Textbook selection: no reference
   5. Career infusion: noted through classroom career awareness activities beginning with sixth grade

New Hampshire

K–8 Minimum Standards
A. Curriculum Content
   1. Grades K–8: traditional influence; recognizing reasonable/unreasonable results interwoven throughout
   2. Grades 9–12: traditional influence
   3. A third document, overlapping the previous curriculum document for grades 5–8, titled "Standards and Guidelines for Middle/Junior High Schools," reviews traditional topics already taught but with the addition of "appreciation of fun and beauty in mathematics"
   4. Core areas: enumerated above
   5. College-preparatory students: a supplementary college- preparatory curriculum is advised
   6. Tasks or activities: no reference
   7. Related topics: no reference
   8. Technology: no reference

B. Teaching Standards
   1. Teaching strategies: suggestions are made on ways to maximize the amount of time a student actually spends paying attention to a learning activity
   2. Connections: no reference
   3. Communication skills: no reference
   4. Student involvement: minimal reference
   5. Student expectations: no reference
   6. Role for minorities/women: no reference
C. Professional Standards
   1. Professional organizations: no reference
   2. Business/government agencies: no reference
   3. Vision: no reference
   4. Textbook selection: no reference
   5. Career infusion: no reference
   6. Special features: a guide for the development of an instructional program, cocurricular program, and supportive services

New Mexico

An Elementary Competency Guide for Grades 1–8
A. Curriculum Content
   1. Grades 1–8: traditional influence
   2. Core areas (known as essential competencies): consists of 265 topics that will be the basis of an achievement/placement test to be administered in grades 3, 5, and 8
   3. College-preparatory students: no references
   4. Activities/tasks: none listed
   5. Related topics: none listed
   6. Technology: reference is made to calculator math, but without elaboration
B. Teaching Standards
   1. Teaching strategies: none listed
   2. Connections: no reference
   3. Communication skills: no reference
   4. Student involvement: no reference
   5. Student expectations: no reference
   6. Role for minorities/women: no reference
C. Professional Standards
   1. Professional organizations: none listed
   2. Business/government agencies: none listed

3. Vision: no reference
4. Textbook selection: no reference
5. Career infusion: none

New York

K–12 Mathematics Curriculum Guides
A. Curriculum Content
   1. Grades K–6: traditional influence
   2. Grades 7–8: no document provided
   3. Grades 9–12: Course I, II, and III—modern/traditional
   4. Core areas: modern/traditional
   5. College-preparatory students: the syllabus/guide quotes from the College Entrance Examination Board's bulletin that all college entrants need to achieve proficiency in use of mathematical techniques, deductive nature of mathematics, calculators and computers, statistics, and more in-depth algebra, geometry, and functions
   6. Tasks or activities: provides many in-depth illustrations of appropriate tasks and activities
   7. Related topics: indicated under the rubric interdisciplinary activities, which includes pure mathematics, music, and art
   8. Technology: two supplementary guides, of 60 and 50 pages respectively, are devoted exclusively to the teaching of computers in grades K–8 and 9–12
B. Teaching Standards
   1. Teaching strategies: presented in a series of supplementary guides—improving reading/study skills in mathematics K–6; ideas for strengthening mathematics skills; suggestions for teaching mathematics using laboratory approaches in grades 1–3; suggestions for teaching mathematics using laboratory approaches in grades 1–6; number and numeration and another for operations; creative problem solving; in addition, strategies involving small-group work, whole-class projects, and written and oral activities are recommended
   2. Connections: references are made to several subject areas, but illustrations are limited to advanced mathematics topics and history of mathematics

3. Communication skills: occasional and general references to the use of communication skills are made within the context of illustrating solutions of certain tasks, but not as a teaching tool
4. Student involvement as doers and thinkers: well-illustrated in the curriculum guide
5. Student expectations: no reference
6. Role for minorities/women: no reference is made to a greater role in mathematics for women and minorities; however, all courses are mandated to include information about disabled and handicapped persons
C. Professional Standards
   1. Professional organizations: noted
   2. Business/government agencies: no reference
   3. Vision: no reference
   4. Textbook selection: no reference
   5. Career infusion: no reference

North Carolina

K–12 Standard Course of Study—Competency Based Curriculum
A. Curriculum Content
   1. Grades K–8: modern/traditional influence
   2. Grades 9–12: the basic program—
      a. non-college-preparatory: traditional courses
      b. college-preparatory: traditional courses
   3. Core areas: supplement the above influences with calculators and computers
   4. College-preparatory students: the guide follows a traditional/modern college preparatory program
   5. Tasks or activities: plentiful throughout the guide
   6. Related topics: occasionally indicated, such as map reading in social studies, consumer mathematics, scale drawings, and insurance tables
   7. Technology: advocated throughout but with no specific suggestions as to how computers and calculators ought to be utilized
B. Teaching Standards
   1. Teaching strategies: no reference

2. Connections: makes connections with other areas of mathematics as well as other curriculum areas
3. Communication skills: no references
4. Student involvement as doers: large number of activities included
5. Student expectations: no reference
6. Role for minorities/women: no reference
C. Professional Standards
   1. Professional organizations: not listed
   2. Business/government agencies: not listed
   3. Vision: not indicated
   4. Textbook selection: adoption process for North Carolina is indicated
   5. Career infusion: noted indirectly in the context of alternative courses of study and alternative tracks

North Dakota

9–12 Mathematics Curriculum Guide
A. Curriculum Content
   1. For K–8 curriculum planning: recommends using the bibliography at end of this guide along with *A Guide for Curriculum Planning K–12,* 4th ed. (DPI, 1983)
   2. Grades 9–12: traditional topics
   3. Core areas: traditional
   4. College-preparatory students: traditional sequence
   5. Tasks or activities: recommended activities are written in the margins of the guide
   6. Related topics: no references
   7. Technology: calculator and computer use is expected
B. Teaching Standards
   1. Teaching strategies: both small- and large-group instruction
   2. Connections: no reference
   3. Communication skills: no reference
   4. Student involvement: many student activities are written in the margins
   5. Student expectations: implies that all students should consider college entry
   6. Role for minorities/women: groups for special consideration are special education and handicapped youngsters; minorities and women are not mentioned
C. Professional Standards
   1. Professional organizations: not listed

2. Business/government agencies: no reference
3. Vision: no grand vision is indicated, only the personal vision of college or employment after graduation
4. Textbook selection: evaluation guidelines are listed
5. Computer software: guidelines are listed
6. Career infusion: no reference

Ohio

9–12 Model Competency-Based Mathematics Program
A. Curriculum Content
1. The model consists of eight strands with objectives grouped by grade levels K-4, 5-8 and 9-12. The objectives are correlated with ability/achievement tests administered in grades 4, 6, 8, 9, and 12.
2. The guide presents its modern/traditional model curriculum in a 214-page document that includes illustrations, activities, diagrams, graphs, charts, and tables.
3. The design of the strands is guided by five assumptions—problem solving and reasoning, curricular sequencing, technology, connections, communication
4. Core areas: modern/traditional influence
5. College-preparatory students: the guide offers an alternative integrated curriculum
6. Related topics: some related areas in mathematics are noted
7. Technology: utilization is noted but not emphasized
B. Teaching Standards
1. Teaching strategies: noted without elaboration
2. Connections: noted but not emphasized
3. Communication skills: noted but not stressed
4. Student involvement as doers and thinkers: strongly advocated
5. Student expectations: high, but student achievement is not emphasized
6. Role for minorities/women: no reference
C. Professional Standards
1. Professional organizations: no reference
2. Business/government agencies: not listed
3. Vision: appears to be along academic lines only

4. Textbook selection: no reference
5. Career infusion: no reference
6. Strong emphasis on the significance of standardized testing and of intervention services in classrooms, school buildings, and school districts

Oklahoma

K–12 Learner Outcomes
A. Curriculum Content
1. Grades 1 and 2: traditional influence
2. Grades 3–5: mathematics learner outcomes not available
3. Grades 6–8: traditional influence of mathematics learner outcomes
5. Grades 9–12: traditional high school courses
6. Tasks or activities: no reference
7. Related topics: no reference
8. Technology: suggests using calculators and computers
B. Teaching Standards
1. Teaching strategies to consider: project work, discussions, cooperative partner or group work, integration of math in other content areas, technology
2. Connections: no reference
3. Communication skills: deemed advisable
4. Student involvement: advisable even though no sample activities are demonstrated
5. Student expectations: no reference
6. Role for minorities/women: no reference
C. Professional Standards
1. Professional organizations: no reference
2. Business/government agencies: no reference
3. Vision: suggests implementing a developmentally appropriate mathematics curriculum
4. Textbook selection: no reference
5. Career infusion: no reference

Oregon

K–11 Mathematics Common Curricular Goals
A. Curriculum Content
1. Grades K-3: traditional mathematics topics plus oral and written communica-

tion skills, as well as appropriate study skills
2. Grades 4–11: same skills as grades K–3
3. Core areas: same as those listed for K–3
4. College-preparatory students: no reference
5. Related topics: no reference
6. Tasks or activities: many
7. Technology: calculator and computer use advised chiefly for solving problems and verification of answers
B. Teaching Standards
   1. Teaching strategies: no reference
   2. Connections: no reference
   3. Communication skills: much emphasis on all aspects of communication skills that could reinforce mathematical skills
   4. Student involvement as doers: noted only indirectly through students' solution of problems and communication activities
   5. Student expectations: no reference
   6. Role for minorities/women: no reference
C. Professional Standards
   1. Professional organizations: no reference
   2. Business/government agencies: no reference
   3. Vision: no reference
   4. Textbook selection: no reference
   5. Career infusion: no reference

Pennsylvania

K–6 Mathematics Content Examples or Activities
A. Curriculum Content
   1. Grades K–3: partial modern/traditional curriculum
   2. Grades 4–6: mostly traditional influence
   3. Core areas: none specified, other than those already indicated above for K–6
   4. College-preparatory students: no reference
   5. Tasks or activities: many and varied, including solutions; many calculator/computer tasks
   6. Related topics: not listed, except for general reference to charts and tables in newspapers and timelines in history

7. Technology: strong endorsement for calculator/computer use
B. Teaching Standards
   1. Teaching strategies: none listed
   2. Connections: insignificant listings
   3. Communication skills: listed only in connection with problem-solving strategies
   4. Student involvement as doers: the major strength of this curriculum document
   5. Student expectations: no reference
   6. Role for minorities/women: no reference
C. Professional Standards
   1. Professional organizations: no reference
   2. Business/government agencies: none listed
   3. Vision: no reference
   4. Textbook selection: no reference
   5. Career infusion: no reference

South Carolina

K–12 Mathematics Framework
A. Curriculum Content
   1. Grades K–12: mostly modern/some traditional strands
   2. Identical strands run through every grouping of grades K–3, 3–6, 6–9 and 9–12
   3. Tasks: the framework offers suggested activities for every grade grouping; these reflect appropriate age, readiness, and ability levels of pupils
   4. Core skills: modern influence
   5. College-preparatory students: suggests following the recommendations of the NCTM's Curriculum Standards
   6. Tasks/activities: offers appropriate individual and group problem-solving activities throughout the curriculum
   7. Related areas: indicated for every topic
   8. Technology: calculator/computer use encouraged for every student at each grade level
B. Teaching Standards
   1. Teaching strategies: advocates a variety of strategies such as project work, group and individual tasks, discussions, practice, teacher exposition
   2. Connections: with other mathematics concepts are suggested

3. Communications skills: strongly endorses sharpening skills in speaking, writing and listening
4. Tasks: many activities that require active student participation
5. Student expectations: not noted
6. Role for minorities/women: high-quality performance is encouraged

C. Professional Standards
1. Professional organizations: membership is implicitly encouraged
2. Business/political/ media partners: notes their importance in the educational process
3. Vision: strong commitment to the vision provided by the standards; Department of Education restructured certification standards to conform to three suggested grade levels—K–4, 5–8, and 9–12
4. Textbook selection: no reference
5. Career infusion: no significant reference

South Dakota

K–12 A Curriculum Guideline for Mathematics
A. Curriculum Content
1. Preschool (the mathematical foundations): describes the learning processes of 1-, 2-, 3-, and 4-year-olds
2. Grades K–3 (primary): lists 16 traditional age-appropriate competencies
3. Grades 4–6 (intermediate): lists 51 traditional competencies
4. Grades 7–9 (junior high): lists 40 traditional competencies including land area/hectare, flow charts, nth roots, and powers
5. Grades 10–12 (senior high): lists 71 modern/traditional competencies
6. Core areas: traditional/modern
7. College-preparatory students: contains a section entitled "Mathematics Required for ..."; in addition to the standard college-preparatory courses, the section recommends informal statistics, consumer mathematics, and computer literacy
8. Tasks or activities: no reference
9. Related topics: no reference
10. Technology: the use of calculators and computers is advocated throughout the curriculum

B. Teaching Standards
1. Teaching strategies: no reference
2. Connections: no reference
3. Communication skills: no reference
4. Student involvement: indicated in connection with problem-solving activities
5. Student expectations: no reference
6. Role for minorities/women: no reference

C. Professional Standards
1. Professional organizations: none listed
2. Business/government agencies: none listed
3. Vision: none indicated
4. Textbook/educational material selection: a section on materials selection describes, in general terms, the criteria for selecting educational materials
5. Career infusion: no reference

Tennessee

K–8 Curriculum Frameworks
A. Curriculum Content
1. Grades K–8: lists 70 traditional topics
2. Grades 9–12: traditional high school mathematics sequence plus an alternate sequence for noncollege-preparatory pupils
3. Core areas: includes the 70 traditional topics for grades K–8 plus the traditional high school mathematics sequence
4. College-preparatory students: no reference
5. Tasks: none listed (see A-7 below)
6. Related topics: none listed
7. Technology: in a prefatory remark, it is noted that Tennessee encourages efforts to integrate computers and calculators with problem-solving and activity-based instruction, even though they do not have the optimal number of these educational tools

B. Teaching Standards
1. Teaching strategies: should include hands-on, activity- based lessons
2. Connections: none listed
3. Communication skills: no reference
4. Student involvement: see A-7 above

5. Student expectations: no reference
6. Role for minorities/women: no reference

C. Professional Standards
1. Professional organizations: no reference
2. Business/government agencies: no reference
3. Vision: no reference
4. Textbook selection: no reference
5. Career infusion: no reference

Utah

K–12 Core Mathematics Curriculum

A. Curriculum Content
1. Grades K–6: traditional influence with emphasis on counting, measuring, and problem solving
2. Grades 7–12: traditional high school sequence
3. Core areas: none are specified, but see K–12 courses above
4. College-preparatory students: no reference
5. Tasks or activities: none indicated
6. Related topics: none indicated
7. Technology: no reference

B. Teaching Standards
1. Teaching strategies: suggests considering self-inquiry, external stimuli, experiences with the physical world, pictorial representations, abstractions, and the environment
2. Connections: no reference
3. Communication skills: stresses value of words and symbols
4. Student involvement: no reference
5. Role for minorities/women: no reference

C. Professional Standards
1. Professional organizations: no reference
2. Business/government agencies: no reference
3. Vision: value of education is noted
4. Textbook selection: no reference
5. Career infusion: no reference
6. Special note: the Utah curriculum writers appear to consider graduation requirements with special concern, although the reason for this concern is not noted

Vermont

K–12 Mathematics Scope and Sequence Framework

A. Curriculum Content
1. Grades K–3: lists 28 traditional grade-appropriate objectives referring to core areas below
2. Grades 4–6: lists 48 traditional grade-appropriate objectives referring to core areas below
3. Grades 7–8: lists 64 traditional grade-appropriate objectives referring to core areas listed below
4. Grades 9–12: lists 53 traditional grade-appropriate objectives referring to core areas listed below
5. Core areas: traditional topics
6. College-preparatory students: no reference
7. Tasks and graded activities: none listed
8. Related topics: none listed
9. Technology: significant use of calculators is noted in the listing of objectives

B. Teaching Standards
1. Teaching strategies: no reference
2. Connections: no reference
3. Communication skills: no reference
4. Student involvement as doers: no activities are listed, but in the area of problem solving, many types of problems are discussed
5. Student expectations: no reference
6. Role for minorities/women: no reference

C. Professional Standards
1. Professional organizations: not listed
2. Business/government agencies: not listed
3. Vision: no reference
4. Textbook selection: no reference
5. Career infusion: no reference

Virginia

K–12 Standards of Learning Objectives

A. Curriculum Content
1. Grades K–8, general math 9, algebra I, consumer math, and geometry: "standards"

of one sentence each are written for each required topic taught in every grade

2. Grades 9–12: the traditional high school course sequence is designated as optional
3. Core areas: are not specified as such
4. College-preparatory students: the only reference is in the advanced-placement course description
5. Tasks or activities: no specific exercises are shown, although descriptions of types of activities are recommended for each grade
6. Related topics: no reference
7. Technology: the use of calculators and computers is strongly advocated, but without specific recommendations

B. Teaching Standards
1. Teaching strategies: no reference
2. Connections: no reference
3. Communication skills: no reference
4. Student involvement: no direct references
5. Student expectations: no reference
6. Role for minorities/women: no reference

C. Professional Standards
1. Professional organizations: no reference
2. Business/government agencies: no reference
3. Vision: no reference
4. Textbook selection: no reference
5. Career infusion: no reference

Washington

K–8 Guidelines for Mathematics Curriculum
A. Curriculum Content
1. Grades K–8: modern/traditional influence
2. Core areas: all grades do the same traditional strands, but with differing expectations and strategies; there is strong feeling to counter the restrictive influence of the basic-skill push of recent years.
3. College-preparatory students: no reference
4. Tasks or activities: lists no specific activities, but does list instructional resources and manipulatives that are task-related for each grade

5. Related topics: no reference
6. Technology: students are advised to use calculators to perform multidigit computations and computers for problem-solving techniques

B. Teaching Standards
1. Teaching strategies: no reference
2. Connections: note A-4 above
3. Communication skills: no reference
4. Student involvement: many references to the activities a teacher should develop
5. Student expectations: no reference
6. Role for minorities/women: no reference

C. Professional Standards
1. Professional organizations: no reference
2. Business/government agencies: no reference
3. Vision: gives a parochial vision that the aim is to develop thoughtful behavior in mathematics
4. Textbook selection: no reference
5. Career infusion: no reference

West Virginia

K–12 Mathematics Instructional Goals and Objectives
A. Curriculum Content
1. Grades K–8: traditional influence
2. Grades 9–12: traditional high school sequence
3. Core areas: all topics listed for K–12 though with different objectives and strategies for each grade level
4. College-preparatory students: adds to the traditional core areas the topics of traditional advanced high school mathematics
5. Tasks: none are suggested
6. Related topics: no references
7. Technology: advocates use of calculators and computers, but no specific applications are given; recommends a computer mathematics course for students

B. Teaching Standards
1. Teaching strategies: no references
2. Connections: the classroom use of connections with other mathematics topics, other areas of study, and the real world is

advocated, but with no specific illustrations
3. Communication skills: advocates their use, but does not provide illustrations
4. Student involvement: no references
5. Student expectations: no references
6. Role for minorities/women: no references

C. Professional Standards
1. Professional organizations: no reference
2. Business/government agencies: no reference
3. Vision: no reference
4. Textbook selection: no reference
5. Career infusion: no reference

Wisconsin

K–12 A Guide to Curriculum Planning in Mathematics

A. Curriculum Content
1. Grades K–3, 4–6, 7–8, and 9–12 with spirally arranged traditional curriculum strands from grade K
2. Core areas: students of each grade level study every strand at their appropriate level of understanding; "The Wisconsin Plan" describes a curriculum for grades 7–12 which consists of an integrated structure that includes some arithmetic, statistics, geometry, and algebra. A more traditional sequence is offered by the "modified" Wisconsin Plan.
5. College-preparatory students: college preparation for mathematics begins with the traditional sequence in grade 9
6. Tasks or activities: none indicated
7. Technology: use of calculators is encouraged for all grades.

B. Teaching Standards
1. Teaching strategies: the only suggestion noted is that individual differences among students ought to be considered as part of every strategy
2. Connections: several noted
3. Communication skills: suggests that reading skills in the mathematics classroom ought to be sharpened
4. Student involvement as doers and thinkers: students will be required to make decisions when solving problems

5. Student expectations: no reference
6. Role for minorities/women: no reference

C. Professional Standards
1. Professional organizations: both national and local professional associations are listed
2. Business/government agencies: no reference
3. Vision: vague vision for the future of mathematics education
4. Textbook selection: criteria are provided
5. Career infusion: is noted

Wyoming

K–12 School Accreditation

A. Curriculum Content
1. No state-mandated grade levels for elementary, junior, or senior high schools
2. All public school students shall meet student performance standards at the level set by the school and district
3. Core areas: standards set by the school and district
4. College-preparatory students: no reference
5. Tasks or activities: none listed
6. Related topics: no reference
7. Technology: reference to technology is made in general terms only—no specific recommendations

B. Teaching Standards
1. Teaching strategies: no reference
2. Connections: no reference
3. Communication skills: no reference
4. Student involvement as doers: no reference
5. Expectations for at-risk students: each district establishes procedures to intervene with such students with parent participation
6. Role for minorities/women: attention given to addressing the needs of gender, ethnic, or socioeconomic groups that are functioning below performance levels

C. Professional Standards
1. Professional organizations: each school has a staff-development plan in which professional growth should be encouraged

2. Business/government agencies: parents are listed as partners in the educational process
3. Vision for the future: no references
4. Textbook selection: no reference
5. Career infusion: the only reference is a general one to "life-skills"

STATE-LEVEL CURRICULUM GUIDELINES: A LISTING

THIS chapter provides bibliographic information on the state curriculum documents discussed in chapter 5. The publications are organized by state; for each state, we have provided the full address for that state's department of education, including the office to contact regarding curriculum publications (if such an office has been specified by the state department). The phone number shown is the best number to use for ordering the publications or for getting further information on the publications. We have also provided the addresses and phone numbers for states whose departments of education do not publish statewide curriculum frameworks. These states may produce curriculum materials on specific topics in mathematics and in other disciplines, but they are not statewide guides as described in chapter 5.

For each publication, the listing provides the full title, document number or ISBN (if available), number of pages, year of publication or reprinting (or "n.d." if no date is available), and price. Pricing is given on those publications for which Kraus had information; note that the prices shown are taken from the department's order form. Shipping and handling are often extra, and some states offer discounts for purchasing multiple copies. If a document is listed in ERIC, its ED number is shown as well.

Alabama

State Department of Education
Gordon Persons Office Building
50 North Ripley Street
Montgomery, AL 36130-3901

Division of Student Instructional Services
Coordinator, Curriculum Development/Courses
of Study
(205) 242-805

Alabama Course of Study: Mathematics
Bulletin 1989, No. 31, 200p., 1989.

Alaska

State Department of Education
Goldbelt Building
P.O. Box F
Juneau, AK 99981

Division of Education Program Support
Administrator, Office of Basic Education
(907) 465-2841

Alaska Elementary Mathematics: Model Curriculum Guide
2d ed., 77p., 1986.

Alaska Secondary Mathematics: Model Curriculum Guide
2d ed., 116p., 1986.

Arizona

State Department of Education
1535 West Jefferson
Phoenix, AZ 85007

Education Services
Instructional Technology (602) 542-2147

Arizona Essential Skills for Mathematics
DDD985, 22p., 1987.

Arkansas

Department of Education
Four State Capitol Mall
Room 304 A
Little Rock, AR 72201-1071

Instructional Services
Coordinator, Curriculum and Assessment
(501) 682-4558

Mathematics, Grades 1-8
80p., n.d.

Mathematics, Grades 9-12. Arkansas Public School Course Content Guide
43p., n.d.

California

State Department of Education
P.O. Box 944272
721 Capitol Mall
Sacramento, CA 95814

California Department of Education
Bureau of Publications
(916) 445-1260

Mathematics Framework for California Public Schools: Kindergarten through Grade Twelve
ISBN 0-8011-0358-4, 160p., 1991.

Colorado

State Department of Education
201 East Colfax Avenue
Denver, CO 80203-1705

Colorado Sample Outcomes and Proficiencies for Elementary, Middle and High School Education
162p., 1991. Includes outcomes and proficiencies for mathematics.

Connecticut

State Department of Education
P.O. Box 2219
165 Capitol Avenue
State Office Building
Hartford, CT 06145

Program and Support Services
Division of Curriculum and Professional
  Development
(203) 566-8113

A Guide to Curriculum Development: Purpose, Practices and Procedures
72p., 1981.

A Guide to Curriculum Development in Mathematics
108p., 1981.

Delaware

State Department of Public Information
Post Office Box 1402
Townsend Building, #279
Dover, DE 19903

Instructional Services Branch
State Director, Instruction Division
(302) 739-4647

Content Standards for Delaware Public Schools
233p., 1986. Includes content standards for mathematics.

Florida

State Department of Education
Capitol Building, Room PL 116
Tallahassee, FL 32301

Curriculum Support Services
Bureau of Elementary and Secondary Education
(904) 488-6547

Curriculum Frameworks for Grades 6-8 Basic Programs. Volume IV: Mathematics and Computer Education
42p., 1990.

Curriculum Frameworks for Grades 9-12 Basic and Adult Secondary Programs. Volume IV: Mathematics and Computer Education
132p., 1990.

Georgia

State Department of Education
2066 Twin Towers East
205 Butler Street
Atlanta, GA 30334

Office of Instructional Programs
Director, General Instruction Division
(404) 656-2412

Georgia's Quality Core Curriculum (K-12)
25-diskette set (AppleWorks version), 17-diskette set (IBM WordStar version), 1989.

Hawaii

Department of Education
1390 Miller Street, #307
Honolulu, HI 96813

Office of Instructional Services
Director, General Education Branch
(808) 396-2502

The Hawaii Department of Education is revising its statewide frameworks; the new publications are scheduled to be available in 1993.

Idaho

State Department of Education
Len B. Jordan Office Building
650 West State Street
Boise, ID 83720

Chief, Bureau of Instruction/School
Effectiveness
(208) 334-2165

Idaho K-8 Course of Study Guide: Mathematics
63p., 1987, updated 1990.

Idaho Secondary Courses of Study Guide: Mathematics
45p., 1991 draft.

Illinois

State Board of Education
100 North First Street
Springfield, IL 62777

School Improvement Administrative
Curriculum Improvement (217) 782-2826

State Goals for Learning and Sample Learning Objectives. Mathematics: Grades 3, 6, 8, 10, 12
4M 7-464B-26 No. 238, 69p., 1986.

Indiana

State Department of Education
Room 229, State House
100 North Capitol Street
Indianapolis, IN 46024-2798

Center for School Improvement and
Performance
Manager, Office of Program Development
(317) 232-9157

Mathematics Proficiency Guide
268p., 1991.

Iowa

State Department of Education
Grimes State Office Building
East 14th & Grand Streets
Des Moines, IA 50319-0146

Division of Instructional Services
Bureau Chief, Instruction and Curriculum
(515) 281-8141

A Guide to Curriculum Development in Mathematics. A Curriculum Coordinating Committee Report
120p., 1987.

Kansas

State Department of Education
120 East Tenth Street
Topeka, KS 66612

Kansas Mathematics Curriculum Standards
99p., 1990.

The Kansas State Department of Education is currently developing mathematics outcomes which will be available mid-1993.

Kentucky

State Department of Education
1725 Capitol Plaza Tower
Frankfort, KY 40601

Office of Instruction
Division of Curriculum and Staff Development
(502) 564-2106

A List of Valued Outcomes for Kentucky's Six Learning Goals. Council on School Performance Standards
6p., n.d.

Louisiana

State Department of Education
P.O. Box 94064
626 North 4th Street
12th Floor
Baton Rouge, LA 70804-9064

Office of Academic Programs
Secondary Education (504) 342-3404
Elementary Education (504) 342-3366

Introduction to Algebra. Curriculum Guide, Grade 8
Bulletin 1802, 88p., 1987.

Mathematics Curriculum Guide, Grades K-8
Bulletin 1609, 169p., rev. 1986.

Geometry Curriculum Guide
Bulletin 1581, 70p., rev. 1984.

Advanced Mathematics Curriculum Guide
Bulletin 1583, 167p., rev. 1984.

Computer Science Curriculum Guide
Bulletin 1610, 23p., rev. 1983.

Computer Literacy Curriculum Guide
Bulletin 1739, 67p., 1985.

Maine

State Department of Education
State House Station No. 23
Augusta, ME 04333

Bureau of Instruction
Director, Division of Curriculum
(207) 289-5928

The Maine State Department of Education does not produce statewide frameworks for K-12 mathematics.

Maryland

State Department of Education
200 West Baltimore Street
Baltimore, MD 21201

Bureau of Educational Development
Division of Instruction, Branch Chief, Arts and
  Sciences
(301) 333-2307

Mathematics. A Maryland Curricular
Framework
42-970 10/90, 42p., 1990.

Massachusetts

State Department of Education
Quincy Center Plaza
1385 Hancock Street
Quincy, MA 02169

School Programs Division
(617) 770-7540

The Massachusetts State Department of Education does not produce statewide frameworks for K-12 mathematics.

Michigan

State Board of Education
P.O. Box 30008
608 West Allegan Street
Lansing, Michigan 48909

Instructional Specialists Program
(517) 373-7248

Model Core Curriculum Outcomes
73p., 1991 working document. Contains Educational Outcomes for K-12 subjects, including outcomes for mathematics and science.

Minnesota

State Department of Education
712 Capitol Square Building
550 Cedar Street
St. Paul, MN 55101

Minnesota Curriculum Services Center
(612) 483-4442

Model Learner Outcomes for Mathematics
Education
E722B, 153p., 1991.

Mississippi

State Department of Education
P.O. Box 771
550 High Street, Room 501
Jackson, MS 39205-0771

Bureau of School Support
Textbook Division (601) 359-2791

Mississippi Curriculum Structure: Mathematics
59p., 1986 (sixth printing 1991).

Missouri

Department of Elementary and Secondary
Education
P.O. Box 480
205 Jefferson Street, 6th Floor
Jefferson City, MO 65102

Center for Educational Assessment
  (314) 882-4694

Core Competencies and Key Skills for Missouri
Schools. Grade 2: Mathematics
30p., 1991.

Core Competencies and Key Skills for Missouri
Schools. Grade 3: Mathematics
30p., 1991.

Core Competencies and Key Skills for Missouri
Schools. Grade 4: Mathematics
35p., 1991.

Core Competencies and Key Skills for Missouri
Schools. Grade 5: Mathematics
35p., 1991.

Core Competencies and Key Skills for Missouri Schools. Grade 6: Mathematics
38p., 1991.

Montana

Office of Public Instruction
106 State Capitol
Helena, MT 59620

Department of Curriculum Services
Curriculum Assistance and Instructional Alternatives
(406) 444-5541

Montana School Accreditation: Standards and Procedures Manual
34p., 1989.

The Montana Office of Public Instruction does not produce other statewide frameworks.

Nebraska

State Department of Education
301 Centennial Mall, South
Post Office Box 94987
Lincoln, NE 68509

The Nebraska State Department of Education does not produce statewide frameworks.

Nevada

State Department of Education
Capitol Complex
400 West King Street
Carson City, Nevada 89710

Instructional Services Division
Director, Basic Education Branch
(702) 687-3136

Elementary Course of Study
65p., 1984. Includes scope and sequence for mathematics.

Nevada Secondary Course of Study. Volume 1: Academic Subjects
05282, 72p., n.d. Includes information on Required Courses in computer education and in mathematics.

New Hampshire

State Department of Education
101 Pleasant Street
State Office Park South
Concord, NH 03301

Division of Instructional Services
General Instructional Services Administrator
(603) 271-2632

Minimum Standards for New Hampshire Public Elementary School Approval, Kindergarten-Grade 8: Working Together
36p., 1987. Includes elementary school curriculum, K-8.

Standards & Guidelines for Middle/Junior High Schools
101p., 1978. Includes information on mathematics.

Standards for Approval of New Hampshire Public High Schools, Grades 9-12
53p., 1984.

New Jersey

Department of Education
225 West State Street, CN 500
Trenton, NJ 08625-0500

Division of General Academic Education
(609) 984-1971

New Jersey High School Graduation Requirements
1 p., 1988.

The New Jersey Department of Education does not produce statewide frameworks.

New Mexico

State Department of Education
Education Building
300 Don Gaspar
Santa Fe, NM 87501-2786

Learning Services Division
Instructional Materials (505) 827-6504

An Elementary Competency Guide for Grades 1-8
88p., 1987, rev. 1990, reprinted 1990. Includes "Competencies by Subject Area" for mathematics and computer literacy.

Graduation Requirements
SBE Regulation No. 90-2, section A.4.3, 12p., 1990. High School graduation requirements.

New York

State Education Department
111 Education Building
Washington Avenue
Albany, NY 12234

The University of the State of New York
The State Education Department
Publications Sales Desk (518) 474-3806

Handbook on Requirements for Elementary and Secondary Schools. Education Law, Rules of the Board of Regents, and Regulations of the Commissioner of Education
140p., 2d ed. 1989.

Mathematics K-6: A Recommended Program for Elementary Schools
91-7143, 71p., reprinted 1991.

Improving Reading/Study Skills in Mathematics K-6
88-8417, 038900, 31p., reprinted 1989.

Suggestions for Teaching Mathematics using Laboratory Approaches, Grades 1-6: 1. Numbers and Numeration
88-8091, 24p., reprinted 1989.

Suggestions for Teaching Mathematics using Laboratory Approaches, Grades 1-6: 2. Operations
88-8092, 28p., reprinted 1989.

Suggestions for Teaching Mathematics using Laboratory Approaches, Experimental Edition. 5. Numbers and Numeration, Operations, Geometry, Measurement
84-6615, 44p., reprinted 1984.

Suggestions for Teaching Mathematics using Laboratory Approaches, Grades 1-6: 6. Probability
89-8365, 029000, 28p., 1990.

Suggestions for Teaching Mathematics using Laboratory Approaches, Grades 1-3: 7. Metric System
88-8415, 41p., reprinted 1989.

Three-Year Sequence for High School Mathematics: Course I
87-7078, 84p., 1988.

Three-Year Sequence for High School Mathematics: Course II
89-7844, 79p., reprinted 1989.

Three-Year Sequence for High School Mathematics: Course III
83p., reprinted 1988.

Teaching Mathematics with Computers, K-8
88-7411, 61p., 1988.

Teaching Mathematics with Computers, 9-12
86-8317, 53p., 1988.

Creative Problem Solving
88-8070, 018900, 48p., 1988, reprinted 1989.

Ideas for Strengthening Mathematics Skills
91-7146, 40p., reprinted 1991.

North Carolina

Department of Public Instruction
Education Building
116 West Edenton Street
Raleigh, NC 27603-1712

Publications Sales Desk (919) 733-4258

North Carolina Standard Course of Study and Introduction to the Competency-Based Curriculum
530p., 1985.

Teacher Handbook: Mathematics, Grades K-12. North Carolina Competency-Based Curriculum
570p., 1985.

North Dakota

State Department of Public Instruction
State Capitol Building, 11th Floor
600 Boulevard Avenue East
Bismarck, ND 58505-0440

Office of Instruction
Supplies (701) 224-2272

Mathematics Curriculum Guide, K-8
150p., 1986.

Ohio

State Department of Education
65 South Front Street, Room 808
Columbus, Ohio 43266-0308

Division of Curriculum, Instruction, and Professional Development
(614) 466-2761

Model Competency-Based Mathematics
214p., n.d.

Oklahoma

Department of Education
Oliver Hodge Memorial Education
Building
2500 North Lincoln Boulevard
Oklahoma City, OK 73105-4599

Instructional Services Division
Curriculum/Instructional Computers
(405) 521-3361

Learner Outcomes Oklahoma State Competencies: Grades 1-12
439p., n.d.

Oregon

State Department of Education
700 Pringle Parkway, S.E.
Salem, OR 97310

Publications Sales Clerk (503) 378-3589

Mathematics Common Curriculum Goals, K-12
59p., n.d. Available only on IBM diskette.

Pennsylvania

Department of Education
333 Market Street, 10th Floor
Harrisburg, PA 17126-0333

Office of Basic Education
Director, Bureau of Curriculum and Instruction
(717) 787-8913

Chapter 5 Curriculum Regulations of the Pennsylvania State Board of Education. Guidelines for Interpretation and Implementation
32p., 1990.

Mathematics Content in Elementary School, Grades K-3
193p., 1988, reprinted 1990

Mathematics Content in Elementary School, Grades 4-6
233p., 1989

Rhode Island

Department of Education
22 Hayes Street
Providence, RI 02908

Division of School and Teacher Accreditation
(401) 277-2617

The Rhode Island Department of Education is currently developing common core of learning goals; availability date is early 1994.

South Carolina

**State Department of Education
1006 Rutledge Building
1429 Senate Street
Columbia, SC 29201**

The South Carolina State Department of Education is revising its statewide frameworks; the revised publications will be issued in 1993.

South Dakota

**Department of Education and Cultural Affairs
435 South Chapelle
Pierre, SD 57501**

Division of Elementary and Secondary Education
Office of Curriculum & Instruction
(605) 773-3261/4670

A Curriculum Guideline for Mathematics, Kindergarten-Twelve
103p., 1981.

Tennessee

**State Department of Education
100 Cordell Hull Building
Nashville, TN 37219**

Curriculum and Instruction (615) 741-0878

Mathematics Curriculum Framework, Grades 9-12
38p., 1991.

Tennessee K-8 Curriculum Frameworks
10p., 1991. Includes framework for mathematics.

Texas

**Texas Education Agency
William B. Travis Building
1701 North Congress Avenue
Austin, TX 78701-1494**

Publications Distribution Office (512) 463-9744

State Board of Education Rules for Curriculum—Essential Elements
AD202101, 545p., 1991. Includes essential elements for mathematics.

Utah

**State Office of Education
250 East 500 South
Salt Lake City, UT 84111**

Division of Operations
Coordinator, Curriculum (801) 538-7774

Mathematics Core Curriculum, Grades K-3
16p., 1987.

Mathematics Core Curriculum, Grades 4-6
28p., 1987.

Mathematics Core Curriculum, Grades 7-12
59p., 1990.

Vermont

**State Department of Education
120 State Street
Montpelier, VT 05602-2703**

Basic Education
Chief, Curriculum and Instruction Unit
(802) 828-3111

Framework for the Development of a Mathematics Scope and Sequence
17 in. x 22 in. folded sheet, n.d.

Virginia

Department of Education
P.O. Box 6-Q, James Monroe Building
Fourteenth & Franklin Streets
Richmond, VA 23216-2060

Instruction and Personnel
Administrative Director of General Education
(804) 225-2730

Standards of Learning Objectives for Virginia Public Schools: Mathematics
35p., rev. 1988.

Standards of Learning Objectives for Virginia Public Schools: Optional Mathematics Courses
51p., 1987.

Washington

Superintendent of Public Instruction
Old Capitol Building
Washington & Legion
Olympia, Washington 98504

Curriculum, Instruction Support and Special
  Services Service Unit
Director, Curriculum Support
(206) 753-6727

Guidelines for K-8 Mathematics Curriculum
56p., n.d.

West Virginia

State Department of Education
1900 Washington Street
Building B, Room 358
Charleston, West Virginia 25305

Bureau of General, Special and Professional
  Education
Director, General Education
(304) 348-7805

Mathematics Instructional Goals and Objectives
95p., 1991.

Wisconsin

State Department of Public Instruction
General Executive Facility 3
125 South Webster Street
Post Office Box 7841
Madison, WI 53707-7841

Publication Sales (608) 266-2188

A Guide to Curriculum Planning in Mathematics
(Bulletin No. 91330), 340p., 1986, rev. 1991.

Wyoming

State Department of Education
2300 Capitol Avenue, 2d Floor
Hathaway Building
Cheyenne, WY 82002

Division of Certification, Accreditation and
  Program Services
Accreditation/Special Services Unit (307)
777-6808

School Accreditation
6p., n.d.

The Wyoming State Department of Education
does not produce other statewide frameworks.

PERFORMANCE ASSESSMENT OVERVIEW

by Elizabeth Badger
Director of Assessment
Commonwealth of Massachusetts Department of Education, Quincy, Massachusetts

TESTS play a major role in American schooling. The National Commission on Testing and Public Policy (1990) has estimated that approximately 127 million standardized tests are administered to school children each year, with 20 million days of teachers' time devoted to such testing. These tests are used to grade students, teachers, school districts, even the nation. Until recently, these standardized, multiple-choice tests have been accepted without question as accurate measures of students' achievement. Within the last few years, however, a major revolution has occurred in beliefs about testing. Educators, test developers, and the public are beginning to question the premise that these tests are valid and reliable measures of performance. They have begun to advocate alternatives that give clearer and more useful information about student achievement.

This "testing revolution" is a direct result of the current reform movement in curriculum and instruction, which stresses the application of knowledge and skills over passive memorization of facts and formula. Advocates contend that active learning cannot be measured well in a multiple-choice format. Reading and recognizing a correct option from a set of incorrect

options may require thoughtful consideration, but it does not require the same creative thinking as the actual retrieval and organization of prior knowledge to solve a problem—for example, knowing how to write an essay is not the same as writing one. It is argued that the latter kind of creative thinking more closely resembles what is needed in today's world and should form the focus of testing programs.

Dissatisfaction with the format of traditional tests is compounded by the belief that these tests do not provide the kind of information necessary for accountability. The current public interest on the effectiveness of schooling has been accompanied by increasing pressure for results that are easily understood by the public. The scaled scores given by standardized, norm-referenced tests (the type of tests that are traditionally available) are useful only if one is interested in comparing students to others (i.e., the "norm"). They carry no inherent meaning and tell little about mathematical performance. Public interest in educational policy requires something more definitive—results that describe what students can and cannot do.

In addition to recognizing the need for tests that reflect what is valued in education, policymakers are also relying upon the powerful influence of tests on teaching to propel the forces of reform. While few would completely abandon multiple-choice tests—they have proven their technical quality and their cost-effectiveness in large-scale testing—emphasis is shifting from tests that are designed to rank schools and students to tests that are developed to measure students' performances vis-à-vis a set of standards.

A New Vision of School Mathematics

As mentioned in Chapter 1, the National Council of Teachers of Mathematics' (NCTM's) *Curriculum and Evaluation Standards for School Mathematics* (1989) and *Professional Standards for Teaching Mathematics* (1991) drastically changed the climate of opinion regarding mathematics. In convincing detail, these publications described a different set of goals, a different curriculum, and a different conception of the role of evaluation in school mathematics.

A Different Set of Goals for School Mathematics

Unlike the proponents of New Math in the 1960s, who focused on the intellectual structure of mathematics itself, the authors of the NCTM Standards were concerned with making school mathematics accessible to all children. The goal of instruction was not to prepare a small group of potential mathematicians but rather to give all students a sense of "mathematical power," which was defined as the ability to solve problems, communicate mathematically, reason, and recognize connections among different mathematical ideas and topics. Whether it concerned shapes or numbers, this vision of mathematics was firmly rooted in the students' world. It contrasted sharply with the traditional notions that stressed a sequence of mainly arithmetic procedures, to be taught in strict sequence. It also contrasted with the notion that the primary goal of schooling is increasing facility in abstraction and the manipulation of symbols.

A Different Curriculum

Until recently, learning mathematics was thought to occur incrementally, in a logical progression from the simple to the complex. Students began with the "basics"—number facts and simple operations on numbers. These were regarded as the initial building blocks, the foundations on which mathematical knowledge rested. New knowledge was built upon this foundation, as new facts and more complex operations were introduced. The mathematics curriculum for each grade rested firmly upon previous learning. Learning itself was thought to take place mainly through drill and practice. The ability to use and manipulate numbers was the primary goal of mathematics instruction. Problems were introduced only to illustrate or reinforce this practice. There was no discussion of context. In contrast to more culturally bound subjects such as language arts, school mathematics was viewed as having the advantage of being almost completely "school-based."

Researchers have argued that this decontextualization has led many students to view mathematics as a set of meaningless rules. They further argue that much of the blame for the poor performance of American students can be traced to the fact that students see little connection between the various mathematical topics, or between mathematics itself and real life. In contrast to this notion of mathematics as an unwieldy collection of facts, formulas, and procedures, the authors of the Standards, along with other mathematics educators, have begun to stress the unifying "big ideas." A network of concepts has replaced the building blocks as a metaphor of how mathematics is learned and should be taught.

A Different Conception of Evaluation

Recognizing the powerful influence that assessment exerts upon the curriculum, the leaders of the reform explicitly linked the two and stressed that assessment must be used to promote good instruction. While the NCTM's Curriculum and Evaluation Standards provided an analysis of the types of behavior that indicated an understanding of mathematics, the Mathematical Sciences Education Board (1991) issued a set of principles and goals for mathematics assessment. The following goals were included:

- Assessment will be aligned with the mathematical knowledge, skills, and processes that the nation needs all of its students to know and be able to do.

- Assessment practices will promote the development of mathematical power for all students.
- A variety of effective assessment methods will be used to evaluate outcomes of mathematics education.
- Mathematics teachers and school administrators will be proficient in using a wide variety of assessment methods for improving the learning and teaching of mathematics.
- The public will become better informed about assessments and assessment practices.

Current National Initiatives Affecting Mathematics Assessment

The policy statements discussed above have been translated into several major initiatives. At the national level, the most far reaching has been a series of proposals to establish national testing systems that use performance measures as well as multiple-choice questions. Two kinds of national examinations have been proposed: a voluntary individual assessment using performance assessment techniques; and an expansion of the National Assessment of Educational Progress, which also includes newly developed performance tasks. In addition, there are other national initiatives that will affect how assessment is viewed in the future.

Prominent among these initiatives is The New Standards Project, based at the Learning Research and Development Center (LRDC) at the University of Pittsburgh and the National Center on Education and the Economy. More than twenty-five states and districts have joined this consortium, which plans to develop an assessment system based on student projects, portfolios of student work, and performance evaluation. The first performance examinations will be in math and language arts at grade 4, and the system is expected to be in place for grades 4, 8, and 10 by 1998.

At the high-school level, the College Board Pacesetter Examinations are being designed as a national, syllabus-driven examination system for all high school students. Based on the NCTM Curriculum Standards, these will be partially multiple-choice and partially open-response examinations. A preliminary syllabus

and some pilot assessments for high school mathematics will be available in late 1993.

Also at the secondary level, the Secretary of Labor's Commission on Achieving Necessary Skills (SCANS) has published several reports on the competencies needed in the workplace. As with the other projects, these reports stress the need for more integrated practical thinking skills, to be measured against well-recognized standards. SCANS advocates multiple measures of assessment, including a student resume of achievement that would begin in the middle grades and lead to a Certificate of Initial Mastery by grade 10.

In addition, Congress will be considering reauthorization of Chapter 1 in 1993, and a major revision of its evaluation component is anticipated. Performance assessment, especially portfolios, is likely to be used as an option or a replacement for norm-referenced multiple-choice testing by states and local districts. In California, a state-funded project is experimenting with the California Learning Record as a means of reporting the progress of Chapter 1 children. The record provides a framework for multifaceted assessments of children's literary and mathematical development. Currently, other states are piloting adaptations of this record or considering alternatives to standardized testing.

Finally, many states and districts are developing their own standards and replacing statewide programs of testing with assessment systems. Twenty-one states currently use performance assessment (e.g., portfolios, constructed response, and hands-on demonstrations) in some subject areas; and an additional nineteen states plan to adopt some or all of these methods (State Education Leader 1992). These states are being assisted in their efforts to develop new assessment methods by the Council of Chief State School Officers, which is establishing consortia of states around specific assessment issues or subject areas.

Establishing a Local Assessment Program

Although national testing programs may appear far removed from the everyday realities of the classroom, the pressure for clearer, more specific information about students' achievement is being felt by local school districts.

Furthermore, the type of assessment that is being advocated by national groups is conceived as an integral part of effective instruction. As a result, schools are beginning to assume the responsibility for designing and carrying out their own assessments of students' achievement. This represents a different role for teachers.

When decisions about students' progress are based on the results of standardized tests that are created by "experts," or when grades are dependent upon the number of correct answers to straightforward questions, there is little need for teachers to be concerned about issues of reliability and validity. However, when teachers are asked to identify and evaluate students' behavior and abilities as they engage in solving problems, they need a deeper understanding of the various components of assessment.

What Is Assessment All About?

Assessment is not testing; assessment is the process of gathering information in order to make a judgment about something. There are two necessary components to this process. One is a standard on which judgments can be based. The other is a range of good, reliable sources of information. The first component concerns validity; the second, reliability.

Validity refers not to the test itself but rather to the inferences drawn from the test results. The question that is asked is not "Is this test providing consistent results?" but "Given the evidence provided by this test, is it possible to make a defensible judgment about the quality that we are measuring?"

Validity is the key issue in the testing reform movement. It is argued that, given the new conception of school mathematics and instruction, we cannot make valid inferences about students' understanding of mathematics on the basis of traditional testing alone. In other words, judgments about students' ability to solve problems or about their understanding of concepts cannot be based solely on their knowledge of formulas or their computational ability.

Reliability is a statistical term that refers to the consistency of results. It is based on the notion that test scores are only estimates of whatever we are measuring and that the more estimates we obtain, the closer we will come to the "true" score. It follows from this that reliability can be increased by increasing the number of questions (or estimates). For example, a test that requires a student to calculate ten proportional reasoning problems is more reliable than one that contains only two such problems, because in the former case a lucky guess or a careless error will have less effect on the final results.

Reliability is also increased by limiting the focus of the questions. Tests that measure a narrow range of abilities (e.g., computation) are more reliable than those that measure broader abilities (e.g., problem solving) because there is less opportunity for extraneous information (i.e., error) to be reflected in the results. For example, multiple-choice tests can be constructed to be extremely reliable because their questions are highly targeted, aimed at measuring single skills or one kind of information. In contrast, alternate types of assessment are less reliable because they are more vulnerable to students' interpretation and style. Furthermore, since knowledge is often closely tied to the particular situation in which it was learned, students often respond differently to tasks that are set in different contexts, despite the fact that they may involve essentially the same concepts. As a result, the context of the problem can effect the reliability of the results.

Finally, the fact that the scores of many alternative types of assessment depend upon teachers' judgment raises other issues of reliability. Interpretation of standards can vary widely among different groups of people. The clearer the criteria for scoring and the more thorough the training of raters, the more reliable their judgments will be.

Both reliability and validity imply a standard or criteria against which students' performance can be judged. Clearly articulated criteria also establish priorities about what to test and the relative weight to be placed on different aspects of behavior (e.g., solution versus strategy in problem solving). Although this may appear to be a formidable requirement, a simple analysis may often be all that is needed. For example, the following questions will help teachers focus their assessment tasks on those skills and content area that they believe are important:

- What kinds of knowledge/processes do I want to measure?
- What response would indicate a good understanding?
- What are the possible misconceptions that I expect?
- What are the implications for instruction?

This kind of analysis does not imply that there will be no surprises, such as the student who shows a unique and unforeseen insight, strategy, or misconception that was not anticipated and is difficult to account for. However, it does ensure that the assessment tasks are focused on meaningful performance and that evidence will be gathered to make a valid judgment about understanding.

What Are "Good" Assessment Tasks?

Tests are only one source of information that can be used in the assessment process. Other sources may consist of observations, interviews, the products of activities or projects, writings, videos, drawings, or other forms of communication. Many of these nontest sources of information come under the general term *performance testing*. The term is meant to stress some type of mental or physical activity that entails more than memorizing or recognizing. It suggests that the task is "real" and requires students to "show" what they can do.

The good assessment task differs from the more usual textbook problem insofar as it challenges students to engage in mathematical thinking. However, there are other features that mark good tasks, such as the following:

- They involve real mathematics.
- They elicit a range of responses.
- They can be solved in a variety of ways.
- They require communication.
- They stimulate the best possible performance; that is, they are engaging and clearly stated.

Developing Tasks

There is no such thing as a perfect task. Professional task developers, however experienced and/or creative, are seldom surprised when a favorite problem fails to elicit the kinds of thinking that it was intended to measure. On the other hand, almost any routine textbook question can be turned into an assessment occasion and a source of information about students' understanding. Depending upon the age and maturity of the students, the following are some starting points for transforming ordinary textbook questions into more challenging and fruitful tasks.

Explain a procedure. Asking students to explain a procedure to a younger child provides insight into the quality of their understanding. There are usually clear distinctions in the quality of explanations. At the basic level, students will simply reiterate the procedure, but those who have a clearer understanding will use examples and analogies from other contexts, such as the student below who was asked to explain $\frac{1}{2} \times \frac{3}{4}$.

"Well, that's $\frac{1}{2}$ of $\frac{3}{4}$. Say you had $\frac{3}{4}$ of a pie left and wanted to divide it in half. You could give $\frac{1}{4}$ to each person, but there would be $\frac{1}{4}$ left over. If you divided that $\frac{1}{4}$ in half, each person would get another $\frac{1}{8}$ of the pie. So in the end, each person would get $\frac{1}{4} + \frac{1}{8}$. That's $\frac{3}{8}$."

As with most good assessment devices, asking for explanations carries an instructional message. Not only does it give teachers valuable information about students' thinking, it forces students to begin to reflect on their own reasoning. When younger children have difficulty expressing their thinking in words, pictures or manipulatives can convey their reasoning. For example, contrast the child who represents 3×5 as:

$$* \; * \; * \quad \times \quad * \; * \; * \; * \; *$$

with one who draws:

$$
\begin{array}{ccccc}
* & * & * & * & * \\
* & * & * & * & * \\
* & * & * & * & *
\end{array}
$$

While the second child understands what multiplication is all about; the first child needs more time to discuss or explore.

Interpret information (written or graphical). Students' ability to read detailed information from charts and graphs often conveys the erroneous impression that they can also interpret that information. However, on a state assessment, surprisingly few eighth or twelfth grade students were able to respond correctly to the following problem, which focused on the shape of the relationship rather than on such features as scale and exact values (Badger 1989).

The two graphs below have not been completed. Draw the lines so that the graphs look the way they would if you had used real numbers. Explain why you drew the graphs the way you did.

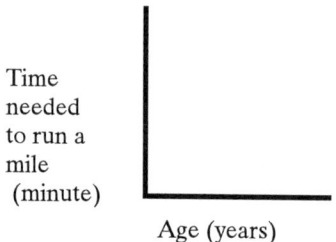

Time
needed
to run a
mile
(minute)

Age (years)

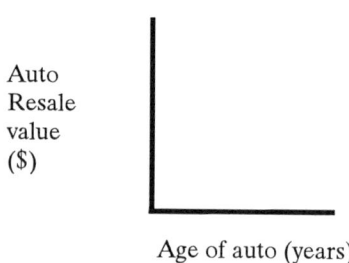

Auto
Resale
value
($)

Age of auto (years)

How the graphs can be evaluated depends upon the students' explanations. For example, when a student compares herself to an older person and states, "The older you get, the more time you need to run a mile," a straight linear relationship is correct. On the other hand, when students refer to the changing pattern from childhood to old age, a curvilinear slope is required.

Write a problem. Students can be asked to write problems that involve computations, ratios, exponents, formulas, and other mathematical procedures and expressions that are often taught in isolation. Again, this activity tells much about their understanding of the meaning of the procedures. For example, the following response to "Write a story problem that would be solved by 12 × 4 = 48" suggests that this fourth-grade student did not understand what was being requested but was only reacting to a set of numbers.

"In 2 days there are 48 hours. In one day there are 24 hours. In one half day there are 12 hours. You know that because 12 × 4 = 48."

Not only do students' problems reveal their ideas about the practical uses of mathematics. For example, when students confuse problem posing with fantasy, they may have little understanding of what a problem consists of. Furthermore, since posing analogous but simpler problems is a good problem-solving strategy, students benefit from the practice involved.

Give an argument. Asking students to take a position and defend it, not only indicates how well they understand the topic, but measures their ability to recognize the relevance of different kinds of data. The following question requires students to express their understanding of averages, as well as to construct an argument.

Roger doesn't believe that adding a constant (the same number) to every student's test score will simply change the average test score by that same amount. Convince Roger that this is true.

Depending upon the student's maturity, an ideal answer would be in generalized form— either an algebraic proof or a narrative example using the simplest and smallest set of numbers (e.g., 2 and 3). Students who rely upon "convincing" examples (such as 84, 93, 87. . .) may produce correct responses but fail to understand the concept of an "average". When this question was asked in a state assessment program (Badger 1989), more than three-quarters of eighth-grade students were unable to make a convincing argument using a worked example, despite their ability to identify the correct average of a set of numbers on a multiple-choice test. Students also showed an underlying misconception of what qualifies as a convincing argument in mathematics. For most students, a convincing argument was one that was specific and personal, replete with realistic examples.

In addition to these textbook/written type of problems, an assessment should include more prolonged tasks, involving both individual and group activities and demanding a synthesis of different skills. Surveys can be used at any level. They involve conceptual analysis (formulating the problem, determining the questions), collecting and organizing the data, analyzing and displaying it, and making inferences and conclusions. Each of the above components of the task can be evaluated separately and/or a holistic score can be given. In addition, computational, social, and graphical skills can be assessed.

There are also investigations that are more theoretical in nature, requiring students to explore problems and present evidence or argue for conjectures. Although this kind of thinking is often associated with older students, the youngest child can begin to explore patterns in numbers and nature. Assessment is made on the basis of the correctness of the examples given, the orderliness of the approach, the clarity with which the findings are presented, and the adequacy of the resulting generalization.

Portfolios

Although at the time of this writing few states are using portfolios to monitor statewide achievement, many teachers are recognizing the benefits of portfolios in the classroom (see end of chapter for details). Although most portfolios focus on the language arts, mathematics portfolios are becoming increasingly popular.

As it is currently used, the term *portfolio* is intended to convey far more than a collection of student work. It suggests a deliberate selection of work samples that are chosen to represent different aspects of performance. When used for evaluation, the resulting portfolio usually represents a collaboration between teacher and students. While the teacher may define the specific or general categories of tasks that will be included, students are often asked to select the actual pieces of work to be included under each category. In fact, student involvement and input is considered crucial in implementing portfolios in the classroom. One practitioner warns teachers to, "keep the focus on student ownership. Unless students become genuinely engaged in the process of compiling portfolios as a means of observing and reflecting on their own learning, they are not likely to respond well to portfolios. Such a situation will leave the teacher doing all of the work of evaluating and wondering if it's really worth all the trouble" (Cooper 1992, 12).

The success of the portfolio as a means of assessment can be attributed to the fact that many of its features closely reflect the aims of the curriculum reform movement. Descriptions of three of these features follow:

Breadth of information provided. Portfolios can provide more valid and reliable information on student achievement than can traditional testing because they provide multiple instances of students' achievement. By accommodating different learning styles, portfolios are able to give a more accurate appraisal of achievement than tests that rely upon a single format.

Instructional potential. By requiring students to make decisions concerning the portfolio (e.g., what to include; how it will be organized), portfolios provide an opportunity for discussion and reflection. As one observer states, "The major virtue of portfolios is that they can be designed to function simultaneously as a teaching tool and as an assessment medium" (Mitchell 1992, 105).

Communication. Since the content of a portfolio will primarily reflect the day-to-day work of the classroom, it carries a clear message about the goals of the curriculum to parents as well as to students. For this reason, it can be used to replace or supplement grades as a means of reporting student progress.

As with any assessment tool, it is important that the function of portfolios be clearly defined. For example, portfolios that are used for program evaluation or monitoring require a greater degree of standardization than those which are used in the classroom to provide students opportunities for reflection and self-evaluation. Despite the enthusiasm engendered by their potential for instruction and assessment, there are unresolved issues associated with portfolios. The primary difficulties concern classroom management, commitment on the part of students, and scoring or grading.

At the high-school level, exhibitions of mastery are one of the "Common Principles" (Sizer 1989) around which members of the Coalition of Essential Schools have organized their reform. In a sense, they epitomize the difference between "knowing how" and "knowing that." Requiring a broad range of competencies, these exhibitions are regarded as the culminating activity of a course or as a prelude to graduation. Modeled on athletic or artistic competitions, they involve students preparing for one or more exemplary tasks that embody the goals of the curriculum. Trained judges score performance against standards of excellence that are known to all participants ahead of time. Fundamental to this notion is a set of clearly defined standards that are known in advance, which form the goals of the project, as well as students' ability to understand and internalize the criteria of genuine competence. (Wiggins 1989).

Scoring Performance Tasks

There are various types of scoring, depending upon the specificity of the information required and the type of questions asked. Scoring ranges from general impressions, called "holistic scoring," to the simple right/wrong decisions that are traditionally used.

Right/wrong. The simplest type of scoring is the traditional one of determining whether or not a student has given the required response. This is most appropriate when testing for knowledge and simple skills that require only recognition or memory for their response. Correct answers are totaled and reported in terms of a percentage or ratio.

Checklists. In a large classroom, where some students demand a great deal of attention, a checklist is a useful device for ensuring that some students are not forgotten. Organizing a matrix of students and skills, with enough space for comments, will tell at a glance which student needs observation and what skills need to be addressed. The very act of drawing up a checklist forces decisions about the essential understanding and skills, as well as the relative importance of different aspects of behavior. However, the main function of checklists is to provide coverage, not insight, and they should not be used as the primary method of evaluation.

Holistic. Holistic scoring is used when an overall judgment of performance is warranted. It has a solid tradition in writing assessment, but can be adapted easily to mathematics assessment, particularly with reference to questions that ask for explanations. It is also useful in scoring problems for which there may be a range of solutions that may vary in terms of quality or completeness (see table 1).

The simplest form of holistic scoring uses a three-point scale of unacceptable, some errors, and acceptable. This range is appropriate for

scoring individual papers within a classroom. However, a large range of scores will also give more reliable results and is more appropriate in judging responses to complex tasks.

Multidimensional scoring. Complex products, such as portfolios, often involve a series of measures, reflecting the different dimensions that are being evaluated. For example, a portfolio can be rated for use of appropriate problem-solving strategies, execution, mathematical language, breadth and types of content, etc. Although the scores on these different dimensions may be highly correlated, scoring and reporting on different dimensions of performance provide useful feedback for instruction. This approach also reinforces the perception that mathematics performance is not unidimensional but entails the integration of different kinds of thinking and knowledge. For parents, it can carry a message about the importance of such aspects of performance as communication and cooperation, which are often neglected in discussions of mathematics achievement.

Analytic. Analytic scoring does not depend on holistic impressions of students' responses, but looks for the presence or absence of specific features. This type of scoring is tightly associated with the requirements of the individual task and is particularly useful for diagnosis and remediation.

What constitutes an adequate level of consistency among rater judgments depends on the purpose of the assessment. Reliability is not a problem for individual teachers who are scoring tasks for their own instructional decisions. In fact, it is sometimes recommended that students score their own work or those of others in order to recognize the criteria for good performance. On the other hand, when important decisions are being made about individual students, reliability is an important consideration, and it is usually suggested that each task be scored at least twice by different raters to ensure that judgments are consistent.

In measuring achievement against school or district standards, inter-rater reliability may be determined by comparing the scores that raters give on an identical sample of responses. If their scores are consistent, it is assumed that the raters will score consistently on the entire

Table 1. Holistic Scoring Rubric		
Josh and his friends ordered two large pizzas. Each pizza was cut into 8 equal slices. Paula ate ⅛ (a slice) of a pizza. Josh ate ⅜ of a pizza. Ron ate ¼ of a pizza. Rhonda ate ½ of a pizza. In your answer booklet, draw a picture to show how much pizza each person ate. Then find out how much pizza was left.		
Level	**Specific scoring**	**General scoring**
1	Incorrect response; no work shown • "I don't know"	Work is completely incorrect, irrelevant, unintelligible, or off-task
2	Minimal understanding; some work correct • 1 or 2 pizzas cut into some pieces	Response demonstrates a minimal understanding of the problem posed but does not suggest a reasonable approach. Although there is some correct work or concepts, the response contains serious misconceptions, major errors, or serious flaws in reasoning.
3	Evidence of conceptual understanding; reasonable approach indicated • student accurately draws amount of pizza each child ate • 1 pizza drawn correctly with accurate pieces and 1 not accurate	Response contains evidence of a conceptual understanding of the problem in that a reasonable approach is indicated. However, on the whole, the response is not well developed. Examples provided do not illustrate the desired conclusions.
4	Clear understanding; acceptable approach • accurate picture of 2 pizzas—incorrect or no answer • correct answer with no drawing • answer "¾ of 1 pizza" with accurate picture	Response demonstrates a clear understanding of the problem and provides an acceptable approach. The response also is generally well developed and coherent but contains minor weaknesses in the development. Examples provided may not completely illustrate the desired conclusions.
5	Complete understanding; appropriate, fully developed solutions • correct answer: ⅜ of 2 pizzas with accurate picture	Response demonstrates a complete understanding of the problem, is correct, and the methods of solution are appropriate and fully developed. Responses are logically sound, clearly written, and do not contain any significant errors. Examples are well chosen and illustrate the desired conclusions.

set of responses. Another method is for individual teachers to score their own class of students and then compare examples of their highest, middle, and lowest scores with responses from other schools. When discrepancies occur, all scores from a class or school are altered accordingly.

Although scoring training typically takes place as part of a large-scale testing program in which teachers are used as raters, the training procedure itself has been found to be a useful vehicle for staff development. In this model, teachers come together to compare a set of student responses and discuss the development of an appropriate scoring rubric. In the process, they are required to reflect upon the standards that they hold for their own students, as well as to consider the mathematical thinking that is indicated by different types of responses. This experience has proven to be useful to teachers, not only in terms of evaluating their students' work, but also in developing better assessment tasks of their own.

Management Issues

Essentially, good assessment tasks do not differ from good instructional tasks; it is the role of the teacher that differs. Consider the following questions: "How many ways can you arrange 16 square tiles to make a rectangle? Is there a rule?"

In an instructional setting, the teacher will have a final destination in mind, such as helping students to understand the connection between geometric and symbolic representations of factors. However, the path to that destination is essentially uncharted. It will be students' responses that determine the course of the discussions and the issues to be explored. Few good teachers will cut off a provocative response just because it is unexpected. Instead, in the give-and-take, the teacher will probably probe, provide clues, and open up other avenues of thought to steer the students gently to some final conclusion.

In contrast, when the same task is used for assessment, the teacher's role changes from conductor and facilitator to observer and evaluator. How many rectangles did the student make? Did the student recognize the same shape when it was rotated? Was a 4 × 4 square rejected because it was "not a rectangle"? Was the student able to relate the shapes to the number system? What kinds of generalizations were made? How well were they expressed?

This more inactive role is often difficult for teachers to assume. Furthermore, in a large class, close observation is often impossible. One way to gather information is to have the students work in groups. Not only does group work facilitate more productive efforts from students, it also facilitates good assessment. Discussion is a necessary condition when complex problems must be solved jointly. And students' conjectures and arguments are windows onto their thinking. Another way is to require some individual written work or product as a follow-up after any group or class

discussion. This allows teachers to check their observations of students' understanding and to assess their ability to transfer the knowledge gained in one situation to another. For example, after small-group work using tiles to construct rectangles, the teacher might ask students to answer the following questions: "What other sets of tiles could you use to make rectangles? Are there any sets of tiles that don't make rectangles? If so, what is the difference between the sets?" Or, for a more advanced group: "How many rectangles can you make with 18 tiles? With 24 tiles? With 30 tiles? With 36 tiles? Do you see a pattern? If so, express it."

These written responses can then be judged according to a scoring criteria that reflects their adequacy. (For a thorough and practical discussion of assessing cooperative problem solving, see Kroll et al. 1991.)

Conclusions

Despite the fact that many of the activities described in this chapter were not in common usage five years ago, a practitioner new to alternative assessment has the support of a large number of resources. Some of the suggested readings that follow attempt to explain the rationale for this radical change in how we view testing; others give practical suggestions for implementation. Although there remain unresolved issues about the validity and reliability of this type of assessment for purposes of high stakes accountability, few would deny its relevance to the classroom and its appropriateness for the classroom teacher.

References

Cooper, Winfield. 1992. "Getting Started with Portfolios—What Does It Take?" *Portfolio News* 3(2): 12–14.

Education Commission of the States. 1992. *State Education Leader.* 11(1): Focus issue on assessment and reform.

Kroll, Diana, et al. 1991. "Cooperative Problem Solving, but What about Grading?" *Arithmetic Teacher* 39(6): 17–23.

Mathematical Sciences Education Board. 1991. *For Good Measure, Principles and Goals for Mathematics Assessment. Report of the Na-*

tional Summit on Mathematics Assessment.
Washington, DC.

———. 1993. *Measuring Up: Prototypes for Mathematics Assessment.* Washington, DC: National Research Council.

National Commission on Testing and Public Policy. 1990. *From Gatekeeper to Gateway: Transforming Testing in America.* Washington, DC.

National Council of Teachers of Mathematics. 1989. *Curriculum and Evaluation Standards for School Mathematics.* Reston, VA.

———. 1991. *Professional Standards for Teaching Mathematics.* Reston, VA.

Office of Technology Assessment. 1992. *Testing in American Schools: Asking the Right Questions.* Washington, DC: U.S. Congress.

Romberg, Thomas, Linda Wilson, and Mamphono Khaketla. 1990. *An Examination of Six Standard Mathematics Tests for Grade Eight.* Madison, WI: National Center for Research in Mathematical Sciences Education.

Sizer, Theodore. 1989. "Diverse Practice, Shared Ideas: The Essential School." In *Organizing for Learning toward the Twenty-First Century.* Reston, VA: National Association of Secondary School Principles.

Wiggins, Grant. 1989. "A True Test: Toward More Authentic and Equitable Assessment." *Phi Delta Kappan.* 70(9): 703—713.

———. 1991. "An Assessment Glossary." Paper presented at The New Standards Project, Snowmass, CO), July 29–August 4, 1991.

Portfolio Networks and Newsletters

Mathematics Portfolio Project
California Assessment Program
P.O. Box 944272
Sacramento, CA 94244-2720
 Teachers in this project have been field-testing portfolios in order to document student growth in mathematics.

Northwest Evaluation Association
5 Centerpointe Drive, Suite 100
Lake Oswego, OR 97035
 The association publishes *Portfolio Assessment Newsletter* three times a year. Its goal is to support an information network for educators interested in portfolios and portfolio assessment.

Northwest Regional Educational Laboratory:
Information Center
101 SW Main Street, Suite 500
Portland, OR 97204
 This clearinghouse disseminates a newsletter, reports on portfolio projects, and provides annotated bibliographies of articles and papers on portfolios.

Portfolio Assessment Clearinghouse
c/o San Dieguito Union High School District
710 Encinitas Boulevard
Encinitas, CA 92024
 Publishes *Portfolio News,* a quarterly publication about portfolios. $25 a year. Also has a network of schools involved in portfolio assessment.

Further Reading

Association for Supervision and Curriculum Development. 1989. *Educational Leadership* 46 (April): Focus issue on assessment.

———. 1992. *Educational Leadership* 49 (May): Focus issue on assessment.

Badger, Elizabeth. 1991. "More Than Testing." *Arithmetic Teacher* 39: n.p.

———. 1989. *On Their Own: Students' Responses to Open-Ended Questions in Mathematics.* Quincy, MA: Massachusetts Department of Education.

California Assessment Program. 1989. *A Question of Thinking: A First Look at Students' Performance on Open-Ended Questions in Mathematics.* Sacramento, CA: California State Department of Education.

Herman, Joan, Pamela Aschbacker, and Lynn Winters. 1992. *A Practical Guide to Alternative Assessment.* Alexandria, VA: Association for Supervision and Curriculum Development.

Hymes, Donald, Ann Chafin, and Peggy Gonder. 1991. *The Changing Face of Testing and Assessment.* Washington, DC: American Association of School Administrators.

Kulm, Gerald, ed. 1990. *Assessing Higher Order Thinking in Mathematics.* Washington, DC: American Association for the Advancement of Science.

Mitchell, Ruth. 1992. *Testing for Learning: How New Approaches to Evaluation Can Improve American Schools.* New York: The Free Press.

Mumme, Judy. 1990. *Portfolio Assessment in Mathematics.* Berkeley, CA: California Mathematics Project.

National Council of Teachers of Mathematics. 1992. *Mathematics Teacher* 83(8): Focus issue on alternative assessment.

———. 1992b. *Arithmetic Teacher* 40 (Sept.): Focus issue on assessment.

Pandey, Tej. 1991. *A Sampler of Mathematics Assessment.* Sacramento, CA: California State Department of Education.

Stenmark, Jean K. 1991. *Mathematics Assessment: Myths, Models, Good Questions, and Practical Suggestions.* Reston, VA: National Council of Teachers of Mathematics.

RECOMMENDED CURRICULUM GUIDES

by Joan Mullee
Ossining Public Schools, Ossining, New York

UCH has been written about the teaching of mathematics in today's classrooms. This chapter includes references to curriculum guides, books, journal articles, and catalogs which represent a variety of specific methods, activities, resources, and current trends in mathematics education. They are grouped according to grade level. At the secondary level, references are also grouped by subject.

General (K-12)

Alabama State Department of Education. 1989.
   Mathematics. Alabama Course of Study: 31.
   Montgomery, AL. 196p. ED 327 399.
This curriculum guide prescribes the minimum course content for all mathematics programs in the public schools of the state of Alabama. It forms the basis for the development, by local education agencies, of all state mathematics programs. Minimum course content for each grade has been organized into seven content areas: number, measurement, geometry, algebra, patterns and functions, logic, and statistics and probability. Instructional objectives and student outcomes related to problem solving, the use of calculators and computers, the practice of estimation and mental arithmetic, and the development of language facility are inter-

woven into every grade level and/or course. Minimum course content for each grade level is presented as student outcomes that consider the developmental characteristics of learners and the developmental nature of mathematical concepts. This guide is divided into the following sections: (1) "Introduction"; (2) "Trends, Issues, and Position Statements"; (3) "The Elementary Program in Mathematics"; and (4) "The Secondary Program in Mathematics." Appended are the time requirements for subject areas and study habits, homework, and student responsibilities.

Artzt, Alice F., and Claire M. Newman. 1991.
   "Equivalence: A Unifying Concept." *Mathematics Teacher* 84(2): 128-32. EJ 421 904.
The concept of how equivalence can relate various segments of a mathematics program and help students develop an appreciation of the unity of mathematics is discussed. Activities and examples of equivalence at all grade levels are included.

Blume, Glendon W., and Robert F. Nicely, Jr., eds. 1991. *A Guide for Reviewing School Mathematics Programs*. Reston, VA: National Council of Teachers of Mathematics. 65p. ED 339 597.
The position statements on curriculum, evaluation, teaching, leadership, staff development,

equity for under-represented groups, and the effective use of technology issued by the National Council of Teachers of Mathematics (NCTM) in the *Curriculum and Evaluation Standards for School Mathematics* and *Professional Standards for Teaching Mathematics* have precipitated major changes in the K-12 mathematics curriculum. To help school districts capitalize on this opportunity for curriculum redesign and enhancement, the NCTM has designed this guide to enable school district personnel to analyze their mathematics programs by identifying and listing critical elements in the areas of goals, curriculum, instruction, evaluation, and teacher and administrator responsibility. The guide is intended to: (1) stimulate critical analysis of content, methodology, assessment, and management issues related to the K-12 mathematics program; (2) identify some of the major desired directions for the K-12 mathematics curriculum; (3) help districts identify discrepancies between what is and what could be; and (4) point out directions for change. The initial section describes how to use the guide, encouraging a cooperative approach by teachers and administrators, and giving several uses that involve all teachers, selected teachers of mathematics, personnel from outside the district, or textbook analysis. After presenting an example of how to complete the forms, the subsequent sections are proposed to evaluate the K-12 mathematics program in the following areas: (1) goals; (2) K-4 curriculum; (3) 5-8 curriculum; (4) 9-12 curriculum; (5) instruction; (6) evaluation; (7) teachers; and (8) administration. The questions "How important is this to you?" and "To what extent does this happen in your setting?" are used to evaluate the critical factors in each of the areas. See *Curriculum and Evaluation Standards for School Mathematics* (ED 304 336).

Fey, James T. 1992. *Calculators in Mathematics Education 1992 Yearbook.* Reston, VA: National Council of Teachers of Mathematics. 256p.
How to use the power of calculators in the teaching and learning of mathematics. Includes ways that calculators can influence the content and the process of mathematics instruction; how calculators have been used successfully for teaching mathematics from primary grades

through college; how calculators have influenced testing in state assessment programs and college admissions; classroom activities that illustrate the new instructional possibilities.

Fong, Ho-Kheong. 1989. "Models for the Integration of Computing into Mathematics Curricula." *Computers and Education* 13(2): 157-66. EJ 398 075.
Presents an alternative model for using the computer as an aid in teaching mathematics. Highlights include a mathematics teaching model; computer software for drill and practice, tutorials, simulation, games, and computer-managed instruction; integrating computers into the mathematics curriculum; and examples of computer activities for mathematics.

Hatfield, Mary M., and Jack Price. 1992. "Promoting Local Change: Models for Implementing NCTM's Curriculum and Evaluation Standards." *Arithmetic Teacher* 39(5): 34-37. EJ 441 989.
Following commentary stating that the most feasible locus for curricular change and instructional reform is the individual school, four models for change that implement the NCTM's curriculum and evaluation standards are presented. These models are not offered as panaceas but rather as examples of efforts currently being utilized which may provide beneficial starting points for individual schools. See *Curriculum and Evaluation Standards for School Mathematics* (ED 304 336).

Kenney, Margaret J., and Christian R. Hirsch, eds. 1991. *Discrete Mathematics across the Curriculum, K-12 1991 Yearbook.* Reston, VA: National Council of Teachers of Mathematics. 248p. ED 334 062.
This yearbook provides the mathematics education community with specific perceptions about discrete mathematics, concerning its importance, its composition at various grade levels, and ideas about how to teach it. Many practical suggestions with respect to the implementation of a discrete mathematics school program are included. A unifying thread throughout the book is that discrete mathematics incorporates many of the recommendations found in the *Curriculum and Evaluation Standards for School Mathematics of the National Council of Teachers of Mathematics* (ED 304 336).

Contents include: "Perspectives and Issues" (rise to prominence, implications, and cautions); "Teaching in Grades K-8" (strengthening traditional curricula through problem-solving approach); "Teaching in Grades 7-12" (appropriateness for both precollege and non-college-oriented students); " Graph Theory" (formulation of real-world applications of algorithmic thinking); "Matrices" (computer-based storage and manipulation of numerical data); "Counting Methods" (enhancement of combinatorial reasoning with manipulatives, geometric concepts, and generating functions); "Recursion, Iteration, and Induction" (spreadsheet programs and computer graphics as tools to explore concepts and applications); "Algorithms" (thinking procedurally without the rigor of computer programming); and "Units, Activities, and Projects" (open-ended, relevant problems lead to debate and discussions among students).

Lee, Carolyn S., comp. 1991. *Mathematics Education Programs that Work: A Collection of Proven Exemplary Educational Programs and Practices in the National Diffusion Network.* Washington, DC: Office of Educational Research and Improvement. 35p. ED 334 081.
The National Diffusion Network (NDN) was established in 1974 through the Office of Educational Research and Improvement upon the belief that there are few difficulties encountered by school systems that have not been solved successfully in some other location. The primary function of the NDN is to disseminate information about a wide array of exemplary educational programs through which a local school system may solve its own unique problems without the necessity of starting from scratch. The term *exemplary program* is conferred only after a project has been approved by the U.S. Department of Education's Joint Dissemination Review Panel, recently renamed the Program Effectiveness Panel. This catalog contains descriptions of the exemplary mathematics education programs in the NDN that are available to school systems and other educational institutions for implementation in their classrooms. Part I of the catalog contains the descriptions of thirteen funded programs for K-12 mathematics instruction. Part II contains the descriptions of five nonfunded programs that offer training and

technical assistance through cost/service agreements negotiated with potential adopters. Part III contains contact details for the fifty state facilitators who can provide further information concerning any particular program, instructional materials and services about specific content areas, or professional development based on needs assessment from the prospective adopter.

"Making Sense of the Future." 1990. *Journal of Computer Assisted Learning* 6(1): 14-33. EJ 409 016.
Describes activities of the Educational Technology Center at Harvard University, which focus on the use of computers and other technology to improve K-12 instruction in science, mathematics, and computing. Topics discussed include curriculum development; students' prior conceptions; the construction of scientific knowledge; the use of software; and recommendations for future developments.

Michigan State Board of Education. 1989. *An Interpretation of: The Michigan Essential Goals and Objectives for Mathematics Education.* Lansing, MI. 255p. ED 316 443.
This is a companion document to the *Michigan Essential Goals and Objectives for Mathematics Education* (ED 295 827). It aims to assist teachers, curriculum specialists, and other educators in their endeavors to improve their K-12 mathematics education programs. The document illustrates the integration of mathematical content with process skills. The six process skills are conceptualization, mental arithmetic, estimation, computation, applications and problem solving, and calculators and computers. The major purpose of this interpretive document is to provide examples and specific information on the content strands, the mathematical processes, and specific objectives. Vocabulary, comments, and examples for each objective are presented at different grade levels.

Michigan State Board of Education. 1988. *Michigan Essential Goals and Objectives for Mathematics Education.* Lansing, MI. 44p. ED 295 827.
This document is designed to assist administrators and teachers in planning, developing, and implementing grades K-9 mathematics programs, and to provide some guidelines for

grades 10-12 instruction. It is intended to provide a philosophical foundation and curricular framework from which educators may construct a comprehensive local program to meet the instructional needs of students. The document illustrates the integration of mathematical content with process skills. The mathematical content strands are: (1) whole numbers and numeration; (2) fractions, decimals, ratio, and percent; (3) measurement; (4) geometry; (5) statistics and probability; (6) algebraic ideas; (7) problem solving and logical reasoning; and (8) calculators. See also *An Interpretation of: The Michigan Essential Goals and Objectives for Mathematics Education*, ED 316 443.

National Council of Teachers of Mathematics. 1989. *Curriculum and Evaluation Standards for School Mathematics*. Reston, VA. 258p. ED 304 336.

This document contains a set of standards for mathematics curricula in North American schools (K-12) and for evaluating the quality of both the curriculum and student achievement. In the introduction, the background, need, and overview of these standards are described. A total of fifty-four standards are divided into four categories: (1) curriculum standards for grades K-4; (2) curriculum standards for grades 5-8; (3) curriculum standards for grades 9-12; (4) evaluation standards. Each group of curriculum standards contains a statement of what mathematics the curriculum should include and a description of the student activities associated with that mathematics. The evaluation standards are presented in three categories: general assessment, student assessment, and program evaluation. Each standard is elaborated on in a focus section followed by a discussion with examples. For a summary of the document, see *ERIC/SMEAC Mathematics Education Digest*, no. 1, 1990 (ED 319 630). See also "A Guide for Reviewing School Mathematics Programs" (ED 339 597), "Promoting Local Change: Models for Implementing NCTM's Curriculum and Evaluation Standards" (EJ 441 989), and "Curriculum and Evaluation Standards for School Mathematics Addenda Series."

National Council of Teachers of Mathematics. 1991. *Curriculum and Evaluation Standards for School Mathematics Addenda Series*. Reston, VA.

The *Curriculum and Evaluation Standards for School Mathematics* provides a vision and a framework for revising and strengthening the K-12 mathematics curriculum in North American schools and for evaluating both the mathematics curriculum and students' progress. The document addresses not only what mathematics students should learn but also how they should learn it. When completed, it is expected that the "Addenda Series" will consist of twenty-two supporting books designed to interpret and illustrate how the vision could be translated into classroom practices. Targeted at mathematics instruction in grades K-6, 5-8, and 9-12, the themes of problem solving, reasoning, communication, and connections are woven throughout the materials, as is the view of assessment as a means of guiding instruction. Activities have been field tested by teachers to ensure that they reflect the realities of today's classrooms.

Kindergarten Book. Addenda Series, Grades K-6. 1991. ED 339 603.

The book explores four areas: (1) "Patterns"; (2) "Number Sense and Operations"; (3) "Making Sense of Data"; and (4) "Geometry and Spatial Sense."

First-Grade Book. Addenda Series, Grades K-6. 1991. ED 339 602.

This book explores the same four areas as the *Kindergarten Book.* (Addenda Series will also include *Second-Grade, Third-Grade, Fourth-Grade, Fifth-Grade,* and *Sixth-Grade* Books.)

Dealing with Data and Chance. Addenda Series, Grades 5-8. 1991. ED 339 616.

Illustrates five themes: (1) data gathering by students; (2) communication; (3) problem solving; (4) reasoning; and (5) connections.

Developing Number Sense. Addenda Series, Grades 5-8. 1991. ED 339 604.

Categories include: (1) "Expanded Activity"; (2) "Reasonableness"; (3) "Relative Size of Numbers"; (4) "Whole Number Computation"; (5) "Fractions"; (6) "Decimals"; (7) "Percents"; and (8) "Measurement and Graph Interpretation." Lists of references and recommended reading are also provided.

Geometry in the Middle Grades. Addenda Series, Grades 5-8. 1992. 88p.
Provides different approaches to some geometry topics and supplies a collection of sample activities.

Patterns and Functions. Addenda Series, Grades 5-8. 1991. ED 339 598.
The five sections are: (1) "Exponents and Growth Patterns"; (2) "Number Theory and Counting Patterns"; (3) "Rational Number Patterns"; (4) "Measurement and Geometric Patterns"; and (5) "Graphs and Functions as Patterns."

Geometry from Multiple Perspectives. Addenda Series, Grades 9-12. 1991. ED 339 615.
The chapters are: (1) "Why Should Geometry Be Considered from Multiple Perspectives?"; (2) "Elements of Contemporary Geometry"; (3) "Triangles from Multiple Perspectives"; (4) "Quadrilaterals from Multiple Perspectives"; (5) "Polygons from Multiple Perspectives"; (6) "Solids: Expanded Perspectives"; (7) "Reasoning About Shapes Using Coordinates and Transformations"; (8) "Congruence from Multiple Perspectives"; (9) "Similarity from Multiple Perspectives"; and (10) "Reasoning, Justification, and Proof."

Connecting Mathematics. Addenda Series, Grades 9-12. 1991. ED 339 617.
The chapters in this book are: (1) "Connecting with Functions"; (2) "Connecting with Matrices;" (3) "Data Analysis: A Context for Connections;" (4) "Building and Using Connections in Reasoning"; and (5) "Building and Using Connections in Problem Solving."

A Core Curriculum: Making Mathematics Count for Everyone. Addenda Series, Grades 9-12. 1992. 152p.
Offers several models for organizing mathematics content and gives suggestions for assessment techniques that focus on problem solving, reasoning, and disposition toward mathematics.

Data Analysis and Statistics Across the Curriculum. Addenda Series, Grades 9-12. 1992. 88p.
Includes examples and activities that illustrate how to integrate statistical concepts into the traditional mathematics curriculum.

National Council of Teachers of Mathematics. 1991. *Professional Standards for Teaching Mathematics.* Reston, VA. 190p.
This book presents a set of standards that promotes a vision of mathematics teaching, evaluating mathematics teaching, the professional development of mathematics teachers, and responsibilities for professional development and support.

National Research Council. 1989. *Everybody Counts: A Report to the Nation on the Future of Mathematics Education.* Washington, DC: National Academy Press. 114p.
A collaborative work of the National Research Council, the Mathematical Sciences Education Board, Board on Mathematical Sciences, and their joint Committee on the Mathematical Sciences in the Year 2000. Presents data about the lack of mathematical skill and study in the U.S.; links mathematical proficiency with job opportunity in a technological age. Calls for increasing the amount and level of mathematics that all students take and for updating the curriculum to reflect current applications of mathematics and the use of new technologies in mathematics. Chapter titles are: "Opportunity ... tapping the power of mathematics"; "Human Resources ... investing in intellectual capital"; "Mathematics ... searching for patterns"; "Curriculum ... developing mathematical power"; "Teaching ... learning through involvement"; "Change ... mobilizing for curricular reform"; "Action ... moving into the 21st century"; "References ... documenting the challenge."

National Research Council. 1990. *Reshaping School Mathematics: A Philosophy and Framework for Curriculum.* Washington, DC: National Academy Press. 74p. Copies may be purchased from the National Academy Press, 2101 Constitution Avenue, NW, Washington, DC 20418.
This book is a response to NCTM's *Curriculum and Evaluation Standards* and the National Research Council's *Everybody Counts.* Chapter titles are: "A Rationale for Change"; "A

Practical Philosophy"; "Redesign from a Technological Perspective"; "Redesign from a Research Perspective"; "A Framework for Change." The latter discusses principles, goals, and enabling conditions.

New York State Education Department. 1989. *Ideas for Strengthening Mathematics Skills.* Albany, NY. 37p. For copies, write to Publications Desk, New York State Education Department, Albany, NY 12234.
This is a book of ideas for helping students who have difficulty learning math. Suggested approaches include manipulative materials, alternate algorithms, games, relevant applications, and diagnosis of computational errors. Examples of each approach are given. The book includes an appendix, "Students with Handicapping Conditions."

Parkrose School District. 1991. *Mathematics Curriculum.* Portland, OR. 338p. Write to Parkrose Public Schools, 10636 NE Prescott Street, Portland, OR 97220.
The guide begins with a list of nine general outcomes for all mathematics students, then shows a chart in which general exit outcomes are correlated to specific mathematical objectives. For each grade level, the guide states teaching emphases and a course statement for each strand that includes a list of essential learning objectives correlated to the list of outcomes, assessment tools/strategies, and materials and resources available.

Smith, Laura B. 1989. *A Catalog of Successful Math Programs across Alabama, Florida, Georgia, Mississippi, North Carolina, and South Carolina. Vol. II.* Research Triangle Park, NC: Southeastern Educational Improvement Lab. 31p. ED 310 001.
This catalog describes exemplary mathematics programs across six southeastern states. The program title, site, content focus, grade level, achievement levels, program description, and address for contact are summarized for each program. The programs were identified through a literature search and through recommendations from mathematics experts. Three summary tables by content focus, grade levels (K-12), and achievement levels are provided. The

content focus includes early childhood mathematics; arithmetic, seventh grade mathematics; eighth grade mathematics; enrichment topics/problem solving; consumer mathematics; general mathematics; algebra I; geometry; algebra II; advanced mathematics; trigonometry; calculus; minority enhancement; and other topics. A total of thirty-eight programs are outlined for six states. Thirteen additional programs are listed as minority/female-focused programs.

Stenmark, Jean Kerr, ed. 1992. *Mathematics Assessment; Myths, Models, Good Questions and Practical Suggestions.* Reston, VA: National Council of Teachers of Mathematics. 67p.
Supplies assessment models adapted from those used by other teachers, as well as step-by-step instructions on how to use portfolios and other assessment techniques.

Suydam, Marilyn N. 1990. "Curriculum and Evaluation Standards for Mathematics Education." *ERIC/SMEAC Mathematics Education Digest No. 1.* Columbus, OH: ERIC Clearinghouse for Science, Mathematics and Environmental Education. 4p. ED 319 630.
This digest reviews the National Council of Teachers of Mathematics' *Curriculum and Evaluation Standards for School Mathematics* (ED 304 336). Topics summarized are: (1) rationale; (2) underlying assumptions; (3) five goals for students; (4) framework for curriculum; (5) standards for each grade cluster; (6) standards for general assessment, student assessment, and program evaluation; (7) suggested change in mathematics instruction; and (8) steps for implementation. Ten references are listed.

Velazquez, Robert. 1990. "Organizing Mathematics Courses for the Gifted in Ontario, Canada." *Gifted Child Today* 13(5): 52-54, EJ 419 982.
This paper discusses methods used to differentiate the presentation of mathematics to gifted students in Ontario, Canada. Activities are described that help students to develop structured inquiry, reinforce categorization skills, develop efficient study habits, and encourage probing and divergent questions.

Elementary

General (K-8)

Franklin, Margaret. 1990. *Add-Ventures for Girls: Building Math Confidence. Elementary Teacher's Guide.* Reno, NV: University of Nevada. 278p. For copies, write to WEEA Publishing Center, Education Development Center, 55 Chapel Street, Newton, MA 02160.

This book describes the attitudes and the practices that cause girls to lose interest in math; it then provides strategies, activities, and resources that teachers can use to help girls overcome barriers and reach their full potential in mathematics.

Hendrickson, A. Dean. 1989. *Meaningful Mathematics. Teacher's Guide to Lesson Plans.* Washington, DC: National Science Foundation.

Mathematics and the use of mathematical thinking should be much more than what has been traditional school arithmetic. Much of the mathematical reasoning can be developed and experienced out of school, particularly in the home. This material is a teacher's guide designed to help parents support what is done with their children in class. Background material for parents is provided. Some underlying principles for teaching mathematics meaningfully are listed. Assessment material is presented.

Kindergarten. ED 316 385. A total of forty-four activities on the following concepts and skills are included: (1) comparing; (2) sorting; (3) ordering; (4) working with patterns; (5) number concept; and (6) arithmetic operations. Songs and verses for kindergarten children are provided.

Level One. ED 316 386. A total of sixty-one lesson plans covering the following topics are included: (1) number concept; (2) number operation; (3) place value; (4) problem solving; and (5) measurement.

Level Two. ED 316 388. A total of fifty-eight activities on the following concepts and skills are included: (1) comparing; (2) counting; (3) classification; (4) using patterns; (5) numbers; (6) number operations; (7) problem solving; (8) place value; (9) equality; (10) fractions; (11) correspondences; (12) geometry; (13) logic; (14) estimation; and (15) measurement.

Level Three. ED 316 390. A total of fifty-two activities on the following concepts and skills are included: (1) computation; (2) numeration; (3) fractions; (4) geometry; (5) arithmetic operations; (6) problem solving; (7) number relations; and (8) logic.

Level Three. Recording Forms and Worksheets for Pupil Use. ED 316 391. This material contains recording forms and worksheets for activities for the Level Three experiences.

Level Four. ED 316 392. A total of twenty-nine activities on the following concepts and skills are included: (1) equality and inequality; (2) use of operations; (3) place value; (4) computation; (5) geometry; (6) logic; (7) ratio; (8) use of calculators; (9) use of LOGO.

Level Five. ED 316 394. A total of thirty-one activities on the following concepts and skills are included: (1) ratio; (2) using operations; (3) problem solving; (4) measurement; (5) fractions; (6) decimals; (7) computations; (8) geometry; (9) word problems; (10) logic; (11) scientific notations; (12) use of calculators; (13) use of LOGO; and (14) probability.

Level Six. ED 316 396. A total of thirty-five activities on the following concepts and skills are included: (1) computation; (2) scientific notations; (3) logic; (4) word problems; (5) ratio; (6) inequality; (7) area; (8) graphing; (9) geometry; (10) problem; (11) estimation; (12) using data; and (13) use of LOGO.

Lamoille North Supervisory Union. 1991. *Mathematics: K-6.* Hyde Park, VT. 78p.

This guide includes a scope and sequence for grades K-6. At each grade level, a chart is presented that lists topics, specific skills, and concepts and activity numbers or suggested manipulative aids. This list is followed by the thirty-seven numbered activities referred to in the chart. Each activity appears on a single page. An activity may be used at more than one grade level.

Mason, Virgyl. 1988. *Kendall Demonstration Elementary School Mathematics Curriculum Guide.* Washington, DC: Pre-College Programs, Gallaudet University. 149p. For information, write Outreach, KDES, Gallaudet Pre-College Programs, Gallaudet University, Washington, DC 20002.

This guide includes a scope and sequence of mathematical objectives for levels 1 through 8, sample activities for each strand, and progress charts for each level. The progress charts are checklists related to the scope and sequence. The guide includes four appendices: a glossary of terms used in the guide, a list of references, a materials resource list, and record keeping forms.

Merttens, Ruth, and Jeff Vass. 1990. *Sharing Math Cultures: IMPACT (Inventing Math for Parents and Children and Teachers)*. n.p. 205p. ED 342 613.

This book describes the Inventing Math for Parents and Children and Teachers (IMPACT) project, a project that involves parents routinely in the primary school curriculum and that includes curriculum materials that are designed to be valid across different cultures. This book describes and promotes educational settings in which the child is not the center or object of teaching, but is the initiator and tutor. It explains how any teacher or school can set up and run a similar initiative. Detailed examples are given of the processes involved and the specific methods used for effective parental involvement. It describes and discusses sample materials for implementing the scheme and outlines a radically new approach to evaluation. It also reflects upon the relationship among curriculum, IMPACT, and primary math. The chapters examine ongoing debates in education, particularly the relationship of the community to the curriculum; the research context of IMPACT; how to set up IMPACT in a school; the classroom and the day-to-day running of IMPACT in schools; appropriate materials which extend classwork so that suitable tasks can be done at home jointly with a parent; and the issues and techniques of evaluation.

New York State Education Department, Bureau of Curriculum Development. 1989. *Creative Problem Solving*. Albany, NY. 60p. ED 327 407. For copies, write Publications Desk, New York State Education Department, Albany, NY 12234.

This publication was prepared to assist school personnel in strengthening their overall mathematics programs. These materials give an overview of a variety of teaching techniques for use in the first eight grades. The illustrative problems are concerned with those topics usually studied at the elementary school level.

The recommended problem-solving approaches are designed to be useful in every subject. The materials in this publication may also prove valuable as a basis for in-service teacher education. Chapters include: (1) "An Enlarged View of Problem Solving"; (2) "How to Create Problems"; (3) "Organizing Facts to Perceive Patterns"; (4) "Some More Attempts at Creating Problems"; (5) "Many, Many, Many"; and (6) "Some Concluding Thoughts on Problem Solving." Appended is a section on students with handicapping conditions.

New York State Education Department, University of the State of New York. 1988. *Teaching Mathematics with Computers, Grades K-8*. Albany, NY. 71p. Copies may be obtained by submitting requests in writing on school stationary to the Publications Sales Desk, New York State Education Department, Room 164 EBA, Washington Avenue, Albany, NY 12234.

This booklet includes suggestions for staff development, hardware, and software. In the section called "Teaching Mathematics with Computers," specific software is reviewed. The reviews are sorted by grade level clusters. This section also discusses LOGO, BASIC, and the use of workstations and test generators.

Rowland, Tim. 1990. "Can Openers." *Mathematics in School* 19(2): 34-37. EJ 410 993.

Describes the *Calculator-Aware Number (CAN)* curriculum developed by the *Primary Initiatives in Mathematics Education (PrIME)* project. Summarizes the CAN curriculum as it relates to learning by children and use by teachers.

Texas Higher Education Coordinating Board. 1990. *Mathematics/Computer Integrated Curriculum*. Austin, TX.

This is a series designed for a program to integrate the teaching and learning of mathematical and computer concepts and skills in the elementary school. Each manual contains a number of lessons, and each of these lessons includes information on the topic, suggested grade level, mathematics concepts and skills, objective, prerequisite skills needed, and activities. There are six manuals.

Grades K-1. ED 323 113. Topics contained in the lessons include: (1) problem solving; (2)

geometry; (3) numbers; (4) measurement; (5) number concepts; (6) addition; (7) comparing; (8) estimating; (9) time; and (10) fractions. Software used for the activities is primarily LOGO.

Grade 2. ED 323 114. Topics contained in the lessons include: (1) problem solving; (2) geometry; (3) numbers; (4) measurement; (5) number concepts; (6) addition; (7) time; (8) patterns; and (9) making inferences and drawing conclusions. Software programs used for the activities are specified for each lesson.

Grade 3. ED 323 115. Topics contained in the lessons include: (1) problem solving; (2) geometry; (3) numbers; (4) measurement; (5) number concepts; (6) addition; (7) time; (8) LOGO; (9) division; (10) fractions; and (11) probability, statistics, and graphing. Software programs used for the activities are specified for each lesson.

Grade 4. ED 323 116. Topics contained in the lessons include: (1) problem solving; (2) geometry; (3) numbers; (4) measurement; (5) number concepts; (6) LOGO; and (7) puzzles. Software programs used for the activities are specified for each lesson.

Grade 5. ED 323 117. Topics contained in the lessons include: (1) problem solving; (2) geometry; (3) numbers; (4) number concepts; (5) statistics; (6) measurement; and (7) probability, statistics, and graphing. Software programs used for the activities are specified for each lesson.

Grade 6. ED 323 118. Topics contained in the lessons include: (1) problem solving; (2) geometry; (3) numbers; (4) number concepts; (5) computer commands; (6) fractions; and (7) graphing. Software programs used for the activities are specified for each lesson.

Thiessen, Diane, et al. 1989. *Elementary Mathematical Methods. 3d ed. New York: Macmillan.* 684p. ED 299 154.
This book contains fifteen chapters: (1) "Teaching Problem Solving"; (2) "Using Calculators and Computers in Elementary School Mathematics"; (3) "Organizing Instruction"; (4) "Teaching Numeration of Whole Numbers"; (5) "Teaching Numeration of Common and Decimal Fractions"; (6) "Teaching Addition and Subtraction of Whole Numbers"; (7) "Teaching Multiplication and Division of Whole Numbers"; (8) "Teaching Addition and Subtraction

of Common and Decimal Fractions"; (9) "Teaching Multiplication and Division of Common and Decimal Fractions"; (10) "Teaching Measurement"; (11) "Teaching Geometry"; (12) "Teaching Rates, Ratios, Proportions and Percents"; (13) "Teaching Statistics and Probability"; (14) "Teaching Integers and Their Operations"; and (15) "Teaching Number Patterns and Theory." The two appendices include discussions on "Calculators and Computers" and "Mainstreaming—Can Individual Needs Be Met?" The chapters in this book reflect the strands taught in the elementary school mathematics curriculum. Throughout each chapter are collections of problem sets that divide the chapter into sections. A number of these problems are similar to lessons that could be used with elementary school children.

Thwaites, G. N. 1989. "Counting." *Mathematics in School* 18(1): 14-15. EJ 389 722.
Discusses a counting system and number operations. Suggests six distinct areas in a "number" subject: one-to-one correspondences; simple counting process; complicated counting process; addition and multiplication; algorithms for the operations; and the decimal system.

Trafton, Paul R., and Albert P. Shulte, eds. 1989. *New Directions for Elementary School Mathematics. 1989 Yearbook.* Reston, VA: National Council of Teachers of Mathematics. 245p. ED 306 112.
This yearbook consists of five sections, with samples of children's work appearing between sections. Part 1, "Perspective on Change in Elementary School Mathematics," describes and discusses broad components of change in elementary school mathematics. The central role of problem solving, communication, and reasoning are highlighted, as is the need for revising our thinking about computation. Part 2, "Children's Reasoning and Strategies: Implications for Teaching," addresses knowledge about children's thinking and reasoning, how they construct ideas, and the importance of incorporating their mathematical understandings into instruction. Part 3, "New Directions in Teaching the Content of the Curriculum," reflects the ideas presented in the previous two sections and offers practical guidance for teachers and curriculum developers. Part 4, "In the Class-

room," captures the spirit of mathematical exploration and the active involvement of students through descriptions of classroom investigations. The yearbook concludes with Part 5, "Perspectives and New Directions in Teaching and Learning," which focuses on important factors that influence the way mathematics is taught and learned and that must be considered in implementing change.

Van De Walle, John A. 1990. *Elementary School Mathematics, Teaching Developmentally*. White Plains, NY: Longman. 475p. ED 317 428.
This is a methods book to help teachers guide children to develop ideas and relationships about mathematics. The methods and activities are designed to get children mentally involved in the construction of those ideas and relationships. Chapters 1-3 discuss what it means to teach mathematics and some of the important variables shaping mathematics education, the general philosophy behind the subtitle "Teaching Developmentally," and the learning of mathematics as a problem-solving process. Chapters 4-19 each address a different part of the elementary mathematics curriculum. Activities (stressed as the most important feature of these chapters), problems for discussion and exploration, and suggested reading references are provided for each of the chapters. The remaining chapters discuss the role of calculators and computers in mathematics; planning lessons, classroom use of materials, cooperative learning groups, homework, and the role of basal textbooks; assessment with an emphasis on diagnosis and how to listen to children; and the special considerations that should be given to children with special needs. The appendices contain summaries of the NCTM Standards for grades K-8, guides for mathematics learning activities, and masters and construction tips.

Pre-K–Grade 4

Aze, Ian. 1989. "Negatives for Little Ones?" *Mathematics in School* 18(2): 16-17. EJ 391 241.
Discusses a primary teaching method for the concept of negative numbers. Describes some activities for teaching the concept.

Baroody, Arthur J. 1989. *A Guide to Teaching Mathematics in the Primary Grades.*

Wakefield, NH: Longwood. 425p. ED 307 110.
This book was written as a tool to help teachers of prekindergarten to grade 3 make use of recent developments in the cognitive psychology of mathematical learning. Chapters included are: (1) "The Nature of Children's Mathematical Thinking"; (2) "Designing Effective Mathematics Instruction"; (3) "Oral Counting"; (4) "Numbering"; (5) "Numerical Relationships"; (6) "Informal Arithmetic"; (7) "Reading and Writing Symbols"; (8) "Concepts and Their Formal Representations"; (9) "Basic Number Combinations"; (10) "Base-Ten Place-Value Skills and Concepts"; (11) "Multidigit Mental Arithmetic"; (12) "Multidigit Written Calculation"; (13) "Geometry and Fractions"; (14) "Epilogue." Chapters 3-6 focus on informal mathematics, while chapters 7-12 focus on formal mathematics. Each chapter describes the learning patterns of the concepts or skills and then presents instructional activities. Suggested developmental sequences of skills and concepts of prekindergarten through grade 3 are appended. An extensive reference list is provided as well as an index of games and activities and an index of skills and concepts.

Fielder, Dorothy R. 1989. "Project Hands-On Math: Making a Difference in K-2 Classrooms." *Arithmetic Teacher* 36(8): 14-16. EJ 394 149.
Outlines the "Project Hands-On Math" developed in Georgia. Describes the five objectives, developmental phases, implementation, and evaluation of the project. Lists concrete manipulative materials used in this project.

Hamilton Township Schools. 1992. *Mathematics Curriculum Guide*. Hamilton Township, NJ. 128p.
This guide is meant to be used for long-range (semester or year) as well as short-range (individual lesson) planning. It includes objectives that must be covered as well as suggested strategies, resources, supplementary materials, and student evaluation components. The main body of the guide presents a chart. Specific objectives are stated on the left while a center column lists any suggestions for strategies and activities as well as corresponding page numbers in the textbook and supplementary materi-

als. The rightmost column lists assessment tools and activities.

Lynchard, Becky, et al. 1989. *Mathematics Teaching Modules. Basic Skills Assessment Program. Grade Three.* Jackson, MS: Mississippi State Department of Education. 143p. ED 312 136.
This document contains teaching modules based on seventeen objectives for grade 3 mathematics. The modules may be used for program evaluation and curricular planning as well as detailed outlines for classroom activities. Each module has eight sections: (1) "Objective"; (2) "Explanation of Objective"; (3) "Prerequisite Skills"; (4) "Vocabulary"; (5) "Lesson Plan Outline" (providing suggestions for introducing the objective, daily activities, and a culminating activity); (6) "Materials" (listing needed manipulatives and exhibits); (7) "Sample Test I"; and (8) "Sample Test II." The appendices include: "Mississippi Basic Skills Mathematics—Grade 3"; "Grade 3—Answer Key to Sample Tests"; and "Diagnostic/Prescriptive Evaluation Sheets" (third grade *Basic Skills Assessment Program* mathematics objectives). For the Grade 5 module, see ED 312 137.

Payne, Joseph N., ed. 1990. *Mathematics for the Young Child.* Reston, VA: National Council of Teachers of Mathematics. 316p. ED 326 393.
This book is for teachers of children in preschool through grade 4. It is designed to help teachers make important decisions about the mathematics curriculum and give them some effective ways to help children attain the mathematical power needed for everyday use and for future careers. Relevant research in each chapter provides a framework and dependable justification for the suggested curriculum and instructional strategies. Instructional activities are interwoven to furnish practical and useful ideas that teachers can modify and expand. Effective teaching, viewed as an interactive activity, is a major goal of this book. The intent is to help teachers understand the spirit of this interaction, the way tasks should be presented, the kind of questions to ask, the way manipulatives should be used to stimulate and develop thinking, and the way children's responses should be used to make instructional decisions. Included are fourteen chapters: (1) "New Directions in Mathematics Education"; (2) "Developing Mathematical Knowledge in the Young Child"; (3) "Developing Problem-Solving Abilities and Attitudes"; (4) "Concepts of Number"; (5) "Place Value and Larger Numbers"; (6) "A Problem-Solving Approach to the Operations"; (7) "Strategies of the Basic Facts"; (8) "Whole Number Computation"; (9) "Fractions and Decimals"; (10) "Geometric Concepts and Spatial Sense"; (11) "Measurement"; (12) "Graphical Representation and Probability"; (13) "Using Microcomputers for Mathematics Learning"; and (14) "Planning for Mathematics Instruction."

Shaw, Jean M. 1990. *Exploring Mathematics: Activities for Concept and Skill Development. Kindergarten-Grade 3.* Glenview, IL: Scott, Foresman. 162p. ED 309 999.
This book presents activities promoting children's understanding of mathematical ideas while at the same time providing the practice children need to refine their skills in mathematics. Chapter topics include: (1) numbers; (2) number facts and operations; (3) problem solving; (4) calculators; (5) fractions; (6) graphing; (7) geometry; and (8) measurement. Teaching materials, many including line drawings, are provided in each chapter. Each of the forty-two activities contains objectives, a list of materials, preparation, time frame, procedures, evaluation, and an extension activity.

Wolfinger, Donna M. 1989. "Mathematics in the Preschool Kindergarten." *Dimensions* 18(1): 5-7. EJ 401 232.
Discusses a mathematics curriculum for children in preschool or kindergarten that emphasizes manipulation. Suggests activities in the following concept areas: observation, classification, comparing and contrasting, seriation, patterning, graphing, informal measurement, and counting.

Grades 5-8

Lynchard, Becky, et al. 1989. *Mathematics Teaching Modules. Basic Skills Assessment Program. Grade Five.* Jackson, MS: Mississippi State Department of Education. 188p. ED 312 137.
This document contains teaching modules based on twenty objectives in grade 5 mathematics.

The modules may be used for program evaluation and curricular planning as well as detailed outlines for classroom activities. The format of the modules and appendices are the same as for Grade 3 (see ED 312 136).

McGehe Consulting. 1989. *Pioneer Math.* Urbana, IL. 249p.
This is a textbook written by middle school students with learning problems. Students worked under the direction of a math teacher and a special education teacher. The guide is illustrated with cartoons and is very project oriented. Most of the pages are workbook-like lessons in basic mathematics, presented in a way that would be suitable for remedial work with upper elementary or middle school students.

Secondary

General (7-12)

Alberta Department of Education. 1989. *Senior High Mathematics (10/13/14): Interim Teacher Resource Manual.* Edmonton. 331p. ED 315 309.
This resource material is intended to assist teachers in translating the intentions of the senior high mathematics program into classroom practice. An overview of the "Course of Studies" for mathematics (10, 13, and 14) is provided at the beginning of each course section in this manual. The overview includes the specific learner expectations, the attitudinal expectations, and the problem-solving expectations for each course. Activities for teaching concepts related to various topics are provided. Comments, process/problem-solving and context, technology integration, and suggestions are provided for each concept. Basic and recommended textbooks are identified for each concept. Learning resource books are listed for each course. Appendices include template sheets, resources, suggestions for using writing assignments in teaching mathematics, guidelines for integrating media, and a list of consultants in the province.

Alberta Department of Education. 1990. *Senior High Mathematics (20/23/24) Interim Teacher Resource Manual.* Edmonton. 298p. ED 325 308.

This *Teacher Resource Manual* (TRM) is designed to help teachers implement the Mathematics (20, 23, and 24) courses. The TRM is a support document that provides helpful information to classroom teachers. An overview of the course of studies for mathematics (20/23 and 24) is provided at the beginning of each course section in the TRM. The overview includes the specific learner expectation, the attitudinal expectations, and the problem-solving expectations for each course. In addition, the mathematics (24) overview includes the specific numeration expectations. For each concept covered in every program, the skills, process/problem-solving context with sample problems, elective suggestions, and resource correlation are included. Templates for a binomial grid and algebra tiles, support resources, samples of student evaluation ideas, ideas for combining writing and mathematics, guidelines for effective media integration, addresses for the Alberta Consumer and Corporate Affairs Offices, and a list of the Alberta mathematics consultants are appended.

Cetorelli, Nancy, et al. *A Review of Selected Microcomputer Software Packages with Lessons for Teaching Mathematics Grades 8-12. A Curriculum Development Project of the Project To Increase Mastery of Mathematics and Science (PIMMS).* Hartford, CT: Connecticut State Board of Higher Education.
The purpose of this project is to help teachers of mathematics get started using the computer to aid the teaching and learning of many mathematical concepts in a meaningful way. To this end, commercially available software packages are described that are versatile, cost effective, and easy to use. A review and several lessons, including homework exercises, have been written for each chapter and are presented in ready-to-use form.
Volume 1. 1987. ED 302 423. Four chapters: (1) "Geometric Supposer: Triangles" (Sunburst); (2) "Graphing Equations" (Conduit); (3) "Discovery Learning in Trigonometry" (Conduit); and (4) "Interactive Experiments in Calculus" (Prentice-Hall).
Volume 2. 1988. ED 302 424. Seven chapters: (1) "Polynomial Practice Using Tiles: Discovering Algebra" (Sunburst); (2) "Interpreting Graphs" (Sunburst/Conduit); (3) "The Geomet-

ric preSupposer" (Sunburst); (4) "Geometric Supposer: Quadrilaterals" (Sunburst); (5) "Superplot" (EduSoft); (6) "Introductory Statistics Software Package" (Addison-Wesley); and (7) "The Whatsit Corporation: Survival Math Skills" (Sunburst).
Volume 3. 1989. ED 309 977. Five chapters: (1) "The Factory"; (2) "Gears: Strategies in Problem Solving"; (3) "Building Perspective"; (4) "The Geometric Supposer: Triangles"; and (5) "The Geometric Supposer: Quadrilaterals." All five software packages reviewed are published by Sunburst Communications.

Dossey, John. 1990. *Discrete Mathematics and the Secondary Mathematics Curriculum.* Reston, VA: National Council of Teachers of Mathematics. 65p. ED 325 367.
Discrete mathematics, the mathematics of decision making for finite settings, is a topic of great interest in mathematics education at all levels. Attention is being focused on resolving the diversity of opinion concerning the exact nature of the subject, what content the curriculum should contain, who should study that material, and how that material should be taught in the classroom. This document details a set of recommendations on the infusion of topics from discrete mathematics into the secondary school (grades 7-12) mathematics curriculum. The suggestions are accompanied by a list of resources, including software, that will aid in the development of specific topics. Modeling, use of technology, algorithmic thinking, recursive thinking, decision making, and mathematical induction as a way of knowing, are six unifying themes that emerge when discrete mathematics is studied and applied. Recommended topics for emphasis in discrete mathematics before algebra, with algebra, and with geometry are included. Appended is an outline for a semester course in discrete mathematics.

EASTCONN Regional Educational Services Center. 1989. *Pre-Tech Mathematics Curriculum.* North Windham, CT. 46p. ED 342 639.
In 1988-1989 the Connecticut Vocational-Technical School System initiated a program for the ongoing review and upgrading of all trade and academic curricula used in the system's schools to insure that each curriculum

is consistent with current standards. After review and analysis, the revised curriculum incorporates the latest thinking of instructors who teach the subject, suggestions from authorities in the mathematics field, and current instructional approaches in the field of education. This curriculum is intended as a plan for learning topics in algebra, trigonometry, and geometry, and as an aid to the mathematics instructor to prepare students for membership in society. The document is separated into several sections: the "Vocational-Technical School Mathematics Sequence"; "Career Relationships and Requirements"; "Instructor's Goals"; "Students' General Competencies"; "A Common Core of Learning"; "Personal Goals for Students"; "Pre-Tech Math History"; and a series of thirty units in three categories—algebra, trigonometry, and geometry. Each unit provides suggested instructional time, unit objectives, and a list of student competencies required of students after completing the unit. The first category contains eleven algebra units discussing the topics of word problems, factoring, fractions, fractional equations, systems of equations, graphing, exponents, radicals, quadratic equations, inequalities, and ratio, proportion, and variation. The second category contains eleven trigonometry units discussing the topics of trigonometric ratios, right triangles, functions of angles of any size, trigonometric identities, radius measure, special angles, composite angle formulas, solving oblique triangles, vectors, and graphs of trigonometric functions. The final category contains eight geometry units dealing with plane figures, measurement, the right triangle, circles, prisms, cylinders, pyramids and cones, and the sphere. Computer software and a textbook are suggested.

Franklin, Margaret. 1990. *Add-Ventures for Girls: Building Math Confidence, Middle School 7-10.* Reno, NV: University of Nevada. 346p. For copies, write to WEEA Publishing Center, Education Development Center, 55 Chapel Street, Newton, MA 02160.
This guide includes strategies, activities, and resources that deal with five major topics: attitudes and math, math relevance, the learning environment, other issues, and mathematics promotion. All of the suggestions are based on research findings, published resources, and

practical ideas from math teachers. They represent sound educational practice and should provide a positive learning environment for both boys and girls.

Lamoille North Supervisory Union. 1991. *Mathematics: 7-12.* Hyde Park, VT. 63p. This curriculum guide includes a scope and sequence chart for Grades 7 and 8 and for each secondary course. Each chart lists skills within strands and indicates the number of weeks that should be spent on a strand. Each skill receives a code (I = introduce, R = reinforce, R,I = reinforce and extend). A course map for grades 9-12 lists skills and indicates how they fit into each of the ten different courses offered.

Natour, Denise M., et al. 1990. *Instructor's Guide to the CCA Basic Skills Mathematics Curriculum.* Urbana, IL: Illinois University, Computer-Based Education Research Lab. 239p. ED 325 332. The curriculum consists of sixteen units covering: (1) addition, subtraction, multiplication, and division of whole numbers; (2) common fractions; (3) mixed numbers; and (4) decimals. Each unit consists of a pretest, assignment of lessons as needed, and a posttest. Paper-and-pencil worksheets that accompany the computer lessons are an integral part of the curriculum. Each unit includes lesson name, computer file where lesson can be found, author(s) and affiliation, objectives, description, interaction, intended audience, and completion time. This curriculum is designed for, and delivered by, the SYS 4 instructional management system. When a student does not earn PREtest credit, SYS 4 can either make an automatic assignment of lessons or alert the instructor to make an assignment. When a student does not earn credit on a POSTtest, intervention is always required. Instructors can choose to have a student redo lessons, do other lessons, work with paper-and-pencil materials, or work with a tutor. Sample computer generated worksheets and an alphabetical list of files used in this curriculum are appended.

New York State Education Department. 1988. *Teaching Mathematics with Computers: Grades 9-12.* Albany, NY. 61p. Copies may be obtained by submitting requests in writing on school stationary to the Publications Sales Desk, New York State Education Department, Room 164 EBA, Washington Avenue, Albany, NY 12234. This report is designed to make suggestions and give examples of how computers may be used to enhance student understanding of mathematical concepts and applications in the secondary school mathematics curricula. Section I describes the various types of software that could be incorporated into the secondary classroom. Section II presents a model for including computer instruction in General Mathematics and the Three-Year Sequence. Section III addresses a NYS Regents Action Plan for including computer mathematics as an option for a local diploma. Section IV proposes advanced computer mathematics courses. Section V provides two tables that can be used to facilitate the selection and use of software within the curriculum. Included in the appendices are two computer programs that teachers might use to demonstrate computer relevance to real-life problems.

"N.S.F. Awards $6 Million for Math Curriculum." 1992. *Education Week* 12 (14 Oct.): 9. States that NSF has granted $6 million to 5 university-based teams to develop a technology-based high school math curriculum. The project is called the Core Plus Mathematics Project; its director is Christian R. Hirsch of Western Michigan University. The project seeks to weave together four major strands into a simplified curriculum for each of the three years of instruction: algebra and functions, statistics and probability, geometry and trigonometry and discrete math, including graphs and algorithm design. The program wants to eliminate placing students in college-bound and non-college-bound tracks, upgrade the mathematics that all students learn and update the curriculum to reflect new developments in mathematics.

Oklahoma State Department of Vocational and Technical Education, Curriculum and Instructional Materials Center. *Mathematics Series: Principles, Concepts, and Applications.* Stillwater, OK. The "General Mathematics" series is intended to introduce students to basic mathematical principles and concepts and to provide the opportunity to apply those principles and concepts. The standard format for each of the units found within the following five modules is: performance objectives; suggested activities; information sheets; tests; answers to tests; and, depending on the specific objectives, there may or may not be assignment sheets; job sheets; transparency masters; and supplements.

Whole Numbers. 1989. ED 312 164. Units include: (1) "Place Value"; (2) "Rounding Whole Numbers"; (3) "Adding Whole Numbers"; (4) "Subtracting Whole Numbers"; (5) "Multiplying Whole Numbers"; (6) "Dividing Whole Numbers"; and (7) "Reviewing Whole Numbers."

Fractions. 1988. ED 303 348. Units included are: (1) "Understanding Fractions"; (2) "Proper, Improper, Mixed, and Equivalent Fractions"; (3) "Converting and Reducing Fractions"; (4) "Adding Fractions"; (5) "Subtracting Fractions"; (6) "Multiplying Fractions"; (7) "Dividing Fractions"; and (8) "Fraction Review."

Decimals and Percents. 1990. ED 312 165. Units include: (1) "Place Values of Decimals"; (2) "Rounding Decimals"; (3) "Adding Decimals"; (4) "Subtracting Decimals"; (5) "Multiplying Decimals"; (6) "Dividing Decimals"; (7) "Converting Percents"; (8) "Solving Problems Using Percents"; and (9) "Reviewing Decimals and Percents."

Measurements. 1988. ED 303 349. Units include: (1) "Linear Measurement"; (2) "Units of Measure"; (3) "Figuring Board Feet"; and (4) "Reviewing Measurements."

Geometry. 1989. ED 310 933. The seven units include: (1) "Geometric Figures"; (2) "Perimeters in Geometry"; (3) "Geometric Areas"; (4) "Circumference and Area of Circles"; (5) "Volume"; (6) "Geometric Angles"; and (7) "Reviewing Geometry."

Prince George's County Public Schools. 1989. *Integrating Computer Software into the Functional Mathematics Curriculum: A Diagnostic Approach.* Upper Marlboro, MD. 891p. ED 312 159.

This curriculum guide was written to provide information on the skills covered in the Maryland Functional Math Test (MFMT) and to outline a process that will allow teachers to integrate fully computer software into their instruction. The materials produced in this directory are designed to assist mildly to moderately handicapped students who will take the MFMT, but may also be helpful to regular education students who are experiencing difficulty with the test.

The first section, "Domain Directory," lists 30 objectives divided into 7 domains on the MFMT. The content scope, question format, teaching strategy, vocabulary, common errors, and task analysis are provided for each of the objectives. The second section, "Assessment Materials," contains tests, answer keys, and skill sheets. The skill sheets are organized by domains, corresponding objectives, and skills. The last section, "Software Materials," provides the software matrices relating specific skills to software programs and a summary reviewing the programs. Appendices include: (1) "MFMT Vocabulary List"; (2) "Student Progress Sheet"; (3) "Computer Software"; (4) "Suggested Assessment Modifications"; (5) "Guidelines for Parents"; (6) "Additional Resources and Supplementary Materials"; and (7) "MFMT List of Domains, Objectives, and Skills."

Swetz, Frank, and J. S. Hartzler, eds. 1991. *Mathematical Modeling in the Secondary School Curriculum.* Reston, VA: National Council of Teachers of Mathematics. 136p. ED 339 601.

Over the past ten years, national conferences and committees investigating the state of American mathematics education have advocated an increased emphasis on problem solving and mathematical modeling situations in the secondary school curriculum. However, little effort has been made to prepare secondary school teachers to use mathematical modeling techniques in their classrooms. The document presents a variety of classroom modeling activities that were developed and classroom tested by the Mathematical Sciences Program of the Pennsylvania State University at Harrisburg at the secondary school level. After an introductory chapter explaining the concept of mathematical modeling and how it differs from problem solving, the bulk of the document is a series of twenty-two classroom activities, followed by an appendix giving teachers' guides for each of the activities, and an extensive bibliography for related mathematical modeling activities. The twenty-two activities can be broken up into the following subject levels, together with the concepts and skills involved: (1) General Mathematics—involving ratio and proportion, Pythagorean formula, the distance formula, and probability; (2) Algebra 1—involving simple graphing, area computation, inequalities, functions and pattern recognition, algebraic operations, and simple programming; (3) Algebra 2—involving linear inequalities and graphing, circle equation, permutations and counting techniques, graphing and programming linear parabolic equations, basic trigonometric functions, velocity and acceleration formulas, exponential functions, and matrix arithmetic; and (4) Precalculus—involving transcendental functions.

See also *Problem Solving Through Critical Thinking, Grades 5-8* (ED 331 721), under Elementary.

Grades 8 and 9

Alberta Department of Education. 1989. *Mathematics: Program of Studies/Curriculum Guide. Grades 8 and 9. Interim: 1989.* Edmonton. 118p. ED 308 064.
This program is for students of grades 8 and 9 who may function a year or two behind their age peers in mathematics. It is designed to develop concepts, skills, and attitudes required for effective computation and problem solving at home, in the classroom, in the workplace, and in the community. Other goals are to develop a positive self-concept, critical and creative thinking skills, ability to use modern technology, and reading skills and other forms of communication required for learning mathematics and solving problems. Concepts taught include number systems and operations; ratio, proportion, and percent; geometry and measurement; data investigation and display; and algebra. Themes are managing your money, world of work, using math at home, and travel and recreation. The curriculum guide includes learning objectives, related life skills, related applications across the curriculum, and suggested strategies and activities for each concept developed. See also *Mathematics 8 and 9: Teacher Resource Manual (Interim—1989).* ED 313 237.

Dawson, Jon, ed. 1989. *Mathematics 8 and 9: Teacher Resource Manual. Interim 1989.*

Edmonton: Alberta Department of Education. 272p. ED 313 237.
A manual to supplement *Mathematics: Program of Studies/Curriculum Guide. Grades 8 and 9 (Interim—1989)*, ED 308 064. This teacher resource manual has been developed to assist classroom teachers in implementing the *Integrated Occupational Mathematics* program for grades 8 and 9. The first chapter of this manual gives an introduction, including resources and scope/sequence. The next two chapters describe the overviews and instructional strategies of the grade 8 and 9 themes. The themes developed at each grade level are classified as managing money; world of work; using math at home; and travel and recreation. The last chapter discusses generic strategies: (1) problem solving; (2) use of technology (including calculators and computers); (3) computational facility and estimation; (4) using a math lab; and (5) evaluation. Resource materials are listed for each of the five topics. List of thirteen references.

Grade 10

Alberta Department of Education. 1990. *Mathematics 16, Program of Studies/Curriculum Guide, Grade 10. Interim—1990.* Edmonton. 91p. ED 325 306.
The Mathematics 16 program provides for the development of essential concepts, skills, and attitudes required for effective computation and problem solving. The program is activity-based and addresses the need for students to be able to transfer and apply specific mathematical concepts and skills to more generalized situations in everyday life and the world of work. A focus on the use of technology throughout the program assists students in developing the ability to use calculators and computers in performing routine tasks more easily replicated by these technologies. Contents include chapters entitled: (1) "Rationale"; (2) "Philosophy"; (3) "Goals for the Mathematics 16 Program"; (4) "Model for the Mathematics 16 Program"; (5) "Interpersonal Skills and the Social Sphere"; (6) "Required and Elective Components"; (7) "Planning"; (8) "Learning Resources for Mathematics 16"; (9) "Methodology"; (10) "Evaluation"; (11) "Scope and Sequence"; (12) "Program of Studies/Presentation of Content"; and (13) "Suggested Options." See also *Mathemat-*

ics 16, Teacher Resource Manual (Interim—1990). ED 325 307.

Alberta Department of Education. 1990. *Mathematics 16, Teacher Resource Manual. Interim—1990.* Edmonton. 315p. ED 325 307.
A manual to supplement *Mathematics 16, Program of Studies/Curriculum Guide, Grade 10 (Interim—1990),* ED 325 306. This manual should be used as a practical planning and instructional tool in translating the intentions of the Mathematics 16 program. The themes include earning money, budgeting and banking skills for the consumer, and math in the workplace. The program's emphases and methodology cover problem solving, the use of technology (calculator and computer), computational facility and estimation, situational and concrete approaches, and assessment and evaluation. An annotated list of learning resources, Alberta Consumer and Corporate Affairs Offices, and the regional offices of education are appended.

Grade 12

Manitoba Department of Education and Training. 1989. *Mathematics (304, 301, 300).* Winnipeg. 991p. ED 325 362.
This guide for grade 12 mathematics has been developed as part of an overall revision of the mathematics program for Manitoba, Canada schools. It is the framework upon which grade 12 teachers may build their mathematics programs. This guide consists of four main sections, including a general overview of the program and three Mathematics (304, 301, 300) course outlines. The outlines are based on the aims of mathematics, which are to foster maximum and continuous growth in three behavioral areas: (1) the cognitive or factual; (2) the affective which includes feelings and perceptions; and (3) the psychomotor or movements. The mathematical content of this program has been chosen to promote the above goals.

Algebra

College Entrance Examination Board, Educational Testing Service. 1990. *Algebridge. Concept Based Instructional Assessment.* Princeton, NJ. 382p. ED 339 600.

A major hurdle for students in studying algebra is the transition from arithmetic to algebra. *Algebridge* is a teaching supplement that integrates assessment and instruction in a student-centered approach, focusing on understanding concepts, not memorizing algorithms. Each unit uses a four-step approach to diagnose and correct student misconceptions. The steps are to *assess* whether a student understands a particular concept by administering the *instructional assessment*; *discuss* with the students the given idea while scoring the assessment to see where misunderstanding lies; *instruct* to clear up conceptual misunderstandings and fill in knowledge gaps using practice sheets and suggested activities; and *reassess* using the *follow-up assessment*. The units include: (1) "Fractions in Expressions and Equations"; (2) "Pattern Recognition and Proportional Reasoning"; (3) "Meaning of Negative Numbers"; (4) "Constructing Numerical Equations"; (5) "Attacking Word Problems Successfully"; (6) "Concept of Variable"; (7) "Concept of Equality and Inequality"; and (8) "Operations on Equations and Inequalities." Each unit includes assessment instruments and practice sheets particular to the concepts discussed. The answer keys to separate reassessment instruments are included in the answer section of each unit.

EASTCONN Regional Educational Services Center. 1989. *Algebra Curriculum.* North Windham, CT. 23p. ED 342 637.
In 1988-89 the Connecticut Vocational-Technical School System initiated a program for the ongoing review and upgrading of all trade and academic curricula used in the system's schools to insure that each curriculum is consistent with current standards. After a review and analysis, the revised curriculum incorporates the latest thinking of instructors who teach the subject, suggestions from authorities in the mathematics field, and current instructional approaches in the field of education. This curriculum is intended as a plan for learning algebra and as an aid to the mathematics instructor to prepare students for membership in society. The document is separated into several sections: the "Vocational-Technical School Mathematics Sequence"; "Career Relationships and Requirements"; "Instructor's Goals"; "Students' General Competencies; A

Common Core of Learning"; "Personal Goals for Students"; and a series of ten algebra units. Each unit provides suggested instructional time, unit objectives and a list of student competencies required of students after completing the unit. The units are: (1) "Signed Numbers and Order of Operations"; (2) "Evaluation of Algebraic Expressions"; (3) "Equations in One Variable"; (4) "Fractional Parts of Numbers"; (5) "Inequalities and More about Exponents"; (6) "Multivariable Equations and Additions of Fractional Equations"; (7) "Percents and Polynomials"; (8) "Graphs of Linear Equations"; (9) "Rational Expressions, Exponents, Substitution Axiom, Complex Fractions, and More Graphing"; and (10) "Substitution, Subsets, Square Roots, Domain, and Additive Property of Inequality." Computer software and a textbook are recommended.

EASTCONN Regional Educational Services Center. 1989. *Algebra with Introduction to Trigonometry Curriculum*. North Windham, CT. 27p. ED 342 638.
In 1988-89 the Connecticut Vocational-Technical School System initiated a program for the ongoing review and upgrading of all trade and academic curricula used in the system's schools to insure that each curriculum is consistent with current standards. After a review and analysis, the revised curriculum incorporates the latest thinking of instructors who teach the subject, suggestions from authorities in the mathematics field, and current instructional approaches in the field of education. This curriculum is intended as a plan for learning algebra with trigonometry and as an aid to the mathematics instructor to prepare students for membership in society. The document is separated into several sections: "The Vocational-Technical School Mathematics Sequence"; "Career Relationships and Requirements"; "Instructor's Goals"; "Students' General Competencies"; a "Common Core of Learning"; "Personal Goals for Students"; and a series of fourteen algebra and trigonometry units. Each unit provides suggested instructional time, unit objectives, and a list of student competencies required of students after completing the unit. The units are: (1) "Radical Expression, Elimination, Complex Fractions"; (2) "Factoring"; (3) "Scientific Notation, Consecutive Integers, and Rational Equations"; (4)

"Graphical Solutions, Equation of a Line"; (5) "Multiplication of Radicals, Division of Polynomials, and System of Equations"; (6) "Solutions of Quadratic Equations by Factoring, Intercept-Slope Method of Graphing"; (7) "Multiplicative Property of Inequality, Uniform Motion Problems, Difference of Two Squares"; (8) "Pythagorean Theorem, Distance, Uniform Motion, Square Roots, and Rounding"; (9) "Absolute Value Inequalities, Rational Equations"; (10) "Equation of a Line through Two Points, Functional Notation"; (11) "Parallel Lines, Equation of a Line with a Given Slope, Radical Slope Equations, Formula"; (12) "Absolute Value Inequalities, Multiplication of Radical Expressions, Linear Inequalities"; (13) "Square Roots (Quotient Theorem), Trinomial Factoring, the Quadratic Formula"; and (14) "Trigonometry of the Right Triangle." A textbook and computer software are suggested.

Hughes, Barnabas. 1988. "First Year Algebra: A Computer Coordinated Curriculum." *Journal of Computers in Mathematics and Science Teaching* 7(4): 23-25. EJ 382 874.
Explores a year-long computer coordinated curriculum for algebra. Discusses the model student and goals of the programs. Lists the curriculum in a weekly format with the topics for each week.

Kysh, Judith. 1991. "Implementing the Curriculum and Evaluation Standards: First Year Algebra." *Mathematics Teacher* 84(9): 715-22. EJ 438 346.
An alternative first year algebra program developed to bridge the gap between the NCTM's Curriculum and Evaluation Standards and institutional demands of schools is described. Increased attention is given to graphing as a context for algebra, calculator use, solving "memorable problems," and incorporating geometry concepts, while deemphasizing rational expressions and radical simplification.

McCreery, Louis. 1989. *Motivators for Algebra One: A Situational Approach. Lesson Plans for the New California Framework and the New National Standards*. Pomona, CA: California State Polytechnic University. 100p. ED 309 960.

This monograph provides activities for beginning algebra students to develop an understanding of the meaning and use of letter symbols, to show applications of algebra in their own lives, and to develop an understanding of mathematical processes through investigation. The activities included were developed by considering the students' mathematical knowledge, interests, maturity level, and everyday living. Several attributes of a student-centered classroom are listed. The lessons fall under four headings: (1) "Bridges from Arithmetic to Algebra"; (2) "Estimation and Conjecture"; (3) "Situational Lessons"; and (4) "Investigations." Learning materials and answers to questions in the lessons are appended.

Geometry

Brutlag, Dan, and Carole Maples. 1992. "Making Connections: Beyond the Surface." *Mathematics Teacher* 85(3): 230-35. EJ 440 174.

An integrated curriculum geometry unit, "Beyond the Surface," is presented and built around four principles promoting students' mental connections. The principles include a significant problem context; assignment of concrete, active tasks; employing tools embodying the mathematical concept; and reflection on established connections through writing, discussions, and projects.

Chazan, Daniel, and Richard Houde. 1989. *How To Use Conjecturing and Microcomputers To Teach Geometry.* Reston, VA: National Council of Teachers of Mathematics. 56p. ED 309 993.

This book describes how teachers have taught students to behave like working mathematicians who conjecture and prove within a community of learners through the use of microcomputers and the *Geometric Supposers* software. The first section discusses the definition and importance of the conjecture, describes inquiry skills and understandings students should develop, and argues against seven commonly accepted myths. The second part answers questions typically asked by teachers about: (1) students' conjecturing; (2) curriculum; (3) ways to merge conjecturing into what students are already doing; (4) laboratory sessions including planning, setting expectations, guiding students as

they explore, and evaluating laboratory work; (5) classroom discussions after the laboratory; (6) traditional laboratory sessions; and (7) whole-group exploration. The third part discusses some questions to be considered by the school community, students, and teachers before using a computer tool. A hypothetical introductory lesson with the *Supposers* is appended.

Posamentier, Alfred S. 1989. "Geometry: A Remedy for the Malaise of Middle School Mathematics." *Mathematics Teacher* 28(9): 678-80. EJ 406 083.

Proposes a new geometry curriculum for motivating middle school students. Discusses the treatment of geometry including visual justifications of geometric phenomena, examination of the properties of various common geometric figures, use of art and architecture, and inspection of geometric transformations. Eleven references are listed.

Texas Education Agency, Division of Curriculum Development. 1990. *Guidelines for Teaching Geometry.* Austin, TX. 62p. ED 334 070.

This guidebook is designed to assist teachers and other local school district personnel in the planning and teaching of the high school geometry course, with essential elements as mandated by the Texas State Board of Education in 1989. Research on student learning in geometry provides the logical rationale for the instructional sequencing and techniques presented in this course. The publication presents the philosophy and the intent of the geometry course and discusses teacher preparation, student credits and prerequisites, the required essential elements of instruction, and the integrated use of technology. Included are sample pupil performance objectives and classroom activities that illustrate how the essential elements of the geometry course can be taught. Local school district personnel are advised to use these guidelines, both suggested and stated, in the development of their own curriculum documents for a geometry course. Four appendices contain a description of the van Hiele levels of mental development, activities to help students achieve the levels, a geometry instructional plan, and sixty-four references.

Business Math

EASTCONN Regional Educational Services Center. 1989. *Business Mathematics Curriculum*. North Windham, CT. 24p. ED 335 497.
This curriculum guide for teaching business mathematics in the Connecticut Vocational-Technical School System is based on the latest thinking of instructors in the field, suggestions from mathematics authorities, and current instructional approaches in education. The curriculum guide consists of six sections: (1) career relationships and requirements; (2) instructor's goals; (3) competencies expected of students; (4) correlations with the common core of learning (matrix); (5) personal goals for students; and (6) course outline (eleven units containing student competencies and essential content, suggested activities, and suggested assessment approaches). Units cover these topics: basic mathematics skills workshop; personal business mathematics; checking and savings accounts; cash purchases/charge accounts/credit cards/loans; auto transportation/housing costs; insurance/investments/recordkeeping; business expenses; purchasing and sales; marketing/warehousing and distribution/services; accounting and accounting records; and financial and information management. Appendices list twenty-two suggested software items, a recommended textbook, and a reference source.

Statistics

Rouncefield, Mary. 1989. "Practical Statistics in the Sixth Form." *Mathematics in School* 18(1): 6-7. EJ 389 720.
Introduces a statistics curriculum emphasizing an instructional sequence of practical, real data, discussion, and model/theory instead of theory, example, and practice. Provides one example and describes the teaching methods of the curriculum.

SOURCE LIST
FOR SPECIAL PROJECTS

*by Kim F. Garnett
Coordinator of Elementary Education
Center for Continuing Education, Saint Leo, Florida

I N 1990, President George Bush and the fifty state governors set six national educational goals. Among these goals was the statement that by the year 2000, American students will rank first in mathematics and science achievement.

In order for American students to meet these goals, a drastic restructuring of mathematics education is needed. As teachers, we must adapt to changing what we are teaching, the methodology by which we are teaching, and our assessment practices. Dossey, Mullis, Lindquist, and Chambers (1988) stated that teachers must utilize "more innovative forms of instruction—such as those involving small group activities, laboratory work, and special projects" if the United States is to maintain its competitive edge throughout the world.

As classroom facilitators, we are in a unique position to help students communicate mathematically, discover connections in mathematics, use reasoning skills, and promote the creative posing and solving of nonroutine types of problems. The National Council of Teachers of Mathematics (NCTM) determined that "the development of each student's ability to solve problems is essential if he or she is to be a productive citizen" (NCTM 1989, 6). The NCTM further emphasized the importance of providing opportunities to communicate in the mathematics classroom by stating, "The ability to read, write, listen, think creatively, and communicate about problems will develop and deepen students' understanding of mathematics" (NCTM 1989, 78).

The NCTM *Curriculum and Evaluation Standards for School Mathematics* (1989) addresses the educational needs of this highly technical age with new societal goals: (1) mathematically literate workers, (2) lifelong learning, (3) opportunity for all, and (4) an informed electorate. The mathematically literate worker must be "prepared to understand the complexities and technologies of communica-

Special acknowledgement and thanks to William L. Farber, Director, Mathematics Resource Center, City College of New York, for contributing additional research and resources.

tion, to ask questions, to assimilate unfamiliar information, and to work cooperatively in teams" (NCTM 1989, 3). These expectations do not correlate with the traditional pedagogy in which students work independently to solve drill and practice exercises. Through classroom projects, teachers will "turn their students on" to the marvels and mysteries of mathematics while simultaneously achieving these goals.

The way in which mathematics is taught is as important as the content that is taught. Students need to view mathematics as an "integrated whole" rather than a collection of isolated topics (e.g., geometry, measurement, statistics, probability, algebra). Mathematics instructional pedagogy will influence whether students value mathematics. The introduction of classroom mathematics projects utilizing supplemental materials will enhance your regular mathematics curriculum by captivating your students as willing participants in the learning process.

This chapter presents a multitude of suggested mathematics classroom project ideas and resources. With creativity, imagination, and collaborative planning with your students, modifications for any project will be easily implemented to tailor your instruction to the needs and interests of your students. You are encouraged to contact any sources listed to supplement your existing mathematics program with the excitement of active learning through classroom projects.

This chapter is divided into eight major strands:
- Books
- Organizations/publications
- Magazines/journals
- Multimedia, technology, and software
- Programs
- Contests/competitions
- Materials/resources
- Catalogs

References

Dossey, J., I. Mullis, M. Lindquist, and D. Chambers. 1988. *Mathematics: Are We Measuring Up?* Princeton, NJ: Educational Testing Service.

National Council of Teachers of Mathematics. 1989. *Curriculum and Evaluation Standards for School Mathematics.* Reston, VA: NCTM.

Books

Select from among the books listed in this section (in alphabetical order by author) ideas and suggestions for any K–12 mathematics classroom. The activities and projects described in these books promote mathematical connections, problem solving, mathematical communication, and higher-order reasoning skills through hands-on, active involvement.

Math Power in School, Math Power at Home, and *Math Power in the Community,* by the American Association for the Advancement of Science (Waldorf, MD: AAAS Books, 1990).
Three-volume set of *Math Power* books filled with ideas for mathematics projects. For further information, contact: American Association for the Advancement of Science, AAAS Books, Dept. A37, P.O. Box 753, Waldorf, MD 20604; (301) 645-5643.

A Guide for Reviewing School Mathematics Programs, ed. by Glendon W. Blume and Robert F. Nicely Jr. (Reston, VA: NCTM, 1991). ISBN 0-87353-334-8.
This guide is designed to enable school district personnel to analyze their mathematics programs and to help point out directions for change. To order, contact: NCTM, 1906 Association Drive, Reston, VA 22091; (800) 235-7566 or (703) 620-9840.

One, Two, Buckle My Shoe: Math Activities for Young Children, by Sam Ed Brown (Mt. Rainier, MD: Gryphon House, 1982). ISBN 0-87659-103-9.
Intended for parents and teachers of young children. Brown advocates allowing children the freedom to explore and interact with their world.

Connections: Linking Manipulatives to Mathematics (Grades 1–6), by Micaelia Randolph Brummett and Linda Holden Charles (Sunny-

vale, CA: Creative Publications). ISBNs: 0-88488-768-5 (grade 1); 0-88488-769-3 (grade 2); 0-88488-770-7 (grade 3); 0-88488-771-5 (grade 4); 0-88488-772-3 (grade 5); 0-88488-773-7 (grade 6).
This program is specifically designed to connect manipulative instruction to various topics in mathematics. Each grade-level resource book contains twenty manipulative lessons focusing on all the major strands.

A Collection of Math Lessons from Grades 1 through 3, by Marilyn Burns and Bonnie Tank (Math Solution, 1988). ISBN 0-941-35501-2.
This book features practical ideas for teaching early childhood mathematics through problem solving. The book's format is classroom-tested lessons demonstrating how to use cooperative group instruction, concrete materials, and student writing to develop children's understanding of mathematical concepts. This book is distributed by Cuisenaire, 12 Church Street, P.O. Box D, New Rochelle, NY 10802.

A Collection of Math Lessons from Grades 3 through 6, by Marilyn Burns (Math Solution, 1987). ISBN 0-941-35500-4.
In this book, the author presents her view of what mathematics teaching ought to be, illustrated with samples of children's work. This book is distributed by Cuisenaire, 12 Church Street, P.O. Box D, New Rochelle, NY 10802.

A Collection of Math Lessons from Grades 6 through 8, by Marilyn Burns (Math Solution, 1990). ISBN 0-941-35503-9.
Chapters from this book include lessons on percents, measuring angles, introducing algebra, patterns and functions, and probability investigations. This book is distributed by Cuisenaire, 12 Church Street, P.O. Box D, New Rochelle, NY 10802.

Mathematical Questions from the Classroom, Part 1 and Part 2, by Richard J. Crouse and Clifford Sloyer (Janson Publications, 1987). ISBN 0-939765-04-7 (set); ISBN 0-939765-02-0 (pt. 1); ISBN 0-939765-03-9 (pt. 2).
This two-part book focuses on the mathematics teacher's ability to respond to student questions accurately and effectively. Part 1 covers topics from junior high school mathematics, algebra I, and geometry. Part II covers algebra II, calculus, elementary functions, probability, and trigonometry.

What Expert Teachers Say about Teaching Mathematics, by Richard Dahlke and Roger Verhey (Palo Alto, CA: Dale Seymour Publications, 1986). ISBN 0-86651-308-6.
This book offers preservice and in-service teachers of K–8 mathematics insights into how experienced teachers manage their teaching of elementary- or middle-level mathematics. In addition, the book is based on various questions covering an array of topics related to the teaching of mathematics and includes selected responses from experienced teachers.

Stories of Excellence: Ten Case Studies from a Study of Exemplary Mathematics Programs, by Mark Driscoll (Reston, VA: NCTM, 1987). ISBN 0-87353-236-8.
This guide discusses ten case studies of what Mark Driscoll and his colleageus describe as exemplary mathematics programs. To order, contact: NCTM, 1906 Association Drive, Reston, VA 22091; (800) 235-7566 or (703) 620-9840.

Kids Are Consumers Too: Real World Mathematics for Today's Classroom, by Jan Fair and Mary Melvin (Reading, MA: Addison-Wesley, 1986). ISBN 0-201-20298-3.
Provides a vast collection of ideas and materials for making the math curriculum real and for helping students become wise consumers.

Handbook of Research on Mathematics Teaching and Learning, ed. by Douglas A. Grouws (New York: Macmillan, 1992). ISBN 0-02-922381-4.
This is a comprehensive handbook that describes the evolution of mathematics education research. It discusses the influence of current research on teaching and the teaching of specific topics. In addition, the book examines

mathematics learning in the classroom versus mathematics in real life.

Explorations for Early Childhood, by Lalie Harcourt (Reading, MA: Addison-Wesley, 1988). ISBN 0-201-19106-7.
This activity-based mathematics program is designed for use in kindergarten and prekindergarten classes. Activities are designed and spiralled to promote repeated exposure to concepts in each of these main strands of primary mathematics: problem solving, number, geometry, and measurement.

Mathematics Project Handbook, 3d ed., by Adrien Hess, Glen Allinger, and Lyle Anderson (Reston, VA: NCTM, 1989). ISBN 0-87353-283-X.
This guide contains ideas on organizing, researching, evaluating, and presenting mathematics projects. In addition, it includes ideas from mathematics fairs, lists of resources, things to construct, and questions for investigations as well as a new list of publishers and references and suggestions for parents, teachers, and students in middle-level grades and high school.

Jaw Breakers and Heart Thumpers (Fresno, CA: AIMS Education Foundation, n.d.).
This book features activities that integrate math and science. Children can measure heart rates, observe patterns in fingerprints, calculate the amount of sugar in bubble gum, and learn to gather and record data for graphing. To order, contact: AIMS Education Foundation, P.O. Box 7766, Fresno, CA 93747.

A Call for Change, by the Mathematical Association of America Committee on the Mathematical Education of Teachers (Washington, DC: MAA, 1991). ISBN 0-88385-072-9.
This book discusses standards common to the preparation of mathematics teachers at all levels: elementary school (K–4), middle school (5–8), and secondary school (9–12).

Reshaping School Mathematics, by the Mathematical Sciences Education Board and the National Research Council (Washington, DC: National Academy Press, 1990). ISBN 0-309-04187-2.
This booklet proposes a framework for reform of school mathematics in the United States. Included are changing perspectives on the need for mathematics, the nature of mathematics, and the learning of mathematics. In addition, the changing roles of calculators and computers in the practice of mathematics are summarized.

Counting on You, by the National Academy of Science (Washington, DC: National Academy Press, 1991). ISBN 0-309-03977-0.
This booklet is directed to school boards, school administrators, parents, college and university faculties, policy makers and government, business, and industry leaders, members of the media, and teachers. The booklet describes why significant change in mathematics is necessary and what steps have been taken thus far to bring about such change on a nationwide basis.

Curriculum and Evaluation Standards for School Mathematics (1989) and *Professional Standards for Teaching Mathematics* (1991), by NCTM (Reston, VA: NCTM). ISBN 0-87353-273-2 and ISBN 0-87353-307-0.
To order, contact: NCTM, 1906 Association Drive, Reston, VA 22091; (800) 235-7566 or (703) 620-9840.

Everybody Counts: A Report to the Nation on the Future of Mathematics Education, by the National Research Council (Washington, DC: National Academy Press, 1990). ISBN 0-309-03977-0.
This report is in response to the urgent national need to revitalize mathematics and science education. It examines mathematics education as all one system, from kindergarten through graduate school. In addition, this report reflects the thinking of seventy leading American mathematics educators and scholars.

From Zero to Infinity, 4th ed., by Constance Reid (Washington, DC: Mathematical Association of America, 1992). ISBN 0-88385-505-4.
This book is a classic and is designed for a young person beginning to be interested in

mathematics. It shows how interesting the everyday natural numbers are. This edition brings up to date those portions pertaining to the fast developing application of computers to the determination of the nature of large numbers.

Lure of the Integers, by Joe Roberts (Washington, DC: Mathematical Association of America, 1992). ISBN 0-88385-505-4.
This book stems from a collection of many years of casual accumulation of numerical facts. Most of the mathematics presented belongs to elementary mathematics in the sense that no deep mathematical expertise is required to follow what is written. References are provided for further study.

"Mathematics Projects." Unit IV of five units in *Learning on Your Own!,* by Phil Schlemmer (West Nyack, NY: Center for Applied Research in Education, 1987). ISBN 0-87628-511-6.
Student problem-solving projects designed to promote individual or small-group learning skills for upper elementary through middle school students.

Elementary School Mathematics: Teaching Developmentally, by John Van de Walle (White Plains, NY: Longman, 1990). ISBN 0-8013-0203-X.
A comprehensive resource book recommended for preservice and in-service instruction, individual teachers, and elementary media centers. In twenty-three chapters, with 378 activities promoting discovery and experimentation. Available through: Longman, 95 Church Street, White Plains, NY 10601-1505.

A Mathematical Mystery Tour: Higher-Thinking Math Tasks, by Mark Wahl (Tucson, AZ: Zephyr Press Learning Materials, 1988). ISBN 0-913705-26-8.
Will motivate even the hardest-to-reach students to learn mathematics. History, design, writing, geometrical constructions, botany, zoology, astronomy, and philosophy are integrated with mathematics. The Cooperative Learning Model is suggested throughout the "tour."

Add-Ventures for Girls: Building Math Confidence, by the Research and Planning Center,

University of Nevada (Newton, MA: Women's Educational Equity ACT Publishing Center, 1990).
Classroom teaching strategies effective with girls are stressed. For information on this and other resources, contact: WEEA Publishing Center, 55 Chapel Street, Newton, MA 02160; (800) 225-3088 or (617) 969-7100.

Preparing Young Children for Math: A Book of Games, by Claudia Zaslavsky (New York: Schocken Books, 1986). ISBN 0-8052-0796-1.
Intended for parents and teachers of young children (ages 2–8). Hands-on active learning is promoted using everyday materials.

Organizations/Publications

Become actively involved with other mathematics educators across the United States by joining organizations that promote improved and meaningful mathematics instruction. Members of organizations listed in this section receive periodicals and/or other published materials that will serve as excellent resources for mathematics projects.

The American Mathematical Society is a nonprofit organization devoted to promoting the interests of mathematical scholarship and research. In addition, the Society sponsors meetings, symposia, seminars, and institutes; provides employment services; and publishes mathematical books and journals. For information, contact: American Mathematical Society, P.O. Box 6248, Providence, RI 02940; (401) 272-9500; (800) 556-7774.

Brown's Directory of Instructional Programs is a resource to teachers and administrators who are looking for a particular program or for those who are researching new programs. For more information, contact: Infinity Impressions, 88 East Main Street, Suite 500, Mendham, NJ 07945.

Thinking through Mathematics, by Edward A. Silver, Jeremy Kilpatrick, and Beth Schlesinger, is one of a series of publications initiated by the College Board's Educational EQuality Project, a ten-year effort to improve the quality of secondary education and to ensure equal access to college for all students. In addition, *Thinking through Mathematics* focuses on how mathematics teaching and learning can be improved by developing more powerful approaches to connecting thinking and mathematics. It is published by the College Entrance Examination Board and can be obtained by contacting: College Board Publications, Box 886, New York, NY 10101-0886.

COMAP (Consortium for Mathematics and Its Applications) is a nonprofit corporation dedicated to the improvement of mathematics education. COMAP oversees a network of thousands of mathematicians, educators, and practitioners from elementary and secondary schools, two-year colleges, industry, and government. Various COMAP publications develop student modules, teacher training programs, television courses, software sets, or level-specific newsletters. For information, contact: COMAP, 60 Lowell Street, Arlington, MA 02174; (617) 641-2600.

The Logo Foundation is a nonprofit education organization for the support of teachers who use Logo. For information regarding curriculum and staff-development projects, workshops, seminars, starting Logo user groups, and published materials, contact: Logo Foundation, 250 W. 57th Street, Suite 2603, New York, NY 10107-2603; (212) 765-4918.

The Math Science Network offers publications and videos (e.g., *Math for Girls and Other Problem Solvers*). For information, contact: Math Science Network, 2727 College Avenue, Berkeley, CA 94705.

The Mathematical Association of America is the largest organization dedicated to primaarily college-level mathematics. The range of concern, however, extends naturally to topics of interest to college-bound high school students. The MAA publishes three journals and a newsletter. For more information, contact: The Mathematical Association of America, 1529 18th Street, NW, Washington, DC 20036; (202) 387-5200.

The Math/Science Network will hold conferences across the country in math and science for girls. Hands-on career-related activities and career panels will be presented. For information, contact: Renata Tervalon, Math/Science Network, 2727 College Avenue, Berkeley, CA 94705; (415) 841-MATH or (415) 841-0201.

The National Association for the Education of Young Children (NAEYC) is an organization for early childhood educators. The NAEYC promotes developmentally appropriate curriculum and materials. As a member, the early childhood educator will receive a subscription to *Young Children*, an excellent periodical that often features a commitment to hands-on mathematics learning. For information, contact: NAEYC, 1834 Connecticut Avenue, NW, Washington, DC 20009-5786; (800) 424-2460 or (202) 232-8777.

The National Council of Teachers of Mathematics (NCTM) is an organization that all K–12 mathematics teachers should consider joining. NCTM members receive the NCTM *News Bulletin* six times yearly; opportunities to attend state, regional, national, and international conferences; opportunities to subscribe to NCTM periodicals (e.g., *Mathematics Teacher, Arithmetic Teacher, Research in Mathematics Education*); a *Catalog of Educational Materials*; and opportunities to purchase a multitude of NCTM publications (e.g., *Curriculum and Evaluation Standards for School Mathematics*, and *Professional Standards for Teaching Mathematics*). For information, contact: National Council of Teachers of Mathematics, 1906 Association Drive, Reston, VA 22091.

School Science and Mathematics is the official journal of the School Science and Mathematics Association (SSMA). It promotes the improvement of teaching mathematics and science, and is published monthly, October through May. SSMA individual membership, which includes the journal, is $25. For information, write to: School Science and Mathematics Association, 126 Life Science Building, Bowling Green State University, Bowling Green, OH 43403-0256.

The Young Astronaut Council sponsors math and science enrichment activities for elementary school children. For information, contact: The Young Astronaut Council, Box 65432, 1211 Connecticut Avenue, NW, Suite 800, Washington, DC 20036.

Magazines/Journals

These magazines and journals will serve as motivating supplements to classroom textbooks in the mathematics classroom. *Info Power* is a student-written magazine for pupils in the middle grades, and *The Nth Degree* is written by secondary students. The teaching magazines listed provide ideas for classroom-tested projects. You might want to submit your own manuscript for publication!

The American Mathematical Monthly, issued ten times per year, contains articles, short notes, and features concerning all aspects of mathematics. It is intended for all who are mathematically inclined. It is both a journal of mathematics and mathematics culture. In addition, the journal contains a problems section, a selection of book reviews, and a comprehensive list of telegraphic reviews. For more information, contact: The Mathematical Association of America, 1529 Eighteenth Street, NW, Washington, DC 20036; (202) 387-5200.

Arithmetic Teacher is the official publication of the National Council of Teachers of Mathematics. This journal, published monthly (except June, July, and August), focuses on mathematical activities through the middle grades. It presents new developments in curriculum, instruction, learning, and teacher education. Contact: The National Council of Teachers of Mathematics, 1906 Association Drive, Reston, VA 22091; (703) 620-9840.

The College Mathematics Journal, published five times a year, contains material for all who are interested in undergraduate mathematics, in enrichment material for high school students, and in using the computer in the classroom. Regular sections include Student Research Projects, Classroom Capsules, Media Highlights, Book Reviews, Problems and Solutions, and Fallacies, Flaws, and Flimflam. For information, contact: The Mathematical Association of America, 1529 Eighteenth Street, NW, Washington, DC 20036; (202) 387-5200.

Consortium, COMAP's quarterly high school newsletter, provides a multitude of ideas for introducing students to applications of mathematics. Each issue contains a pull-out section designed for classroom use. This section can be photocopied and distributed to students. For further information, contact: Consortium for Mathematics and Its Applications (COMAP), Inc., 60 Lowell Street, Arlington, MA 02174; (617) 641-2600.

The Elementary Mathematician features an activity-based format and contains problem-exploration activities designed for elementary grades. In addition, the journal features pull-

out sections, activity sheets, cross-curriculum extensions, and mini-lessons. For further information, contact: COMAP, Inc., Suite 210, 57 Bedford Street, Lexington, MA 02173; (617) 862-7878.

Games Junior is a magazine of puzzles and mathematical recreations. Contact: Games Junior, P.O. Box 2028, Harlan, IA 51593.

The World Almanac *Info Power* has 32-page themed issues designed for grades 5-8. Each issue is interdisciplinary in nature and requires students to ask questions, think for themselves, and learn collaboratively. A 16-page teaching guide accompanies each issue. *Info Power* could be used as a springboard for a variety of classroom math projects. For information, call Sundance for a free sample: (800) 343-8204.

The *Instructor* teaching magazine is published nine times annually. It is an excellent resource for classroom-tested mathematics projects, materials, and teacher resources. For information, contact: Scholastic, Inc., 351 Garver Road, P.O. Box 2700, Monroe, OH 45050-2700; (800) 544-2917.

D. C. Heath's *Making the Case for Math: A Special Report on Elementary Mathematics in the 1990s* is a 32-page magazine that describes math instruction for the 1990s according to the new standards. For a free copy, contact: D. C. Heath & Company, Making the Case for Math, 70 Mansell Court, Suite 100, Roswell, GA 30076.

The Instructor: Middle Years is a magazine for middle-level educators that features multidisciplinary classroom projects to motivate middle-grades learners. For information, contact: *Middle Years*, Instructor Magazine, P.O. Box 53896, Boulder, CO 80322-3896.

Mathematics Magazine, published five times a year, provides informal mathematical articles covering a variety of interesting topics in mathematics. The level of interest is designed for undergraduate and graduate students. For information, contact: The Mathematical Association of America, 1529 18th Street, NW, Washington, DC 20036; (202) 387-5200.

Mathematics Teacher is published monthly (except June, July, and August) and is devoted to the improvement of mathematics instruction in the junior high schools, senior high schools, two-year colleges, and teacher education colleges. For information, contact: The National Council of Teachers of Mathematics, 1906 Association Drive, Reston, VA 22091; (703) 620-9840.

The Nth Degree is a secondary mathematics journal written mostly by secondary school students. Encourage your students to submit manuscripts on applications, historical background, computer programs, graphics, and problem solving to: Tim McNamara, Faculty Editor, *The Nth Degree*, 1250 Amherst Street, Buffalo, NY 14216; (716) 875-8212.

School Science and Mathematics is published monthly from May to October. For more information, contact: School Science and Mathematics Association, 126 Life Science Building, Bowling Green State University, Bowling Green, Ohio 43403-0256.

Teaching Pre-K–8 is a magazine published eight times annually. An excellent resource for classroom-tested mathematics projects, materials, and teacher resources. Regular departments include "Teaching Math" and "Math Reproducible." For information, contact: *Teaching Pre-K–8*, P.O. Box 54805, Boulder, CO 80323-4805; (800) 678-8793.

Multimedia, Technology, and Software

In this technological age, it is imperative that educators provide students with opportunities to interact with technology in multimedia classrooms. The instructional resources listed in this section promote real-life simulation activities and projects using technology.

Algebra for Everyone. Videotape providing teachers with useful techniques for active learning in small groups. Exploration and experimentation are encouraged as students are given the tools needed to understand algebra rather than the use of traditional approaches. For information, contact the National Council of Teachers of Mathematics, 1906 Association Drive, Reston, VA 22091; (800) 235-7566 or (703) 620-9840.

"Designing for Discovery: Interactive Multimedia-learning Environments at Bank Street College." Technical report (19 pages) that portrays multimedia classrooms characterized by interactions of students, teacher,

materials, and the world. Students are encouraged to solve problems cooperatively through active engagement. To order, contact: Bank Street College Bookstore, 610 West 112th Street, New York, NY 10025. Ask for Technical Report No. 15; $3 prepaid.

A Look at Children's Thinking. Two 25-minute videotapes (+ 27p. study guide) exploring how young children think in mathematics. Demonstrations with kindergarten and first grade students show the distinctions necessary in assessing what students understand about quantities and the relationships between numbers. For further information, contact: Educational Enrichment, 2221 West Lindsey, P.O. Box 1524, Norman, OK 73070.

Mathematics for the Middle School. Three 20-minute videotapes, created by Marilyn Burns, focusing on specific aspects of mathematics instruction in grades 6–8. Throughout each tape, teachers model ways of engaging classes of middle school students in problem solving, small-group learning, the use of manipulatives, and open communication. Contact: Cuisenaire Company of America, Inc., 12 Church Street, Box D, New Rochelle, New York 10802.

Mathematics with Manipulatives. Six 20-minute videotapes, created by Marilyn Burns, showing how manipulatives materials can be used in K-6 classrooms. Contact: Cuisenaire Company of America, Inc., 12 Church Street, Box D, New Rochelle, New York 10802.

"Square One TV." Award-winning show for television that motivates children for mathematics. Also available are teacher's guides: *Square One TV Goes to School, Square One TV Curriculum Connections, Square One TV Game Shows,* and *Square One TV Mathnet Mysteries.* Teachers are legally permitted to tape "Square One TV" for in-school classroom use for up to three years. PBS broadcast times are in local listings. For further information, contact: Children's Television Workshop, GPO Box 5373, New York, NY 10087-5373, Attn: Schools AAT592.

Teaching Mathematics with Calculators: A National Workshop, presented by the Mathematical Association of America and the National Council of Teachers of Mathematics. This videotape contains two parts: "The Graphing Calculator: Building New Models" (running time: 24:25) and "The Fractions Calculator: Old Things New Ways" (running time: 17:58). This videotape is an introduction to the use of the TI-81 Graphics Calculator and the TI-Explorer Fractions Calculator, both of which are manufactured by Texas Instruments Inc. To obtain this videotape: (800) TI-CARES.

The Theorem of Pythagoras is the first in a series of modules designed to use computer animation to help instructors teach basic concepts in mathematics. This module consists of a 15-minute videotape and a workbook to guide students through the video, elaborating on the important ideas. The modules are designed to support material for existing courses in high school. Information concerning this project can be obtained by writing to the project office at the following address: Project MATHEMATICS!, California Institute of Technology, Pasadena, CA 91125; (818) 356-6345. To order, contact either: The Mathematical Association of America, 1529 Eighteenth Street, NW, Washington, DC 20036; (202) 387-5200—or the National Council of Teachers of Mathematics, 1906 Association Drive, Reston, VA 22091; (703) 620-9840.

Visions of Exploration. Multimedia educational program that integrates science, math, social studies, language arts, fine arts, and career education. Contact: Steve Anderson, *USA Today*, at (703) 276-5872.

WINGS for Learning offers multiple media math modules with its *Journeys in Mathematics*, designed for grades 4–6. Students learn by doing. Computer software, print materials, and manipulatives are used as students work individually on small-group collaborative projects and whole-class sharing and discovery. These modules are ideal for cooperative-learning classrooms. For further information, contact: Journeys in Mathematics module, WINGS for Learning, 1600 Green Hills Road, P.O. Box 660002, Scotts Valley, CA 95067-0002; (800) 321-7511 or (408) 438-5502.

Educational Electronics stocks basic calculators to pocket computers for the classroom from: Texas Instruments, Casio, Sharp, Educator, and others. Contact: Educational Electronics, 70 Finnell Drive, Weymouth Landing, MA 02188.

Texas Instruments offers a wide variety of classroom calculators for K–12 students and overhead calculators for all models. For example: TI-108 (8-digit solar); TI Math Mate (correct order of operations); Math Explorer (fractions, integer division with whole-number remainders); TI-30 Challenger (trigonometric and scientific functions); TI-34 (one-variable statistics, trigonometric functions, four-number bases); TI-81 (interactive graphics analyses). For more information, call: (800) TI-CARES.

Computer-Intensive Algebra (CIA) is an algebra curriculum that uses computer and calculator technology to explore mathematics using real-world situations. For further information, contact: James T. Fry, Dept. of Mathematics, University of Maryland, College Park, MD 20742; or M. Kathleen Heid, Pennsylvania State University, 171 Chambers Building, University Park, PA 16802.

The Kidware 2 "Learning Center" approach is designed for the early elementary classroom in which a computer center and other learning centers are used for thematic teaching. For further information, contact: Kidware Teacher's Software Guides, Mobius Corporation, 405 North Henry Street, Alexandria, VA 22314, Attn: Suzanne Thouvenelle; (703) 684-2911.

For a free Math Kit containing eight Logo activities designed for K–8 students, contact: Terrapin Software, 400 Riverside Street, Portland, ME 04103; (207) 878-8200.

"Counters: An Action Approach to Counting and Arithmetic" for the Apple II (48K required) includes one disk, one back-up, and a 22p. teacher's guide. This software furnishes an environment for young children learning to count meaningfully from one through nine. Sections of addition and subtraction are designed as applications of counting. For further information, contact: Wings For Learning, 1600 Green Hills Road, P.O. Box 660002, Scotts Valley, CA 95067.

"Graphing and Probability Workshop, Grades 3–8" for the Apple (requires 128K) includes two disks and a 128p. and 63p. manual. This software presents activities based on NCTM's Curriculum Standards, and uses technology to involve students in mathematical lessons. For further information, contact: Scott Foresman and Company, 1900 E. Lake Avenue, Glenview, IL 60025.

"LOGO Geometry" for the Apple II (requires 128K) includes one disk each for grades

K-1, 2-3, and 4-6. The program is designed to enhance the understanding of geometry. The program also develops knowledge of the uses of LOGO and writing procedures for the computer. For information, contact: Silver Burdett and Ginn, Simon and Schuster Education Group, 250 James Street, Morristown, NJ 07960-1918.

"Mathematica" is a system of doing mathematics by computer. It is a powerful mathematical tool appropriate for use by high school students to advanced graduate students. In addition, "Mathematica" supports numerical, symbolic, and graphical computation. It can be used as both an interactive problem-solving environment and a modern, high-level programming language. On many computer systems, "Mathematica" also provides interactive documents that mix text, animated graphics, and sound. For information, contact: Wolfran Research, Inc., 100 Trade Center Drive, Champaign, Illinois 61820-7237; (217) 398-0700.

Programs

Two of the programs listed in this section address the NCTM societal goal in this highly technical age—opportunity for all. Current statistics indicate that females do not frequently enter science and technology careers because they are not as mathematically prepared as males. The other programs described promote classroom projects through active engagement.

The Dwight D. Eisenhower Mathematics and Science National Programs and other FIRST Office Mathematics and Science Projects is a collection of abstracts that describes in broad terms the scope and objectives of previous FIRST (Fund for the Improvement and Reform of Schools and Teaching) Program Grants in mathematics and science. Names and addresses of recipients are included. For more information, contact: Becky Wilt, Eisenhower National Mathematics and Science Program, U.S. Department of Education, 555 New Jersey Avenue, NW, Suite 522, Washington, DC 20208-5524, ATT: Contests/Competitions; (202) 219-1496.

Family Math is a nationally recognized parental involvement project that features par-

ents and children working together, valuing mathematics through problem exploration and hands-on activities. The range of activities is applicable for children from 5 to 18 years old. An accompanying publication, *Family Math,* serves as a resource guide for the program. This publication is also available in a Spanish version, *Matematica para la familia.* For more information, contact: Lawrence Hall of Science, University of California, Berkeley, California 94720, Attn: Family Math.

How High the Sky? How Far the Moon? An Educational Program for Girls and Women in Math and Science is a program designed for K-12 females. Contact: Women's Educational Equity Publishing Center, Education Development Center, 55 Chapel Street, Newton, MA 02160; (800) 225-3088 or (617) 969-7100.

The International Space Year is a worldwide occasion to recognize our future in the space age. Some educational opportunities for school involvement include projects, contests, curricula, exhibitions, and movies. Updated quarterly is the *International Space Year Educational Activities Catalog,* which provides information and the names of contacts. For information, contact: Educational Affairs Division, NASA Headquarters, Code XE, Washington, DC 20546.

Macmillan Early Skills Manipulatives is a subscription program, designed for preschool through grade 2, that helps young children build thinking skills using manipulatives. Every two months a new set of early skills manipulatives and teaching ideas are sent, including pattern blocks, "court and sort," "geoboards," "dinosaur counters," puzzles, and "attribute tiles." For further information, contact: Macmillan Early Skills Manipulatives, Dept. 3-ZY5, P.O. Box 938, Hicksville, NY 11802.

Math Matters: Kids Are Counting on You is a project kit sponsored by the National PTA. The major focus is on parental involvement in mathematics. Major funding for this project was provided by the Exxon Education Foundation. To obtain the kit or for further information, contact: The National PTA, 700 North Rush Street, Chicago, IL 60611-2571.

Operation SMART (Science, Math, and Relevant Technology) is a program that introduces females to math-related careers through active-learning opportunities. Contact: Evelyn

Roman-Lazen, Operation SMART, Girls Incorporated, 30 East 33d Street, New York, NY 10016; (212) 689-3700.

Project AIMS is a nonprofit, tax-exempt, educational foundation that promotes activities and staff development integrating math and science. Books, posters, newsletters, and special events are featured items. For information, contact: AIMS Education Foundation, P.O. Box 7766, Fresno, CA 93747.

Contests/Competitions

Trying to succeed in a competitive world is a reality for most Americans. Not only is it exciting for students to prepare for and participate in competitive situations, it also helps to prepare them for the future. The contests listed in this section emphasize the creative solving of nonroutine, multiple-step problems—a skill that the National Assessment of Educational Progress (NAEP) continuously reports is underdeveloped in American students.

The annual "Mathcounts" competition is a problem-solving national contest. The winners receive scholarship awards, Space Camp awards, computers, etc. For further information, contact: Mathcounts, 1420 King Street, Alexandria, VA 22314; (703) 684-2828.

Mathematical Olympiad Contest Problems for Children, by George Lenchner (Oceanside, NY: Glenwood Publications). ISBN 0-9626662-0-3. Collection of nonroutine problems designed for interschool competition for elementary children. To order, contact: Mathematical Olympiad Contest Problems for Children, Glenwood Publications, 125 Merle Avenue, Oceanside, NY 11572.

Mathematics Contests: A Handbook for Mathematics Educators, by David R. Johnson and James R. Margenau, describes a wide variety of mathematics competitions designed for students of all ages. In addition, it outlines specific procedures for creating a mathematics competition in a school. *Mathematics Contests: A Guide for Involving Students and School,* by Frederick O. Flener, presents competitive formats along with a multitude of categories and sample questions. The book also offers suggestions on contest rationales, procedures and

strategies for starting a contest, and effective coaching. For further information on either of these books, contact: The National Council of Teachers of Mathematics, 1906 Association Drive, Reston, VA 22091; (703) 620-9840.

The annual "Odyssey of the Mind" competition consists of seven-member teams of students who are judged according to the problem-solving creativity, presentations, and solutions. The four divisions of the competition include K–6, 6–8, 9–12, and college. For further information, contact: Odyssey of the Mind, P.O. Box 27, Glassboro, NJ 08028; (609) 881-1603.

The annual "USA Mathematical Olympiad (USAMO)" is an American mathematics competition. The winners may go on to compete in the "International Mathematical Olympiad (IMO)." For information, contact: Walter Mientka, Executive Director, American Mathematics Competitions, Dept. of Mathematics, University of Nebraska, Lincoln, NE 68588-0322.

Materials/Resources

The materials and resources cited in this section provide ideas and suggestions for mathematics activities and projects. Active engagement, hands-on learning, and critical thinking are encouraged.

The Fabric of Mathematics: A Resource Book for Teachers, by Mary Laycock and Gene Watson, is a comprehensive resource for elementary school mathematics educators. Included in the book are teacher references, an annotated bibliography of children's references, visuals, activities, games, and manipulatives. *The Tapestry of Mathematics: A Resource Book for Secondary School Mathematics Teachers,* by Mary Laycock and Connie Johnson, lists many of the alternatives currently available for teaching secondary mathematics. Topics include wide world applications, models and games, enrichment activities, and student and teacher references. *Weaving Your Way from Arithmetic to Mathematics,* by Mary Laycock and Peggy McLean, is written for K–8 teachers and will assist in evaluating exploratory, manipulative, or abstract levels of thinking for topics involved in counting, such as place value, the four operations, fractions, decimals, proportions and percent, integers, and real numbers. For information on any of these three books, contact: Activity Resources Company, 24872 Calaroga Avenue, Hayward, CA 94545.

Googolplex is a multifaceted, mathematical, building manipulative focusing on three-dimensional geometry. The pieces consist primarily of an equilateral triangle, square, pentagon, and circle, and are joined by double-hinged connectors that allow individual pieces and whole constructions to rotate a full 360 degrees. For more information, contact: Arlington-Hews, Inc., Box 23798, Vancouver Airport P.O., Richmond, British Columbia V7B 1X9 Canada.

Calendar Math is a collection of monthly math activities and projects designed to promote positive parental involvement in the mathematics program. For information, contact: Peter Saarimaki, Coordinator of Math, Board of Education for the City of Toronto, 155 College Street, Toronto, Ontario M5T 1P6 Canada.

Math Learning Centers for the Primary Grades is a resource book that provides the primary teacher with detailed instructions for promoting creative, independent learning with problem solving, graphing activities, manipulative projects, etc. For further information, contact: Center for Applied Research in Education, Book Distribution Center, Route 59 at Brookhill Drive, West Nyack, NY 10995-9900; (800) 288-4745.

Marcy Cook, an internationally renowned educational consultant, provides teacher in-service workshops/seminars on "Hands-on Math for Active Learning: Grades K–8." She provides elementary and middle-level teacher participants with materials, lessons, and activities during an active day of learning. A catalog of motivational math materials is also available. Contact: Marcy Cook, Math Specialist-Consultant for Elementary Schools, P.O. Box 5840, Balboa Island, CA 92662-5840; (714) 673-5912.

Exploring Fractions and Decimals with Manipulatives, by Don Balka is a teacher resource for grades 3–8. The "Didax Fraction Kit" should be used in conjunction with the resource book. The activities are geared toward individuals, dyads, and small groups. For information, contact: Didax Education Resources (Ref. 41-N1002), One Centennial Drive, Peabody, MA 01960.

The John and Mable Ringling Museum of Art provides a *Mathematics Teaching Books Catalog* that offers inexpensive materials and books ($4.95 to $27.95) for K–12 mathematics teachers. Mathematics resource files, mathematical models, curiosities and puzzles, and art with a mathematical flavor are among the listed topics. Contact: The John and Mable Ringling Museum of Art, 5401 Bay Shore Road, Sarasota, FL 34243, Attn: Sales Dept.; (813) 355-5101.

For a free catalog of model airplane products designed for grades 4–12 mathematics classrooms, contact: Midwest Products Co., Inc., School Division, P.O. Box 564, Hobart, IN 46342; (800) 348-3497 or (219) 942-1134.

"From the File Treasury" is a collection of the finest teaching ideas and activities from *The Arithmetic Teacher*. The collection is printed on color-coded 4″ × 6″ file cards. Collection costs $21.50 and may be ordered from: From the File Treasury, National Council of Teachers of Mathematics, 1906 Association Drive, Reston, VA 22091.

Caddyrack is a portable, stackable, and mobile rack system that can be used as a complete activity center for mathematics manipulatives, calculators, and other materials. The Caddyrack is available in various sizes according to individual needs. For more information, contact: SPECTRUM Educational Supplies Limited, 125 Mary Street, Aurora, Ontario L4G 1G3 Canada.

For a complete list of learning materials that promote whole-brain learning, creative and critical thinking, and self-awareness, a free catalog is available from: Zephyr Press, P.O. Box 13448, Tucson, AZ 85732-3448; (602) 745-9199.

Catalogs

The following list, arranged alphabetically by publisher, includes excellent catalogs of books, software, manipulatives, supplemental materials, additional activity resources, and more. for your classroom math projects.

Teaching Tools for Mathematics (Academic Industries, Inc., P.O. Box 428, Baychester Station, Bronx, NY 10469).

Open to the Magic of Math (Activity Resources Co., Inc., P.O. Box 4875, Hayward, CA 94540; (510) 782-8172).

A Catalog of Books and Manipulatives for Pre-School and K–6 (Classic School Products, P.O. Box 160066, Altamonte Springs, FL 32716-0066.

COMAP Catalogue (Consortium for Mathematics and Its Applications (COMAP), Inc., 60 Lowell Street, Arlington, MA 02174; (617) 641-2600).

Middle Grades 1992 Mathematics Catalog: Grades 4 through 9 (Creative Publications Order Department, 5040 West 111th Street, Oak Lawn, IL 60453; (800) 624-0822).

Creative Publications: K–12 Mathematics and Language Arts (Creative Publications Order Department, 5040 West 111th Street, Oak Lawn, IL 60453; (800) 624-0822).

Dale Seymour Publications: K–8 Educational Materials (Dale Seymour Publications, P.O. Box 10888, Palo Alto, CA 94303-0879).

Dale Seymour Publications: Secondary Mathematics (Dale Seymour Publications, P.O. Box 10888, Palo Alto, CA 94303-0879).

Delta Education: Hands-on Math Catalog (Delta Education, Inc., P.O. Box 915, Hudson, NH 03051).

Delta Education: Hands-on Learning Catalog (Delta Education, Inc., P.O. Box 915, Hudson, NH 03051; (800) 442-5444).

Edmund Scientific: Annual Reference Catalog for Optics, Science, and Education (Edmund Scientific Co., 101 E. Gloucester Pike, Barrington, NJ 08007-1380; (609) 573-6250 or (609) 547-3488).

EduCALC: Calculators and Informators (EduCALC, 27953 Cabot Road, Laguna Niguel, CA 92677. Customer Service: (800) 677-7001).

ETA Mathematics Catalog: Math Manipulators, Versa-Tiles and Science Materials (Educational Teaching Aids, 199 Carpenter Avenue, Wheeling, IL 60090; (708) 520-2500).

ETA Mathematics Catalog: A Universe of Math Manipulatives (Educational Teaching Aids, 620 Lakeview Parkway, Vernon Hills, IL 60061).

Gamco Industries, Microcomputer Software (Grades 1–12) (Gamco Industries, Inc., P.O. Box 310J5, Big Spring, TX 79721. U.S. toll-free: (800) 351-1404; Texas toll-free: (800) 447-1516).

Thinking + Mathematics + Problem Solving (Books, Journals, and Software) (Lawrence Erlbaum Associates, Inc., 365 Broadway, Hillsdale, NJ 07642; (201) 666-4110 or (800) 926-6579).

Longman Catalog: Education (Longman Publishing Group, 10 Bank Street, White Plains, NY 10606-1951; (914) 993-5000).

Mathematical Association of America Publications Catalogue (Mathematical Association of America, 1529 18th Street NW, Washington, DC 20036; (202) 387-5200).

Merit: Software (Merit, P.O. Box 392-D, New York, NY 10024; (212) 267-7437).

NCTM: Catalog of Educational Materials (National Council of Teachers of Mathematics, 1906 Association Drive, Reston, VA 22091-1593; (703) 620-9840).

NMSA: Resource Catalogue (National Middle School Association, 4807 Evanswood Drive, Columbus, OH 43229-6292; (614) 848-8211).

Selected Materials for Mathematics: Elementary, Intermediate, Secondary (Opportunities for Learning, Inc., 20417 Nordhoff Street, Dept. MA987, Chatsworth, CA 91311; (818) 341-2535).

Rigby: Big Books and More! (Rigby, P.O. Box 797, Crystal Lake, IL 60014; (800) 822-8661).

Routledge Education Bookshelf (Routledge, 29 West 35th Street, New York, NY 10001-2291; (212) 244-6412).

Summit Learning: K–8 Math Resources (Summit Learning, P.O. Box 493, Fort Collins, CO 80522).

Sunburst: New Directions in Education (software) (Sunburst Communications, 101 Castleton Street, Pleasantville, NY 10570-3498; (800) 628-8897 or (914) 747-3310).

Sundance: Pre-K–6 Catalog (Sundance Publishers & Distributors, Pharos Books, P.O. Box 1326, Littleton, MA 01460).

Universal Education: Mathematics Materials Catalog (Universal Education Math Department, 320 S. Eldorado, Mesa, AZ 85202; (800) 248-3764. *Eastern Sales Office*: P.O. Box 163, Groveland, MA 01834).

WINGS for Learning/Sunburst Communications: Science and Mathematics for High School and College (WINGS for Learning/ Sunburst Communications, 1600 Green Hills Road, P.O. Box 660002, Scotts Valley, CA 95067-0002; (800) 321-7511 or (408) 438-5502).

WINGS for Learning/Sunburst Communications: Educational Computer Courseware (K–12) (WINGS for Learning/Sunburst Communications, 1600 Green Hills Road, P.O. Box 660002, Scotts Valley, CA 95067-0002; (800) 321-7511 or (408) 438-5502).

RECOMMENDED TRADE BOOKS

by Gavrielle Levine
Assistant Professor
C. W. Post Campus of Long Island University, Long Island, New York

A LTHOUGH trade books can make an important contribution to students' learning of mathematics, they are often an underutilized resource. Either as alternative sources of new information or as opportunities to review information in another way, they offer students options in the mathematics learning environment. The National Council of Teachers of Mathematics (NCTM) has identified five new goals for students learning mathematics (1989). Trade books can play an important role in achieving these goals.

To accomplish the first goal, that students learn to value mathematics, the NCTM suggests that students participate in many and varied mathematical experiences. Trade books provide easy access to a broad range of mathematical ideas and applications by presenting mathematical concepts in a variety of contexts and cultures.

The second goal, to increase students' confidence in their ability to do mathematics, can be accomplished in part as students appreciate the many ways in which they can be successful in mathematics. Trade books present topics in mathematics that may not be included in the mathematics curriculum or that may offer more information about those topics that are.

As a result, students can appreciate how successful they can be in these diverse content areas. Beyond this, they can realize how frequently they interact mathematically within their environment.

The third goal, that students become problem solvers, can be accomplished as students solve the problems and puzzles presented in trade books. Because solutions often require integrating many facets of mathematical knowledge, many of these problems need to be completed over long periods of time. This becomes an important lesson to contrast with our "fast-food" culture. Students can learn the value of investing time and effort to solve a worthwhile problem.

The fourth goal, that students learn to communicate mathematically, reflects the importance of learning the symbols of mathematics, as well as reading, writing, and speaking the language of mathematics. Storybooks allow young students to read the language of mathematics, while historical or more technical books enable older students to accomplish this goal. All students can use reference sources to investigate and research terms, symbols, and concepts that are not familiar. Trade books do a wonderful job of putting mathematics into interesting and meaningful contexts, showing that communicating with, through, and about mathematics helps to make sense of the world in meaningful ways.

The fifth goal, that students learn to reason mathematically, can be achieved in at least two ways using trade books. Most books of mathematics problems include discussions of solutions as well as problems. Students can create their own solutions to problems and then examine solutions that others have developed. In this way, they can expand their problem-solving repertoire. In addition, biographies of people who used mathematics in their careers reveal their approaches to solving problems.

When selecting mathematics trade books, the following questions serve as a helpful guide:

1. How accurate is the information presented in the book? Is it current? Does it reflect current thinking about the topic?
2. Are the illustrations and text unambiguous? Are terms defined clearly and accurately? Are concepts presented clearly?
3. Are the concepts, text, and illustrations appropriate for the grade level and interests of the students?
4. Does the book represent males, females, ethnic groups, and minority groups in nonstereotypical roles?
5. Do you think that this book will appeal to students?

These five criteria were used to select the books that have been included in this bibliography.

Current professional activity reflects educators' interest in incorporating trade books in mathematics education. Several recent publications include lesson-plan suggestions that incorporate mathematics trade books in the elementary school curriculum (Burns 1991; Griffiths and Clyne 1991; Rommel 1991). Papers presented at a recent national mathematics conference (the National Council of Teachers of Mathematics Annual Conference in April 1992) discussed the relationship between mathematics and literature (Carey 1992; Larson 1992). Teachers may choose to combine reading a book with a related classroom activity so that students experience the mathematical concept in an active way.

This chapter is organized by grade level, and within that by topic. The four grade levels are early childhood, elementary school, middle school, and secondary school. Students who are more advanced conceptually or who would benefit from review of earlier concepts should be encouraged to read trade books at the appropriate levels. Several of the reference books appear in more than one grade-level listing because they are intended for use by a wider span of student levels.

Early Childhood

Many important concepts related to mathematics can be introduced to young children. The topics included in this section introduce counting (forward and backward), number concepts, geometry, and measurement. As important as the selection of topics is the manner in which they are presented. To match the child's developmental level, books included here present topics in a highly visual and physical way. Often, teachers will find that the books serve as a springboard for developing related classroom activities.

Two contexts appear most frequently in these early childhood books. One is the familiar environment: people, places, and things that are likely to be well known to young children. The other is the world of fantasy: talking animals, castles and princesses, and magic powers. Both contexts are likely to engage children so that they enjoy learning mathematics now and for years to come.

Several early childhood books are available in Spanish as well as in English. When available, the Spanish title is included.

Counting Books

The introduction of counting activities represents an important part of the early childhood mathematics curriculum. Children learn the sounds of number names, begin to associate them with a quantity, and hear them in a counting sequence. Hearing number names repeated in stories and rhymes, counting the objects on a page, and singing the counting sequences are inviting ways to familiarize children with numbers. Books included in this section introduce numbers up to twenty.

Anno's Counting Book, by Mitsumoso Anno (New York: HarperCollins, 1986).
A beautifully illustrated counting book of numbers from one to twelve.

Count-A-Saurus, by Nancy Blumenthal (Riverside, NJ: Four Winds, 1989).
Illustrations of prehistoric animals and numbers combine in a counting rhyme.

Counting Book 1 to 10, by Cyndy Szekeres (Racine, WI: Western).
A sturdy counting book. This book is most appropriate for children between the ages of two and five.

Count on Calico Cat, by Donald Charles (Chicago: Children's Press, 1974).
This book is also available in Spanish: *Cuenta con gato galano.* A counting book with the appealing calico cat character.

Count with Us: It's a Story Tray!, by Charles Reasoner (New York: McClanahan, 1991).
A collection of fourteen miniature board books, each presenting pictures of a different number from zero to ten, as well as +, -, =.

Dancing in the Moon, by Fritz Eichenberg (New York: Harcourt Brace Jovanovich, 1983).
A collection of illustrated counting rhymes using numbers from one to twenty.

Farm Counting Book, by Jane Miller (New York: Simon & Schuster, 1992).
A book featuring clearly presented photographs of farm animals. Each page displays the numerals and the number of animals written in words.

Fish Eyes: A Book You Can Count On, by Lois Ehlert (New York: Voyager, 1990).
Bold, colorful drawings of fish that children can count and associate with the numeral on the page. Smaller dark type, in the lower right corner, invites the introduction of addition when appropriate by including the idea of adding "me" to the current number of fish. A teacher's guide is available from the publisher.

The Fraggles Counting Book, by Harry Ross (Chicago: Children's Press, 1988).
A counting book to ten using groups of Fraggle characters doing familiar activities.

Hidden Numbers, by Stephen Holmes (New York: Harcourt Brace Jovanovich, 1991).
An interactive counting book of numbers to ten. Children can lift the flaps and then count the many sets of hidden objects.

How Many Bugs in a Box?, by David A. Carter (New York: Simon & Schuster, 1988).
A pop-up counting book of brightly colored bugs to count from one to ten. Behind each flap is a descriptive phrase that includes the number of bugs, and pictures of the bugs pop out and can be moved.

How Many Monsters? Learning about Counting, by Joanne Wylie and David Wylie (Chicago: Children's Press, 1985).
This book is also available in Spanish: *¿Cuantos monstruos? Un cuento de numeros.* A story about friendly monsters is used to introduce concepts of counting.

Magic Monkey: A Fun Book of Numbers, by Neil Morris (Minneapolis: Carolrhoda, 1990).
A counting book to ten centered on the adventures of a monkey.

A More or Less Fish Story: Learning about Counting, by Joanne Wylie and David Wylie (Chicago: Children's Press, 1984).
This book is also available in Spanish: *Un cuento de peces, mas o menos.* A funny story about fish is used to introduce counting.

Mother Earth's Counting Book, by Andrew Clements and Lonni Sue Johnson (New York: Picture Book Studio, 1992).
A colorful counting book that blends mathematics with art and science.

My First Book of Counting, by Chuck Murphy (New York: Scholastic, 1992).
This book uses a lift-the-flap format to reveal pictures to count.

My First 1 2 3 Book, by Sebastian Conran (Riverside, NJ: Aladdin, 1988).

Brightly colored dots are shown on each page so that children can associate the numeral with the number of objects.

Nicky, 1–2–3, by Cathryn Falwell (New York: Clarion, 1991).
A counting book presenting pictures of familiar objects.

One Big Bear, by Dick Dudley and Keith Moseley (New York: Barron, 1988).
A pop-up counting storybook that introduces counting from one to ten.

One, Five, Many, by Kveta Pacovska (New York: Clarion, 1990).
A colorful counting book of numbers from one to ten, presenting both the numbers and pictures of objects to count. Flaps, pullouts, a mirror, and punch-out counting pieces make this book ideal for active learning.

1 Is One, by Tasha Tudor (Riverside, NJ: Macmillan, 1956).
A counting book using verse and charming pictures.

One Little Elephant, by Colin West (Chicago: Children's Press, 1988).
A counting rhyme.

One Red Rooster, by Kathleen Sullivan Carroll (Boston: Houghton Mifflin, 1992).
A boldly colored counting and rhyming book that displays numbers to ten and animals to count. On each double page, as the next number is introduced, pictures of the previous numbers of animals are shown in the background. In this way, as children move from one number to the next, they can re-count the previous number of animals.

1 2 3, by Sara Lynn and Rosalinda Kightley (Boston: Little, Brown, 1986).
A colorfully illustrated counting book.

1 2 3, by Jan Pienkowski (New York: Simon & Schuster, 1987).
A miniature board counting book of bold illustrations of toys and animals to ten.

1, 2, 3, by Yoshi (New York: Picture Book Studio, 1991).

A beautifully painted counting book to ten with animals to count for each number. An animal stampede creates a compelling sense of drama as well as an opportunity for forward counting.

One, Two, Three: An Animal Counting Book, by Marc Brown (Boston: Little, Brown, 1976).
Each page displays the numeral, written number, and countable animals to represent the integer in an unambiguous way. Numbers to twenty are included.

1 2 3 I Can Count, by Lynn N. Grundy (Auburn, ME: Ladybird Books).
A colorfully illustrated counting book for numbers from one to ten. Each double page presents the numeral, written number name, and boldly colored picture of counting objects.

One Yellow Lion, Matthew Van Fleet (New York: Penguin, 1992).
A foldout counting book of animals and colors, to the number ten. The numbers and corresponding drawings are unambiguously presented, which makes this book ideal to use when initially introducing numbers.

Percival's Party: A Story about Numbers, by Julia Hynard (Chicago: Children's Press, 1983).
Drawings of animal characters are used to introduce numbers.

Richard Scarry's Counting Book, by Richard Scarry (Racine, WI: Western, 1990).
A beautifully illustrated counting book.

Teddy Bears 1 to 10, by Susanna Gretz (Riverside, NJ: Four Winds, 1986).
Counting to ten with teddy bear pictures.

Ten for Dinner, by Jo Ellen Bogart (New York: Scholastic, 1989).
A birthday-theme counting book of children, to groups of ten, not in sequential order.

Ten Little Animals, by Laura J. Coats (Riverside, NJ: Macmillan, 1990).
A story of a boy's ten stuffed animals who insist on jumping on his bed at bedtime.

Ten Little Ducks, by Franklin Hammond (New York: Scholastic, 1987).
A counting book to ten of ducks engaged in familiar activities.

This Old Man, by Tony Ross (Riverside, NJ: Aladdin, 1990).
A counting board book with lively drawings to illustrate the counting song.

Who Wants One?, by Mary Serfozo (Riverside, NJ: Margaret McElderry, 1989).
A little girl, dressed as a magician, leads her younger brother through the numbers one to ten.

Counting Backward

Once children are comfortable with the names of numbers and the sequence of counting numbers forward, they can be introduced to the pattern of counting backward. Children should be encouraged to count the objects, animals, or people presented in the illustrations.

Bea's 4 Bears, by Martha Weston (New York: Clarion, 1992).
A counting-backward and then -forward story about Bea and her four bears. During the course of the story, she loses her bears, one at a time, then finds them.

Five Little Monkeys Sitting in a Tree, by Eileen Christelow (New York: Clarion, 1991).
A well-illustrated counting backward story from five. Five monkeys, on a picnic with their mother, fall into the river and try to avoid tangling with the crocodile.

How Many Feet in the Bed?, by Diane J. Hamm (New York: Simon & Schuster, 1991).
A little girl counts the feet in her parents' bed as more and more people climb into it and then leave. Includes numbers to ten.

Mouse Count, by Ellen Walsh (New York: Harcourt Brace Jovanovich, 1991).
A counting-forward and -backward book, using the numbers from one to ten. Ten mice are captured by a snake and then escape.

Ten Little Lambs, by Dick Dudley and Keith Moseley (New York: Barron, 1988).

A pop-up counting storybook that introduces counting down from ten to one as well as the concept of opposites.

Ten Little Mice, by Joyce Dunbar (New York: Harcourt Brace Jovanovich, 1990).
A reverse counting book illustrated with drawings of countable mice in a complex nature scene. Because of the complexity of the illustrations, this book is most appropriate once children are familiar with the number–object correspondence. A teacher's guide is available from the publisher.

Number Books

The association between number and quantity, as well as the preliminary introduction of arithmetic number operations—such as addition and subtraction—can begin once children become familiar with number names. The books included in this section introduce beginning number concepts in a variety of ways. Carefully designed illustrations and pictures accurately represent mathematical relationships and appeal to children. Many teachers find it best to teach these important early number concepts using a variety of formats, to reach as wide a range of children as possible.

Magic Monkey: A Fun Book of Numbers, by Neil Morris (Minneapolis: Carolrhoda, 1991).
An easy-to-read text with beautiful illustrations introduces ideas of numbers.

1 + 1 Take away Two!, by Michael Berenstain (Racine, WI: Western).
A picture book that introduces the concept of subtraction.

The Twins, Two by Two, by Catherine Anholt and Laurence Anholt (Cambridge, MA: Candlewick, 1992).
Real and fantasy worlds are engaged in a story about imaginative twins at bedtime, to demonstrate the notion of "two."

What Comes in 2's, 3's, and 4's?, by Suzanne Aker (New York: Simon & Schuster, 1990).
A counting book with pictures of familiar objects to demonstrate the numbers two, three, and four.

Geometry

Our environment is comprised of geometric shapes. Young children enjoy recognizing shapes in their everyday life. Pictures and drawings included in many of the books listed feature children's environments and highlight geometric shapes found in them. Teachers can develop related lessons to help children begin to describe the characteristics of shapes and to compare and contrast different shapes.

Blue Bug's Treasure, by Virginia Poulet (Chicago: Children's Press, 1976).
A beautifully illustrated picture book about sizes and shapes.

Calico Cat Looks at Shapes (formerly *Calico Cat Looks Around*), by Donald Charles (Chicago: Children's Press, 1975).
This book is also available in Spanish: *Mira las formas con gato galano*. A book about recognizing geometric shapes with Calico Cat.

Circles, by Mavis Smith (Boston: Little, Brown, 1991).
A board book of pictures that transform circular shapes into new objects.

Crescents, by Mavis Smith (Boston: Little, Brown, 1991).
A board book of pictures that transform crescent shapes into new objects.

A Fishy Shape Story: Learning about Shapes, by Joanne Wylie and David Wylie (Chicago: Children's Press, 1984).
This book is also available in Spanish: *Un cuento de peces y sus formas*. A funny story about fish is used to introduce ideas about shapes.

Muppet Babies Shape Machine, by Bonnie Worth (Chicago: Children's Press, 1988).
A shape identification book.

Rummage Sale: A Fun Book of Shapes and Colors, by Neil Morris (Minneapolis: Carolrhoda, 1991).
An easy-to-read illustrated text introduces ideas of shapes and colors.

Shapes, by Sara Lynn and Rosalinda Kightley (Boston: Little, Brown, 1986).
A colorful collection of pictures of shapes.

Shapes, by John J. Reiss (Riverside, NJ: Bradbury, 1974).
A colorfully illustrated book of shapes including the oval, circle, and triangle.

The Shapes Game, by Paul Rogers (New York: Henry Holt, 1990).
A collection of riddles that are fun to say, along with boldly colored graphics to introduce the shapes.

Shapes in Nature, by Judy Feldman (Chicago: Children's Press, 1991).
A book of nature pictures that reveal geometric shapes.

Shape Space, by Cathryn Falwell (New York: Clarion, 1992).
A young gymnast dances her way among rectangles, triangles, circles, and squares for a lesson in geometry. This book is colorfully illustrated.

Spirals, Curves, Fanshapes and Lines, by Tana Hoban (New York: Greenwillow, 1992).
A collection of beautiful photographs that highlight these geometric shapes.

Squares, by Mavis Smith (Boston: Little, Brown, 1991).
A board book of pictures that transform square shapes into new objects.

Surprise, Surprise, by Paul Rogers (New York: Scholastic, 1992).
A rhyming-text and lift-the-flap format create an interactive way of introducing counting, shapes, sizes, and colors.

Triangles, by Mavis Smith (Boston: Little, Brown, 1991).
A board book of pictures that transform triangular shapes into new objects.

Measurement

Questions such as "Who is bigger?" and "How much later can I stay up?" have an important role in a young child's life. Measurement concepts can be introduced to children utilizing trade books to create a meaningful and playful environment.

A Big Fish Story: Learning about Size, by
  Joanne Wylie and David Wylie (Chicago:
  Children's Press, 1983).
This book is also available in Spanish: *Un
cuento de un pez grande*. A funny story about
fish introduces concepts of size.

Calico Cat at School, by Donald Charles
  (Chicago: Children's Press, 1981).
A day at school in the life of Calico Cat shows
familiar activities with the time presented on a
clock.

*Do You Know Where Your Monster Is To-
  night?: Learning about Time*, by Joanne
  Wylie and David Wylie (Chicago: Children's
  Press, 1984).
This book is also available in Spanish: *¿Sabes
donde esta tu monstruo esta noche? Un cuento
sobre la hora*. A story about friendly monsters
is used to introduce concepts of time.

Holly and Harry: A Fun Book of Sizes, by Neil
  Morris (Minneapolis: Carolrhoda, 1991).
A whimsical and easy-to-read text introduces
ideas of size.

Linda's Late: A Fun Book of Time, by Neil
  Morris (Minneapolis: Carolrhoda, 1991).
A playful and easy-to-read text introduces
ideas of time.

Little Monster: Learning about Size, by Joanne
  Wylie and David Wylie (Chicago: Children's
  Press, 1985).
This book is also available in Spanish: *El
pequeño monstruo: Un cuento de tamaños*. A
story about friendly monsters is used to
introduce concepts of size.

Elementary School

Children in elementary school generally have
developed informal ideas about mathematics
from their observations and experiences in the
world (Ginsburg 1989). Some of these ideas
may be more accurate than others. The
classroom teacher has the opportunity to assess
the accuracy of children's informal mathematics
information, to correct misconceptions, and to
teach correct ideas. The books listed may prove
fruitful vehicles both for assessing children's

knowledge and for extending it. The topics
included in this section are counting (forward
and backward), number concepts, geometry,
measurement, time, money, problem solving
and puzzles, and reference books.

Counting Books

Elementary school children enjoy reading
counting books. Those books that are most
appealing often introduce large numbers and
provide children with an opportunity to associate numbers with objects as well as to use a
number to name a set of objects (ten elephants).
Counting forward, or counting on, forms the
basis for the development of addition skills.
High-interest graphics can include the familiar,
the fantastic, and the perceptually challenging
("Find the number hidden in this picture").
Classroom activities can extend the material
and themes presented in these books.

Counting can provide an opportunity for
planning interdisciplinary lessons. One series of
counting books introduces numbers as spoken
in several countries throughout the world.
These books can become the basis for combined mathematics and social studies lessons
about people in other countries. The idea that
everyone counts, but that the sounds of that
counting can be different, may be an important
aspect of children's multicultural education.

Count and See, by Tana Hoban (New York:
  Macmillan, 1972).
A photographic counting book that includes the
numeral, word, corresponding set of dots, and
pictures of the objects to be counted.

Count on Clifford, by Norman Bridwell (New
  York: Scholastic, 1987).
A counting book featuring the popular character
Clifford the Big Red Dog. This book is most
suitable for students through grade three.

Count Your Way through the Arab World, by
  Jim Haskins (Minneapolis: Carolrhoda,
  1987).
This book combines counting with the culture
and history of the Arab world. Students through
grade four will enjoy this book.

Count Your Way through Africa, by Jim
  Haskins (Minneapolis: Carolrhoda, 1989).

This book combines counting with the culture and history of Africa. Students through grade four will enjoy this book.

Count Your Way through Canada, by Jim
   Haskins (Minneapolis: Carolrhoda, 1989).
This book combines counting with the culture and history of Canada. Students through grade four will enjoy this book.

Count Your Way through China, by Jim
   Haskins (Minneapolis: Carolrhoda, 1987).
This book combines counting with the culture and history of China. Students through grade four will enjoy this book.

Count Your Way through Germany, by Jim
   Haskins (Minneapolis: Carolrhoda, 1990).
This book combines counting with the culture and history of Germany. Students through grade four will enjoy this book.

Count Your Way through India, by Jim Haskins
   (Minneapolis: Carolrhoda, 1990).
This book combines counting with the culture and history of India. Students through grade four will enjoy this book.

Count Your Way through Israel, by Jim
   Haskins (Minneapolis: Carolrhoda, 1990).
This book combines counting with the culture and history of Israel. Students through grade four will enjoy this book.

Count Your Way through Italy, by Jim Haskins
   (Minneapolis: Carolrhoda, 1990).
This book combines counting with the culture and history of Italy. Students through grade four will enjoy this book.

Count Your Way through Japan, by Jim
   Haskins (Minneapolis: Carolrhoda, 1987).
This book combines counting with the culture and history of Japan. Students through grade four will enjoy this book.

Count Your Way through Korea, by Jim
   Haskins (Minneapolis: Carolrhoda, 1989).
This book combines counting with the culture and history of Korea. Students through grade four will enjoy this book.

Count Your Way through Mexico, by Jim
   Haskins (Minneapolis: Carolrhoda, 1989).
This book combines counting with the culture and history of Mexico. Students through grade four will enjoy this book.

Count Your Way through Russia, by Jim
   Haskins (Minneapolis: Carolrhoda, 1987).
This book combines counting with the culture and history of Russia. Students through grade four will enjoy this book.

From One to One Hundred, by Teri Sloat (New
   York: Dutton, 1991).
A beautifully illustrated counting book of selected numbers to one hundred. The key, at the bottom of the page, indicates how many of which pictures can be found on the page.

How Much Is a Million?, by David M.
   Schwartz (New York: Scholastic, 1987).
A magician introduces students to the world of large numbers. This book is most appropriate for students through grade four, and is also available in Big Book format.

If You Made a Million, by David M. Schwartz
   (New York: Scholastic, 1989).
This beautifully illustrated book introduces the concept of large numbers. This book is most appropriate for students through grade four. A teacher's guide is available.

Numbers, by John J. Reiss (Riverside, NJ:
   Bradbury, 1971).
A colorfully illustrated counting book from one to twenty, then ten to one hundred, and then one thousand.

Out for the Count, by Katherine Cave (New
   York: Simon & Schuster, 1991).
A counting book about an adventurous dream, in rhyme, with illustrations of "creatures" (pythons, penguins, and bats) in numbers to one hundred.

Over in the Meadow, by Olive A. Wadsworth
   (New York: Scholastic, 1985).
An Appalachian counting rhyme featuring animal mothers and babies. Both the book and the cassette are available. This book is most appropriate for students through grade two.

Pigs from 1 to 10, by Arthur Geisert (Boston: Houghton Mifflin, 1992).
A story, counting, and puzzle book using the numbers zero to nine. On each double page, all ten piglets are hidden in the etchings.

26 Letters and 99 Cents, by Tana Hoban (New York: Scholastic, 1987).
A beautiful book of color photographs showing the alphabet and money to be counted. This book is most appropriate for students through the second grade.

Counting Backward

Once children are familiar with the counting-forward pattern, counting backward can be introduced. Counting backward is conceptually related to subtraction.

Roll Over!, by Merle Peek (New York: Clarion, 1981).
A backward counting book based on the song of ten sleepy animals falling, one by one, out of bed. The written music is included.

Ten, Nine, Eight, by Molly Bang (New York: Penguin, 1985).
A colorfully illustrated counting-backward story of a girl at bedtime. This book is most suitable for children through grade two and is available in a Big Book format.

Waiting for Sunday, by Carol Blackburn (New York: Scholastic, 1986).
The story of a little boy counting down to his birthday is used to introduce counting backward. This book is most suitable for children through the second grade.

Number Concepts

Introduction to the concepts underlying number operations is crucial for children's success in mathematics computation and problem solving. Many of the books included in this section highlight mathematical relationships in a clear way.

Addition Annie, by David Gisler (Chicago: Children's Press, 1991).
An illustrated storybook introducing addition for beginning readers.

Bunches and Bunches of Bunnies, by Louise Mathews (New York: Scholastic, 1978).
An introduction to multiplication using collections of bunnies holding sets of familiar objects.

A Cache of Jewels, by Ruth Heller (New York: Scholastic, 1991).
A book of collective nouns (a "parcel" of penguins; a "gam" of whales). Although this book is not, strictly speaking, a mathematics book, it introduces descriptive language about groups. Teachers may choose to use the material in this type of book to develop interdisciplinary lessons including mathematics. This book is also available in Big Book format.

Colors, Shapes, Words, and Numbers, by Alan Snow (Chicago: Children's Press, 1989).
A large book introducing numbers, addition, subtraction, and shapes with colorful illustrations.

The Doorbell Rang, by Pat Hutchins (New York: William Morrow, 1989).
The story of sharing grandma's cookies among friends as more and more arrive is used to introduce division.

Eating Fractions, by Bruce McMillan (New York: Scholastic, 1992).
Pictures of fractional parts are shown using photographs of food. Recipes are included. Children through third grade are most likely to enjoy this book.

Henry and the Boy Who Thought Numbers Were Fleas, by Marjorie Kaplan (Riverside, NJ: Four Winds, 1991).
A story about a city dog who has attended college mathematics classes. The dog helps a boy from Indiana who is having difficulty with multiplication. This book is most suitable for students in grades four through six.

Mathematics, by Irving Adler (New York: Doubleday, 1990).
This book presents in a lively way the relationships between mathematics and nature, art and music. This book also includes mathematical games and computer activities.

Me + Math = Headache, by Lee Wardlaw
(Summerland, CA: Red Hen, 1986).
A story about a third-grade boy who does not
see the usefulness of math. Through the course
of his experiences, he comes to recognize
math's utility in his daily life.

My First Number Book, by Marie Heinst (New
York: Dorling Kindersley, 1992).
A beautifully illustrated book of counting,
sorting, matching, and comparing number
activities. These activities are used to introduce
number concepts.

Number Art, by Leonard E. Fisher (Riverside,
NJ: Four Winds, 1982).
An informative picture book presenting the
history and design of numerals for thirteen
notation systems.

Numbers, by Philip Carona (Chicago: Children's Press, 1982).
An illustrated book intended to answer students' questions about numbers.

One, Two, One Pair!, by Bruce McMillan
(New York: Scholastic, 1992).
A collection of color photographs is used to
explore the idea of pairs. This book is most
appropriate for children through second grade.

One, Two, Three, and Four. No More?, by
Catherine Gray (Boston: Houghton Mifflin,
1988).
Verse is used to introduce addition and
subtraction concepts using the numbers one
through four.

Roman Numerals, by David Adler (New York:
HarperCollins, 1977).
This book presents an introduction to Roman
numerals.

Rooster's Off To See the World, by Eric Carle
(New York: Picture Book Studio, 1987).
A story about animals is used to introduce
mathematics concepts. This book is most
appropriate for use by children through third
grade.

Sea Squares, by Joy N. Hulme (New York:
Hyperion, 1991).
A beautifully illustrated book, with a sea
theme, that introduces the concept of squared
numbers.

The Stingy Baker, by Janet Greeson
(Minneapolis: Carolrhoda, 1990).
An illustrated retelling of a story accounting for
the origins of the "baker's dozen."

Tillie Tiger's Times Tables, by Pat Paris (New
York: Barron, 1990).
This book provides an opportunity for students
to practice multiplication facts to one hundred.
The student can move a wheel, which is bound
inside each page, to see a new problem.

Zero: Is It Something? Is It Nothing?, by
Claudia Zaslavsky (New York: Franklin
Watts, 1989).
A lively book presenting many demonstrations
of the concept of zero. For example, there are
zero elephants at a child's birthday party, zero
can be important in a street address, and
astronauts count down to zero.

Geometry

One of the activities that elementary school-
age children find most exciting is identifying
shapes in their environment. As children's
abilities to conserve quantities become more
developed, they realize that shapes do not
necessarily change when they are viewed from
a different perspective, or even when their
position is changed. This leads to the pleasure
of recognizing a familiar object in an unfamiliar perspective. Books such as *Topsy-Turvies:
Pictures To Stretch the Imagination*, by
Mitsumoso Anno, provide the opportunity to
find shapes in various environments. *Tangrams:
330 Puzzles*, by Ronald Read, encourages
children to rearrange tangram shapes to create
new designs.

Circles, Triangles, and Squares, by Tana
Hoban (Riverside, NJ: Macmillan, 1974).
A collection of photographs of familiar children's activities is used to demonstrate basic
geometric concepts.

Fun with Math: Shapes and Solids, by Lakshmi
Hewavisenti (New York: Gloucester, 1991).

A book of activities and games using geometry. For example, several shapes are traced onto graph paper to compare their sizes.

If You Look around You, by Fulvio Testa (New York: Dial, 1987).
This beautifully illustrated book identifies shapes in drawings of various aspects of children's environments.

Lines, by Philip Yenawine (New York: Delacorte, 1991).
This book explores the variety of lines in paintings. It provides an opportunity to develop an interdisciplinary lesson that integrates mathematics and art.

Magic Tricks for Children, by Len Collis (New York: Barron, 1989).
A collection of instructions for performing optical illusions.

Pancakes, Crackers, and Pizza: A Book of Shapes, by Marjorie Eberts and Margaret Gisler (Chicago: Children's Press, 1984).
An illustrated easy-to-read book about shapes. It is most suitable for students through second grade.

Shapes, by Philip Yenawine (New York: Delacorte, 1991).
This book uses paintings to help students appreciate that different shapes elicit different responses. It provides an opportunity to develop an interdisciplinary lesson that integrates mathematics and art.

Shapes, Shapes, Shapes, by Tana Hoban (New York: Greenwillow, 1986).
A collection of photographs highlighting geometric shapes in our environment.

Tangrams: 330 Puzzles, by Ronald Read (New York: Dover, 1978).
A collection of challenging problem-solving pattern puzzles and their solutions, constructed using tangram pieces.

Topsy-Turvies: Pictures To Stretch the Imagination, by Mitsumoso Anno (New York: Philomel, 1989).
A beautiful collection of perception-challenging drawings.

Measurement

As children's conservation and number skills develop, they utilize them in measurement situations. The books in this section introduce children to measuring quantities to solve a problem, following directions, reading maps, and comparing the sizes of objects. A teacher may wish to create related classroom activities to extend children's understanding of these measurement concepts. Many measurement topics lend themselves to interdisciplinary units. For example, map reading provides an opportunity to develop a mathematics–social studies unit.

Bear's New House, by Annie Cobb (Westwood, NJ: Silver Press, 1991).
An illustrated storybook that integrates measurement, following plans, and building to scale to solve problems.

Detective Duckworth to the Rescue, by Annie Cobb (Westwood, NJ: Silver Press, 1991).
An illustrated story introducing map-reading skills to solve problems.

Fun with Math: Measuring, by Lakshmi Hewavisenti (New York: Gloucester, 1991).
A book of imaginative activities and games using measurement. For example, the construction of a scale to measure paper clips is made from two thread spools and a straw.

Mouse's Birthday Party, by Annie Cobb (Westwood, NJ: Silver Press, 1991).
An illustrated story introducing map-reading skills (looking for landmarks) to solve problems.

Squirrel's Treasure Hunt, by Annie Cobb (Westwood, NJ: Silver Press, 1991).
An illustrated storybook including direction-following skills to solve problems about a scavenger hunt.

Time

Telling time is one of the most challenging and motivating tasks for an elementary school–age child. Children are introduced to the calendar in kindergarten, and to telling time in various stages (to the hour, half and quarter hour, and minutes), using both analog and digital clocks after that. Several books included in this list describe the history of the calendar and timekeeping devices. Others are storybooks that children can read to appreciate the importance of being able to tell time and to develop time-telling skills.

Brendan's Best-Timed Birthday, by Deborah Gould (Riverside, NJ: Macmillan, 1988).
This book tells the story of a boy who receives a digital stopwatch for his birthday and uses it to time the last-minute party preparations.

Calendar Art, by Leonard E. Fisher (Riverside, NJ: Four Winds, 1987).
This book describes how different timetables were developed and used to calculate days, weeks, months, and years.

Clocks! How Time Flies, by Siegfried Aust (Minneapolis: Lerner Books, 1991).
This book describes the history of timekeeping devices and provides instructions for making them.

The Cuckoo-Clock Cuckoo, by Annegert Fuchshuber (Minneapolis: Carolrhoda, 1988).
The twenty-four–hour adventures of a curious cuckoo are accompanied by clocks showing the time.

My First Book of Time, by Claire Llewellyn (New York: Dorling Kindersley, 1992).
Beautiful photographs, games, puzzles, questions, and answers can introduce concepts related to telling time and the seasons. A glossary of terms and a foldout clock with movable hands are included.

Ripley's Believe It or Not! Mind Teasers: Hours, Days, and Years, by Robert R. Ripley (Mankato, MN: Capston, 1991).
A book of amazing facts about clocks, calendars, and time presented in a cartoon format. The concepts and language are most appropriate for students in grades four and above.

Time, by Jan Pienkowski (New York: Simon & Schuster, 1980).
A miniature board book that shows analog clock times and drawings of what occurs at that time. This book is most appropriate for use with young elementary children.

Time, by Feenie Ziner and Elizabeth Thompson (Chicago: Children's Press, 1982).
A well-illustrated book that presents explanations about confusing concepts related to time.

Money

Children at very young ages learn to appreciate the value of money. Nonetheless, they often become confused about the relative value of coins. It is, after all, counterintuitive that a small coin like a dime should be worth more than a larger nickel or penny. The books in this section are intended to clarify some of the confusions about money.

Alexander, Who Used To Be Rich Last Sunday, by Judith Viorst (New York: Atheneum, 1989).
A story about Alexander and his money. Although he loses his money quickly, he learns to appreciate the value of money.

Money, by Benjamin Elkin (Chicago: Children's Press, 1983).
A well-illustrated book that presents explanations for some of the confusing concepts related to money.

Money Madness$$! Garfield Learns about Money, by Jim Davis (Racine, WI: Western, n.d.).
An introduction to the value and use of money based on the popular Garfield character.

Problem Solving and Puzzles

What could be more fun than solving puzzles that use recently learned skills? Puzzle books allow children to practice the skills that they have acquired in a relaxing yet challenging way. Sometimes, students enjoy solving puzzles so much that they think that mathematics means playing games! From a teacher's perspective, problem solving continues to be one of the most important goals of mathematics learning.

Anno's Math Games, by Mitsumoso Anno (New York: Philomel, 1987).
A beautifully drawn collection of mathematics puzzles.

Anno's Math Games II, by Mitsumoso Anno (New York: Philomel, 1989).

Another volume of beautifully drawn mathematics puzzles and mazes.

Anno's Math Games III, by Mitsumoso Anno (New York: Philomel, 1991).
A beautifully illustrated collection of mazes and puzzles intended to encourage thinking and abstracting.

Card Games for Children, by Len Collis (New York: Barron, 1989).
A collection of instructions for playing forty-one card games.

Fun with Math: Counting, by Lakshmi Hewavisenti (New York: Gloucester, 1991).
A book of activities and games using addition and subtraction. For example, children are encouraged to add the digits in a telephone number or find pairs of playing cards that add to ten.

Fun with Math: Problem Solving, by Lakshmi Hewavisenti (New York: Gloucester, 1991).
A book of creative activities and games using problem solving. For example, an activity using fractions instructs students to fill several glasses with different heights of water and then listen to the music the glasses make when tapped.

Reference

An assortment of mathematics reference books is available. Children can use these to explore answers to their own questions, as well as to find answers to questions teachers ask. A biography of Albert Einstein, a scientist who used mathematics in his career, is also included. Teachers may choose to include mathematics-related biographies when introducing a unit on the lives of famous people.

Albert Einstein, by Ibi Lepscky (New York: Barron, 1982).
A biography of the childhood of a scientist who used mathematics in his career.

First Math Dictionary, by Richard W. Dyches and Jean M. Shaw (New York: Franklin Watts, 1991).
A colorfully illustrated dictionary of mathematics terms, with simple verbal as well as visual definitions and examples of the concepts.

The Information Please Kids' Almanac, by Alice Siegel and Margo McLoone Basta (Boston: Houghton Mifflin, 1992).
A collection of difficult-to-find information and fun facts that can be used as the basis of statistical activities.

Mathematics Encyclopedia, by Leslie Foster (Chicago: Rand McNally, 1986).
A colorfully illustrated collection of stories and folktales that describe mathematical principles.

Math Yellow Pages, by Marjorie Frank (Nashville, TN: Incentive Publications, 1988).
An alphabetical list of explanations and definitions of mathematical terms and concepts. Measurement-conversion charts and skills checklists are also included.

What Are You Figuring Now? A Story about Benjamin Banneker, by Jeri Ferris (Minneapolis: Carolrhoda, 1988).
The biography of an eighteenth-century African American who used mathematics to survey land and predict eclipses.

Women in Space: Reaching the Last Frontier, by Carole S. Briggs (Minneapolis: Lerner, 1988).
A biography of Dr. Sally K. Ride, the first woman astronaut, who used mathematics in her career.

Middle School

Middle school students have met and mastered many ideas of arithmetic. Their interests now turn to refining these ideas and appreciating when and how to use their skills. Therefore, books included in this section reflect a shift to more applied topics. The topics included in this section are number concepts, probability and statistics, geometry, money, problem solving and puzzles, and reference material.

One of the challenges for students in the middle school grades is to appreciate their unique capabilities. Mathematics in particular seems to be one area where many students lose self-confidence. To address this, teachers can introduce a wide variety of mathematical skills and ideas that will enable students to recognize and appreciate their abilities.

Many of the books included in this section may also be appropriate for secondary school students.

Number Concepts

Middle school students are generally familiar with arithmetic operations. Discovering why these familiar operations "work" can expand their understanding of mathematics.

Fascinating Fibonaccis, by Trudi H. Garland (Palo Alto, CA: Seymour Publications, 1987).
This book demonstrates the existence of the Fibonacci sequence of numbers in nature. Some of the mathematical properties of Fibonacci numbers are discussed.

Number Treasury, by Stanley Bezuszka and Margaret Kenney (Boston: Boston College Mathematics, 1971).
A description of various number-theory issues, such as prime and composite numbers, and digital patterns.

The Phantom Tollbooth, by Norton Juster (New York: Knopf, 1988).
Travel with Milo through the fantastic Kingdom of Wisdom and see what happens in Dictionopolis, where only words matter, and in Digitopolis, where numbers prevail. This book may also be appropriate for upper elementary students as well as secondary students.

Probability and Statistics

Statistics provide another description of numerical information. Assessing their meaning is as important as calculating the relationships. Games of chance are an interesting application of probability theory.

Do You Wanna Bet?, by Jean Cushman (New York: Clarion, 1991).
Several concepts of chance are introduced through scenes from the everyday lives of two children at home and at school. They roll dice, make predictions about the weather, participate in a raffle, examine baseball statistics, and play cards to understand probability. A bibliography of additional books for students and teachers is included.

Reading the Sports Page: A Guide to Understanding Sports Statistics, by Jeremy R. Feinberg (New York: New Discovery, 1992).
An explanation of the many uses of statistics in sports, including how to read and interpret charts and tables to understand players' averages and scores.

Geometry

Middle school students are intrigued by visual illusions and perceptual contradictions. The graphics of M. C. Escher provide surprising visual challenges, as do the geometric designs found in Islamic art.

Adventures with Impossible Figures, by Bruno Ernst (New York: Parkwest Publications, 1987).
A fantastic collection of photographs and drawings of figures that could never exist in the real world.

Arabic Geometrical Pattern and Design, by J. Bourgoin (New York: Dover, 1973).
This book presents pictures of many patterns based on geometric figures that are found in Islamic art.

Can You Believe Your Eyes?, by J. R. Block and H. E. Yuker (New York: Gardner, 1989).
This book displays and then explains more than 250 visual illusions and visual oddities.

The Eighth Book of Tan, by Sam Loyd (New York: Dover, 1968).
A collection of tangram puzzles and solutions.

Geometric Concepts in Islamic Art, by Issam El-Said and Ayse Parman (Essex, UK: Scorpion Publishers, 1990).
A collection of pictures that explain the geometric basis of Islamic art.

A History of Pi, by Petr Beckman (Boulder, CO: Golem, 1977).
This book describes the historical background of pi. The discussions reveal the social and historical conditions during which its precision progressed and when it did not.

The Magic Mirror of M. C. Escher, by Bruno Ernst (New York: Parkwest Publications, 1987).
A collection of the drawings and etchings of M. C. Escher, which display many visual contradictions. Biographical information about Escher is also included.

Money

One high-interest application of mathematics for middle school students is money and its uses. These books are ideal for developing interdisciplinary mathematics–social studies lessons. Several books demonstrate the role of money in economic systems. A novel, from the Archie series, demonstrates the importance of money in a teenager's life.

Economics and the Consumer, by M. Barbara Killen (Minneapolis: Lerner, 1990).
An introduction to the ideas of consumer economics, including chapters about supply and demand, smart shopping, and technology and the consumer.

Eyewitness Books: Money, by Joe Cribb (New York: Alfred A. Knopf, 1990).
A beautifully illustrated collection of brief and intriguing articles about the history of money as well as international forms of money.

The Fortunate Fortunes: Business Successes that Began with a Lucky Break, by Nathan Aaseng (Minneapolis: Lerner, 1989).
Brief stories of the fortuitous development of familiar and popular products, such as Kleenex and Kellogg's Corn Flakes.

From Rags to Riches: People Who Started Businesses from Scratch, by Nathan Aaseng (Minneapolis: Lerner, 1990).
Brief biographies of people who started businesses such as Procter & Gamble, JCPenney, and Hershey.

Good-Bye Millions, by Michael J. Pellowski (New York: Hyperion, 1992).
Veronica Lodge, friend of Archie at Riverdale High, learns about earning money and economizing in this teenage novel.

Money and Financial Institutions, by Jane E. Gungaum (Minneapolis: Lerner, 1990).

A description of financial institutions, such as banks and the Federal Reserve System, as well as an explanation of how the money supply works.

The Problem Solvers: People Who Turned Problems into Products, by Nathan Aasang (Minneapolis: Lerner, 1989).
Brief biographies of inventors who created familiar products such as mass-produced baby food, the dishwasher, and the Jacuzzi.

The Stock Market, by Robin R. Young (Minneapolis: Lerner, 1991).
An introduction to the principles of the stock market, including its history, and an explanation of stocks and bonds.

Taxes and Government Spending, by Andrea Lubov (Minneapolis: Lerner, 1990).
An explanation of important principles governing the American tax system, how governments allocate funds, and the government deficit.

Problem Solving and Puzzles

Number concepts and relationships are effectively mastered through a variety of challenging problems and puzzles. The books listed include problems of several different types. Some will be solved quickly, whereas others will require a larger investment of time and thought. Sometimes, students will want to work together to solve a more difficult problem. This cooperative learning style has several benefits. In addition to learning to work together, students must find the language to communicate their ideas to another person.

The Big Book of Puzzlers, by Karen C. Anderson (New York: Disney, 1992).
This book contains a diverse collection of mathematical puzzles based on Disney film themes.

The Book of Think: Or How To Solve a Problem Twice Your Size, by Marilyn Burns (Boston: Little, Brown, 1976).
The exercises, brain teasers, and puzzles in this book are intended to provide understanding of the various conditions that determine success or failure in problem solving.

Crossmatics: A Challenging Collection of Cross-Number Puzzles, by A. Dudley (Palo Alto, CA: Seymour Publications, 1990).
A selection of problems using a crosswordlike grid for recording the answers.

The I Hate Mathematics Book, by Marilyn Burns (Boston: Little, Brown, 1975.
This book introduces mathematical concepts and terms through the process of experimentation and discovery with puzzles, riddles, magic tricks, and brain teasers.

Mathematical Fun, Games, and Puzzles, by Jack Frohlichstein (New York: Dover, 1962).
A collection of puzzles and games that demonstrate how mathematics is used in everyday life.

Mathematical Challenges for the Middle Grades: From "The Arithmetic Teacher," ed. William D. Jamski (Reston, VA: NCTM, 1990).
A collection of problems and puzzles from *The Arithmetic Teacher* based on topics such as geometry, probability, number theory, and computation.

Mathematical Olympiad Contest Problems for Children, by George Lenchner (New York: Glenwood, 1990).
A collection of 250 problems drawn from the Mathematical Olympiads, covering topics such as series, variation, combinations and permutations, magic squares, and sequences. The problem-solving process is discussed.

Math for Boys, by Carole Marshe (Bath, NC: Gallopade, 1990).
A collection of mathematical stories and puzzles intended to be of high interest to boys. Subjects include the value of birthday presents, girlfriends, a pizza party, and a paycheck. The language is written for a male reader.

Math for Girls, by Carole Marshe (Bath, NC: Gallopade, 1990).
A collection of mathematical stories and puzzles intended to be of high interest to girls. Subjects include rock & roll songs, fudge recipes, a shopping spree, and a social calendar. The language is written for a female reader.

Math for Smarty Pants, by Marilyn Burns (Boston: Little, Brown, 1982).
The mathematical ideas, puzzles, and tricks in this volume are intended to reveal the many different ways in which people can be mathematically capable.

Math Fun: Test Your Luck, by Rose Wyler and Mary Elting (New York: Julia Messner, 1992).
A collection of intriguing problems based on familiar experiences.

Math Fun: With a Pocket Calculator, by Rose Wyler and Mary Elting (New York: Julia Messner, 1992).
A collection of puzzles, problems, and brain teasers that can be fun to solve using a calculator.

Merlin Book of Logic Puzzles, by Margaret C. Edmiston (New York: Sterling Publications, 1992).
A collection of logic problems and their solutions. The theme of the problems is the Merlin legend.

The Mirror Puzzle Book, by Marion Walter (New York: Parkwest Publications, 1985).
A collection of nonverbal puzzles. Mirrors are used to teach reflective symmetry.

New Puzzles in Logical Deduction, by George J. Summers (New York: Dover, 1968).
A collection of fifty logic puzzles and step-by-step solutions. The problems are presented in a mystery-story format.

Paper Capers: An Amazing Array of Games, Puzzles, and Tricks, by Jack Botermans (New York: Henry Holt, 1986).
This book provides suggestions for creating brain teasers and puzzles from familiar shapes and materials.

Sideways Arithmetic from Wayside School, by Louis Sachar (New York: Scholastic, 1989).
A set of fifty hilarious brain teasers that use mathematics skills. This set of problems is appropriate for students in grades three to twelve.

Reference

Middle school students enjoy the independence of searching out answers to questions and discovering as yet "untaught" ideas. This collection of reference books provides the resources for students to investigate mathematics questions as they would in any other content area.

Albert Einstein, by Karin Ireland (Westwood, NJ: Silver Burdett, 1989).
A biography of a famous scientist who used mathematics as a tool in his work.

Buckminster Fuller, by Robert R. Potter (Westwood, NJ: Silver Burdett, 1990).
A biography of a visionary architect who used mathematics in his work.

The Crescent Dictionary of Mathematics, by W. Karush (Palo Alto, CA: Seymour Publications, 1987).
A reference tool providing summaries and illustrations of mathematical concepts and relationships.

Galileo and the Magic Numbers, by Sidney Rosen (Boston: Little, Brown, 1958).
A biography of Galileo that demonstrates the importance of mathematics to his discoveries.

The Information Please Kids' Almanac, by Alice Siegel and Margo McLoone Basta (Boston: Houghton Mifflin, 1992).
A collection of difficult-to-find information and fun facts that can be used as the basis of statistical activities.

Mathematics Encyclopedia, by Leslie Foster (Chicago: Rand McNally, 1986).
A colorfully illustrated collection of stories and folktales that describe mathematical principles.

Math Equals, by Teri Perl (Reading, MA: Addison-Wesley, 1978).
A collection of brief biographies of nine women mathematicians from the fourth through the twentieth centuries.

Math Yellow Pages, by Marjorie Frank (Nashville, TN: Incentive Publications, 1988).
An alphabetical list of explanations and definitions of mathematical terms and concepts. Measurement conversion charts and a skills checklist are also included.

The Silver Burdett Mathematical Dictionary, by R. E. Jason Abdelnoor (Westwood, NJ: Silver Burdett, 1979).
A dictionary of mathematical terms and concepts that uses illustrations and examples to present clear explanations.

Webster's New World/Crescent Mathematics Dictionary, by W. Karush (New York: Prentice Hall, 1989).
A revised and updated version of the *Crescent Dictionary of Mathematics.* In addition to summaries and illustrations of mathematical concepts and relationships, computer terms are included. This book is available in a paperback edition.

Secondary School

Trade books can introduce secondary students to new realms of mathematics. Some of these new areas may stimulate further investigation or have career potential. Students are often surprised to discover the breadth of mathematics. The books included in this section are intended to expand students' awareness of what mathematics can be.

Motivational issues are very important for this age group, particularly with respect to mathematics. Many students, often females, lose self-confidence about their mathematics ability. Once given the opportunity to avoid mathematics courses, they do. Experience and research have shown that this self-exclusion can prevent them from entering many professional fields. Reading trade books may be a nonthreatening way to reestablish self-confidence in mathematics. Students may find themselves interested in topics that they are surprised to discover are mathematics-related.

The topics included in this section are: number concepts, geometry, computer science, money, problem solving, and reference materials. Many of the books included in the middle school section may also be appropriate for secondary students.

Number Concepts

These trade books introduce the secondary student to new mathematical ideas.

Graphs and Their Uses, by Oystein Ore (Washington, DC: Mathematics Association, 1990).
This book discusses many types of graphs and their practical applications (such as scheduling football games) as well as more esoteric applications (such as solving ancient puzzles).

How To Lie with Statistics, by Darrell Huff and Irving Geis (New York: Norton, 1954).
This seasoned volume demonstrates the deceptive possibilities of familiar statistical presentations, and the importance of examining them carefully.

How To Think about Statistics, by John L. Phillips, Jr. (New York: W. H. Freeman, 1988).
This book introduces the logic of statistics, as well as several of its applications.

Invitation to Number Theory, by Oystein Ore (Washington, DC: Mathematics Association, 1967).
This book takes a historical look at the development of number theory. Many principles of number theory are explored.

Number Theory and Its History, by Oystein Ore (New York: Dover, 1988).
This book describes the history and several concepts of number theory. For example, the theory of prime numbers and counting are discussed.

What Is Calculus About?, by W. W. Sawyer (Washington, DC: Mathematics Association, 1961).
This book introduces many of the ideas of calculus. It describes the historical development of mathematics that led to the need for and the development of calculus. Topics such as acceleration, areas, volumes, and integrals are discussed.

Geometry

The books in this section describe applications of geometric principles, as well as the theory underlying geometric relations. Many of the books of visual illusions listed for middle school students may be of interest to secondary students as well.

The Dot and the Line: A Romance in Lower Mathematics, by Juster Norton (New York: Random, 1977).
A sensible straight line, in love with a perfect dot, practices transforming himself into complex shapes to impress the dot and win her love.

An Eye for Fractals: A Graphic and Photographic Essay, by Michael McGuire (Reading, MA: Addison-Wesley, 1990).
A collection of more than one hundred photographs of fractal geometry. Explanations of these fractals are included.

Flatland: A Romance of Many Dimensions, by Edwin A. Abbott (Princeton: Princeton University Press, 1991).
A charming fantasy of our experience of the world if we could perceive only two dimensions instead of three.

Geometric Design, by Dale Seymour (Palo Alto, CA: Seymour Publications, n.d.).
A guide to constructing complex geometric figures from primitive shapes such as triangles, squares, and pentagons.

The Golden Section, by G. E. Runion (Palo Alto, CA: Seymour Publications, 1990).
Demonstrations of the ratio of the golden section in nature, art, and architecture, along with a set of problem-solving activities.

Mind Sights, by Roger N. Shepard (New York: W. H. Freeman, 1990).
A collection of original visual illusions and ambiguous figures.

Pyramid, by David Macaulay (Burlington, MA: Houghton Mifflin, 1975).
With very clear text and illustrations, the step-by-step construction of an Egyptian pyramid is explained. Illustrations show the complete design as well as sectional views. The author has produced a series of similar books dealing with other architectural structures. These books include *Castle, Cathedral, City, Mill, Unbuilding,* and *Underground.*

Sphereland, by Dionys Burger and trans. by Cornelie J. Rheinboldt (New York: HarperCollins, 1983).
A sequel to *Flatland* in which a Hexagon living in a two-dimensional world accepts the concept of a third dimension, explores the possibility that his plane is curved and expanding, and begins to believe that a fourth dimension also exists.

Tilings and Patterns: An Introduction, by Branko Grunbaum and G. C. Shephard (New York: W. H. Freeman, 1986).
This book is an encyclopedic treatment of the history and theory of patterns.

Visual Illusions: Their Causes, Characteristics, and Applications, by Matthew Luckiesh (New York: Dover, 1965).
This book includes explanations of many familiar visual illusions and their uses in lighting, design, camouflage, and architecture.

Computer Science

The area of computer science is an exciting extension of mathematics. Several books have been included to introduce some of the ideas behind computers, as well as some of the people involved in developing this field.

Chaos, Fractals, and Dynamics: Computer Experiences in Mathematics, by Robert L. Devaney (New York: Science TV, 1989).
An introduction, including guided computer graphics experiments (using BASIC), fractals, the Mandelbrot set, and Julia sets.

The Cybernetics Group, by Steve J. Heims (Cambridge, MA: MIT Press, 1991).
A group biography of the cybernetics group that met at MIT in the 1940s and 1950s to try to develop new applications of the current mathematics and science theories for the understanding of human behavior.

Glory and Failure, by Michael Lindgren (Cambridge, MA: MIT Press, 1990).
A nontechnical description of various approaches to the beginning of automated computing. The works of Johann Muller, Charles Babbage, and Jeorg and Edvard Sheutz are described.

John Van Neumann and the Origins of Modern Computing, by William Aspray (Cambridge, MA: MIT Press, 1990).
A biography of the developer of early computer science theory.

Money

Secondary students are aware of the role that money plays in their present lives and in their future. The books included in this section describe various systems in which money is used. Students may be motivated to learn more about these systems and realize the need for some of the mathematics that they are studying in the classroom. Several books identify the issues related to earning money from the perspective of the secondary student.

Banking, by Nancy Dunnan (Westwood, NJ: Silver Burdett, 1990).
This book introduces students to the way in which the banking system works. Concepts such as savings and interest, loans, money management, and different types of investments offered by banks are discussed.

Collectibles, by Nancy Dunnan (Westwood, NJ: Silver Burdett, 1990).
This book describes how to begin collecting "collectibles." Suggestions for selecting a type of collectible, where to look for purchases, and when to sell the collection are included.

College Cash, by Van Hutchinson (New York: Harcourt Brace Jovanovich, 1988).
A practical guide for high school and college students who are interested in starting their own businesses.

Entrepreneurship, by Nancy Dunnan (Westwood, NJ: Silver Burdett, 1990).
This book offers practical advice for turning a hobby into a business. Developing a business plan, getting financing, finding clients, and keeping business records are discussed.

Get Real, by James Tenuto and Susan Schwartzwald (New York: Harvest, 1992).
A financial guide for high school students who are collegebound. Practical information is included, such as how to balance a checkbook, how to apply for a credit card, and how to pay the rent.

The Stock Market, by Nancy Dunnan
    (Westwood, NJ: Silver Burdett, 1990).
This book presents a description of a stock,
explains why stocks are issued, and recommends ways of organizing a financial notebook.
Investment suggestions are made.

Problem Solving

Problem solving can be a challenging and
engaging way of integrating information that
has been learned over time. Frequently, topics
are taught in segmented time frames. Many
problems demand the combination of assorted
bits of information and perspectives to arrive at
a solution. And solving problems and puzzles
can be fun!

Can You Solve These?, by David Wells (New
    York: Parkwest Publications, 1985).
This is a set of three collections of puzzles
from England. The use of a calculator is likely
to be helpful when solving some of the
problems.

The Divine Proportion: A Study of Mathematical Beauty, by H. E. Huntley (New York:
    Dover, 1970).
Descriptions of mathematical relationships in
familiar patterns and shapes, as well as more
esoteric issues, calculations, and problems.

Exploratory Problems in Mathematics, by
    Frederick W. Stevenson (Reston, VA:
    NCTM, 1992).
A collection of open-ended problems and
suggestions on how to approach the solutions.

The Great Book of Math Teasers, by Robert
    Muller (New York: Sterling Publications,
    1989).
A collection of puzzles for people with a strong
mathematics background, as well as for those
without one.

Logic Puzzles, by Ruth Minshull (Northport,
    MI: SAA Publications, 1980).
A group of fifty-one logic problems of
different difficulty levels, with solution hints.

The Master Book of Mathematical Recreations,
    by Fred Schuh (New York: Dover, 1969).
This book explains the mathematics behind
puzzles, games, and card tricks.

Math and Logic Games, by Franco Agostini
    (New York: HarperCollins, 1986).
A collection of puzzles, riddles, paradoxes, and
problems based on the theories of famous
mathematicians.

Mathematical Bafflers, by Angela Dunn (New
    York: Dover, 1980).
A collection of challenging problems that
require facility with algebra, geometry, logic,
and probability.

Mathematical Brain Benders, by Stephen Barr
    (New York: Dover, 1982).
A selection of problems from topics such as
logic, topology, geometry, and origametry. For
several of the problems, some knowledge of
algebra and geometry would be helpful.

Mathematical Puzzling, by A. Gardiner (New
    York: Oxford University Press, 1988).
A collection of approximately two hundred
mathematical puzzles covering a wide range of
topics. For example, problems using Fibonacci
numbers, magic squares, and polygons are in
this book.

Mathematics, Magic, and Mystery, by Martin
    Gardner (New York: Dover, 1976).
This collection of problems uses mathematics
tricks to demonstrate laws of probability,
number theory, and sets.

Mathematics Project Handbook, 3d ed., by
    Adrien Hess, Glenn Allinger, and Lyle
    Anderson (Reston, VA: NCTM, 1989).
A collection of ideas for developing and
evaluating mathematics projects.

Second Book of Mathematical Bafflers, by
    Angela Dunn (New York: Dover, 1983).
Another collection of challenging mathematical
problems for which facility with algebra,
geometry, logic, and probability will be useful.

Reference

This collection of reference materials is
intended to serve several functions. Some of the
materials can be used by students to answer
questions about mathematics concepts. Other
selections may spark the interest of students in
unfamiliar aspects of mathematics, such as

cybernetics and fractals. Still others may reveal to students how mathematicians "do" mathematics and its impact on the life of the mathematician in a professional as well as a personal way.

Ada: A Life and Legacy, by Dorothy Stein (Cambridge, MA: MIT Press, 1985).
A biography of Ada Lovelace, who assisted Charles Babbage in developing the first computing machine.

The Crescent Dictionary of Mathematics, by W. Karush (Palo Alto, CA: Seymour Publications, 1987).
A reference tool providing summaries and illustrations of mathematical concepts and relationships.

Engineering in the Mind's Eye, by Eugene S. Ferguson (Cambridge, MA: MIT Press, 1992).
The author argues that engineering is a combination of intuition and nonverbal thinking as much as of computations and equations.

The Mathematical Tourist: Snapshots of Modern Mathematics, by Ivars Peterson (New York: W. H. Freeman, 1989).
Written for the nonmathematician, this book explores mathematical concepts in a nontechnical way.

Mathematics Appreciation, by Theoni Pappas (San Carlos, CA: Wide World Publishing/ Tetras House, 1988).
This book reveals the relationship of mathematics with other subject areas, as well as the beauty of mathematics. Discussions of topics such as the golden rectangle, non-Euclidean geometry, optical illusions, and magic squares are included.

Mathematics Illustrated Dictionary: Facts, Figures, and People, by Jeanne Bendick (New York: Franklin Watts, 1989).
A comprehensive resource that explains mathematical terms and concepts. Brief biographies of mathematicians are included as well.

Math Equals, by Teri Perl (Reading, MA: Addison-Wesley, 1978).

A collection of brief biographies of nine women mathematicians from the fourth through the twentieth centuries.

The Silver Burdett Mathematical Dictionary, by R. E. Jason Abdelnoor (Westwood, NJ: Silver Burdett, 1979).
A dictionary of mathematical terms and concepts that uses illustrations and examples to present clear explanations.

Stand and Deliver, by Nicholas Edwards (New York: Scholastic, 1988).
A novel based on the true story of an effective high school mathematics teacher who inspired his students to find success in mathematics.

Webster's New World/Crescent Mathematics Dictionary, by W. Karush (New York: Prentice Hall, 1989).
A revised and updated version of the *Crescent Dictionary of Mathematics*. In addition to summaries and illustrations of mathematical concepts and relationships, computer terms are included. This book is available in a paperback edition.

Women in Mathematics, by Lynn M. Osen (Cambridge, MA: MIT Press, 1974).
A collection of biographies of female mathematicians that starts with Hypatia (370–415), goes through the Renaissance, and ends with Emmy Noether (1882–1935). Their stories parallel the growth of mathematics. These biographies reveal the human as well as professional dimensions of their lives.

The World of Mathematics, by James R. Newman (Redmond, WA: Microsoft, 1988).
An accessible four-volume compilation of mathematics literature through the centuries which includes diverse entries from authors such as Archimedes, René Descartes, Bertrand Russell, and George Polya.

References

Burns, Marilyn. 1991. *Mathematics and Literature*. Mathematics Solutions, distr. by Cuisenaire.

Carey, Deborah A. 1992. "Children's Literature: Providing Context for Problem Solving." Paper presented at the Annual NCTM meeting, Nashville, TN.

Ginsburg, Herbert P. 1989. *Children's Arithmetic: How They Learn It and How You Teach It*. Austin, TX: Pro-ed.

Griffiths, Rachel, and Margaret Clyne. 1991. *Books You Can Count On*. Portsmouth, NH: Heinemann.

Larson, Carol N. 1992. "Children's Literature and Mathematics." Paper presented at the Annual NCTM meeting, Nashville, TN.

National Council of Teachers of Mathematics. 1989. *Curriculum and Evaluation Standards for School Mathematics*. Reston, VA.

Rommel, Carol A. 1991. *Integrating Beginning Math and Literature*. Nashville, TN: Incentive Publications.

CURRICULUM MATERIAL PRODUCERS

T This chapter provides information on publishers and producers of mathematics materials, books, supplementary materials, software, and other items. For some of the larger publishers, we have provided a listing of mathematics series and book titles. For other companies, we provide a description of products. Much of the information in this chapter is based on the publishers' catalogues; for more details, you should contact the publishers and producers directly. The addresses and phone numbers given are for the offices that will supply catalogues and other promotion material; note that these phone numbers are not for the editorial offices.

Addison-Wesley Publishing Company
Jacob Way
Reading, MA 01867
800-447-2226

Elementary Mathematics

Addison-Wesley Mathematics© 1991 (series)
Grades K-8. A basal mathematics program. Texts, teacher's editions, workbooks, tests, teaching aids, record keeping forms, computer management system, implementation package (available in Spanish)

Math in Stride (series)
Grades 1-6. Hands-on activity program with workbooks, teacher's editions, blackline masters

Challenge: A Program for the Mathematically Talented (series)
Grades 3-6. Texts, workbooks, teacher's editions, manipulative kit

The Middle Grades Mathematics Project (series)
Grades 5-8. Using *Mouse and Elephant: Measuring Growth* covers area, perimeter, surface area and volume. Sourcebooks, activity books

Explorations (series)
Grades K-2. Texts, teacher's resource books, manipulative kits, Ted the Puppet

Other materials for elementary in part include: *Mathematics Their Way* with blackline masters, manipulatives kits, materials kits; *Workjobs*, activity centered early childhood math concepts; *Kids Are Consumers Too! Real World Mathematics for Today's Classroom, Baseball: A Game of Numbers, Build Your Own Polyhedra, Teaching Problem Solving Strategies, Mathematics Games for Fun and Practice, Cooperative Learning in Mathematics: A Handbook for Teachers, Teaching Mathematics to Children with Special Needs*

Middle/High School Mathematics

Addison-Wesley Pre-Algebra (series)
Grades 7–12. Texts, teacher's editions, resource binders, solutions manuals, overhead transparencies, software for Apple/IBM, *Problem-Solving Experiences in Pre-Algebra*

Addison-Wesley Algebra
Grades 8–12. Texts, resource packages, teacher's edition solutions manuals (available in Spanish)

Algebra I: Expressions, Equations and Applications (series)
Grades 8–12. Texts, teacher's edition, resource book, software, solutions manuals, *Problem-Solving Experiences in Algebra*

Addison-Wesley Algebra and Trigonometry
Grades 10–12. Text, teacher's edition, resource package, solutions manual, *Making Practice Fun*, Apple/IBM software

Addison-Wesley Geometry (series)
Grades 9–12. Text, teacher's edition, resource package, Apple/IBM software, *Problem-Solving Experiences in Geometry*

Informal Geometry
Grades 9–12. Text, teacher's edition, resource binder, laboratory manual

Essentials of Mathematics
Grades 8–12. Texts, teacher's edition, resource book, overhead transparencies, Apple/IBM software

Consumer Mathematics
Grades 9–12. Texts, teacher's edition, resource package, computer activities

Other materials for secondary in part include: *Mathematics Resources: Problem-Solving Experiences in Mathematics* (series), *Exploring Mathematics on the TI–81™ Graphing Calculator, Math Motivators* (series), *MasterGrapher/3D Grapher*. The Foerster Series includes *Algebra I, Algebra and Trigonometry, Trigonometry,* and *Precalculus with Trigonometry.* The Demana/Waits Series includes *Precalculus Mathematics* and *Transition to College Mathematics.* Other titles: *Elements of Calculus & Analytic Geometry, Calculus, Calculus and Analytic Geometry, Elementary Statistics, Using Statistics, Topics in Discrete Mathematics* (series), *Chaos, Fractals and Dynamics: Computer Experiments in Mathematics*

Euclid's Toolbox, geometry software for Macintosh and IBM

AGS

Publishers' Building
Circle Pines, MN 55014-1796
800-328-2560

Grades 6–12; adult. *Basic Mathematics Skills, LifeSkills Mathematics Series, Mathematics for Consumers, Mathematics for Business: Preparing Students for the Business World, Read It! Solve It! Math You Should Know, KeyMath Teach and Practice (TAP), Early Steps, Activity Pacs* include texts, teacher's guides, workbooks

AIMS (Activities Integrating Math & Science Education Foundation)

P.O. Box 8120
Fresno, CA 93747
209-255-4094

Historical Connections: Resources for Using History of Mathematics in the Classroom
Biographical information on ten mathematicians, anecdotes, quotes, activity sheets

American Association for the Advancement of Science

Dept. A62, P.O. Box 753
Waldorf, MD 20604
301-645-5643

The Sourcebook for Science, Mathematics & Technology Education, 1992
Information about people, programs, and organizations involved in improving education

American School Publishers

SRA School Group
P.O. Box 5380
Chicago, IL 60680-5380
800-843-8855

The Random House Achievement Program in Mathematics (series)
Grades 1–6. Cooperative learning approach with texts, teacher's guides

Spotlight on Math (series)
Grades 3–8. Skill books, workbooks, applications books, teacher's guide

Grades 1–8. Supplementary materials include: *Homework Math* (series) activities (available in Spanish); *Practicing Math Applications, Practicing Math, Practicing Problem Solving, Learning Math/Mastering Math/Extending Math Skills* (drill-and-practice kits); *Gaining Math skills, Mathematics Laboratory* (drill-and-practice cards); *The Random House Audio Mathematics Program*, a multimedia program using audiocassettes

Amsco School Publications, Inc.
315 Hudson Street
New York, NY 10013-1085
212-675-1221

Grades 7–12. Texts include *Mastering Fundamental Mathematics, Algebra I Review Guide*

Barnell-Loft
SRA School Group
P.O. Box 5380
Chicago, IL 60680-5380
800-843-8855

Grades pre-K–6. Supplementary programs for key concepts in math

William K. Bradford Publishing Co.
310 School Street
Acton, MA 01720
800-421-2009

Grades K–12. Software to supplement graphing, algebra, geometry, trigonometry, probability, calculus & precalculus, applied mathematics; competency exams

Wm. C. Brown Publishers
2460 Kerper Boulevard
Dubuque, IA 52001
800-228-0459

Mathematics for Elementary School Teachers: A Conceptual Approach
Grades K–6. Text, teacher's manual, test items, software

Mathematics for Elementary Teachers: An Activity Approach
Grades K–6. Text, 48 material cards/teacher's manual.

Primarily textbooks for college-age students. Include *Preparation for Algebra, Prealgebra: College Preparatory Mathematics, Beginning Algebra, Principles of Intermediate Algebra with Applications, Essentials of Precalculus, Algebra and Trigonometry, Finite Mathematics, Calculus for Management, Social and Life Sciences, Statistics, Business Mathematics in a Changing World*

Cambridge University Press
40 West 20th Street
New York, NY 10011
212-924-3900

Board Games Round the World
Grades K–12. Mathematical games

Carolina Biological Supply Company
2700 York Road
Burlington, NC 27215
800-334-5551

Grades K–12. Plastic coins, geometric forms, probability kits, number boards, abacuses, measure sets

Casio, Inc.
570 Mt. Pleasant Avenue
P.O. Box 7000
Dover, NJ 07801
201-361-5400

Grades K–12. Regular, scientific, and graphing calculators

Chariot Software Group
3659 India Street, Suite 100C
San Diego, CA 92103
800-242-7468

Grades K–12. Macintosh software for teaching algebra, geometry, decimals, fractions. Includes games and art

Children's Television

Schools D1992A
GPO Box 5373
New York, NY 10087-5373
212-595-3456

The Square One Activity Booklet
Grades 3-6. Book and teacher's guide of
puzzles and activities

COMAP, Inc. (Consortium for Mathematics and Its Applications)

Suite 210
57 Bedford Street
Lexington, MA 02173
800-77-COMAP

Going Bananas!
Grades K-3. Multidisciplinary units involving
problem-solving, cooperative learning, writing

High School Mathematics and Its Applications
(HiMAP) Modules
Grades 7-12. Lesson plans, group and
individual exercise, transparencies, worksheets

Big Math Attack Packs
Grades 7-12. Self-contained lessons:
Sociopoliticopack, architecture/designpack,
businesspack, calcpack, enviropack, mediapack,
statpack

Computer Science Press

W. H. Freeman & Company
41 Madison Avenue
New York, NY 10010
801-973-4660

Mathematics: A Human Endeavor. A Book for
Those Who Think They Don't Like the
Subject
Grades 9-12. Text, teacher's guide, transpar-
ency masters

Elementary Algebra
Grades 9-12. Text, teacher's guide, test
masters, transparency masters

Geometry
Grades 9-12. Text, teacher's guide with
solutions, test and transparency masters

CONDUIT

The University of Iowa
Oakdale Campus
Iowa City, IA 52242
800-365-9774

Grades 7-12. IBM and Macintosh software for
teaching algebra, trigonometry, mathematical
relationships

Contemporary Books, Inc.

Department F92
180 North Michigan Avenue
Chicago, IL 60601
800-621-1918

Series for ESL, ABE, & special-needs students
include *Number Sense: Discovering Basic Math
Concepts, Real Numbers: Developing Thinking
Skills in Math, Number Power: The Real World
of Adult Math, Lifeskills: Developing Consumer
Competence*

Corwin Press, Inc.

P.O. Box 2526
Newbury Park, CA 91319-8526
805-499-9774

Mathematics Programs: A Guide to Evaluation
Grades K-12. Program evaluation

Creative Publications

5040 West 11th Street
Oak Lawn, IL 60453
800-624-0822

Grades pre-K-6. Series titles include:
*Connections™ Program, Mathematics on Their
Way™, Explorations™, Decimal Factory™*.
Teacher's resource books. Manipulatives,
enrichment materials, games, calculators

Critical Thinking Press & Software
Midwest Publications
P.O. Box 448
Pacific Grove, CA 93950
800-458-4849

Grades K-12. Remedial-average-gifted-at risk. Activity books, resource books, software, tests

Cuisenaire Co. of America, Inc.
P.O. Box 5026
White Plains, NY 10602-5026
800-237-3142

About Teaching Mathematics: A K–8 Resource
Grades K-8. Text, black-line masters, bibliography

Math by All Means
Grade 3. Replaces textbook instruction. Teacher's manual

Grades K-8. Videotapes include: *Mathematics with Manipulatives, Mathematics for Middle School, Mathematics: Teaching for Understanding.* Many manipulatives for the overhead projector (thermometers, clocks, counters, coins), manipulatives for the classroom, calculators, models for graphing, classifying, geometry, measurement. Resources for problem solving, grade-level kits

Curriculum Associates, Inc.
5 Esquire Road
North Billerica, MA 01862-2589
800-225-0248
508-667-8000

Solve™ Action Problem Solving (series)
Grades 4-adult ed. Workbook, teacher's guide

Figure It Out (series)
Grades 2-6. Workbooks, teacher's guide

Enright Computation Series
Grades 4-adult. Practice books, answer guides, booster packs, teacher's resource manual

Supplementary elementary materials include Apple II software *Factmaster* for addition, subtraction, division and multiplication, math books in Spanish

Dale Seymour Publications
P.O. Box 10888
Palo Alto, CA 94303-0879
800-USA-1100

Problem Solving Experiences in Mathematics-Secondary (series)
Grades 8-12. Covers algebra, geometry, general math, prealgebra

Mathematical Investigations (series)
Grades 8-12 gifted. Life math

The Stella Octangula and Platonic Solids: Visual Geometry Project
Grades 7-12. Activity book, video, manipulative kit,

Other secondary titles include *The Art of Problem Solving, Problem Solving: A Basic Mathematics Goal, Teaching Problem Solving: What, Why & How.*

Supplementary materials include puzzles, contest books, recreational math, sophisticated manipulatives, calculators, test books, resources

DDL Books, Inc.
6521 N.W. 87th Avenue
Miami, FL 33178
800-63-LIBRO

Grades K-8. Dedicated to dual literacy. Math books in Spanish such as *SM Matematicas* (series), *Matematicas Cometa* (series), *Matematicas* (series)

Delmar Publishers, Inc.
2 Computer Drive West
P.O. Box 15015
Albany, NY 12212-5015
800-347-7707

Early Childhood. *Math and Science for Young Children, Experiences in Math for Young Children*

Delta Education
P.O. Box M
Nashua, NH 03060-6012
800-258-1302

Delta Mathematics Program: A Complete K-6
Mathematics Program
Grades K-6. Manipulative kits, teacher's
guides, printed materials, challenge activities,
cooperative learning experiences

Supplementary elementary materials include
manipulatives, number boards, games, kits

Didax, Inc.
One Centennial Drive
Peabody, MA 01960
800-458-0024

Preschool, elementary, special needs. Unifix
Structural Mathematics Materials, manipulatives
to use with *Unifix Mathematics Activities*
(series). Shapes, patterns, geometry. Materials
for overhead transparencies. Weighing, measuring, estimating manipulatives.

DLM
P.O. Box 4000
One DLM Park
Allen, TX 75002
800-527-4747

Numbers in Rhyme
Grades Pre-K-1. Big books, small books.

Math Predictable Storybooks
Grades K-1. Storybooks, teacher's guide,
blackline masters

On the Button in Math
Grades Pre-K-1. Resource book, but-
tons, laces

Supplementary material for early childhood and
elementary math students include hands-on
activities mats, manipulatives, puzzles,
problem-solving kits

Edmark
P.O. Box 3218
Redmond, WA 98073-3218
800-362-2890

Grades K-6. Supplementary and enrichment
books, software, audiocassettes, manipulatives

Educational Activities, Inc.
P.O. Box 392
Freeport, NY 11520
800-645-3739

Grades L-adult. Software for Apple/Macin-
tosh/MS-DOS, voice-interactive programs,
support materials

Entry Publishing, Inc.
P.O. Box 20277
New York, NY 10025
800-736-1405

Reading-disabled grades 5-adult. Teacher's
guides, workbooks, novels, software, audio-
and videocassettes, bilingual materials including
Autoskill Mathematics Program and *Math Talk:
The Wonderful Problems of Fizz & Martina*

ETA
620 Lakeview Parkway
Vernon Hills, IL 60061
800-445-5985

Grades K-12. Math manipulatives, videos,
calculators, puzzles, games

Everyday Learning Corporation
P.O. Box 1479
Evanston, IL 60204-1479
708-866-0702

Everyday Mathematics
Grades K-3. Text, teacher's manual, lesson
guide, *Minute Math,* Student Activity Aids,
blackline masters

*Using Mathematics and Science To Explore
Our World*
Grades 1-3.

Grades K–12. Calculators, *Everything Math Deck*, counters, enrichment materials

Fearon/Janus/Quercus
500 Harbor Boulevard
Belmont, CA 94002
800-877-4283

Practical Arithmetic Series
Special education/remedial programs. Texts

Math in Action
Special education/remedial programs. Texts, teacher's guide, resource books

Ferranti Educational Systems, Inc.
801 S. 18th Street
Columbia, PA 17512
717-684-4398

Software for students and teachers

Focus Media, Inc.
839 Stewart Avenue
P.O. Box 865
Garden City, NY 11530
800-645-8989

Grades pre-K–12. Software for Apple/MS-DOS/Macintosh/Commodore, support materials

Gamco Industries, Inc.
P.O. Box 1862N1
Big Spring, TX 79721-1911
800-351-1404

Grades K–12. Software for Apple/MS-DOS/Macintosh, support materials, games, calculator activities

Glencoe/Macmillan/McGraw-Hill
P.O. Box 543
Blacklick, OH 43004-0543
800-334-7344

Merrill Pre-Algebra: A Transition to Algebra
Grades 7–12. Text, teacher's edition, software, tests, resources, games, workbook, lab manual, solutions manual (available in Spanish)

Merrill Algebra (series)
Grades 7–12. *Algebra 1, Algebra 2 with Trigonometry* include texts, teacher's edition, test, resources, workbooks, lab manuals, solutions manuals (available in Spanish)

Merrill Mathematics Connections: Essentials and Applications
Grades 9–12. Texts, teacher's edition, resources, lab manual, software (available in Spanish)

Merrill Advanced Mathematical Concepts
Grades 9–12. Text, teacher resource, teacher's edition, solutions manual

Mastering Essential Mathematics Skills
Grades 9–12 for minimum competency. Text, teacher's edition

Globe Book Company
Simon and Schuster
4350 Equity Drive
P.O. Box 2649
Columbus, OH 43216

800-848-9500
Globe High School Mathematics (Series)
Grades 9–12. Texts, teacher's editions, teacher's resource books

Mathematics Workshop (series)
Grades 9–12. Includes basic skills, exam preparation, exploring careers, problem solving. Texts, teacher's editions

GoodYearBooks
1900 East Lake Avenue
Glenview, IL 60025
800-628-4480

Grades K–8. Math motivators

J. L. Hammett Co.
P.O. Box 9057
Braintree, MA 02184-9704
800-333-4600

Grades K-12. Manipulatives, UNIFIX structural mathematics materials, enrichment materials

Harcourt Brace Jovanovich, Inc.
School Department
6277 Sea Harbor Drive
Orlando, FL 32821-9989
800-CALL-HBJ

Mathematics Plus (series)
Grades K-8. Text, *Math Fun Magazines*, teacher's editions, teacher's resource, testing program, practice workbooks, *Reaching All Students,* reteaching workbooks, enrichment workbooks, *A Day with No Math* including a read along cassette, literature lap books, big books, multicultural projects, *Mathematics from around the World,* manipulative kits, interactive teaching packs, calculators, software

Mathematics Unlimited
Grades K-8. Texts, teacher's editions, teacher's resources, practice workbooks, reteaching workbooks, enrichment workbooks, posters, big books, *The Best Problems Ever, Unlimited Challenges for Problem Solvers,* activity books, manipulatives, in-service series videotapes, software, calculators, tests

Hartley Courseware
133 Bridge Street
Dimondale, MI 48821
800-247-1380

Grades pre-K-12. Software for Apple/MS-DOS/Macintosh/Commodore, support materials

D. C. Heath
125 Spring Street
Lexington, MA 02173
800-334-3284

Heath Mathematics Connections (series)
Grades K-8. Texts, manipulatives, skill pads, charts, teacher's edition, overhead transparencies, gameboards, counters, little books, assessments (available in Spanish)

New Ways in Numbers (series)
Grades K-8. Texts, teacher's editions, tests, copymasters

Heath General Mathematics
Grades 6-9. Texts, teacher's editions, workbook, visual aids, manipulatives, software

Heath Pre-Algebra
Grades 7-12. Texts, teacher's edition, resource file, workbook, study guide, transparencies, software

Heath Algebra 1
Grades 7-12. Text, teacher's edition, resource binder, software, transparencies

Heinemann Boynton/Cook
361 Hanover Street
Portsmouth, NH 03801-3959
800-541-2086

Teacher's resource books for teaching mathematics, the relationship of math and literature, math and writing. Includes *Read Any Good Math Lately?*

Hispanic Books Distributors, Inc.
1665 West Grant Road
Tucson, AZ 85745
800-634-2124

Grades 3-6. *Coleccion Mi Primera Matematica* (series) for multiplying, dividing, measuring and weighing.

Holt, Rinehart, and Winston
6277 Sea Harbor Drive
Orlando, FL 32821-9989
800-225-5425

Holt Pre-Algebra
Grades 6-9. Text, teacher's edition, teacher's resource, transparencies, overhead manipulatives, software, tests

Holt Introductory Algebra (series)
Grades 9-11. Texts, teacher's edition, teacher's resources, manipulatives, transparencies

Algebra (series)
Grades 9–12. *Holt Algebra 1* and *Holt Algebra with Trigonometry* include texts, teacher's edition, teacher's resources, manipulatives, transparencies

Houghton Mifflin
Department J
One Beacon Street
Boston, MA 02108-9971
800-323-5663

The Mathematics Experience (series)
Grades K–3.

HRM Video
175 Tompkins Avenue
Pleasantville, NY 10570
914-767-7496

Grades 4–12. Videos presenting problems and problem-solving strategies ranging from simple arithmetic to geometry

IBM Direct
PC Software Department 829
One Culver Road
Dayton, NJ 08810
800-222-7257

Pre-K–adult. Software for IBM, support materials, multimedia materials including CD-ROM, videodiscs, videocassettes. In developing math skills, algebra, geometry

I/CT (Instructional/ Communications Technology, Inc.)
10 Stepar Place
Huntington Station, NY 11746
516-549-3000

Grades K–adult. Software for Apple/MS-DOS and support materials, filmstrips, books, activity books, audiocassettes. In basic math facts, computer power

Intellimation
Library for the Macintosh
Department 2SCK
P.O. Box 219
Santa Barbara, CA 93116-9954
800-346-8355

Grades 7–12. Software for Macintosh, multimedia programs. In mastering calculus, algebra, statistics, geometry

Interact
Box 997-H92
Lakeside, CA 92040
800-359-0961

Grades K–12. Simulation programs. Student guides, teacher's guides, supplementary materials including software. Topics include developing calculator skills, money management, metric measurement

Janson Publications, Inc.
Dept. H9, P.O. Box 860
Dedham, MA 02026-0011
800-322-MATH

Contemporary Precalculus through Applications: Functions, Data Analysis and Matrices
Grades 11–14. Text, answer key, teacher's manual, assessment package, *Graphing Calculator Precalculus Manual*, workshop video

Contemporary Applied Mathematics
Grade 7–14. Text, teacher's manual

Mathematical Modeling in the Secondary School Curriculum
Grades 9–12. Teaching resource

Mathagrams
Grades 7–12. Integers to fractals, prealgebra to advanced math.

Experiments in Mathematics for the Curious
Grades 7–12.

Middle Grades Modules: Probability, Graphing, Data, and More
Grades 5–8. Each module forms a 2- or 3-day lesson.

Algebridge: Educational Testing Service and The College Board
Grades 7–12. Student booklets, teacher manual

Middle Grades Mathematics Project (series)
Grades 5-8. Five titles

Grades 7-College. Mathematical texts and teacher's resources

JC Cassettes
Box 73
Calumet, OK 73014
405-893-2293

Grades 1-12. Teach math by cassette

Jostens Learning Corporation
7878 North 16th Street, Suite 100
Phoenix, AZ 85020-4402
800-422-4339

Grades K-8. Computer software *Basic Learning System/Math*

Journal Films
930 Pitner Avenue
Evanston, IL 60202
800-325-5448

Grades K-12. Elementary videomath series includes *Movement, Cubes, Size, Hexagons.* Primary mathematics series includes *Division, Counting, Base and Place Value, Numbers All around Us.* Intermediate math videos include *Fractions, Introduction to Decimals.* Secondary mathematics films include *Statistics, Triangle and Square Numbers, Fractions and Percentages, Symmetry*

Judy/Instructo
4424 West 78th Street
Bloomington, MN 55435
800-832-5228

Grades pre-K-2. Educational materials for math including desk tapes, sorting boxes, picture cards, puzzles, alphabet wall charts

Kaplan
P.O. Box 609
Lewisville, NC 27023-0609
800-324-2014

Grades K-6. Manipulatives, calculators, counters, games, enrichment materials

Key Curriculum Press
2512 Martin Luther King Jr. Way
P.O. Box 2304
Berkeley, CA 94702
800-338-7638

Creating Miracles: A Story of Student Discovery
Grades K-12. Teacher's resource

Discovering Geometry: An Inductive Approach
Grades 9-12. Text, teacher's guide, answer key, teacher's resource book, test and exams, *Mathercise D*

Visual Geometry Project
Grades 9-12. Three dimensional geometry in *The Platonic Solids* and *The Stella Octangula* videos, activity book, manipulatives kit

The Geometer's Sketchpad
Grades 9-12. Macintosh software

Lakeshore Learning Materials
2695 East Dominguez Street
P.O. Box 6261
Carson, CA 90749
800-421-5354

Grades 2-7. Skill-building textbooks, workbooks, activity books, teacher's guides, and software in beginning and practical math, prealgebra, money management. Calculators

Longman Publishing Group
95 Church Street
White Plains, NY 10601-1566
800-447-2226

Grades 4-12. Problem-solving games and skills workbooks

McDougal, Littell & Company
P.O. Box 8000
St. Charles, IL 60174
800-225-3809

Gateways to Algebra and Geometry: An
  Integrated Approach (series)
Grades 7-12. Text, teacher's edition, resource
file, workbooks, quizzes, solutions manual,
transparencies, software

Algebra: An Integrated Approach (series)
Grades 8-12. Includes *Algebra 1, Algebra 2
and Trigonometry.* Text, teacher's edition,
resource file, workbooks, quizzes, solutions
manual, software

Geometry for Enjoyment and Challenge
Grades 9-12. Text, teacher's edition, resource
file, quizzes, solutions manual, transparencies,
software

Daily Mathematics: Critical Thinking and Problem Solving (series)
Grades 1-8. Workbooks, teacher's manual,
transparencies (available in Spanish)

Macmillan/McGraw-Hill School Division
220 East Danieldale Road
DeSoto, TX 75115
800-224-1111

Mathematics in Action. I Can! Math Activity
  Program (series)
Grades K-8. Teacher's editions, jumbo books,
activity books, literature books, computer work-
shop, floormats, workmats, read-aloud selec-
tions, math songs, posters, *Critical Thinking,
Calculator Workshop, Computer Software
Workshop, Home Involvement* (available in
Spanish)

Marilyn Burns Education Assoc.
150 Gate 5 Road, Suite 1001
Sausalito, CA 94965
415-332-4181

About Teaching Mathematics: A K–8 Resource
Grades K-8. A teaching resource

Math Teachers Press, Inc.
5100 Gamble Drive, Suite 398
Minneapolis, MI 55416
800-852-2435

Grades K-12. Manipulatives, enrichment materials, calculators

Mathcad
P.O. Box 120
Buffalo, NY 14207-0120
800-289-3023

Grades 9-12. Software for complex math
concepts

Mathematical Association of America
1529 18th Street, NW
Washington, DC 20036
202-387-5200

Student Research Projects in Calculus
Grades 10-college. Five calculus teachers
present over 100 projects

Mathematical Sciences Education Board
818 Connecticut Avenue, NW, Suite 500
Washington, DC 20006
800-624-6242

Math Matters: Kids are Counting on You
Grades K-6 (available in Spanish)

Mathematics Criterion Center
P.O. Box 2061
Evanston, WY 82931-1061
800-334-3173

Grades 7-12. Teacher resource packages,
multiple-choice test packages, student
workbooks, solutions manuals. In basic math,
algebra, geometry, precalculus, calculus

MECC
6160 Summit Drive North
Minneapolis, MN 55430-4003
800-685-MECC

Grades Pre-K–adult. Software for Macintosh/ Apple/MS-DOS, multimedia materials. In geometry, *Exploring Chaos*, counting & measuring

Media Materials Publishing
Department 920301
1821 Portal Street
Baltimore, MD 21224
800-638-1010

Grades K–adult special needs students. Includes *Basic Mathematic Skills*, *Life Skills Mathematics*, *Mathematics for Consumers*, flash cards, games, counters

Micrograms Publishing
1404 North Main Street
Rockford, IL 61103
800-338-4726

Grades K–6. Software for Apple. Includes *Oliver's Crosswords*, *Friendly Figures*, *Time and Money*, *Mathosaurus*, *Wild West Math*

Mimosa Publications
P.O. Box 26609
San Francisco, CA 94126
800-443-7389

Moving Into Math
Grades K–3. Student booklets, enrichment materials, manipulative kits, games, problem solving cards, teacher's resource book, staff development program

Minimath Projects
5120 Farwell Avenue
Skokie, IL 60077
708-677-3130

String design kits and polyhedron kits.

Modern Curriculum Press
13900 Prospect Road
Cleveland, OH 44136
800-321-3106

Excel in Mathematics (series)
Grades 1–8. Texts, tests, keys

Modern Curriculum Press Mathematics
Grades K–6. Texts, teacher's editions

Problem Solving in Math
Grades K–6. Workbooks on a variety of subjects

Modern Educational Resource Guide
5000 Park Street North
St. Petersburg, FL 33709
800-243-6877

Grades K–12. Videos, videodiscs, CD-ROMs, software including *Algebra, Calculus, Pre-Calculus, Math Patrol, Square One*

Nasco
901 Janesville Avenue
P.O. Box 901
Fort Atkinson, WI 53538-0901
414-563-2446

Grades K–12. Manipulatives, games, enrichment materials, calculators, videotapes, software

National Council of Teachers of Mathematics
1906 Association Drive
Reston, VA 22091
703-620-9840

Mathematics for the Young Child
Grades preschool–5. Teacher resource

Activities for Junior High School and Middle School Mathematics: Readings from "The Arithmetic Teacher" and "The Mathematics Teacher"
Grades 6–9. Teacher resource

Algebra for Everyone
Grades 9–12 underachievers. Teacher resource followed by *Algebra for Everyone In-Service Handbook*

Applications of Secondary School Mathematics, Readings from "The Mathematics Teacher"
Grades 9-12. Teacher resource

Grades K-12. Teacher resources, theory, enrichment programs, research results. Magazines for teachers, *Arithmetic Teacher* (K-8), *Mathematics Teacher* (grades 7-14), *Journal of Research in Mathematics Education*

National Women's History Project
7738 Bell Road
Windsor, CA 95492-8518
707-383-6000

Outstanding Women in Mathematics and Science
Grades 5-adult

Open Court Publishing Company
315 Fifth Street
Peru, IL 61354
800-435-6850

Open Court Real Math™ (series)
Grades K-8. Student books, workbooks, games, manipulatives, practice sheets, teacher's resource sets

Orange Cherry/Talking Schoolhouse Software
P.O. Box 390, Dept. S
Pound Ridge, NY 10576-0390
800-672-6002

Grades pre-K-8. Software for Apple/Macintosh/IBM/Tandy, support materials, multimedia programs in math problem solving, story problems, beginning concepts

The Peoples Publishing Group, Inc.
P.O. Box 70
365 West Passaic Street
Rochelle Park, NJ 07662
800-822-1080

Math for the Real World
Grades 6-adult. Text, teacher's guide for remedial mathematics

Phoenix Learning Resources
468 Park Avenue South
New York, NY 10016
800-221-1274

Learning Skills Series, Arithmetic
Grades 2-adult. Worktexts, placement tests, teacher's manual

Essential Math Skills
Grades 2-adult. Text, teacher's edition, tests

Learning Basic Arithmetic (series)
Grades K-3. Remedial workbook series with teacher's manual and placement tests

Prentice-Hall
School Division of Simon & Schuster
Englewood Cliffs, NJ 07632-9940
800-848-9500

Algebra (series)
Grades 8-12. *Algebra 1, Algebra 2 with Trigonometry* include teacher's edition, teacher's resource book, connections, transparencies, tests

Informal Geometry
Grades 9-12. Text, teacher's edition, teacher's resources, transparencies, overhead manipulatives

Geometry
Grades 9-12. Text, teacher's edition, teacher's resources, transparencies, overhead manipulatives, tests, solutions manual

Connections (series)
Grades 9-12. *Algebra 1 Connections, Algebra 2 with Trigonometry Connections, Geometry Connections.* Texts, transparencies, manipulatives, calculator books, software, dictionary, *SuperCourse for the SAT*

Grades 9-12. Electives and advanced placement texts include *Statistics: A First Course, Algebra & Trigonometry, Calculus & Analytic Geometry, Excursions in Modern Mathematics*

The Psychological Corporation
Harcourt Brace Jovanovich, Inc.
555 Academic Court
San Antonio, TX 78204-2498
800-228-0752

Grades 2-8. Math assessment

Queue, Inc.

338 Commerce Drive
Fairfield, CT 06430
800-232-2224

Grades K-12. Software for Apple/IBM/
Macintosh, support materials, multimedia
materials.

S&S

P.O. Box 513
Colchester, CT 06415-0513
800-243-9232

Grades K-12. Manipulatives

Saxon Publishers

1320 West Lindsey
Norman, OK 73069
800-284-7019

Saxon's Primary Mathematics Series
Grades K-3. Text, scripted lessons, teacher's
manual, homework pages, masters, factcards,
manipulatives

Grades 4-12. *Math 54, Math 65, Math 76,
Math 87, Algebra 1/2, Algebra 1, Algebra 2*

Scholastic Inc.

2931 East McCarty Street
Jefferson City, MO 65101-9968
800-325-6149

Grades 1-10. *Math Shop Series*, software

ScottForesman

1900 East Lake Avenue
Glenview, IL 60025
800-554-4411

The University of Chicago School Mathematics
Project: Secondary Component Materials
(series)
Grades 7-12. Titles include *Transition Mathematics, Algebra, Geometry, Advance Algebra,
Functions, Statistics & Trigonometry,
Precalculus & Discrete Mathematics* and
include texts, teacher's editions, manipulatives,
teacher resources (tests, lesson, teaching aids),
solution manuals

The Explorer Series
Grades 7-12. Computer software for UCSMP
classes. Includes *GeoExplorer, GraphExplorer,
StatExplorer*

Exploring Mathematics (series)
Grades 6-8. Texts, teacher's editions, teacher's
resources, workbooks (practice, reteaching,
enrichment, problem solving, critical thinking
and answer keys)

Exploring Mathematics Software (series)
Grades 3-8. *Graphing & Probability Workshop, Geometry Workshop, Number Sense
Workshop, Fraction Workshop, Spreadsheet
Workshop*

Standard Japanese textbooks, translated, for
grades 7-9, and Russian textbooks, translated,
for grades 1-3

Silver Burdett & Ginn

4350 Equity Drive
P.O. Box 2649
Columbus, OH 43216
800-848-9500

Silver Burdett & Ginn Mathematics: Exploring
    Your World
Grades K-8. Texts, consumable workbooks,
teacher's edition, teacher support system (available in Spanish)

Grades 6-12. Supplemental titles include
*Visual Science: Computers & Mathematics, The
Silver Burdett Mathematical Dictionary*, plus
manipulatives, software, and videos

Tom Snyder Productions, Inc.

80 Coolidge Hill Road
Watertown, MA 02172
800-342-0236

Grades 1-8. Videotapes for program
enrichment

Society for Visual Education, Inc.
Dept. JT
1345 Diversey Parkway
Chicago, IL 60614-1299
800-829-1900

Grades 4-8. Videocassettes, filmstrips, software, videodiscs, and book cassettes to enhance program

Sorpis West
1140 Boston Avenue
Longmont, CO 80501
303-651-2829

Grades K-12. *Math Plus* is an entire curriculum.

South-Western Publishing Co.
5101 Madison Road
Cincinnati, OH 45227
800-543-7972

Pre-Algebra
Grades 9-12. Textbook, supplement, tests, software

Geometry for Decision Making
Grades 9-12. Textbook, supplement, tests, software, teacher's resources

Computer Mathematics: Structured BASIC with Math Applications
Grades 9-12. Textbook, supplement, software, teacher's manual

Spring Branch Software, Inc.
P.O. Box 342
Manchester, IA 52057
319-927-6537

Grades 9-college. *MacNumerics for the Macintosh Computer*

SRA School Group
P.O. Box 5380
Chicago, IL 60680-5380
800-843-8855

Grades 1-9. Supplementary programs include *Mathematics Laboratory* (skill, project, and

activity cards), *Computer Drill and Instruction: Mathematics* (software to reinforce curriculum)

Steck-Vaughn Company
P.O. Box 26015
Austin, TX 78755
800-531-5015

Early Math (series)
Grades K-2. Readiness set, workbooks

Steck-Vaughn Working with Numbers (series)
Grades 1-12. Workbooks, teacher's guides, answer keys, calculator applications

Mathematics Skills Books (series)
Grades 4-8. Booklets, answer key, teacher's edition

Counters and Seekers (series)
Grades 1-2. Books, parent involvement component, teacher's guide

Sunburst Communications
39 Washington Avenue, Box 40
Pleasantville, NY 10570-3498
800-431-1934

Grades 9-12. Texts include *Elementary Algebra, a Guided Inquiry*; *Algebra II/Trigonometry, a Guided Inquiry*; *Geometry, a Guided Inquiry*

Teacher Ideas Press
Attn. Dept. 400
P.O. Box 3988
Englewood, CO 80155-3988
800-237-6124

Grades K-6. Resource books for teachers include *Math through Children's Literature: Activities that Bring NCTM Standards Alive*

Tricon Mathematics
Box 146
Mt. Pleasant, MI 48804
517-772-2811

Grades K-12. Software, calculators, manipulatives, enrichment materials

Tutorsystems

Woodmill Corporate Center
5153 West Woodmill Drive
Wilmington, DE 19808
800-545-7766

Grades 3–adult. *Stat Complete Series* software for growth in reading, math, and grammar

Ventura Educational Systems

3440 Brokenhill Street
Newbury Park, CA 91320
800-336-1022

Grades K–12. *Elementary Math Series, Geometry Concepts*
(IBM/Macintosh instructional software). Teacher productivity tools, staff development programs

Wadsworth School Group

10 Davis Drive
Belmont, CA 94002-3098
800-831-6996

Fundamentals with Elements of Algebra: A
    Bridge to College Mathematics
Grades 9–12. Text, answer book, instructor's manual, transparencies

Essential Mathematics with Geometry
Grades 9–12. Text, answer book, videotapes, testing system

Intermediate Algebra
Grades 9–12. Text, instructor's manual, solutions manual, algebra tutor software, testing system

CABRI Geometry Software
Grades 9–12. Macintosh and IBM software, user guide

Grades 9–college. Texts in areas of technical mathematics, applied mathematics, trigonometry, precalculus, analytic geometry, calculus, finite mathematics, discrete mathematics, linear algebra, differential equations, abstract algebra/geometry, statistics, general interest. Software in most areas

Waterford Institute

1480 East 9400 South
Sandy, UT 84092
801-572-1172

Grades K–6. *Mental Math Games* software to improve arithmetic skills

Watten/Poe Teaching Resource Center

P.O. Box 1509
14023 Catalina Street
San Leandro, CA 94577
800-833-3389

Grades K–8. Integrated curriculum materials, teacher's resource books, geoboards, manipulatives, geometry kits, resource books, book/audiocassette packages, chalkboards

Weaver Instructional Systems

Grades 6161 28th Street, SE
Grand Rapids, MI 49506
616-942-2891
Math Efficiency System

Grades 2–6. Computer-assisted math system, individualized, multi-level program

Weekly Reader Corp.

3000 Cindel Drive
P.O. Box 8037
Delran, NJ 08075
800-446-3355

Math Word Problem-Solving Program (series)
Grades 1–6. Skills books, teacher's guide, duplicating masters

Table & Graph Skills
Grades 2–6. Skills books, teacher's guide, duplicating masters

WINGS for Learning/Sunburst

1600 Green Hills Road
P.O. Box 660002
Scotts Valley, CA 95067-0002
800-321-7511

Grades 6–college. Software for Apple/Macintosh/IBM/Tandy, multimedia materials includes *Number Connections, Subtract with Balancing Bear, Journey in Mathematics, Computational Games, How the West Was One + Three x Four*

World Book Educational Products

101 Northwest Point Boulevard
Elk Grove Village, IL 60007
708-290-5300

Early World of Learning
Preschool. Reviews size, colors, shapes, numbers, reading readiness, position & direction, time, listening & sequencing, motor skills, social-emotional development

The Wright Group

19201 120th Avenue, NE
Bothell, WA 98011-9512
800-523-2371

Understanding Mathematics (series)
Grades 5–7. Whole language approach to math through poetry. Titles include *Soldiers, How Many?, Ten Little Goblins,* and *Skittles* with teacher's manuals, resource books, blackline masters

STATEWIDE TEXTBOOK ADOPTION

THERE are twenty-two states that have statewide adoption of textbooks and other instructional materials: Alabama, Arizona, Arkansas, California, Florida, Georgia, Idaho, Indiana, Kentucky, Louisiana, Mississippi, Nevada, New Mexico, North Carolina, Oklahoma, Oregon, South Carolina, Tennessee, Texas, Utah, Virginia, and West Virginia.

The policies and procedures for textbook adoption are similar in all twenty-two states, with some minor variations.

Textbook Advisory Committee

In general, the state board of education is responsible for developing guidelines and criteria for the review and selection of textbooks and for appointing members to a textbook advisory committee. However, in a few states, the appointment of committee members is the responsibility of the governor or of the Commissioner of Education.

The textbook advisory committee is usually composed of educators, lay citizens, and parents, and can have from nine to twenty-seven appointees, depending upon the state. Membership is weighted, however, toward individuals who are educators: elementary and secondary teachers in the subject areas in which textbooks are to be adopted, instructors of teacher education and curriculum from local universities and colleges, school administrators, and school board members. Lay citizens, in

order to sit on the committee, should be interested in and conversant with educational issues. An effort is made to select appointees who reflect the diversity of their state's population, and therefore decisions about appointments are often made with the purpose of having a wide representation of ethnic backgrounds and geographical residence within the state.

Adoption Process

The textbook and instructional materials adoption process takes approximately twelve months.

Once the textbook advisory committee is formed, the members conduct an organizational meeting to formulate policy on such issues as adoption subjects and categories; standards for textbook evaluation, allocation of time for publisher presentations, and location of regional sites for such; sampling directions for publishers; and publisher contact. The committee may appoint subcommittees, made up of curriculum and/or subject specialists, to assist them in developing criteria for evaluating instructional materials.

After these procedural matters are agreed upon, the committee issues an official textbook call or invitation to textbook publishers to submit their books. This document provides the publisher with adoption information and subject area criteria, which can be either the curriculum framework or essential skills list. Those publishers interested in having their materials considered for adoption submit their intention to bid, which shows the prices at which the publishers will agree to sell their material during the adoption period. Publishers usually bid current wholesale prices or lowest existing contract prices at which textbooks or other instructional materials are being sold elsewhere in the country.

If their bid has been accepted by the committee, the publishers submit sample copies of their textbooks for examination. The committee then hears presentations by the publishers. This meeting allows the publisher to present the texts submitted for adoption and to answer any questions the committee may have on the material. After publisher presentations, the textbooks are displayed in designated areas throughout the state for general public viewing. The committee then holds public hearings (usually two) which provide citizens with the opportunity to give an opinion on the textbooks offered for adoption. After much discussion and evaluation, the committee makes recommendations for textbook adoption to the state board of education.

When the board of education approves the committee's recommendations, it negotiates the contract with the chosen publishers and disseminates the list of instructional materials to the school districts. The school districts will then make their textbook selections from this list. A few states also allow their school districts to use materials for the classroom that are not on the adoption list.

Textbook and Instructional Materials

There are two categories of instructional materials: basal and supplementary. Basal, or basic, materials address the goals, objectives, and content identified for a particular subject. Supplementary materials, used in conjunction with the basic text, enhance the teaching of the subject.

Instructional materials may include all or some of the following: hardcover books, softcover books, kits, workbooks, dictionaries, maps and atlases, electronic/computer programs, films, filmstrips, and other audiovisual materials.

The textbook adoption period generally runs from four to six years (California, the exception, has an eight-year contract period for K-8 only). The grade levels for adoption are usually K-12, with the following subject areas: English/language arts, social studies, foreign languages, English as a Second Language, science, mathematics, fine arts, applied arts, health education, physical education, vocational education, driver education, technology education, special education, home economics.

Textbooks and instructional materials are ultimately judged by how well they reflect the state curriculum framework and/or esential skills objectives. Materials are rated on the following criteria: organization, accuracy, and currency of subject content; correlation with grade level requirements for the subject; adaptability for students with different abilities, backgrounds, and experiences; types of teacher aids provided; author's background and training; physical features; and cost.

In addition, some states have social content requirements that textbooks have to meet. For instance, textbooks should be objective in content and impartial in interpretation of the subject, and should not include offensive language or illustrations. American values (defined as democracy, the work ethic, and free enterprise), culture, traditions, and government should be presented in a positive manner. Respect for the individual's rights, and for the cultural and racial diversity of American society, can also be addressed in the text. Finally, some states declare that textbooks should not condone civil disorder, lawlessness, or deviance.

Kraus thanks the personnel we contacted at the state departments of education for their help in providing the states' textbook adoption lists.

List of Textbooks

Of the twenty-two states that have statewide adoption of instructional materials, all adopt books for mathematics. Their materials are listed below; the title, grade level, publisher, and copyright date are provided.

Alabama

Mathematics, Elementary (Grades One through Eight), Basic

Addison-Wesley Mathematics: Grades 1–8
Addison-Wesley, 1991 (Termination Year: 1997)

Mathematics Unlimited: Grades R/1–8
Harcourt, 1991 (Termination Year: 1997)

Houghton Mifflin Mathematics: Grades K–8
Houghton, 1991 (Termination Year: 1997)

Mathematics in Action: Grades K–8
Macmillan, 1991 (Termination Year: 1997)

Math 65, an Incremental Development:
Grades 5–12
Saxon, 1987 (Termination Year: 1997)

Math 76, an Incremental Development:
Grades 6–12
Saxon, 1985 (Termination Year: 1997)

Exploring Mathematics: Grades KR–8
ScottForesman, 1991 (Termination Year: 1997)

Mathematics: Exploring Your World:
Grades K–8
Silver, 1991 (Termination Year: 1997)

Algebra, Basic

Addison-Wesley Prealgebra: Grades 7–12
Addison-Wesley, 1990 (Termination Year: 1997)

Addison-Wesley Algebra: Grades 8–12
Addison-Wesley, 1990 (Termination Year: 1997)

Addison-Wesley Algebra & Trigonometry:
Grades 10–12
Addison-Wesley, 1990 (Termination Year: 1997)

Merrill Prealgebra: A Problem-Solving Approach: Grade 7
Glencoe, 1989 (Termination Year: 1997)

Merrill Algebra One: Grades 9–12
Glencoe, 1990 (Termination Year: 1997)

Merrill Algebra Two with Trigonometry:
Grades 9–12
Glencoe, 1990 (Termination Year: 1997)

Heath Prealgebra: Grades 9–12
Heath, 1990 (Termination Year: 1997)

Heath Algebra I: Grades 9–12
Heath, 1990 (Termination Year: 1997)

Heath Algebra II: Grades 9–12
Heath, 1990 (Termination Year: 1997)

Holt Prealgebra: Grades 9–12
Holt, 1986 (Termination Year: 1997)

HBJ Algebra 1: Grades 9–12
Holt, 1990 (Termination Year: 1997)

HBJ Algebra 2 with Trigonometry:
Grades 9–12
Holt, 1990 (Termination Year: 1997)

Pre-Algebra: An Accelerated Course:
Grades 7–8
Houghton, 1988 (Termination Year: 1997)

Essentials for Algebra: Concepts & Skills:
Grade 9
Houghton, 1989 (Termination Year: 1997)

Elementary Algebra: Grades 9–12
Houghton, 1988 (Termination Year: 1997)

Basic Algebra: Grades 9–12
Houghton, 1990 (Termination Year: 1997)

Algebra: Structure & Method: Grades 9–12
Houghton, 1990 (Termination Year: 1997)

Algebra & Trigonometry: Structure & Method:
Grades 9–12
Houghton, 1990 (Termination Year: 1997)

Algebra 1, an Integrated Approach:
Grades 9–12
McDougal, 1991 (Termination Year: 1997)

Algebra 2 & Trigonometry: Grades 9–12
McDougal, 1991 (Termination Year: 1997)

Algebra I: Grades 9–12
Prentice Hall, 1990 (Termination Year: 1997)

Algebra II with Trigonometry: Grades 9–12
Prentice Hall, 1990 (Termination Year: 1997)

Algebra 1/2, an Incremental Development:
Grades 7–8
Saxon, 1983 (Termination Year: 1997)

Algebra I, an Incremental Development:
Grades 7–12
Saxon, 1990 (Termination Year: 1997)

Algebra II, an Incremental Development:
Grades 9–12
Saxon, 1984 (Termination Year: 1997)

*University of Chicago School Mathematics
Project: Algebra*: Grades 9–12
ScottForesman, 1990 (Termination Year: 1997)

*University of Chicago School Mathematics
Project: Advanced Algebra*: Grades 9–12
ScottForesman, 1990 (Termination Year: 1997)

Algebra 1: Grades 9–12
Houghton, 1989 (Termination Year: 1997)

Basic—Advanced

Algebra 1: Grades 9–12
Houghton, 1989 (Termination Year: 1997)

Basic—Low Level

Merrill Algebra Essentials: Grades 9–12
Glencoe, 1988 (Termination Year: 1997)

Unified Geometry, Basic

Addison-Wesley Geometry: Grades 9–12
Addison-Wesley, 1990 (Termination
Year: 1997)

Geometry: Grades 9–12
Glencoe, 1990 (Termination Year: 1997)

Geometry: Grades 9–12
Holt, 1991 (Termination Year: 1997)

Geometry: Grades 9–12
Houghton, 1990 (Termination Year: 1997)

Discovering Geometry: An Inductive Approach:
Grades 9–12
Key Curriculum, 1989 (Termination
Year: 1997)

Geometry for Enjoyment & Challenge:
Grades 9–12
McDougal, 1991 (Termination Year: 1997)

Geometry: Grades 9–12
Prentice Hall, 1990 (Termination Year: 1997)

*University of Chicago School Mathematics
Project: Geometry*: Grades 9–12
ScottForesman, 1991 (Termination Year: 1997)

ScottForesman Geometry: Grades 9–12
ScottForesman, 1990 (Termination Year: 1997)

Basic—Low Level

Basic Geometry: Grades 9–12
Houghton, 1990 (Termination Year: 1997)

Supplementary

Merrill Informal Geometry: Grades 9–12
Glencoe, 1988 (Termination Year: 1997)

General Mathematics, Basic

Essentials of Mathematics: Grades 8–12
Addison-Wesley, 1989 (Termination
Year: 1997)

Merrill Integrated Mathematics: Grades 7–8
Glencoe, 1991 (Termination Year: 1997)

Merrill General Mathematics: Grades 9–12
Glencoe, 1987 (Termination Year: 1997)

Merrill Applications of Mathematics:
Grades 9–12
Glencoe, 1988 (Termination Year: 1997)

Globe High School Mathematics: Grades 9–12
Globe, 1989 (Termination Year: 1997)

Heath General Mathematics: Grades 9–12
Heath, 1989 (Termination Year: 1997)

Practical Mathematics: Skills & Concepts:
Grades 9–12
Holt, 1989 (Termination Year: 1997)

Mathematics: Structure & Method: Grades 7–8
Houghton, 1988 (Termination Year: 1997)

Essentials for High School Mathematics:
Grades 9–12
Houghton, 1989 (Termination Year: 1997)

Applications of High School Mathematics:
Grades 9–12
Houghton, 1990 (Termination Year: 1997)

Prentice Hall Refresher Mathematics:
Grades 9–12
Prentice Hall, 1989 (Termination Year: 1997)

Mathematics in Life: Grades 9–12
ScottForesman, 1989 (Termination Year: 1997)

University of Chicago School Mathematics Project: Transition Mathematics: Grade 8
ScottForesman, 1990 (Termination Year: 1997)

Supplementary—Low Level

Arithmetic Skills: Grades 9–12
Amsco, 1988 (Termination Year: 1997)

Mathematics Workshop: Grades 7–12
Globe, 1989 (Termination Year: 1997)

Essential Math Skills: Grades 9–12
Phoenix, 1990 (Termination Year: 1997)

Supplementary—Remedial

Achieving Competence in Mathematics:
Grades 9–12
Amsco, 1987 (Termination Year: 1997)

Consumer Mathematics, Basic

Addison-Wesley Consumer Mathematics:
Grades 9–12
Addison-Wesley, 1989 (Termination
Year: 1997)

Consumer Mathematics: Grades 9–12
Glencoe, 1989 (Termination Year: 1997)

Practical Mathematics: Consumer Applications:
Grades 9–12
Holt, 1989 (Termination Year: 1997)

Prentice Hall Consumer Mathematics:
Grades 9–12
Prentice Hall, 1989 (Termination Year: 1997)

Consumer & Career Mathematics:
Grades 9–12
ScottForesman, 1989 (Termination Year: 1997)

Dollars & Sense: Problem Solving Strategies in Consumer Mathematics: Grades 9–12
South-Western, 1989 (Termination Year: 1997)

Mathematics—Advanced Mathematics

Elements of Calculus & Analytic Geometry:
Grades 11–12
Addison-Wesley, 1989 (Termination
Year: 1997)

Precalculus Mathematics: A Graphing Approach: Grades 11–12
Addison-Wesley, 1990 (Termination
Year: 1997)

Merrill Advanced Mathematical Concepts:
Grades 9–12
Glencoe, 1991 (Termination Year: 1997)

Merrill Precalculus Mathematics: Grades 9–12
Glencoe, 1988 (Termination Year: 1997)

Precalculus: Grades 9–12
Heath, 1989 (Termination Year: 1997)

Calculus with Analytic Geometry: Grades 9–12
Heath, 1990 (Termination Year: 1997)

HBJ Advanced Mathematics: A Preparation for Calculus: Grades 9–12
Holt, 1988 (Termination Year: 1997)

HBJ Calculus with Analytic Geometry:
Grades 9–12
Holt, 1989 (Termination Year: 1997)

Advanced Mathematics: A Precalculus Course:
Grades 9–12
Houghton, 1989 (Termination Year: 1997)

Introductory Analysis: Grades 9–12
Houghton, 1988 (Termination Year: 1997)

Precalculus Mathematics: Grades 9–12
Prentice Hall, 1989 (Termination Year: 1997)

Calculus & Its Applications: Grades 9–12
Prentice Hall, 1990 (Termination Year: 1997)

Advanced Mathematics, an Incremental Development: Grades 9-12
Saxon, 1989 (Termination Year: 1997)

Supplementary

Trigonometry: Grades 11-12
Addison-Wesley, 1990 (Termination Year: 1997)

Exploring Data: Grades 7-12
Dale Seymour, 1986 (Termination Year: 1997)

Exploring Probability: Grades 7-12
Dale Seymour, 1987 (Termination Year: 1997)

Art & Techniques of Simulation: Grades 7-12
Dale Seymour, 1987 (Termination Year: 1997)

Exploring Surveys & Information From Samples: Grades 7-12
Dale Seymour, 1987 (Termination Year: 1997)

Finite Mathematics: Grades 9-12
Houghton, 1988 (Termination Year: 1997)

Trigonometry with Applications: Grades 9-12
Houghton, 1990 (Termination Year: 1997)

Modern Analytic Geometry: Grades 9-12
Houghton, 1989 (Termination Year: 1997)

Introductory Statistics & Probability: A Basis for Decision Making: Grades 9-12
Houghton, 1988 (Termination Year: 1997)

Analysis of Elementary Functions: Grades 9-12
Houghton, 1990 (Termination Year: 1997)

Probability & Statistics: Grades 10-12
South-Western, 1989 (Termination Year: 1997)

Precalculus with Unit Circle Trigonometry: Grades 9-12
West, 1990 (Termination Year: 1997)

Mathematics/Computer Literacy Skills, Supplementary

Computer Mathematics: Structured Basic with Math Applications: Grades 10-12
South-Western, 1989 (Termination Year: 1997)

Understanding Programming in Basic IBM/ MS-DOS: Grades 9-12
West, 1991 (Termination Year: 1997)

Understanding Programming in Basic Apple: Grades 9-12
West, 1991 (Termination Year: 1997)

Understanding Pascal: Turbo: Grades 9-12
West, 1989 (Termination Year: 1997)

Fundamentals of Pascal: Understanding Programming & Problem Solving: Grades 9-12
West, 1990 (Termination Year: 1997)

Microcomputers: Concepts, Skills & Applications: Grades 8-12
West, 1991 (Termination Year: 1997)

Computers in Business: Information Processing: Grades 9-12
West, 1991 (Termination Year: 1997)

Working with Application Software: Appleworks Version: Grades 7-12
West, 1991 (Termination Year: 1997)

Computing Today: Grades 7-10
West, 1991 (Termination Year: 1997)

Developing Computer Skills Using PFS: First Choice: Grades 7-12
West, 1991 (Termination Year: 1997)

Developing Computer Skills Using Appleworks: Grades 7-12
West, 1991 (Termination Year: 1997)

Arizona

Mathematics

Heath Mathematics: Grades K-8 (also in Spanish)
Heath, 1988 (Termination Year: 1994)

Mathematics Today: Grades K-8 (also in Spanish)
Harcourt, 1987 (Termination Year: 1994)

Mathematics Unlimited: Grades K-8 (also in Spanish)
Harcourt, 1988 (Termination Year: 1994)

Houghton Mifflin Mathematics: Grades K–8
(also in Spanish)
Houghton, 1987 (Termination Year: 1994)

Real Math: Grades K–8
Open Court, 1987 (Termination Year: 1994)

Prealgebra

Prealgebra: Grades 7–12
Addison-Wesley, 1991 (Termination Year: 1994)

Heath Prealgebra: Grades 7–12
Heath, 1986 (Termination Year: 1994)

Prealgebra: Grades 7–12
Harcourt, 1986 (Termination Year: 1994)

Holt Prealgebra: Grades 7–12
Holt, 1986 (Termination Year: 1994)

Houghton Mifflin Mathematics: Structure & Method: Grades 7–12
Houghton, 1988 (Termination Year: 1994)

Prealgebra: Grades 7–12
Houghton, 1988 (Termination Year:1994)

Algebra

Addison–Wesley Algebra: Grades 8–12
Addison-Wesley, 1988 (Termination Year: 1994)

Heath Algebra: Grades 7-12
Heath, 1987 (Termination Year: 1994)

HBJ Algebra: Grades 7-12
Holt, 1987 (Termination Year: 1994)

Holt Algebra I: Grades 7-12
Holt, 1986 (Termination Year: 1994)

Basic Algebra: Grades 7-12
Houghton, 1988 (Termination Year: 1994)

General Mathematics

General Mathematics: A Fundamentals Approach: Grades 7-12
Addison-Wesley, 1986 (Termination Year: 1994)

Heath General Mathematics: Grades 7-12
Heath, 1985 (Termination Year: 1994)

Unified Math

Houghton Mifflin Unified Mathematics: Grades 6
Houghton, 1985 (Termination Year: 1994)

Mathematics

Challenge: A Program for the Mathematically Talented: Grades 3–6
Addison-Wesley, 1986-87 (Termination Year: 1994)

Mathematics Their Way: Grades K–2
Addison-Wesley, 1976 (Termination Year: 1994)

Explorations: Grades 3–6
Addison-Wesley, 1986 (Termination Year: 1994)

The Middle Grades Math Project: Grades 5–8
Addison-Wesley, 1986 (Termination Year: 1994)

Getting Ready for Algebra: Grades
Heath, 1988 (Termination Year: 1994)

Happy Ways to Number: Grades
Holt, 1987 (Termination Year: 1994)

Mastering Math: Grades K–8
Random House, 1977 (Termination Year: 1994)

Extending Math Skills: Grades K–8
Random House, 1977 (Termination Year: 1994)

Practicing Problem Solving: Grades K–8
Random House, 1984 (Termination Year: 1994)

Practicing Math Applications: Grades K–8
Random House, 1984 (Termination Year: 1994)

Spotlight on Math: Grades K–8
Random House, 1984 (Termination Year: 1994)

Problem Solving in Math: Grades K–8
Scholastic, (Termination Year: 1994)

Working with Numbers: Grades K-8
Steck-Vaughn, 1990 (Termination Year: 1994)

Arkansas

Mathematics, Grades 1-6

Addison-Wesley Mathematics: Grades 1-6
Addison-Wesley, 1987 (Termination
Year: 1993)

Mathematics Today: Grades 1-6
Harcourt, 1987 (Termination Year: 1993)

Mathematics Unlimited: Grades 1-6
Harcourt, 1987 (Termination Year: 1993)

Houghton Mifflin Mathematics: Grades 1-6
Houghton, 1987 (Termination Year: 1993)

McGraw-Hill Mathematics: Grades 1-6
Macmillan, 1987 (Termination Year: 1993)

Macmillan Mathematics: Grades 1-6
Macmillan, 1987 (Termination Year: 1993)

Merrill Mathematics: Grades 1-6
Glencoe, 1987 (Termination Year: 1993)

Invitation to Mathematics: Grades 1-6
Scott, Foresman, 1987 (Termination
Year: 1993)

Silver Burdett & Ginn Mathematics:
Grades 1-6
Silver, 1987 (Termination Year: 1993)

The Spectrum Mathematics Series: Grades 1-6
Glencoe, 1984 (Termination Year: 1993)

MCP Mathematics: Grades 1-6
Modern Curriculum, 1987 (Termination
Year: 1993)

Problem Solving in Math: Grades 1-6
Modern Curriculum, 1983 (Termination
Year: 1993)

Mathematics, Grades 7-8

Addison-Wesley Mathematics: Grade 7-8
Addison-Wesley, 1987 (Termination
Year: 1993)

Mathematics Today: Grades 7-8
Harcourt, 1987 (Termination Year: 1993)

Heath Mathematics: Grades 7-8
Heath, 1987 (Termination Year: 1993)

Mathematics Unlimited: Grades 7-8
Holt, 1987 (Termination Year: 1993)

Houghton Mifflin Mathematics: Grades 7-8
Houghton, 1987 (Termination Year: 1993)

Mathematics Structure & Method Course I:
Grades 7-8
Houghton, 1988 (Termination Year: 1993)

Laidlaw Mathematics: Grades 7-8
Glencoe, 1987 (Termination Year: 1993)

Laidlaw Mathematics, Series 2000: Grades 7-8
Glencoe, 1987 (Termination Year: 1993)

McGraw-Hill Mathematics: Grades 7-8
Macmillan, 1987 (Termination Year: 1993)

Merrill Mathematics: Grades 7-8
Glencoe, 1987 (Termination Year: 1993)

Math 76, An Incremental Developer: Grade 7
Saxon, 1985 (Termination Year: 1993)

Invitation to Mathematics: Grades 7-8
Scott, Foresman, 1987 (Termination
Year: 1993)

Silver Burdett & Ginn Mathematics:
Grades 7-8
Silver, 1987 (Termination Year: 1993)

Basic Algebra

Algebra In Easy Steps: Grades 9-12
Prentice Hall, 1982 (Termination Year: 1993)

Basic Algebra: Grades 8-10
Houghton, 1988 (Termination Year: 1993)

Merrill Algebra Essentials: Grades 8-10
Glencoe, 1988 (Termination Year: 1993)

Mathematics, Grades 7-12, Algebra I

Algebra: Grades 9-12
Addison-Wesley, 1986 (Termination
Year: 1993)

HBJ Algebra I: Grades 7-12
Holt, 1987 (Termination Year: 1993)

Heath Algebra I: Grades 7-12
Heath, 1987 (Termination Year: 1993)

Holt Algebra I: Grades 7-12
Holt, 1986 (Termination Year: 1993)

Algebra Structure & Method: Grades 8-11
Houghton, 1988 (Termination Year: 1993)

Algebra 1: Grades 8-11
Houghton, 1986 (Termination Year: 1993)

Laidlaw Algebra I: Grades 8-12
Glencoe, 1987 (Termination Year: 1993)

Merrill Algebra One: Grades 8-10
Glencoe, 1986 (Termination Year: 1993)

Algebra I: Grades 9-12
Saxon, 1981 (Termination Year: 1993)

ScottForesman Algebra I: Grades 9-12
Scott, Foresman, 1984 (Termination
Year: 1993)

Harper & Row Algebra: Grades 7-12
Glencoe, 1985 (Termination Year: 1993)

Prealgebra (Math II)

Prealgebra: Grades 7-9
Addison-Wesley, 1987 (Termination
Year: 1993)

Algebra: An Introductory Course: Grades 7-12
Amsco, 1987 (Termination Year: 1993)

Prealgebra: Skills/Problem Solving/Applications: Grades 7-12
Harcourt, 1986 (Termination Year: 1993)

Heath Prealgebra: Grades 7-10
Heath, 1986 (Termination Year: 1993)

Holt Prealgebra: Grades 7-12
Holt, 1986 (Termination Year: 1993)

Essentials for Algebra: Concepts & Skills:
Grade 9
Houghton, 1986 (Termination Year: 1993)

Preparing to Use Algebra: Grades 7-10
Glencoe, 1986 (Termination Year: 1993)

*Merrill Prealgebra: A Problem Solving
Approach*: Grades 7-9
Glencoe, 1986 (Termination Year: 1993)

Algebra 1/2: Grades 8-9
Saxon, 1983 (Termination Year: 1993)

ScottForesman Algebra: Grades 9-12
ScottForesman, 1987 (Termination Year: 1993)

Algebra II

Algebra & Trigonometry: Grades 10-12
Addison-Wesley, 1986 (Termination
Year: 1993)

HBJ Algebra 2 with Trigonometry:
Grades 7-12
Harcourt, 1987 (Termination Year: 1993)

Heath Algebra II with Trigonometry:
Grades 9-12
Heath, 1987 (Termination Year: 1993)

Holt Algebra 2 with Trigonometry:
Grades 7-12
Holt, 1986 (Termination Year: 1993)

Algebra & Trigonometry Structure & Method:
Grades 9-12
Houghton, 1988 (Termination Year: 1993)

Laidlaw Algebra 2 with Trigonometry:
Grades 9-12
Glencoe, 1987 (Termination Year: 1993)

Merrill Algebra Two with Trigonometry:
Grades 9-12
Glencoe, 1986 (Termination Year: 1993)

Algebra II: Grades 10-12
Saxon, 1984 (Termination Year: 1993)

ScottForesman Algebra II: Grades 9-12
Scott, Foresman, 1984 (Termination
Year: 1993)

Two with Trigonometry: Grades 7-12
Glencoe, 1985 (Termination Year: 1993)

Geometry

Informal Geometry: Grades 9–12
Addison-Wesley, 1986 (Termination Year: 1993)

Geometry: An Informal Approach:
Grades 10–12
Prentice Hall, 1986 (Termination Year: 1993)

HBJ Geometry: Grades 7–12
Harcourt, 1987 (Termination Year: 1993)

Holt Geometry: Grades 7–12
Holt, 1986 (Termination Year: 1993)

Basic Geometry: Grades 9–11
Houghton, 1988 (Termination Year: 1993)

Laidlaw Geometry: Grades 9–12
Glencoe, 1987 (Termination Year: 1993)

Merrill Informal Geometry: Grades 9–11
Glencoe, 1988 (Termination Year: 1993)

Merrill Geometry: Grades 10–11
Glencoe, 1987 (Termination Year: 1993)

ScottForesman Geometry: Grades 9–12
Scott, Foresman, 1984 (Termination Year: 1993)

Harper & Row Geometry: Grades 7–12
Glencoe, 1986 (Termination Year: 1993)

Trigonometry

Trigonometry: Grades 9–12
Harcourt, 1987 (Termination Year: 1993)

Trigonometry with Applications: Grades 10–12
Houghton, 1987 (Termination Year: 1993)

ScottForesman Trigonometry: Grades 9–12
Scott, Foresman, 1983 (Termination Year: 1993)

Calculus

Elements of Calculus & Analytic Geometry:
Grades 9–12
Addison-Wesley, 1981 (Termination Year: 1993)

Calculus with Analytic Geometry: Grades 10–12
Heath, 1986 (Termination Year: 1993)

Advanced Mathematics

Precalculus with Trigonometry: Grades 11–12
Addison-Wesley, 1987 (Termination Year: 1993)

HBJ Advanced Mathematics: A Preparation for Calculus: Grades 9–12
Harcourt, 1988 (Termination Year: 1993)

Advanced Mathematics: A Precalculus Course:
Grades 11–12
Houghton, 1987 (Termination Year: 1993)

ScottForesman Geometry: Grades 9–12
Scott, Foresman, 1987 (Termination Year: 1993)

Harper & Row Geometry: Grades 7–12
Glencoe, 1986 (Termination Year: 1993)

Trigonometry

Trigonometry: Grades 9–12
Harcourt, 1987 (Termination Year: 1993)

Trigonometry with Applications: Grades 10–12
Houghton, 1987 (Termination Year: 1993)

ScottForesman Trigonometry: Grades 9–12
Scott, Foresman, 1983 (Termination Year: 1993)

Elements of Calculus & Analytic Geometry:
Grades 9–12
Addison-Wesley, 1981 (Termination Year: 1993)

Introductory Analysis: Grades 11–12
Houghton, 1988 (Termination Year: 1993)

Merrill Advanced Mathematical Concepts:
Grades 11–12
Glencoe, 1986 (Termination Year: 1993)

Merrill Precalculus: Grades 11–12
Glencoe, 1988 (Termination Year: 1993)

Geometry Trigonometry, Algebra III:
Grades 11–12
Saxon, 1985 (Termination Year: 1993)

ScottForesman Precalculus: Grades 9–12
Scott, Foresman, 1987 (Termination Year: 1993)

General Math 9-12 (Math I)

General Mathematics: Grades 9–12
Addison-Wesley, 1986 (Termination Year: 1993)

Refresher Mathematics: Grades 7–12
Prentice Hall, 1986 (Termination Year: 1993)

Achieving Competence in Mathematics: Grades 7–12
Amsco, 1987 (Termination Year: 1993)

General Mathematics: Skills/Problem Solving/ Applications: Grades 9–12
Harcourt, 1987 (Termination Year: 1993)

Heath General Mathematics: Grades 7–12
Heath, 1985 (Termination Year: 1993)

Fundamentals of Math: Skills & Applications: Grades 9–10
Houghton, 1986 (Termination Year: 1993)

Mathematics Skills for Daily Living: Grades 9–12
Glencoe, 1986 (Termination Year: 1993)

Merrill General Mathematics: Grades 9–12
Glencoe, 1986 (Termination Year: 1993)

Mathematics in Life: Grades 9–12
Scott, Foresman, 1987 (Termination Year: 1993)

General Math III (Consumer Mathematics)

Applying Mathematics: Grades 9–12
Addison-Wesley, 1986 (Termination Year: 1993)

Business Mathematics: Grades 9–12
Glencoe, 1987 (Termination Year: 1993)

Consumer Mathematics: Grades 9–12
Glencoe, 1986 (Termination Year: 1993)

Essentials of Mathematics: Consumer, Career Skills, Applications: Grades 9–12
Harcourt, 1987 (Termination Year: 1993)

Business Mathematics: Grades 9–12
Houghton, 1986 (Termination Year: 1993)

Consumer Mathematics: Grades 9–12
Houghton, 1988 (Termination Year: 1993)

Applying Mathematics in Daily Living: Grades 9–12
Glencoe, 1986 (Termination Year: 1993)

Merrill Applications of Mathematics: Grades 9–12
Glencoe, 1988 (Termination Year: 1993)

Consumer & Career Mathematics: Grades 9–12
Scott, Foresman, 1987 (Termination Year: 1993)

Mathematics for the College Boards: Grades 10–12
Amsco, 1985 (Termination Year: 1993)

Basic Math Skills for Today's Living: Grades 7–12
Hammond, 1986 (Termination Year: 1993)

The Spectrum Mathematics Series: Grades 7–8
Glencoe, 1984 (Termination Year: 1993)

Mastering Essential Mathematics Skills: Grades 9–12
Glencoe, 1988 (Termination Year: 1993)

Mathematics for Business: Grades 7–12
Media Materials, 1986 (Termination Year: 1993)

Sound Foundations: A Practical Mathematics Simulation: Grades 7–12
South-Western, 1987 (Termination Year: 1993)

Occupational Mathematics: Grades 9–12
South-Western, 1986 (Termination Year: 1993)

Mastering Computational Skills: Grades 7–12
Scott, Foresman, 1984 (Termination Year: 1993)

Essential Mathematics for Life: Grades 9–12
Scott, Foresman, 1985 (Termination Year: 1993)

Elementary Algebra, a Guided Inquiry: Grade 9
Sunburst, 1984 (Termination Year: 1993)

Algebra II/Trigonometry, a Guided Inquiry:
Grades 11-12
Sunburst, 1986 (Termination Year: 1993)

Geometry, a Guided Inquiry: Grade 10
Sunburst, 1984 (Termination Year: 1993)

California

State-Adopted Basic Mathematics Instructional Materials

Addison-Wesley Mathematics: Grades K-8
Addison-Wesley, 1989 (Termination
Year: 1993)

Laidlaw Algebra: Grade 8
Glencoe, 1987 (Termination Year: 1993)

Heath Mathematics: Grades K-8
Heath, 1988 (Termination Year: 1993)

Heath Math Worlds: Grades 4-8
Heath, 1986 (Termination Year: 1993)

Mathematics Unlimited: Grades K-8
Harcourt, 1988 (Termination Year: 1993)

Algebra, Structure & Method: Grade 8
Houghton, 1986 (Termination Year: 1993)

Algebra 1: Grade 8
Houghton, 1986 (Termination Year: 1993)

Real Math: Grades K-8
Open Court, 1987 (Termination Year: 1993)

Invitation to Mathematics: Grades K-8 (also in Spanish)
ScottForesman, 1988 (Termination Year: 1993)

Silver Burdett & Ginn Mathematics:
Grades K-8
Silver, 1988 (Termination Year: 1993)

Florida

Mathematics Basal Programs

Addison-Wesley Mathematics: Grades K-8
Addison-Wesley, 1985 (Termination
Year: 1992)

Mathematics Today: Grades K-8 (also in Spanish)
Harcourt, 1985 (Termination Year: 1992)

Heath Mathematics: Grades K-8
Heath, 1985 (Termination Year: 1992)

Houghton Mifflin Mathematics: Grades K-8
Houghton, 1985 (Termination Year: 1992)

Riverside Mathematics: Grades K-8
Houghton, 1987 (Termination Year: 1992)

Series M: Macmillan Mathematics: Grades K-8 (also in Spanish)
Macmillan, 1985 (Termination Year: 1992)

Invitation to Mathematics: Grades K-8 (also in Spanish)
ScottForesman, 1985 (Termination Year: 1992)

Prealgebra

Prealgebra: Grades 9-12
Glencoe, 1986 (Termination Year: 1992)

Prealgebra: An Accelerated Course:
Grades 9-12
Glencoe, 1985 (Termination Year: 1992)

Preparing to Use Algebra: Grades 9-12
Glencoe, 1986 (Termination Year: 1992)

Prealgebra: Skills/Problem Solving/Applications: Grades 9-12
Harcourt, 1986 (Termination Year: 1992)

Heath Prealgebra: Grades 9-12
Heath, 1986 (Termination Year: 1992)

Holt Prealgebra: Grades 9-12
Holt, 1986 (Termination Year: 1992)

Prealgebra, Grades 9-12

Addison-Wesley Prealgebra: Grades 7-12
Addison-Wesley, 1991 (Termination
Year: 1992)

Merrill Prealgebra: A Problem-Solving Approach: Grades 7-12
Glencoe, 1989 (Termination Year: 1992)

Heath Prealgebra: Grades 7-12
Heath, 1990 (Termination Year: 1992)

Prealgebra: An Accelerated Course:
Grades 7-12
Houghton, 1988 (Termination Year: 1992)

University of Chicago School Mathematics Project: Transition Mathematics: Grades 7-12
ScottForesman, 1990 (Termination Year: 1993)

Algebra

Addison-Wesley Algebra: Grades 9-12
Addison-Wesley, 1990 (Termination
Year: 1992)

Merrill Algebra One: Grades 9-12
Glencoe, 1990 (Termination Year: 1992)

Algebra: Structure & Method: Grades 9-12
Houghton, 1990 (Termination Year: 1992)

Prentice Hall Algebra: Grades 9-12
Prentice Hall, 1990 (Termination Year: 1992)

University of Chicago School Mathematics Project: Algebra: Grades 9-12
ScottForesman, 1990 (Termination Year: 1993)

Algebra Honors, Grades 9-12

Algebra 1: Expressions, Equations & Applications: Grades 9-12
Addison-Wesley, 1990 (Termination
Year: 1992)

Algebra 1: Grades 9-12
Houghton, 1989 (Termination Year: 1992)

Algebra 1: An Integrated Approach:
Grades 9-12
McDougal, 1991 (Termination Year: 1992)

Algebra II, Grades 9-12

Addison-Wesley Algebra & Trigonometry:
Grades 10-12
Addison-Wesley, 1990 (Termination
Year: 1992)

HBJ Algebra 2 with Trigonometry:
Grades 9-12
Holt, 1990 (Termination Year: 1992)

Algebra & Trigonometry: Structure & Method:
Grades 9-12
Houghton, 1990 (Termination Year: 1992)

Prentice Hall Algebra 2 with Trigonometry:
Grades 9-12
Prentice Hall, 1990 (Termination Year: 1992)

University of Chicago School Mathematics Project: Advanced Algebra: Grades 9-12
ScottForesman, 1990 (Termination Year: 1993)

Algebra II Honors, Grades 9-12

Algebra & Trigonometry: Functions & Applications: Grades 10-12
Addison-Wesley, 1990 (Termination
Year: 1992)

Algebra 2 & Trigonometry: Grades 9-12
Houghton, 1989 (Termination Year: 1992)

Algebra 2 & Trigonometry: Grades 9-12
McDougal, 1991 (Termination Year: 1992)

Precalculus

Precalculus with Trigonometry: Functions & Applications: Grades 11-12
Addison-Wesley, 1987 (Termination
Year: 1992)

Precalculus: Grades 10-12
Heath, 1988 (Termination Year: 1992)

Introductory Analysis: Grades 10-11
Houghton, 1991 (Termination Year: 1992)

Before Calculus: Grades 9-12
ScottForesman, 1989 (Termination Year: 1992)

Precalculus with Unit Circle Trigonometry:
Grades 9-12
West, 1990 (Termination Year: 1992)

Calculus, Advanced Placement Calculus AB

Elements of Calculus & Analytic Geometry:
Grades 12
Addison-Wesley, 1981 (Termination
Year: 1992)

Calculus with Analytic Geometry: Grades 12
Heath, 1986 (Termination Year: 1992)

Advanced Placement Calculus BC

Calculus & Analytic Geometry: Grades 12
Addison-Wesley, 1984 (Termination
Year: 1992)

Computer Applications

Using Computers in Mathematics:
Grades 9-12
Addison-Wesley, 1986 (Termination
Year: 1992)

General Mathematics, Applying Basic Skills in Mathematics

*Applying Mathematics, a Consumer/Career
Approach*: Grades 9-12
Addison-Wesley, 1986 (Termination
Year: 1992)

Basic Skills in Mathematics

Mathematics Skills for Daily Living:
Grades 9-12
Glencoe, 1986 (Termination Year: 1992)

Holt Math 1000: Grades 9-12
Holt, 1984 (Termination Year: 1992)

Business Mathematics I

Business Mathematics: Grades 9-12
Glencoe, 1986 (Termination Year: 1992)

Applied Business Mathematics: Grades 9-12
South-Western, 1985 (Termination Year: 1992)

Consumer Mathematics

Consumer Mathematics: Grades 9-12
Glencoe, 1986 (Termination Year: 1992)

Mathematics for Daily Living: Grades 9-12
Glencoe, 1986 (Termination Year: 1992)

Mathematics for Today's Consumer:
Grades 9-12
Glencoe, 1982 (Termination Year: 1992)

*Essentials of Mathematics: Consumer & Career
Skills & Applications*: Grades 9-12
Harcourt, 1983 (Termination Year: 1992)

Consumer & Career Mathematics:
Grades 9-12
ScottForesman, 1985 (Termination Year: 1992)

Fundamental Mathematics I

Fundamentals of Mathematics: skills and Applications: Grades 9-12
Glencoe, 1986 (Termination Year: 1992)

Fundamental Mathematics II

Applying Mathematics in Daily Living:
Grades 9-12
Glencoe, 1986 (Termination Year: 1992)

General Mathematics I

*General Mathematics, a Fundamental
Approach*: Grades 9-12
Addison-Wesley, 1986 (Termination
Year: 1992)

Refresher Mathematics: Grades 9-12
Prentice Hall, 1986 (Termination Year: 1992)

Mathematics for the real World: Grades 9-12
Glencoe, 1982 (Termination Year: 1992)

*General Mathematics: Skill/Problem Solving/
Application*: Grades 9-12
Harcourt, 1982 (Termination Year: 1992)

Heath General Mathematics: Grades 9-12
Heath, 1985 (Termination Year: 1992)

Mathematics in Life: Grades 9-12
ScottForesman, 1985 (Termination Year: 1992)

General Mathematics II

Mathematics for Everyday Life: Grades 9-12
Glencoe, 1982 (Termination Year: 1992)

Holt General Mathematics: Grades 9-12
Holt, 1982 (Termination Year: 1992)

General Mathematics III

Essentials for Algebra: concepts & Skills:
Grades 9-12
Glencoe, 1986 (Termination Year: 1992)

Analytic Geometry, Grades 9-12

Modern Analytic Geometry: Grades 9-12
Houghton, 1989 (Termination Year: 1992)

Geometry

Geometry with Applications & Problem Solving: Grades 9-12
Addison-Wesley, 1984 (Termination Year: 1992)

Geometry: Grades 9-12
Glencoe, 1985 (Termination Year: 1992)

Merrill Geometry: Grades 9-12
Glencoe, 1987 (Termination Year: 1992)

HBJ Geometry: Grades 9-12
Holt, 1987 (Termination Year: 1992)

Holt Geometry: Grades 9-12
Holt, 1986 (Termination Year: 1992)

ScottForesman Geometry: Grades 9-12
ScottForesman, 1987 (Termination Year: 1992)

Geometry Honors

Geometry: Grades 9-12
Addison-Wesley, 1982 (Termination Year: 1992)

Informal Geometry

Addison-Wesley Informal Geometry: Grades 9-12
Addison-Wesley, 1986 (Termination Year: 1992)

Basic Geometry: Grades 9-12
Glencoe, 1984 (Termination Year: 1992)

Geometry: An Informal Approach: Grades 9-12
Prentice Hall, 1986 (Termination Year: 1992)

Integrated Mathematics, Grades 9-12

Integrated Mathematics, Course 3: Grades 11
Amsco, 1991 (Termination Year: 1992)

Merrill Integrated Mathematics, Course 3: Grades 9-12
Glencoe, 1991 (Termination Year: 1992)

Houghton Mifflin Unified Mathematics: Grades 9-12
Houghton, 1991 (Termination Year: 1992)

Liberal Arts Mathematics, Grades 9-12

Merrill Integrated Mathematics, Course I: Grades 9-12
Glencoe, 1991 (Termination Year: 1992)

Houghton Mifflin Unified Mathematics: Grades 9-12
Houghton, 1991 (Termination Year: 1992)

Mathematical Analysis, Grades 9-12

Merrill Advanced Mathematical Concepts: Grades 9-12
Glencoe, 1991 (Termination Year: 1992)

Advanced Mathematics: A Precalculus Course: Grades 9-12
Houghton, 1989 (Termination Year: 1992)

M/J Mathematics 1, 2, 3 Functional Skills

The Spectrum Mathematics Series: Grades 5-8
Glencoe, 1984 (Termination Year: 1992)

Mastering Computational Skills: Grades 6-8
ScottForesman, 1984 (Termination Year: 1992)

M/J Mathematics 2, Advanced

Introductory Algebra 1: Grades 6-8
Harcourt, 1982 (Termination Year: 1992)

Elementary Algebra, Part 1: Grade 7
Houghton, 1985 (Termination Year: 1992)

Mathematics: Structure & Method: Grade 7
Houghton, 1985 (Termination Year: 1992)

M/J Mathematics 3, Advanced

Elementary Algebra, Part 2: Grade 8
Glencoe, 1985 (Termination Year: 1992)

Mathematics: Structure & Method Course 2: Grade 8
Glencoe, 1985 (Termination Year: 1992)

Introductory Algebra 2: Grades 7-12
Harcourt, 1982 (Termination Year: 1992)

Probability & Statistics with Applications

Using Statistics: Grades 9-12
Addison-Wesley, 1985 (Termination
Year: 1992)

Remedial/Compensation Mathematics

Mastering Essential Mathematics Skills:
Grades 9-12
Glencoe, 1984 (Termination Year: 1992)

Trigonometry, Grades 9-12

Trigonometry: Functions & Applications:
Grades 11-12
Addison-Wesley, 1990 (Termination
Year: 1992)

Trigonometry & Its Applications: Grades 9-12
Glencoe, 1990 (Termination Year: 1992)

Trigonometry with Applications: Grades 10-12
Houghton, 1990 (Termination Year: 1992)

Prentice Hall Trigonometry: Grades 11-12
Prentice Hall, 1990 (Termination Year: 1992)

Fundamentals of Trigonometry: Grades 9-12
ScottForesman, 1990 (Termination Year: 1992)

Georgia

Elementary Mathematics

Addison-Wesley Mathematics: Grades K-8
Addison-Wesley, 1987 (Termination
Year: 1993)

Mathematics Today: Grades K-8
Harcourt, 1987 (Termination Year: 1993)

Heath Mathematics: Grades K-8
Heath, 1988 (Termination Year: 1993)

Mathematics Unlimited: Grades K-8
Harcourt, 1987 (Termination Year: 1993)

Houghton Mifflin Mathematics: Grades K-8
Macmillan, 1987 (Termination Year: 1993)

Macmillan Mathematics: Grades K-8
Macmillan, 1987 (Termination Year: 1993)

McGraw-Hill Mathematics: Grades K-8
Macmillan, 1987 (Termination Year: 1993)

Merrill Mathematics: Grades K-8
Glencoe, 1987 (Termination Year: 1993)

ScottForesman Invitation to Mathematics:
Grades K-8
ScottForesman, 1987 (Termination Year: 1993)

Great Beginnings: Grades K-2
ScottForesman, 1992 (Termination Year: 1993)

Silver Burdett & Ginn Mathematics:
Grades K-8
Silver, 1987 (Termination Year: 1993)

Algebra

Addison-Wesley Prealgebra: Grades 7-12
Addison-Wesley, 1991 (Termination
Year: 1993)

Addison-Wesley Algebra 1: Grades 8-12
Addison-Wesley, 1988 (Termination
Year: 1993)

Addison-Wesley Algebra & Trigonometry:
Grades 10-12
Addison-Wesley, 1988 (Termination
Year: 1993)

*Merrill Prealgebra: A Problem-Solving
Approach*: Grades 7-9
Glencoe, 1986 (Termination Year: 1993)

Merrill Algebra Essentials: Grades 9-12
Glencoe, 1988 (Termination Year: 1993)

Merrill Algebra 1: Grades 8-10
Glencoe, 1986 (Termination Year: 1993)

Merrill Algebra 2 with Trigonometry:
Grades 10-12
Glencoe, 1986 (Termination Year: 1993)

Preparing to Use Algebra: Grades 7-9
Glencoe, 1986 (Termination Year: 1993)

Harper & Row Algebra 1: Grades 9-12
Glencoe, 1985 (Termination Year: 1993)

Harper & Row Algebra II with Trigonometry:
Grades 9-12
Glencoe, 1985 (Termination Year: 1993)

Laidlaw Algebra 1: Grades 8-12
Glencoe, 1987 (Termination Year: 1993)

Laidlaw Algebra 2 with Trigonometry:
Grades 9-12
Glencoe, 1987 (Termination Year: 1993)

Prealgebra: Skills/Problem Solving/Applications: Grades 7-12
Holt, 1986 (Termination Year: 1993)

HBJ Algebra 1: Grades 8-12
Holt, 1987 (Termination Year: 1993)

HBJ Algebra 2 with Trigonometry:
Grades 9-12
Holt, 1987 (Termination Year: 1993)

HBJ Introductory Algebra: Grades 8-12
Holt, 1988 (Termination Year: 1993)

Heath Algebra: Grades 9-12
Heath, 1987 (Termination Year: 1993)

Heath Prealgebra: Grades 9-12
Heath, 1986 (Termination Year: 1993)

Holt Prealgebra: Grades 7-12
Holt, 1986 (Termination Year: 1993)

Holt Algebra 1: Grades 9-12
Holt, 1986 (Termination Year: 1993)

Holt Algebra 2 with Trigonometry:
Grades 9-12
Holt, 1986 (Termination Year: 1993)

Mathematics: Structure & Method: Grades 7-8
Houghton, 1988 (Termination Year: 1993)

Elementary Algebra: Grades 8-12
Houghton, 1988 (Termination Year: 1993)

Prealgebra: An Accelerated Course: Grade 7
Houghton, 1988 (Termination Year: 1993)

Algebra: Structure & Method: Grades 8-12
Houghton, 1988 (Termination Year: 1993)

Algebra & Trigonometry Structure & Method:
Grades 9-12
Houghton, 1986 (Termination Year: 1993)

Algebra 1: Grades 8-12
Houghton, 1986 (Termination Year: 1993)

Algebra 2 & Trigonometry: Grades 9-12
Houghton, 1986 (Termination Year: 1993)

Essentials for Algebra: Concepts & Skills:
Grades 7-12
Houghton, 1986 (Termination Year: 1993)

ScottForesman Prealgebra: Grades 7-9
ScottForesman, 1987 (Termination Year: 1993)

ScottForesman Algebra: Grades 9-12
ScottForesman, 1984 (Termination Year: 1993)

*University of Chicago School Mathematics
Project: Transition Mathematics*: Grades 7-12
ScottForesman, 1990 (Termination Year: 1993)

*University of Chicago School Mathematics
Project: Algebra*: Grades 9-12
ScottForesman, 1990 (Termination Year: 1993)

*University of Chicago School Mathematics
Project: Advanced Algebra*: Grades 9-12
ScottForesman, 1990 (Termination Year: 1993)

Geometry

*Addison-Wesley Geometry with Applications &
Problem Solving*: Grades 9-12
Addison-Wesley, 1984 (Termination
Year: 1993)

Addison-Wesley Informal Geometry:
Grades 7-12
Addison-Wesley, 1986 (Termination
Year: 1993)

Merrill Geometry: Grades 10-11
Glencoe, 1987 (Termination Year: 1993)

Harper & Row Geometry: Grades 10-11
Glencoe, 1986 (Termination Year: 1993)

Laidlaw Geometry: Grades 10-12
Glencoe, 1987 (Termination Year: 1993)

HBJ Geometry: Grades 9-12
Holt, 1987 (Termination Year: 1993)

Holt Geometry: Grades 9-12
Holt, 1986 (Termination Year: 1993)

Basic Geometry: Grades 9-12
Houghton, 1988 (Termination Year: 1993)

Geometry: Grades 9-12
Houghton, 1988 (Termination Year: 1993)

ScottForesman Geometry: Grades 10-12
ScottForesman, 1987 (Termination Year: 1993)

University of Chicago School Mathematics Project: Geometry: Grades 9-12
ScottForesman, 1991 (Termination Year: 1993)

Trigonometry

Trigonometry: Grades 10-12
Harcourt, 1987 (Termination Year: 1993)

Trigonometry with Applications: Grades 11-12
Houghton, 1987 (Termination Year: 1993)

ScottForesman Trigonometry: Grades 11-12
ScottForesman, 1983 (Termination Year: 1993)

General Mathematics

General Mathematics I: Grades 9-12
Addison-Wesley, 1986 (Termination Year: 1993)

Applying Mathematics: Grades 9-12
Addison-Wesley, 1986 (Termination Year: 1993)

Achieving Competence in Mathematics:
Grades 9-12
Amsco, 1987 (Termination Year: 1993)

Merrill General Mathematics: Grades 9-12
Glencoe, 1987 (Termination Year: 1993)

Consumer Mathematics: Grades 9-12
Glencoe, 1988 (Termination Year: 1993)

Merrill Applications of Mathematics:
Grades 9-12
Glencoe, 1988 (Termination Year: 1993)

Mathematics Skills for Daily Living:
Grades 9-12
Glencoe, 1986 (Termination Year: 1993)

Applying Mathematics in Daily Living:
Grades 9-12
Glencoe, 1986 (Termination Year: 1993)

Globe High School Mathematics: Grades 9-12
Globe, 1989 (Termination Year: 1993)

General Mathematics: Skills/Problem Solving/ Applications: Grades 9-12
Harcourt, 1987 (Termination Year: 1993)

Essentials of Mathematics: Consumer/Carer Skills/Applications: Grades 9-12
Harcourt, 1987 (Termination Year: 1993)

Heath General Mathematics: Grades 9-12
Heath, 1985 (Termination Year: 1993)

Holt Essential Mathematics: Grades 9-12
Holt, 1987 (Termination Year: 1993)

Fundamentals of Mathematics: Skills & Applications: Grades 9-12
Houghton, 1986 (Termination Year: 1993)

Basic Mathematic Skills: Grades 9-12
Media Materials, 1988 (Termination Year: 1993)

Mathematics for Consumers: Grades 9-12
Media Materials, 1983 (Termination Year: 1993)

Refresher Mathematics: Grades 8-10
Prentice Hall, 1986 (Termination Year: 1993)

ScottForesman Mathematics in Life:
Grades 9-12
ScottForesman, 1987 (Termination Year: 1993)

Consumer & Career Mathematics:
Grades 9-12
ScottForesman, 1987 (Termination Year: 1993)

Advanced Mathematics

Transition to College Mathematics:
Grades 9-12
Addison-Wesley, 1984 (Termination Year: 1993)

Using Computers in Mathematics:
Grades 9-12
Addison-Wesley, 1986 (Termination
Year: 1993)

*Precalculus with Trigonometry: Functions &
Applications*: Grades 11-12
Addison-Wesley, 1987 (Termination
Year: 1993)

Calculus & Analytic Geometry: Grades 11-12
Addison-Wesley, 1984 (Termination
Year: 1993)

Using Statistics: Grades 9-12
Addison-Wesley, 1985 (Termination
Year: 1993)

Merrill Advanced Mathematical Concepts:
Grades 11-12
Glencoe, 1986 (Termination Year: 1993)

*HBJ Advanced Mathematics: A Preparation for
Calculus*: Grades 10-12
Holt, 1988 (Termination Year: 1993)

*Computer Literacy with an Introduction to
Basic Programming*: Grades 7-12
Harcourt, 1986 (Termination Year: 1993)

Computer Programming in the Basic Language:
Grades 11-12
Harcourt, 1987 (Termination Year: 1993)

Precalculus: Grades 10-12
Heath, 1985 (Termination Year: 1993)

Calculus with Analytic Geometry: Grades
10-12
Heath, 1986 (Termination Year: 1993)

Calculus with Analytic Geometry: Grades
10-12
Heath, 1986 (Termination Year: 1993)

*Introductory Statistics and Probability: A Basis
for Decision Making*: Grades 11-12
Houghton, 1988 (Termination Year: 1993)

Advanced Mathematics: A Precalculus Course:
Grades 11-12
Houghton, 1987 (Termination Year: 1993)

Introductory Analysis (Advanced): Grades
11-12
Houghton, 1988 (Termination Year: 1993)

Modern Analytic Geometry: Grades 11-12
Houghton, 1986 (Termination Year: 1993)

Calculus with Analytic Geometry: Grades
11-12
Prentice Hall, 1986 (Termination Year: 1993)

Statistics: A First Course: Grades 11-12
Prentice Hall, 1986 (Termination Year: 1993)

ScottForesman Precalculus: Grades 11-12
ScottForesman, 1987 (Termination Year: 1993)

Precalculus: Grades 10-12
West, 1987 (Termination Year: 1993)

Advanced Mathematics

*Precalculus Mathematics: A Graphing
Approach*: Grades 11-12
Addison-Wesley, 1990 (Termination
Year: 1993)

Idaho

Mathematics, Basic

Addison-Wesley Mathematics: Grades K-8
Addison-Wesley, 1991 (Termination
Year: 1993)

Math, Their Way: Grades K-2
Addison-Wesley, 1976 (Termination
Year: 1993)

Explorations: Grades K-2
Addison-Wesley, 1988 (Termination
Year: 1993)

Merrill Mathematics: Grades 7-12
Glencoe, 1987 (Termination Year: 1993)

Merrill General Mathematics: Grades K-8
Glencoe, 1987 (Termination Year: 1993)

Merrill Applications of Mathematics:
Grades 7-12
Glencoe, 1988 (Termination Year: 1993)

Computational Estimation: Grades 6-8
Dale Seymour, 1987 (Termination Year: 1993)

Merrill Integrated Mathematics: Grades 8-12
Glencoe, 1991 (Termination Year: 1993)

Mathematics Unlimited: Grades K-8
Harcourt, 1987 (Termination Year: 1993)

Mathematics Today: Grades K-8
Harcourt, 1987 (Termination Year: 1993)

General Mathematics: Skills/Problem Solving/ Applications: Grades K-8
Harcourt, 1987 (Termination Year: 1993)

HBJ Fundamentals of Math: Grades 7-12
Holt, 1989 (Termination Year: 1993)

Heath Mathematics: Grades K-8
Heath, 1988 (Termination Year: 1993)

Heath General Mathematics: Grades 7-12
Heath, 1989 (Termination Year: 1993)

Houghton Mifflin Mathematics: Grades K-8
Houghton, 1989 (Termination Year: 1993)

Mathematics. Structure & Method: Grades 7-8
Houghton, 1988 (Termination Year: 1993)

Happy Ways to Numbers: Grades K
Houghton, 1989 (Termination Year: 1993)

Holt Practical Math: Skills and Concepts:
Grades 7-12
Holt, 1989 (Termination Year: 1993)

Holt Practical Math: Consumer Application:
Grades 7-12
Holt, 1989 (Termination Year: 1993)

Mathematics Unlimited: Grades K-8
Holt, 1988 (Termination Year: 1993)

Holt Essential Mathematics: Grades 7-12
Holt, 1986 (Termination Year: 1993)

Macmillan Mathematics: Grades K-8
Macmillan, 1987 (Termination Year: 1993)

Mathematics in Action: Grades K-8
Macmillan, 1991 (Termination Year: 1993)

McGraw-Hill Mathematics: Grades K-8
Macmillan, 1987 (Termination Year: 1993)

Real Math: Grades K-8
Open Court, 1991 (Termination Year: 1993)

Moving into Math Series: Grades K-2
Rigby Education, 1990 (Termination
Year: 1993)

Read and Count Books: Grades K
Rigby Education, 1987 (Termination
Year: 1993)

Read Together Problem-Solving Books:
Grades K-2
Rigby Education, 1990 (Termination
Year: 1993)

Silver Burdett & Ginn Mathematics:
Grades K-8
Silver, 1988 (Termination Year: 1993)

Mathematics: Exploring Your World:
Grades K-8
Silver, 1991 (Termination Year: 1993)

Invitation to Mathematics: Grades K-8
ScottForesman, 1988 (Termination Year: 1993)

Exploring Mathematics: Grades K-8
ScottForesman, 1991 (Termination Year: 1993)

Supplemental

SOLVE Action Problem Solving: Grades 4-12
Curriculum Associates, 1987 (Termination
Year: 1993)

The Art & Techniques of Simulation:
Grades 7-12
Dale Seymour, 1987 (Termination Year: 1993)

Math 65: Grades 5-6
Saxon, 1987 (Termination Year: 1993)

Math 76: Grades 6-7
Saxon, 1986 (Termination Year: 1993)

SRA Mathematics: Grades 6
Science Research, 1989 (Termination
Year: 1993)

*Remarkable Math: Money/Word Problems/
Computation*: Grades 1-4
Science Research, 1985 (Termination
Year: 1993)

Schoolhouse Mathematics 1, 2, 3: Grades 1-4
Science Research, 1988 (Termination
Year: 1993)

Mathematics Laboratory: Grades 5
Science Research, 1988 (Termination
Year: 1993)

Basic Essentials of Math: Grades 5-9
Steck-Vaughn, 1986 (Termination Year: 1993)

Mastering Math: Grades 1-6
Steck-Vaughn, 1988 (Termination Year: 1993)

Working with Numbers: Grades 1-12
Steck-Vaughn, 1987 (Termination Year: 1993)

Focusing on Computers (Math): Grades 7-12
Steck-Vaughn, 1986 (Termination Year: 1993)

Prealgebra/Algebra/Geometry, Basic

*Merrill Prealgebra: A Problem Solving
Approach*: Grades 7-12
Glencoe, 1989 (Termination Year: 1993)

Merrill Algebra Essentials: Grades 7-12
Glencoe, 1988 (Termination Year: 1993)

Merrill Algebra One: Grades 7-12
Glencoe, 1990 (Termination Year: 1993)

*Prealgebra: Skills/Problem Solving/Applica-
tions*: Grade 8
Harcourt, 1986 (Termination Year: 1993)

HBJ Introductory Algebra 1 & 2: Grades 7-12
Holt, 1988 (Termination Year: 1993)

HBJ Algebra 1: Grades 7-12
Holt, 1990 (Termination Year: 1993)

HBJ Algebra 2 with Trigonometry:
Grades 7-12
Holt, 1990 (Termination Year: 1993)

Getting Ready for Algebra: Grades 7-12
Heath, 1988 (Termination Year: 1993)

Heath Prealgebra: Grades 7-12
Heath, 1987 (Termination Year: 1993)

Heath Algebra 1: Grades 7-12
Heath, 1987 (Termination Year: 1993)

Heath Algebra 2: Grades 9-12
Heath, 1987 (Termination Year: 1993)

Algebra Structure & Method: Grades 9-12
Houghton, 1990 (Termination Year: 1993)

Holt Prealgebra: Grades 7-12
Holt, 1986 (Termination Year: 1993)

Algebra I: Grades 7-12
Holt, 1986 (Termination Year: 1993)

Algebra II with Trigonometry: Grades 7-12
Holt, 1986 (Termination Year: 1993)

Geometry: Grades 7-12
Houghton, 1991 (Termination Year: 1993)

Integrated Mathematics: Grades 8-12
McDougal, 1991 (Termination Year: 1993)

Algebra 1, an Integrated Approach:
Grades 8-12
McDougal, 1991 (Termination Year: 1993)

Algebra 2 & Trigonometry: Grades 9-12
McDougal, 1991 (Termination Year: 1993)

Geometry for Enjoyment and Challenge:
Grades 9-12
McDougal, 1991 (Termination Year: 1993)

Algebra 1/2: Grades 7-10
Saxon, 1990 (Termination Year: 1993)

Algebra I: Grades 8-12
Saxon, 1990 (Termination Year: 1993)

Algebra II: Grades 9-12
Saxon, 1984 (Termination Year: 1993)

ScottForesman Prealgebra: Grades 7-9
ScottForesman, 1987 (Termination Year: 1993)

*University of Chicago School Mathematics
Project: Transition Math*: Grades 7-12
ScottForesman, 1991 (Termination Year: 1993)

University of Chicago School Mathematics Project: Algebra: Grades 8-12
ScottForesman, 1990 (Termination Year: 1993)

Geometry: Grades 8-12
Freeman, 1987 (Termination Year: 1993)

Supplemental

Algebra: Grades 6-12
Greenhouse, 1987 (Termination Year: 1993)

Consumer Mathematics, Basic

Addison-Wesley Consumer Mathematics:
Grades 9-12
Addison-Wesley, 1989 (Termination
Year: 1993)

Essentials of Mathematics: consumer, Career Skills, Applications: Grades 7-12
Harcourt, 1987 (Termination Year: 1993)

HBJ Consumer Mathematics: Grades 7-12
Holt, 1989 (Termination Year: 1993)

Holt Practical Math: Consumer Applications:
Grades 7-12
Holt, 1989 (Termination Year: 1993)

Consumer and Career Mathematics:
Grades 7-12
ScottForesman, 1989 (Termination Year: 1993)

Supplemental

Developing Consumer Attitudes: Grades 7-12
South-Western, 1986 (Termination Year: 1993)

Statistics

Quantitative Literacy Series: Grades 7-12
Dale Seymour, 1986 (Termination Year: 1993)

Exploring Data: Grades 7-12
Dale Seymour, 1986 (Termination Year: 1993)

Exploring Probability: Grades 7-10
Dale Seymour, 1987 (Termination Year: 1993)

When Are We Ever Gonna Have To Use This:
Grades 7-12
Dale Seymour, 1988 (Termination Year: 1993)

Elementary Statistics: Grades K-6
Heath, 1986 (Termination Year: 1993)

Kentucky

Elementary Mathematics (K-4)

Addison-Wesley Mathematics: Grades K-4
Addison-Wesley, 1991 (Termination
Year: 1997)

Mathematics Unlimited: Grades K-4
Harcourt, 1991 (Termination Year: 1997)

Houghton Mifflin Mathematics: Grades K-4
Houghton, 1991 (Termination Year: 1997)

Mathematics in Action: Grades K-4
Macmillan, 1991 (Termination Year: 1997)

Exploring Mathematics: Grades K-4
ScottForesman, 1991 (Termination Year: 1997)

Mathematics: Exploring Your World:
Grades K-4
Silver, 1991 (Termination Year: 1997)

Real Math: Grades K-4
Open Court, 1991 (Termination Year: 1997)

Middle School Mathematics (5-8)

Addison-Wesley Mathematics: Grades 5-8
Addison-Wesley, 1991 (Termination
Year: 1997)

Mathematics Unlimited: Grades 5-8
Harcourt, 1991 (Termination Year: 1997)

Houghton Mifflin Mathematics: Grades 5-8
Houghton, 1991 (Termination Year: 1997)

Mathematics in Action: Grades 5-8
Macmillan, 1991 (Termination Year: 1997)

Exploring Mathematics: Grades 5-8
ScottForesman, 1991 (Termination Year: 1997)

Mathematics: Exploring Your World:
Grades 5-8
Silver, 1991 (Termination Year: 1997)

Real Math: Grades 5-8
Open Court, 1991 (Termination Year: 1997)

Math 54: Grade 5
Saxon, 1990 (Termination Year: 1997)

Math 65: Grades 5–6
Saxon, 1987 (Termination Year: 1997)

Math 76: Grade 6-8
Saxon, 1985 (Termination Year: 1997)

Algebra 1/2: Grade 8
Saxon, 1990 (Termination Year: 1997)

Prealgebra (7-8)

Addison-Wesley Prealgebra: Grades 7–8
Addison-Wesley, 1991 (Termination
Year: 1997)

*Merrill Prealgebra: A Problem Solving
Approach*: Grades 7–8
Glencoe, 1989 (Termination Year: 1997)

Holt Prealgebra: Grades 7–8
Harcourt, 1986 (Termination Year: 1997)

Mathematics: Structure & Method: Grades 7–8
Houghton, 1988 (Termination Year: 1997)

Prealgebra: An Accelerated Course:
Grades 7–8
Houghton, 1988 (Termination Year: 1997)

*University of Chicago School Mathematics
Project: Transitional Math*: Grades 7–8
ScottForesman, 1990 (Termination Year: 1997)

Algebra 1/2: Grades 7–8
Saxon, 1990 (Termination Year: 1997)

Algebra I (8)

Merrill Algebra One: Grade 8
Glencoe, 1990 (Termination Year: 1997)

Algebra 1: Grade 7-10
Houghton, 1989 (Termination Year: 1997)

Algebra: Structure & Method: Grade 8
Houghton, 1990 (Termination Year: 1997)

Algebra 1: Grade 8
Prentice Hall, 1990 (Termination Year: 1997)

*University of Chicago Mathematics Project:
Algebra*: Grade 8
ScottForesman, 1990 (Termination Year: 1997)

Algebra 1, an Integrated Approach: Grade 8
McDougal, 1991 (Termination Year: 1997)

Algebra 1: Grade 8
Saxon, 1990 (Termination Year: 1997)

Integrated Mathematics (8)

Merrill Integrated Math: Grade 8
Glencoe, 1991 (Termination Year: 1997)

HBJ Introductory Algebra 1: Grade 8
Holt, 1988 (Termination Year: 1997)

Houghton Mifflin Unified Math: Grade 8
Houghton, 1991 (Termination Year: 1997)

Prealgebra (9-10)

Addison-Wesley Prealgebra: Grades 9–10
Addison-Wesley, 1991 (Termination
Year: 1997)

*Merrill Prealgebra: A Problem Solving
Approach*: Grades 9–10
Glencoe, 1989 (Termination Year: 1997)

Holt Prealgebra: Grades 9–10
Harcourt, 1986 (Termination Year: 1997)

Essentials for Algebra: Concepts & Skills:
Grades 9–10
Houghton, 1986 (Termination Year: 1997)

*University of Chicago Mathematic Project:
Transition Math*: Grades 9–10
ScottForesman, 1990 (Termination Year: 1997)

Algebra 1/2: Grades 9–10
Saxon, 1990 (Termination Year: 1997)

Fundamental Mathematics (9)

Essentials of Mathematics: Grade 9
Addison-Wesley, 1989 (Termination
Year: 1997)

Merrill General Mathematics: Grade 9
Glencoe, 1987 (Termination Year: 1997)

Practical Mathematics: Skills & Concepts:
Grade 9
Holt, 1989 (Termination Year: 1997)

Essentials for High School Math: Grade 9
Houghton, 1989 (Termination Year: 1997)

Applications of High School Math: Grade 9
Houghton, 1990 (Termination Year: 1997)

Refresher Mathematics: Grade 9
Prentice Hall, 1989 (Termination Year: 1997)

Mathematics in Life: Grade 9
ScottForesman, 1989 (Termination Year: 1997)

Essential Math (9-10)

Essentials of Mathematics: Grades 9-10
Addison-Wesley, 1989 (Termination Year: 1997)

Merrill Applications of Mathematics:
Grades 9-10
Glencoe, 1988 (Termination Year: 1997)

Heath General Mathematics: Grades 9-10
Heath, 1989 (Termination Year: 1997)

Practical Mathematics: Skills & Concepts:
Grades 9-10
Holt, 1989 (Termination Year: 1997)

Applications of High School Math:
Grades 9-10
Houghton, 1990 (Termination Year: 1997)

Essentials for High School Math: Grades 9-10
Houghton, 1990 (Termination Year: 1997)

Refresher Mathematics: Grade 9
Prentice Hall, 1989 (Termination Year: 1997)

Mathematics in Life: Grades 9-10
Harcourt, 1989 (Termination Year: 1997)

Informal Geometry

Addison-Wesley Informal Geometry:
Grades 10-12
Addison-Wesley, 1986 (Termination Year: 1997)

Merrill Informal Geometry: Grades 10-12
Glencoe, 1988 (Termination Year: 1997)

Basic Geometry: Grades 10-12
Houghton, 1990 (Termination Year: 1997)

Discovering Geometry: An Individual Approach: Grades 10-12
Key Curriculum, 1989 (Termination Year: 1997)

Basic Integrated Math I (9-12)

Heath General Mathematics: Grades 9-12
Heath, 1989 (Termination Year: 1997)

Basic Integrated Math III (11-12)

Advanced Mathematics: Grades 11-12
Saxon, 1989 (Termination Year: 1997)

Technical Applied Math (9-12)

Career Mathematics: Grades 9-12
Houghton, 1989 (Termination Year: 1997)

Consumer Math (10-12)

Consumer Mathematics: Grades 10-12
Addison-Wesley, 1989 (Termination Year: 1997)

Consumer Mathematics: Grades 10-12
Glencoe, 1989 (Termination Year: 1997)

Practical Mathematics: Consumer Applications:
Grades 10-12
Holt, 1989 (Termination Year: 1997)

Consumer Mathematics: Grades 10-12
Prentice Hall, 1989 (Termination Year: 1997)

Dollars & Sense: Problem Solving Strategies:
Grades 10-12
South-Western, 1989 (Termination Year: 1997)

Business Math (10-12)

Applied Business Mathematics: Grades 10-12
South-Western, 1990 (Termination Year: 1997)

Basic Algebra (9-12)

Algebra I: Grades 9-12
Addison-Wesley, 1990 (Termination Year: 1997)

Merrill Algebra Essentials: Grades 9-12
Glencoe, 1988 (Termination Year: 1997)

Holt Algebra I: Grades 9-12
Holt, 1986 (Termination Year: 1997)

Basic Algebra: Grades 9–12
Houghton, 1990 (Termination Year: 1997)

Algebra 1: Grades 9–12
Saxon, 1990 (Termination Year: 1997)

Algebra I (9-12)

Addison-Wesley Algebra: Grades 9–12
Addison-Wesley, 1990 (Termination Year: 1997)

Merrill Algebra One: Grades 9–12
Glencoe, 1990 (Termination Year: 1997)

Heath Algebra I: Grades 9–12
Heath, 1990 (Termination Year: 1997)

Holt Algebra 1: Grades 9–12
Holt, 1986 (Termination Year: 1997)

HBJ Algebra 1: Grades 9–12
Holt, 1990 (Termination Year: 1997)

Algebra: Structure & Method: Grades 9–12
Houghton, 1990 (Termination Year: 1997)

Algebra 1: Grades 9–12
Houghton, 1989 (Termination Year: 1997)

Algebra 1: Grades 9–12
Prentice Hall, 1990 (Termination Year: 1997)

University of Chicago School Mathematics Project: Algebra: Grades 9–12
ScottForesman, 1990 (Termination Year: 1997)

Algebra 1, an Integrated Approach:
Grades 9–12
McDougal, 1991 (Termination Year: 1997)

Algebra 1: Grades 9–12
Saxon, 1990 (Termination Year: 1997)

Integrated Mathematics 1 (9-12)

Merrill Integrated Math: Grades 9–12
Glencoe, 1991 (Termination Year: 1997)

HBJ Introductory Algebra 1: Grades 9–12
Holt, 1988 (Termination Year: 1997)

Houghton Mifflin Unified Math: Grades 9–12
Houghton, 1991 (Termination Year: 1997)

Integrated Mathematics II (9-12)

Merrill Integrated Math: Grades 9–12
Glencoe, 1991 (Termination Year: 1997)

HBJ Introductory Algebra 2: Grades 9–12
Holt, 1988 (Termination Year: 1997)

Houghton Mifflin Unified Math: Grades 9–12
Houghton, 1991 (Termination Year: 1997)

Integrated Mathematics III (9-12)

Merrill Integrated math: Grades 10–12
Glencoe, 1991 (Termination Year: 1997)

Houghton Mifflin Unified Math: Grades 10–12
Houghton, 1991 (Termination Year: 1997)

Advanced Mathematics: Grades 10–12
Saxon, 1989 (Termination Year: 1997)

Algebra II (10-12)

Addison-Wesley Algebra & Trigonometry:
Grades 10–12
Addison-Wesley, 1990 (Termination Year: 1997)

Merrill Algebra Two with Trigonometry:
Grades 10–12
Glencoe, 1990 (Termination Year: 1997)

Heath Algebra II: Grades 10–12
Heath, 1990 (Termination Year: 1997)

HBJ Algebra 2 with Trigonometry:
Grades 10–12
Holt, 1990 (Termination Year: 1997)

Algebra & Trigonometry: Structure & Method:
Grades 10–12
Houghton, 1990 (Termination Year: 1997)

Algebra 2 with Trigonometry: Grades 10–12
Prentice Hall, 1990 (Termination Year: 1997)

University of Chicago School Mathematics Project: Algebra: Grades 10–12
ScottForesman, 1990 (Termination Year: 1997)

Algebra 2 & Trigonometry: Grades 10–12
McDougal, 1991 (Termination Year: 1997)

Algebra II: Grades 10–12
Saxon, 1984 (Termination Year: 1997)

Algebra II—Accelerated (9-10)

Algebra & Trigonometry: Grades 9-10
Addison-Wesley, 1990 (Termination Year: 1997)

Merrill Algebra Two with Trigonometry: Grades 9-10
Glencoe, 1990 (Termination Year: 1997)

Algebra 2 and Trigonometry: Grades 9-10
Houghton, 1989 (Termination Year: 1997)

University of Chicago School Mathematics Project: Algebra: Grades 9-10
ScottForesman, 1990 (Termination Year: 1997)

Algebra 2 & Trigonometry: Grades 9-10
McDougal, 1991 (Termination Year: 1997)

Algebra II: Grades 9-10
Saxon, 1984 (Termination Year: 1997)

Addison-Wesley Informal Geometry: Grades 10-12
Addison-Wesley, 1986 (Termination Year: 1997)

Merrill Informal Geometry: Grades 10-12
Glencoe, 1988 (Termination Year: 1997)

Basic Geometry: Grades 10-12
Houghton, 1990 (Termination Year: 1997)

ScottForesman Geometry: Grades 10-12
ScottForesman, 1990 (Termination Year: 1997)

Basic Geometry (10-12)

Discovering Geometry: An Individual Approach: Grades 10-12
Key, 1989 (Termination Year: 1997)

Geometry

Addison-Wesley Geometry: Grades 10-12
Addison-Wesley, 1990 (Termination Year: 1997)

Geometry: Grades 10-12
Glencoe, 1990 (Termination Year: 1997)

Geometry: Grades 10-12
Holt, 1991 (Termination Year: 1997)

Geometry: Grades 10-12
Houghton, 1990 (Termination Year: 1997)

Geometry: Grades 10-12
Prentice Hall, 1990 (Termination Year: 1997)

University of Chicago School Mathematics Project: Geometry: Grades 10-12
ScottForesman, 1991 (Termination Year: 1997)

ScottForesman Geometry: Grades 10-12
ScottForesman, 1990 (Termination Year: 1997)

Geometry for Enjoyment & Challenge: Grades 10-12
McDougal, 1991 (Termination Year: 1997)

Discovering Geometry: An Individual Approach: Grades 10-12
Key Curriculum, 1989 (Termination Year: 1997)

Geometry Accelerated (9-10)

Geometry: Grades 9-10
Glencoe, 1990 (Termination Year: 1997)

Geometry: Grades 9-10
Houghton, 1990 (Termination Year: 1997)

University of Chicago School Mathematics Project: Geometry: Grades 9-10
ScottForesman, 1991 (Termination Year: 1997)

Geometry for Enjoyment & Challenge: Grades 10-12
McDougal, 1991 (Termination Year: 1997)

Applied Math (11-12)

Career Mathematics: Grades 11-12
Houghton, 1989 (Termination Year: 1997)

Precalculus (11-12)

Precalculus: A Graphing Approach: Grades 11-12
Addison-Wesley, 1990 (Termination Year: 1997)

Merrill Precalculus Mathematics: Grades 11-12
Glencoe, 1988 (Termination Year: 1997)

Precalculus: Grades 11-12
Heath, 1989 (Termination Year: 1997)

Advanced Mathematics: A Precalculus Course:
Grades 11-12
Houghton, 1989 (Termination Year: 1997)

Precalculus Mathematics: A Problem Solving Approach: Grades 11-12
Prentice Hall, 1989 (Termination Year: 1997)

Advanced Mathematics: Grades 11-12
Saxon, 1989 (Termination Year: 1997)

Advanced Topics in Math (11-12)

Elements of Calculus & Analytic Geometry:
Grades 11-12
Addison-Wesley, 1989 (Termination
Year: 1997)

Merrill Advanced Mathematical Concepts:
Grades 11-12
Glencoe, 1991 (Termination Year: 1997)

Introductory Analysis: Grades 11-12
Houghton, 1988 (Termination Year: 1997)

Probability & Statistics (11-12)

Introduction to Statistics & Probability: A Basis for Decision Making: Grades 11-12
Houghton, 1988 (Termination Year: 1997)

Probability & Statistics: Grades 11-12
South-Western, 1989 (Termination Year: 1997)

AP Calculus AB (12)

Calculus with Analytic Geometry: Grade 12
Heath, 1990 (Termination Year: 1997)

HBJ Calculus with Analytic Geometry:
Grade 12
Holt, 1989 (Termination Year: 1997)

Calculus with Analytic Geometry: Grade 12
Prentice Hall, 1987 (Termination Year: 1997)

Calculus: Grade 12
Saxon, 1988 (Termination Year: 1997)

AP Calculus BC (12)

Elements of Calculus & Analytic Geometry:
Grade 12
Addison-Wesley, 1988 (Termination
Year: 1997)

Functional Math I (9)

Basic Mathematics Skills: Grade 9
Media Materials, 1991 (Termination
Year: 1997)

Functional Math II (10)

Life Skills Mathematics: Grade 10
Media Materials, 1989 (Termination
Year: 1997)

Functional Math III (11)

Mathematics for Consumers: Grade 11
Media Materials, 1983 (Termination
Year: 1997)

Functional Math IV (12)

Mathematics for Business: Grade 12
Media Materials, 1988 (Termination
Year: 1997)

Computer Application (9-12)

PC Power: Microcomputer Applications:
Grades 9-12
Glencoe, 1991 (Termination Year: 1997)

Microcomputers: Concept Skills and Applications: Grades 9-12
West, 1991 (Termination Year: 1997)

Developing Computer Skills Using Appleworks:
Grades 9-12
West, 1991 (Termination Year: 1997)

Developing Computer Skills Using PFS First Choice: Grades 9-12
West, 1991 (Termination Year: 1997)

Computer Program (9-12)

Structured Basic: Grades 9-12
South-Western, 1989 (Termination Year: 1997)

Computer Math Structured Basic with Math:
Grades 9-12
South-Western, 1989 (Termination Year: 1997)

Understanding Structured Programming:
Grades 9-12
West, 1991 (Termination Year: 1997)

Basic Programming Today: Grades 9-12
West, 1990 (Termination Year: 1997)

Fundamentals of Pascal: Understanding Programming & Problem Solving: Grades 9–12
West, 1990 (Termination Year: 1997)

Understanding Pascal: Turbo Version:
Grades 9–12
West, 1990 (Termination Year: 1997)

Computer Science with Pascal for Advanced Placement: Grades 9–12
West, 1989 (Termination Year: 1997)

Louisiana

Mathematics K-8

Addison-Wesley Mathematics: Grades K–8
Addison-Wesley, 1991 (Termination Year: 1998)

Mathematics Unlimited: Grades K–8
Harcourt, 1991 (Termination Year: 1998)

Heath Mathematics Connections: Grades 1–8
Heath, 1992 (Termination Year: 1998)

Houghton Mifflin Mathematics: Grades K–8
Houghton, 1991 (Termination Year: 1998)

Mathematics in Action: Grades K–8
Macmillan, 1991 (Termination Year: 1998)

Exploring Mathematics: Grades K–8
ScottForesman, 1991 (Termination Year: 1998)

Mathematics: Exploring Your World:
Grades K–8
Houghton, 1991 (Termination Year: 1998)

Computing Today: Grades 7–8
West, 1991 (Termination Year: 1998)

Mathematics 7-12, Introduction to Algebra 7-12

Prealgebra: Grades 7–8
Addison-Wesley, 1991 (Termination Year: 1998)

Merrill Prealgebra: A Problem Solving Approach: Grades 8–12
Glencoe, 1989 (Termination Year: 1998)

Heath Prealgebra: Grades 8–12
Heath, 1990 (Termination Year: 1998)

Algebra 1/2: Grades 7–12
Saxon, 1990 (Termination Year: 1998)

Algebra I

Addison-Wesley Algebra: Grades 8–12
Addison-Wesley, 1990 (Termination Year: 1998)

Algebra: Grades 8–12
Addison-Wesley, 1990 (Termination Year: 1998)

Merrill Algebra One: Grades 8–12
Glencoe, 1990 (Termination Year: 1998)

HBJ Algebra One: Grades 8–12
Holt, 1990 (Termination Year: 1998)

Elementary Algebra: Grades 8–12
Houghton, 1988 (Termination Year: 1998)

Algebra: Structure & Method: Grades 8–12
Houghton, 1990 (Termination Year: 1998)

Algebra I, an Integrated Approach:
Grades 8–12
McDougal, 1991 (Termination Year: 1998)

Algebra I: Grades 8–12
Prentice Hall, 1990 (Termination Year: 1998)

Algebra I: An Incremental Development:
Grades 9–12
Saxon, 1990 (Termination Year: 1998)

University of Chicago School Mathematics Project: Algebra: Grades 8–12
ScottForesman, 1990 (Termination Year: 1998)

Algebra II & Trigonometry

Algebra & Trigonometry: Grades 9–12
Addison-Wesley, 1990 (Termination Year: 1998)

Addison-Wesley Algebra & Trigonometry:
Grades 9–12
Addison-Wesley, 1990 (Termination Year: 1998)

Merrill Algebra Two with Trigonometry:
Grades 8–12
Glencoe, 1990 (Termination Year: 1998)

HBJ Algebra 2 with Trigonometry:
Grades 7–12
Holt, 1990 (Termination Year: 1998)

Algebra & Trigonometry: Structure & Method:
Grades 8–12
Houghton, 1990 (Termination Year: 1998)

Algebra 2 & Trigonometry: Grades 8–12
McDougal, 1991 (Termination Year: 1998)

Algebra 2 with Trigonometry: Grades 9–12
Prentice Hall, 1990 (Termination Year: 1998)

Saxon Publishers Algebra 2: Grades 9–12
Saxon, 1984 (Termination Year: 1998)

*University of Chicago School Mathematics
Project: Advanced Algebra*: Grades 9–12
ScottForesman, 1990 (Termination Year: 1998)

Geometry

Addison-Wesley Geometry: Grades 9–12
Addison-Wesley, 1990 (Termination
Year: 1998)

Discovering Geometry, an Inductive Approach:
Grades 9–12
Key Curriculum, 1989 (Termination
Year: 1998)

Geometry for Enjoyment & Challenge:
Grades 9–12
McDougal, 1991 (Termination Year: 1998)

Geometry: Grades 8–12
Prentice Hall, 1990 (Termination Year: 1998)

*University of Chicago School Mathematics
Project: Geometry*: Grades 10–12
ScottForesman, 1991 (Termination Year: 1998)

Integrated Algebra/Geometry

Practical Mathematics: Skills & Concepts:
Grades 7–12
Holt, 1989 (Termination Year: 1998)

*University of Chicago School Mathematics
Project: Transition Mathematics*: Grades 8–12
ScottForesman, 1990 (Termination Year: 1998)

Trigonometry

Trigonometry: Grades 11–12
Addison-Wesley, 1991 (Termination
Year: 1998)

Trigonometry with Applications: Grades 11–12
Houghton, 1990 (Termination Year: 1998)

ScottForesman Trigonometry: Grades 11–12
ScottForesman, 1989 (Termination Year: 1998)

General Math

Heath General Mathematics: Grades 8–12
Heath, 1989 (Termination Year: 1998)

Mathematics in Life: Grades 9–12
ScottForesman, 1989 (Termination Year: 1998)

Math II

Merrill Applications of Mathematics:
Grades 8–12
Glencoe, 1988 (Termination Year: 1998)

Merrill Advanced Mathematical Concepts:
Grades 8–12
Glencoe, 1991 (Termination Year: 1998)

*HBJ Advanced Mathematics: A Preparation for
Calculus*: Grades 9–12
Holt, 1989 (Termination Year: 1998)

Advanced Mathematics: Grades 10–12
Saxon, 1989 (Termination Year: 1998)

Calculus

Precalculus with Trigonometry: Grades 11–12
Addison-Wesley, 1987 (Termination
Year: 1998)

Elements of Calculus & Analytic Geometry:
Grades 9–12
Addison-Wesley, 1989 (Termination
Year: 1998)

Calculus with Analytic Geometry: Grades
10–12
Heath, 1990 (Termination Year: 1998)

Merrill Precalculus Mathematics: Grades 8–12
Glencoe, 1988 (Termination Year: 1998)

HBJ Calculus with Analytic Geometry:
Grades 9–12
Holt, 1989 (Termination Year: 1998)

Precalculus Mathematics: A Problem Solving Approach: Grades 11–12
Prentice Hall, 1989 (Termination Year: 1998)

Calculus with Trigonometry & Analytic Geometry: Grades 11–12
Saxon, 1988 (Termination Year: 1998)

Business Mathematics

Applied Business Mathematics: Grades 9–12
South-Western, 1990 (Termination Year: 1998)

Consumer Math

Consumer Mathematics: Grades 8–12
Glencoe, 1989 (Termination Year: 1998)

Consumer Mathematics: Grades 9–12
Addison-Wesley, 1992 (Termination Year: 1998)

Practical Mathematics: Consumer Applications:
Grades 7–12
Holt, 1989 (Termination Year: 1998)

Consumer Mathematics: Grades 9–12
Prentice Hall, 1989 (Termination Year: 1998)

Consumer & Career Mathematics:
Grades 9–12
ScottForesman, 1989 (Termination Year: 1998)

Dollars & Sense: Problem Solving strategies in Consumer Mathematics: Grades 9–12
South-Western, 1989 (Termination Year: 1998)

Computer Science

Computer Programming in Basic: Grades 9–12
Houghton, 1991 (Termination Year: 1998)

Understanding Structured Programming in BASIC: Grades 9–12
West, 1991 (Termination Year: 1998)

Fundamentals of Pascal: Grades 9–12
West, 1990 (Termination Year: 1998)

Computer Science with Pascal for Advanced Placement Students: Grades 10–12
West, 1989 (Termination Year: 1998)

Understanding Pascal: Turbo Version:
Grades 9–12
West, 1990 (Termination Year: 1998)

Probability & Statistics

Probability & Statistics: Grades 9–12
South-Western, 1989 (Termination Year: 1998)

Mississippi

Mathematics 1-6 (Average to Accelerated)

Mathematics Today: Grades 1–6
Harcourt, 1987 (Termination Year: 1995)

Mathematics Unlimited: Grades 1–6
Holt, 1988 (Termination Year: 1995)

Macmillan Mathematics: Grades 1–6
Macmillan, 1987 (Termination Year: 1995)

McGraw-Hill Mathematics: Grades 1–6
Macmillan, 1987 (Termination Year: 1995)

Invitation to Mathematics: Grades 1–6
ScottForesman, 1988 (Termination Year: 1995)

Silver Burdett & Ginn Mathematics:
Grades 1–6
Silver, 1987 (Termination Year: 1995)

Mathematics 1-6 (Alternate—Basal Program for the Slow Learner)

Spectrum Mathematics: Grades 1–6
Glencoe, 1984 (Termination Year: 1995)

General Mathematics 7-8 (Average to Accelerated)

Mathematics Unlimited: Grades 7–8
Holt, 1988 (Termination Year: 1995)

Macmillan Mathematics: Grades 7–8
Macmillan, 1987 (Termination Year: 1995)

McGraw-Hill Mathematics: Grades 7–8
Macmillan, 1987 (Termination Year: 1995)

Merrill Mathematics: Grades 7–8
Glencoe, 1987 (Termination Year: 1995)

Invitation to Mathematics: Grades 7–8
ScottForesman, 1988 (Termination Year: 1995)

Silver Burdett & Ginn Mathematics:
Grades 7–8
Silver, 1987 (Termination Year: 1995)

Mathematics 7-8 (Alternate—Basal Program for Slow Learners)

Spectrum Mathematics: Grades 7–8
Glencoe, 1984 (Termination Year: 1995)

Math 65, an Incremental Development:
Grades 7–8
Saxon, 1987 (Termination Year: 1995)

Math 76, an Incremental Development:
Grades 7–8
Saxon, 1985 (Termination Year: 1995)

General Math I

Essentials of Mathematics: Grades 7–8
Addison-Wesley, 1989 (Termination
Year: 1995)

Heath General Mathematics: Grades 7–8
Heath, 1989 (Termination Year: 1995)

Practical Mathematics: Skills and Concepts:
Grades 7–8
Harcourt, 1989 (Termination Year: 1995)

Merrill General Mathematics: Grades 7–8
Glencoe, 1987 (Termination Year: 1995)

Prentice Hall Refresher Mathematics:
Grades 7–8
Prentice Hall, 1989 (Termination Year: 1995)

General Math II

Applying Mathematics: Grades 7–8
Addison-Wesley, 1986 (Termination
Year: 1995)

Applying Mathematics in Daily Living:
Grades 7–8
Glencoe, 1986 (Termination Year: 1995)

Essentials of Mathematics: Consumer/Career Skills/Applications: Grades 7–8
Harcourt, 1989 (Termination Year: 1995)

HBJ Fundamentals of Mathematics:
Grades 7–8
Holt, 1989 (Termination Year: 1995)

Merrill Applications of Mathematics:
Grades 7–8
Glencoe, 1988 (Termination Year: 1995)

Consumer Mathematics

HBJ Consumer Mathematics: Grades 7–8
Holt, 1989 (Termination Year: 1995)

Practical Mathematics: Consumer Applications:
Grades 7–8
Harcourt, 1989 (Termination Year: 1995)

Consumer Mathematics: Grades 7–8
Glencoe, 1989 (Termination Year: 1995)

Prentice Hall Consumer Mathematics:
Grades 7–8
Prentice Hall, 1989 (Termination Year: 1995)

Consumer and Career Mathematics:
Grades 7–8
ScottForesman, 1989 (Termination Year: 1995)

Fundamental Mathematics

Heath Prealgebra: Grades 7–8
Heath, 1986 (Termination Year: 1995)

Prealgebra: Skills/Problem Solving/Applications: Grades 7–8
Harcourt, 1986 (Termination Year: 1995)

Holt Prealgebra: Grades 7–8
Harcourt, 1986 (Termination Year: 1995)

Merrill Prealgebra: A Problem Solving Approach: Grades 7–8
Glencoe, 1989 (Termination Year: 1995)

ScottForesman Prealgebra: Grades 7–8
ScottForesman, 1987 (Termination Year: 1995)

College Preparatory Mathematics 7-12, Algebra I

Heath Algebra: Grades 7–12
Heath, 1987 (Termination Year: 1995)

HBJ Algebra: Grades 7–12
Holt, 1987 (Termination Year: 1995)

Holt Algebra: Grades 7–12
Holt, 1986 (Termination Year: 1995)

Algebra: Structure & Method: Grades 7–12
Houghton, 1988 (Termination Year: 1995)

Merrill Algebra One: Grades 7–12
Glencoe, 1986 (Termination Year: 1995)

Algebra II

Heath Algebra II: Grades 7–12
Heath, 1987 (Termination Year: 1995)

HBJ Algebra 2 with Trigonometry:
Grades 7–12
Harcourt, 1987 (Termination Year: 1995)

Holt Algebra II with Trigonometry:
Grades 7–12
Holt, 1986 (Termination Year: 1995)

Algebra & Trigonometry: Structure & Method:
Grades 7–12
Houghton, 1988 (Termination Year: 1995)

Merrill Algebra Two with Trigonometry:
Grades 7–12
Glencoe, 1986 (Termination Year: 1995)

Algebra III

Algebra & Trigonometry: Functions & Applications: Grades 7–12
Addison-Wesley, 1984 (Termination
Year: 1995)

Analysis of Elementary functions: Grades 7–12
Houghton, 1984 (Termination Year: 1995)

College Algebra: Grades 7–12
West, 1986 (Termination Year: 1995)

Unified Geometry

HBJ Geometry: Grades 7–12
Holt, 1987 (Termination Year: 1995)

Holt Geometry: Grades 7–12
Holt, 1986 (Termination Year: 1995)

Geometry: Grades 7–12
Houghton, 1988 (Termination Year: 1995)

Merrill Geometry: Grades 7–12
Prentice Hall, 1987 (Termination Year: 1995)

ScottForesman Geometry: Grades 7–12
ScottForesman, 1987 (Termination Year: 1995)

Analytic Geometry

Analytic Geometry: Grades 7–12
Addison-Wesley, 1986 (Termination
Year: 1995)

Modern Analytic Geometry: Grades 7–12
Houghton, 1989 (Termination Year: 1995)

Advanced Mathematics

HBJ Advanced Mathematics: A Preparation for Calculus: Grades 7–12
Holt, 1988 (Termination Year: 1995)

Advanced Mathematics: A Precalculus Course:
Grades 7–12
Houghton, 1989 (Termination Year: 1995)

Merrill Advanced Mathematical Concepts:
Grades 7–12
Glencoe, 1986 (Termination Year: 1995)

Algebra & Trigonometry: Grades 7–12
West, 1986 (Termination Year: 1995)

College Algebra & Trigonometry: Grades 7–12
West, 1988 (Termination Year: 1995)

Precalculus

Precalculus with Trigonometry: Grades 7–12
Addison-Wesley, 1987 (Termination
Year: 1995)

Introductory Analysis: Grades 7–12
Houghton Mifflin, 1988 (Termination
Year: 1995)

Merrill Precalculus Mathematics: Grades 7–12
Glencoe, 1988 (Termination Year: 1995)

Precalculus Mathematics: Grades 7–12
Prentice Hall, 1989 (Termination Year: 1995)

ScottForesman Precalculus: Grades 7–12
ScottForesman, 1987 (Termination Year: 1995)

Calculus

Calculus & Analytic Geometry: Grades 7-12
Addison-Wesley, 1988 (Termination
Year: 1995)

Calculus with Analytic Geometry: Grades 7-12
Heath, 1986 (Termination Year: 1995)

HBJ Calculus with Analytic Geometry:
Grades 7-12
Holt, 1989 (Termination Year: 1995)

Calculus with Analytic Geometry: Grades 7-12
Prentice Hall, 1987 (Termination Year: 1995)

*An Introduction to Calculus: Methods &
Applications*: Grades 7-12
West, 1986 (Termination Year: 1995)

Trigonometry

Trigonometry: Grades 7-12
Harcourt, 1987 (Termination Year: 1995)

Trigonometry with Applications: Grades 7-12
Houghton, 1987 (Termination Year: 1995)

*Plane Trigonometry: A Problem-Solving
Approach*: Grades 7-12
Prentice Hall, 1989 (Termination Year: 1995)

ScottForesman Trigonometry: Grades 7-12
ScottForesman, 1988 (Termination Year: 1995)

Trigonometry: Grades 7-12
West, 1988 (Termination Year: 1995)

Nevada

Elementary Mathematics

Real Math: Grades 7-8
Open Court, 1991 (Termination Year: not
given)

Mathematics, Their Way: Grades K-2
Addison-Wesley, 1976 (Termination
Year: 1995)

Math in Stride: Grades 3-6
Addison-Wesley, 1989 (Termination
Year: 1995)

Explorations: Grades K-2
Addison-Wesley, 1988 (Termination
Year: 1995)

*Developing Number Concepts Using Unifix
Cubes*: Grades K-3
Addison-Wesley, 1971 (Termination
Year: 1995)

Heath Mathematics Connections: Grades K-6
Heath, 1992 (Termination Year: 1995)

Addison-Wesley Mathematics: Grades K-6
Addison-Wesley, 1989 (Termination
Year: 1993)

Houghton Mifflin Mathematics: Grades K-6
Houghton, 1989 (Termination Year: 1993)

Algebra 1/2: Grades 7
Saxon, 1983 (Termination Year: 1993)

Math 65: Grades 7
Saxon, 1987 (Termination Year: 1993)

Math 76: Grades 8
Saxon, 1985 (Termination Year: 1993)

Exploring Mathematics: Grades K-6
ScottForesman, 1991 (Termination Year: 1994)

Mathematics: Exploring Your World:
Grades K-6
Silver, 1991 (Termination Year: 1994)

Silver, Burdett & Ginn Mathematics:
Grades K-6
Silver, 1988 (Termination Year: 1994)

Mathematics in Action: Grades 6-8
Macmillan, 1991 (Termination Year: not given)

Mathematics Unlimited: Grades 7-8
Harcourt, 1991 (Termination Year: not given)

Secondary Mathematics

Algebra I: Grades 8-12
Addison-Wesley, 1990 (Termination
Year: 1995)

Glencoe Algebra: Grades 9-12
Glencoe, 1990 (Termination Year: 1995)

HBJ Algebra: Grades 9–12
Holt, 1990 (Termination Year: 1995)

Addison-Wesley Algebra: Grades 8–12
Addison-Wesley, 1990 (Termination Year: 1995)

Algebra: Structure & Method: Grades 8–12
Houghton, 1990 (Termination Year: 1995)

Algebra I, an Integrated Approach:
Grades 8–12
McDougal, 1991 (Termination Year: 1995)

*University of Chicago School Math Project:
Algebra*: Grades 8–12
ScottForesman, 1990 (Termination Year: 1995)

Addison-Wesley Prealgebra: Grades 7–12
Addison-Wesley, 1987 (Termination Year: 1993)

Addison-Wesley Informal Geometry:
Grades 9–12
Addison-Wesley, 1986 (Termination Year: 1993)

Heath Prealgebra: Grades 9–12
Heath, 1986 (Termination Year: 1993)

Basic Algebra: Grades 9–12
Houghton, 1988 (Termination Year: 1993)

Basic Geometry: Grades 9–12
Houghton, 1988 (Termination Year: 1993)

Merrill Informal Geometry: Grades 9–12
Glencoe, 1988 (Termination Year: 1993)

*Merrill Prealgebra: A Problem Solving
Approach*: Grades 9–12 Glencoe, 1989 (Termination Year: 1993)

Prentice Hall Refresher Mathematics:
Grades 9–12
Prentice Hall, 1989 (Termination Year: 1993)

Berkey Calculus: Grades 9–12
Saunders College Publishing, 1988 (Termination Year: 1993)

Geometry: A Guided Inquiry: Grades 9–12
Sunburst, 1987 (Termination Year: 1993)

Introductory Algebra: Grades 9–10
Harcourt, 1988 (Termination Year: 1994)

Discovery Geometry: An Inductive Approach:
Grades 9–12
Key Curriculum, 1989 (Termination Year: 1994)

Mathematics Today: Grades K–8
Harcourt, 1987 (Termination Year: not given)

Consumer Mathematics: Grades 9–12
Addison-Wesley, 1989 (Termination Year: not given)

Precalculus with Trigonometry and Applications: Grades 9–12 Addison-Wesley, 1987 (Termination Year: not given)

Calculus & Analytic Geometry: Grades 9–12
Addison-Wesley, 1987 (Termination Year: not given)

Elementary Statistics: Concepts and Methods:
Grades 9–12
Addison-Wesley, 1986 (Termination Year: not given)

Understandable Statistics: Concepts & Methods: Grades 9–12
Heath, 1987 (Termination Year: not given)

Calculus: Grades 9–12
Holt, 1987 (Termination Year: not given)

Advance Mathematics: A Precalculus Course:
Grades 9–12
Houghton, 1987 (Termination Year: not given)

Consumer Mathematics: Grades 9–12
Houghton, 1986 (Termination Year: not given)

Introductory Analysis: Grades 9–12
Houghton, 1988 (Termination Year: not given)

Merrill Advanced Mathematical Concepts:
Grades 9–12
Glencoe, 1986 (Termination Year: not given)

Merrill Precalculus: Grades 9–12
Glencoe, 1988 (Termination Year: not given)

Mathematics for Today's Consumer:
Grades 9-12
Glencoe: 1982 (Termination Year: not given)

Statistics: A First Course: Grades 9-12
Prentice Hall, 1986 (Termination Year: not given)

Calculus with Analytic Geometry: Grades 9-12
Prentice Hall, 1987 (Termination Year: not given)

Consumer & Career Mathematics:
Grades 9-12
ScottForesman, 1987 (Termination Year: not given)

Algebra One: Grade 8
Glencoe, 1990 (Termination Year: not given)

Geometry: Grades 8-12
Houghton, 1992 (Termination Year: not given)

Discovering Geometry: An Inductive Approach:
Grades 8-12
Houghton, 1992 (Termination Year: not given)

Geometry for Enjoyment and Challenge:
Grades 8-12
McDougal, 1991 (Termination Year: not given)

ScottForesman Geometry: Grades 8-12
ScottForesman, 1990 (Termination Year: not given)

University of Chicago School Mathematics Project: Geometry: Grades 8-12
ScottForesman, 1991 (Termination Year: not given)

Addison-Wesley Geometry: Grades 8-12
Addison-Wesley, 1992 (Termination Year: not given)

New Mexico

Advanced Mathematics

Merrill Precalculus Mathematics: Grades 11-12
Glencoe, 1988 (Termination Year: 1994)

Merrill Advanced Mathematical Concepts:
Grades 11-12
Glencoe, 1986 (Termination Year: 1994)

Calculus with Analytic Geometry: Grades 10-12
Heath, 1986 (Termination Year: 1994)

HBJ Advanced Math: A Preparation for Calculus: Grades 9-12
Harcourt, 1988 (Termination Year: 1994)

Trigonometry: Grades 9-12
Holt, 1987 (Termination Year: 1994)

Advanced Mathematics: Grades 11-12
Saxon, 1989 (Termination Year: 1994)

Advanced Math: A Precalculus Course:
Grades 11-12
Houghton, 1987 (Termination Year: 1994)

Introductory Analysis: Grades 11-12
Houghton, 1988 (Termination Year: 1994)

Trigonometry with Applications: Grades 10-12
Houghton, 1987 (Termination Year: 1994)

Precalculus Computer Activity: Grades 11-12
Houghton, 1987 (Termination Year: 1994)

ScottForesman Trigonometry: Grades 11-12
ScottForesman, 1983 (Termination Year: 1994)

ScottForesman Precalculus: Grades 11-12
ScottForesman, 1987 (Termination Year: 1994)

Learning Basic Math & Business Math Using the Calculator: Grades 9-12
South-Western, 1985 (Termination Year: 1994)

Precalculus: Grades 10-12
West, 1987 (Termination Year: 1994)

Trigonometry: Grades 9-12
West, 1985 (Termination Year: 1994)

Transition to College Math: Grades 9-12
Addison-Wesley, 1984 (Termination Year: 1994)

Calculus & Analytic Geometry: Grades 11-12
Addison-Wesley, 1984 (Termination
Year: 1994)

Using Statistics: Grades 9-12
Addison-Wesley, 1985 (Termination
Year: 1994)

Trigonometry: Functions & Applications:
Grades 11-12
Addison-Wesley, 1984 (Termination
Year: 1994)

Precalculus with Trigonometry: Grades 11-12
Addison-Wesley, 1984 (Termination
Year: 1994)

Analytic Geometry: Grades 10-12
Addison-Wesley, 1986 (Termination
Year: 1994)

Techniques of Calculus: Grades 9-12
Amsco, 1983 (Termination Year: 1994)

Mathematics for the College Boards:
Grades 9-12
Amsco, 1985 (Termination Year: 1994)

Real & Complex Analysis: Grades 11-12
Glencoe, 1987 (Termination Year: 1994)

Calculus with Analytic Geometry: Grades
11-12
Glencoe, 1987 (Termination Year: 1994)

Calculus with Analytic Geometry: Grades
10-12
Prentice Hall, 1987 (Termination Year: 1994)

Algebra

Elementary Algebra: Grades 9-12
Freeman, 1979 (Termination Year: 1994)

Introductory Algebra: An Interactive Approach:
Grades 7-10
Glencoe, 1987 (Termination Year: 1994)

Harper & Row Algebra One: Grades 8-11
Glencoe, 1985 (Termination Year: 1994)

*Harper & Row Algebra Two with Trigonome-
try*: Grades 8-11
Glencoe, 1985 (Termination Year: 1994)

*Graphing in the Coordinate Plane Computer
Courseware*: Grades 7-10
Glencoe, 1987 (Termination Year: 1994)

Laidlaw Algebra 1: Grades 7-10
Glencoe, 1987 (Termination Year: 1994)

Laidlaw Algebra 2 with Trigonometry:
Grades 7-10
Glencoe, 1987 (Termination Year: 1994)

Preparing to Use Algebra: Grades 7-10
Glencoe, 1986 (Termination Year: 1994)

*Merrill Prealgebra: Problem Solving
Approach:* Grades 7-9
Glencoe, 1986 (Termination Year: 1994)

Merrill Algebra One: Grades 8-10
Glencoe, 1986 (Termination Year: 1994)

Merrill Algebra Two: with Trigonometry:
Grades 8-10
Glencoe, 1986 (Termination Year: 1994)

Merrill Algebra Essentials: Grades 8-10
Glencoe, 1988 (Termination Year: 1994)

Heath Algebra I: Grades 7-12
Heath, 1987 (Termination Year: 1994)

Heath Algebra II: Grades 7-12
Heath, 1987 (Termination Year: 1994)

Heath Prealgebra: Grades 7-10
Heath, 1988 (Termination Year: 1994)

Getting Ready for Algebra: Grades 7-10
Heath, 1988 (Termination Year: 1994)

Holt Prealgebra: Grades 9-12
Holt, 1986 (Termination Year: 1994)

Holt Algebra 1: Grades 9-12
Holt, 1986 (Termination Year: 1994)

Holt Algebra 2 with Trigonometry:
Grades 9-12
Holt, 1986 (Termination Year: 1994)

HBJ Algebra 1: Grades 7-12
Harcourt, 1987 (Termination Year: 1994)

HBJ Algebra 2 with Trigonometry:
Grades 7–12
Harcourt, 1987 (Termination Year: 1994)

Prealgebra: Skills/Problem Solving/Applications: Grades 7–12
Holt, 1986 (Termination Year: 1994)

HBJ Introductory Algebra 1: Grades 7–10
Harcourt, 1988 (Termination Year: 1994)

HBJ Introductory Algebra II: Grades 7–10
Harcourt, 1988 (Termination Year: 1994)

Algebra II, Incremental Development:
Grades 10–11
Saxon, 1984 (Termination Year: 1994)

Algebra I, an Incremental Development:
Grades 9–10
Saxon, 1983 (Termination Year: 1994)

Algebra 1/2: Grade 8
Saxon, 1990 (Termination Year: 1994)

Basic Algebra: Grades 8–10
Houghton, 1988 (Termination Year: 1994)

Prealgebra: An Accelerated Course: Grade 7
Houghton, 1988 (Termination Year: 1994)

Algebra Structure & Method: Grades 8–11
Houghton, 1988 (Termination Year: 1994)

Algebra & Trigonometry: Structure & Method:
Grades 9–12 Houghton, 1988 (Termination
Year: 1994)

Elementary Algebra: Grades 8–11
Houghton, 1988 (Termination Year: 1994)

Algebra 1: Grades 8–11
Houghton, 1986 (Termination Year: 1994)

Algebra 2: Grades 9–12
Houghton, 1986 (Termination Year: 1994)

Essentials for Algebra: Concepts & Skills:
Grade 9
Houghton, 1986 (Termination Year: 1994)

ScottForesman Algebra First Course:
Grades 9–12
ScottForesman, 1984 (Termination Year: 1994)

ScottForesman Algebra: Second Course:
Grades 9–12
ScottForesman, 1984 (Termination Year: 1994)

College Algebra & Trigonometry: Grades 9–12
West, 1985 (Termination Year: 1994)

Algebra & Trigonometry: Grades 9–12
West, 1986 (Termination Year: 1994)

Algebra & Trigonometry: Functions & Applications: Grades 9–12
Addison-Wesley, 1984 (Termination
Year: 1994)

Algebra 1: Expressions, Equations & Applications: Grades 9–12
Addison-Wesley, 1984 (Termination
Year: 1994)

Addison-Wesley Algebra I: Grades 8–12
Addison-Wesley, 1988 (Termination
Year: 1994)

Addison-Wesley Algebra II: Grades 10–12
Addison-Wesley, 1988 (Termination
Year: 1994)

Prealgebra: Grades 9–12
Addison-Wesley, 1991 (Termination
Year: 1994)

Algebra: An Introductory Course: Grades 9–12
Amsco, 1986 (Termination Year: 1994)

Computer Mathematics

Computer Literacy Through Applications:
Grades 7–8
Glencoe, 1986 (Termination Year: 1994)

Computers & Math: Grades 7–12
Scholastic, 1986 (Termination Year: 1994)

Computer Math, Basic Programming with Applications: Grades 9–12
Heath, 1985 (Termination Year: 1994)

Computer Literacy with Introduction to Basic Programming: Grades 7-12
Holt, 1986 (Termination Year: 1994)

Computer Programming in the Basic Language: Grades 9-12
Holt, 1987 (Termination Year: 1994)

Computer Programming in the Pascal Language: Grades 9-12
Holt, 1988 (Termination Year: 1994)

Pascal: An Introduction to Methodical Programming: Grades 9-12
Houghton, 1985 (Termination Year: 1994)

Introduction to Apple II: Grades 7-9
Houghton, 1984 (Termination Year: 1994)

Computer Programming in Basic: Grades 9-12
Houghton, 1986 (Termination Year: 1994)

Introduction to TRS-80 Basic: Grades 7-9
Houghton, 1985 (Termination Year: 1994)

Lotus 1-2-3 Quick: Grades 9-12
South-Western , 1990 (Termination Year: 1994)

Applications Using the Personal Computer: Grades 9-12
South-Western, 1988 (Termination Year: 1994)

Advanced Structures Basic Manual: Grades 10-12
South-Western, 1985 (Termination Year: 1994)

Appleworks-Integrated Applications for Microcomputers: Grades 10-12
South-Western, 1991 (Termination Year: 1994)

Understanding Fortran: Grades 10-12
West, 1987 (Termination Year: 1994)

Turbo & Apple Pascal: Programming & Problem Solving: Grades 9-12
West, 1987 (Termination Year: 1994)

Understanding & Using Supercalc III: Grades 9-12
West, 1986 (Termination Year: 1994)

Understanding & Using Appleworks: Grades 7-12
West, 1986 (Termination Year: 1994)

Complete Basic Programming: Grades 9-12
West, 1987 (Termination Year: 1994)

Introduction to Computers & Basic Programming: Grades 9-12
West, 1986 (Termination Year: 1994)

Turbo Pascal Programming Today: Grades 9-12
West, 1987 (Termination Year: 1994)

Introduction to Data Structures with Pascal: Grades 10-12
West, 1986 (Termination Year: 1994)

Understanding Fortran: Grades 10-12
West, 1985 (Termination Year: 1994)

Apple Logo Time: Grades 2-3
West, 1986 (Termination Year: 1994)

Building Basic Skills: Grades 4-6
West, 1986 (Termination Year: 1994)

Building Logo Skills: Terrapin: Grades 4-6
West, 1986 (Termination Year: 1994)

Building Logo Skills: Apple: Grades 4-6
West, 1985 (Termination Year: 1994)

Fundamentals of Basic Programming: A Structured Approach: Grades 4-7
West, 1987 (Termination Year: 1994)

Understanding Basic: A Structured Approach: Grades 9-12
West, 1985 (Termination Year: 1994)

Understanding Pascal: A Problem Solving Approach: Grades 9-12
West, 1985 (Termination Year: 1994)

Computer Science with Pascal for Advanced Placement Students: Grades 9-12
West, 1985 (Termination Year: 1994)

Terrapin Logo Time: Grades 2–3
West, 1986 (Termination Year: 1994)

Apple Pascal: A Problem Solving Approach:
Grades 9–12
West, 1985 (Termination Year: 1994)

*Fundamentals of Pascal: Programming &
Problem Solving*: Grades 9–12
West, 1985 (Termination Year: 1994)

Understanding & Using Lotus: Grades 9–12
West, 1986 (Termination Year: 1994)

Understanding & Using Dbase III:
Grades 9–12
West, 1986 (Termination Year: 1994)

Understanding & Using Framework:
Grades 9–12
West, 1986 (Termination Year: 1994)

Complete Logo Programming: Apple:
Grades 7–9
West, 1986 (Termination Year: 1994)

Complete Logo Programming: Terrapin:
Grades 7–9
West, 1986 (Termination Year: 1994)

Advanced Basic: A Structured Approach:
Grades 9–12
West, 1986 (Termination Year: 1994)

*Basic Programming Today: A Structured
Approach*: Grades 9–12 West, 1986 (Termination
Year: 1994)

Strictly Structured Basic: Grades 9–12
West, 1986 (Termination Year: 1994)

Approaching Macintosh: Grades 7–12
Addison-Wesley, 1986 (Termination
Year: 1994)

Addison-Wesley Computer Literacy:
Grades 7–11
Addison-Wesley, 1986 (Termination
Year: 1994)

Using Computers in Mathematics:
Grades 9–12
Addison-Wesley, 1986 (Termination
Year: 1994)

Basic: A Problem-Solving Approach:
Grades 10–12
Addison-Wesley, 1988 (Termination
Year: 1994)

*Computer Applications, Using Word Process-
ing*: Grades 6–9
Addison-Wesley, 1986 (Termination
Year: 1994)

*Problem Solving & Structured Programming in
Pascal*: Grades 11–12
Addison-Wesley, 1985 (Termination
Year: 1994)

Appleworks: Applications & Activities:
Grades 9–12
Addison-Wesley, 1988 (Termination
Year: 1994)

Introduction to Basic: Grades 7–9
Addison-Wesley, 1988 (Termination
Year: 1994)

Computers: Tools for an Information Age:
Grades 10–12
Addison-Wesley, 1986 (Termination
Year: 1994)

*Microcomputer Applications PC-Write, Calc,
File*: Grades 11–12
Brown, 1987 (Termination Year: 1994)

Personal Productivity w/Lotus 1-2-3:
Grades 11–12
Brown, 1987 (Termination Year: 1994)

Personal Productivity w/Dbase III:
Grades 11–12
Brown, 1987 (Termination Year: 1994)

Personal Productivity w/ WordStar:
Grades 11–12
Brown, 1987 (Termination Year: 1994)

Calculator Math: Grades 5–10
Fearon, 1980 (Termination Year: 1994)

Consumer Mathematics

Consumer Mathematics: Grades 9-12
Glencoe, 1988 (Termination Year: 1994)

Essentials of Math: Consumer/Career Skills/ Applications: Grades 9-12
Holt, 1987 (Termination Year: 1994)

Consumer & Career Mathematics:
Grades 9-12
ScottForesman, 1987 (Termination Year: 1994)

Applying Math: A Consumer/Career Approach:
Grades 9-12
Addison-Wesley, 1986 (Termination Year: 1994)

Pacemaker Practical Math Series:
Grades 7-12
Fearon, 1983 (Termination Year: 1994)

Elementary Mathematics

Modern Curriculum Press Mathematics:
Grades K-6
Modern Curriculum, 1987 (Termination Year: 1994)

Silver Burdett Mathematics: Grades K-6
Silver, 1987 (Termination Year: 1994)

Heath Mathematics: Grades K-6
Heath, 1988 (Termination Year: 1994)

Mathematics Today: Grades K-6
Holt, 1987 (Termination Year: 1994)

Mathematics Hoy: Grades K-3
Holt, 1986 (Termination Year: 1994)

Mathematics Unlimited: Grades K-6 (also in Spanish)
Holt, 1987 (Termination Year: 1994)

Macmillan Mathematics: Grades K-6
Macmillan, 1987 (Termination Year: 1994)

McGraw-Hill Mathematics: Grades K-5, 7-8
Macmillan, 1987 (Termination Year: 1994)

Merrill Mathematics: Grades K-6
Glencoe, 1987 (Termination Year: 1994)

Laidlaw Mathematics, Series 2000:
Grades K-6
Glencoe, 1987 (Termination Year: 1994)

Building Skills in Mathematics: Grades K-6
Science Research, 1988 (Termination Year: 1994)

How to Use Bank Accounts: Grades 4-6 (suitable for special education)
Science Research, 1986 (Termination Year: 1994)

How To Understand & Manage Your Time:
Grades 4-6 (suitable for special education)
Science Research, 1986 (Termination Year: 1994)

How To Manage Your Money: Grades 4-6 (suitable for special education)
Science Research, 1986 (Termination Year: 1994)

How To Use Measurements: Grades 4-6 (suitable for special education)
Science Research, 1986 (Termination Year: 1994)

Developing Key Concepts for Solving Problems: Grades K-6
Science Research, 1987 (Termination Year: 1994)

Corrective Mathematics: Grades 4-6 (suitable for special education)
Science Research, 1981 (Termination Year: 1994)

Houghton Mifflin Math: Grades K-6 (also in Spanish)
Houghton, 1987 (Termination Year: 1994)

Riverside Mathematics: Grades K-6
Houghton, 1987 (Termination Year: 1994)

Mastering Computational Skills, Workbook:
Grades 3-6
Harcourt, 1984 (Termination Year: 1994)

ScottForesman Invitation to Mathematics:
Grades 1-6
ScottForesman, 1987 (Termination Year: 1994)

ScottForesman Invitation to Mathematics, Big Book: Grades K ScottForesman, 1985 (Termination Year: 1994)

Explorations, Activity Book: Grades K-2
Addison-Wesley, 1986 (Termination Year: 1994)

Challenge: Grades 3-6
Addison-Wesley, 1987 (Termination Year: 1994)

Addison-Wesley Math: Grades K-6 (also in Spanish)
Addison-Wesley, 1987 (Termination Year: 1994)

McGraw-Hill Mathematics: Grades 1, 5-6
Macmillan, 1987 (Termination Year: 1994)

Practice Workbook: Grades 3-6
Open Court, 1985 (Termination Year: 1994)

Real Math, Readiness Book: 1, 2, 3 Go: Grade 1
Open Court, 1981 (Termination Year: 1994)

Real Math: Grades K-8
Open Court, 1991 (Termination Year: 1994)

Pacemaker Arithmetic Program Plus, Complete Program: Grades K-6
Fearon, 1988 (Termination Year: 1994)

General Mathematics

Math Study Skills Program: Grades 6-10
NASSP, 1980 (Termination Year: 1994)

Solve Action Problem Solving: Grades 4-Ad (suitable for special education)
Curriculum Associates, 1985 (Termination Year: 1994)

Enright Computation Series: Grades 4-9
Curriculum Associates, 1985 (Termination Year: 1994)

Maintenance Book for Whole Numbers: Grades 4-9 (suitable for special education)
Curriculum Associates, 1986 (Termination Year: 1994)

Mathematics: A Human Endeavor:
Grades 9-12
Freeman, 1982 (Termination Year: 1994)

Fundamentals of Mathematics: Grades 7-12
Glencoe, 1986 (Termination Year: 1994)

Spectrum Mathematics Series: Grades 3-8
Glencoe, 1984 (Termination Year: 1994)

Mathematics Skills for Daily Living:
Grades 9-12
Glencoe, 1986 (Termination Year: 1994)

Applying Mathematics in Daily Living:
Grades 9-12
Glencoe, 1986 (Termination Year: 1994)

Mastering Essential Mathematics Skills:
Grades 9-12
Glencoe, 1984 (Termination Year: 1994)

Merrill General Mathematics: Grades 9-12
Glencoe, 1987 (Termination Year: 1994)

Merrill Applications of Mathematics:
Grades 9-12
Glencoe, 1988 (Termination Year: 1994)

Heath General Mathematics: Grades 9-12
Heath, 1989 (Termination Year: 1994)

Holt Essential Mathematics: Grades 9-12
Holt, 1987 (Termination Year: 1994)

General Math: Skills/Problem Solving/Applications: Grades 9-12
Holt, 1987 (Termination Year: 1994)

Laidlaw Mathematics: Grade 8
Glencoe, 1987 (Termination Year: 1994)

Math 76, Incremental Development:
Grades 6-7
Saxon, 1986 (Termination Year: 1994)

Math 65, an Incremental Development:
Grades 5-6
Saxon, 1985 (Termination Year: 1994)

Refresher Mathematics: Grades 7-12
Prentice Hall, 1986 (Termination Year: 1994)

Fundamentals of Math: Skills & Applications:
Grades 9-10
Houghton, 1986 (Termination Year: 1994)

ScottForesman Mathematics in Life:
Grades 9-12
ScottForesman, 1987 (Termination Year: 1994)

Essential Mathematics for Life: Grades 9-12
ScottForesman, 1985 (Termination Year: 1994)

Dollars & Sense: Problem Solving Strategies in Consumer Math: Grades 7-12
South-Western, 1989 (Termination Year: 1994)

Sound Foundations: Practical Mathematics Simulations: Grades 7-12
South-Western, 1987 (Termination Year: 1994)

Occupational Mathematics: Grades 9-12
South-Western, 1986 (Termination Year: 1994)

Essentials of Mathematics: Grades 9-12
Addison-Wesley, 1989 (Termination Year: 1994)

Achieving Competence in Mathematics:
Grades 9-12
Amsco, 1987 (Termination Year: 1994)

Geometry

Harper & Row Geometry: Grades 9-11
Glencoe, 1986 (Termination Year: 1994)

Laidlaw Geometry: Grades 10-12
Glencoe, 1987 (Termination Year: 1994)

Merrill Geometry: Grades 10-11
Glencoe, 1987 (Termination Year: 1994)

Merrill Informal Geometry: Grades 9-11
Glencoe, 1988 (Termination Year: 1994)

Holt Geometry: Grades 9-12
Holt, 1986 (Termination Year: 1994)

HBJ Geometry: Grades 7-12
Harcourt, 1987 (Termination Year: 1994)

Geometry: An Informal Approach:
Grades 9-12
Prentice Hall, 1986 (Termination Year: 1994)

Basic Geometry: Grades 9-11
Houghton, 1988 (Termination Year: 1994)

Geometry: Grades 9-11
Houghton, 1988 (Termination Year: 1994)

ScottForesman Geometry: Grades 9-12
ScottForesman, 1987 (Termination Year: 1994)

Informal Geometry: Grades 9-12
Addison-Wesley, 1986 (Termination Year: 1994)

Geometry for Enjoyment & Challenge:
Grades 9-12
McDougal, 1988 (Termination Year: 1994)

Middle School

Silver Burdett & Ginn Mathematics:
Grades 7-8
Silver, 1987 (Termination Year: 1994)

Heath Math: Grades 7-8
Heath, 1988 (Termination Year: 1994)

Mathematics Today: Grades 7-8
Holt, 1987 (Termination Year: 1994)

Mathematics Unlimited: Grades 7-8
Holt, 1987 (Termination Year: 1994)

Macmillan Math: Grades 7-8
Macmillan, 1987 (Termination Year: 1994)

Merrill Math: Grades 7-8
Glencoe, 1987 (Termination Year: 1994)

Laidlaw Math: Grades 7-8
Glencoe, 1987 (Termination Year: 1994)

Building Skills in Mathematics: Grades 7-8
Science Research, 1988 (Termination Year: 1994)

Houghton Mifflin Math: Grades 7-8
Houghton, 1987 (Termination Year: 1994)

Math Structure & Method: Grades 7-8
Houghton, 1988 (Termination Year: 1994)

Riverside Mathematics: Grades 7-8
Houghton, 1987 (Termination Year: 1994)

ScottForesman Mastering Computational Skills:
Grades 7-9
ScottForesman, 1984 (Termination Year: 1994)

ScottForesman Prealgebra: Grades 7-9
ScottForesman, 1987 (Termination Year: 1994)

ScottForesman Invitation to Mathematics:
Grades 7-8
ScottForesman, 1987 (Termination Year: 1994)

The Middle Grades Math Project: Grades 5-8
Addison-Wesley, 1986 (Termination
Year: 1994)

Addison Wesley Mathematics: Grades 7-8
Addison-Wesley, 1987 (Termination
Year: 1994)

Real Math: Grades 7-8 (suitable for special
education)
Open Court, 1991 (Termination Year: 1994)

Practicing Math Applications: Grades 7-8
American School Publishers, 1984 (Termination
Year: 1994)

Practicing Problem Solving: Grades 7-8
American School Publishers, 1984 (Termination
Year: 1994)

Spotlight on Math: Grades 7-8
American School Publishers, 1983 (Termination
Year: 1994)

Scoring High in Math: Grades 6-8
American School Publishers, 1986 (Termination
Year: 1994)

Mathematics for Daily Living: Grades 7-10
American School Publishers, 1986 (Termination
Year: 1994)

North Carolina

Mathematics, Elementary

Explorations 1: Grades 1-2
Addison-Wesley, 1988 (Termination
Year: 1997)

Mathworks: Grades 1-2
Houghton, 1992 (Termination Year: 1997)

Great Beginnings: Grades 1-2
ScottForesman, 1992 (Termination Year: 1997)

Addison-Wesley Mathematics: Grades 1-8
Addison-Wesley, 1991 (Termination
Year: 1997)

Mathematics Plus: Grades 1-8
Harcourt, 1992 (Termination Year: 1997)

The Mathematics Experience: Grades 1-8
Houghton, 1992 (Termination Year: 1997)

Heath Mathematics Connections: Grades 1-8
Heath, 1992 (Termination Year: 1997)

Mathematics in Action: Grades 1-8
Macmillan, 1992 (Termination Year: 1997)

Exploring Mathematics: Grades 1-8
ScottForesman, 1991 (Termination Year: 1997)

Silver Burdett & Ginn Mathematics:
Grades 1-8
Silver, 1992 (Termination Year: 1997)

Structure and Method: Grades 7-8
Houghton, 1992 (Termination Year: 1997)

Prealgebra

Merrill Pre-Algebra: Grade 7
Glencoe, 1992 (Termination Year: 1997)

Heath Pre-Algebra: Grade 7
Heath, 1992 (Termination Year: 1997)

Holt Pre-Algebra: Grade 7
Holt, 1992 (Termination Year: 1997)

Prentice Hall Pre-Algebra: Grade 7
Prentice Hall, 1992 (Termination Year: 1997)

*University of Chicago School Math Project:
Transition Math*: Grade 7
ScottForesman, 1992 (Termination Year: 1997)

Advanced Mathematics

Merrill Advanced Mathematical Concepts:
Grades 9-12
Glencoe, 1991 (Termination Year: 1997)

Advanced Math Precalculus with Discrete Math & Data Analysis: Grades 9-12
Houghton, 1992 (Termination Year: 1997)

Precalculus Mathematics: Grades 9-12
Prentice Hall, 1991 (Termination Year: 1997)

University of Chicago School Math Project: Pre-Calculus & Discrete Math: Grades 9-12
ScottForesman, 1992 (Termination Year: 1997)

Advanced Placement Calculus

Elements of Calculus & Analytic Geometry: Grades 9-12
Addison-Wesley, 1989 (Termination Year: 1997)

Calculus with Analytic Geometry: Grades 9-12
Heath, 1990 (Termination Year: 1997)

HBJ Calculus with Analytic Geometry: Grades 9-12
Holt, 1989 (Termination Year: 1997)

Calculus with Analytic Geometry: Grades 9-12
Prentice Hall, 1987 (Termination Year: 1997)

Calculus of a Single Variable: Grades 9-12
Wadsworth, 1991 (Termination Year: 1997)

Single Variable Calculus: Grades 9-12
Wadsworth, 1991 (Termination Year: 1997)

Probability and Statistics

Statistics: A First Course: Grades 9-12
Prentice Hall, 1991 (Termination Year: 1997)

University of Chicago School Math Project: Function Statistics & Trigonometry: Grades 9-12
ScottForesman, 1992 (Termination Year: 1997)

Trigonometry

Trigonometry and its Applications: Grades 9-12
Glencoe, 1990 (Termination Year: 1997)

Trigonometry with Applications: Grades 9-12
Houghton, 1990 (Termination Year: 1997)

Prentice Hall Trigonometry: Grades 9-12
Prentice Hall, 1990 (Termination Year: 1997)

Introductory Algebra

HBJ Introductory Algebra: Grades 9-12
Holt, 1988 (Termination Year: 1997)

Algebra I

Addison-Wesley Algebra: Grades 9-12
Addison-Wesley, 1992 (Termination Year: 1997)

Merrill Algebra 1: Applications & Connections: Grades 9-12
Glencoe, 1992 (Termination Year: 1997)

Merrill Algebra Essentials: Grades 9-12
Glencoe, 1988 (Termination Year: 1997)

Holt Algebra 1: Grades 9-12
Holt, 1992 (Termination Year: 1997)

Algebra: Structure & Method: Grades 9-12
Houghton, 1992 (Termination Year: 1997)

Algebra I, an Integrated Approach: Grades 9-12
McDougal, 1991 (Termination Year: 1997)

Prentice Hall Algebra I: Grades 9-12
Prentice Hall, 1990 (Termination Year: 1997)

University of Chicago School Math Project: Algebra: Grades 9-12
ScottForesman, 1990 (Termination Year: 1997)

Algebra II

Addison-Wesley Algebra & Trigonometry: Grades 9-12
Addison-Wesley, 1992 (Termination Year: 1997)

Merrill Algebra 2 with Trigonometry: Applications & Connections: Grades 9-12
Glencoe, 1992 (Termination Year: 1997)

Holt Algebra with Trigonometry: Grades 9-12
Holt, 1992 (Termination Year: 1997)

Algebra and Trigonometry: Structure and Method: Grades 9-12
Houghton, 1992 (Termination Year: 1997)

Algebra II and Trigonometry: Grades 9-12
McDougal, 1991 (Termination Year: 1997)

Prentice Hall Algebra II: Grades 9–12
Prentice Hall, 1990 (Termination Year: 1997)

*University of Chicago School Math Project:
Advanced Algebra*: Grades 9–12
ScottForesman, 1990 (Termination Year: 1997)

Advanced Algebra

Algebra and Trigonometry: Grades 9–12
Prentice Hall, 1991 (Termination Year: 1997)

College Algebra: Grades 9–12
Prentice Hall, 1991 (Termination Year: 1997)

Consumer Mathematics

Consumer Mathematics: Grades 9–12
Addison-Wesley, 1992 (Termination
Year: 1997)

Practical Mathematics: Consumer Applications:
Grades 9–12
Holt, 1989 (Termination Year: 1997)

Prentice Hall Consumer Mathematics:
Grades 9–12
Prentice Hall, 1989 (Termination Year: 1997)

Consumer & Career Mathematics:
Grades 9–12
ScottForesman, 1989 (Termination Year: 1997)

General Mathematics

Addison-Wesley Pre-Algebra: Grades 9–12
Addison-Wesley, 1992 (Termination
Year: 1997)

Essentials of Mathematics: Grades 9–12
Addison-Wesley, 1992 (Termination
Year: 1997)

Merrill Applications of Math: Grades 9–12
Glencoe, 1988 (Termination Year: 1997)

*Merrill Mathematics Connections: Essentials
and Applications*: Grades 9–12
Glencoe, 1992 (Termination Year: 1997)

Applications of High School Mathematics:
Grades 9–12
Houghton, 1992 (Termination Year: 1997)

Essentials for High School Mathematics:
Grades 9–12
Houghton, 1989 (Termination Year: 1997)

*Mathematical Connections: A Bridge to Alge-
bra & Geometry*: Grades 9–12
Houghton, 1992 (Termination Year: 1997)

Refresher Mathematics: Grades 9–12
Prentice Hall, 1989 (Termination Year: 1997)

Geometry

Addison-Wesley Geometry: Grades 9–12
Addison-Wesley, 1992 (Termination
Year: 1997)

Glencoe Geometry: Grades 9–12
Glencoe, 1990 (Termination Year: 1997)

Holt Geometry: Grades 9–12
Holt, 1991 (Termination Year: 1997)

Houghton Mifflin Geometry: Grades 9–12
Houghton, 1992 (Termination Year: 1997)

Prentice Hall Geometry: Grades 9–12
Prentice Hall, 1990 (Termination Year: 1997)

Scott Foresman Geometry: Grades 9–12
ScottForesman, 1990 (Termination Year: 1997)

*University of Chicago School Mathematics
Project: Geometry*: Grades 9–12
ScottForesman, 1991 (Termination Year: 1997)

Analytic Geometry

Modern Analytic Geometry: Grades 9–12
Houghton, 1989 (Termination Year: 1997)

Technical Mathematics

Career Mathematics: Grades 9–12
Houghton, 1989 (Termination Year: 1997)

Oklahoma

Mathematics

Mathematics in Action: Grade K-8
Macmillan, 1992 (Termination Year: 1997)

Mathematics Plus: Grades K-8
Harcourt, 1992 (Termination Year: 1997)

Addison-Wesley Mathematics: Grades K-8
Addison-Wesley, 1991 (Termination
Year: 1997)

*Silver Burdett & Ginn Mathematics: Exploring
Your World*: Grades K-8
Silver, 1992 (Termination Year: 1997)

*Houghton Mifflin: The Mathematics Experi-
ence*: Grades K-8
Houghton, 1992 (Termination Year: 1997)

Exploring Mathematics: Grades K-8
ScottForesman, 1991 (Termination Year: 1997)

Great Beginnings Core System: Grades K-2
ScottForesman, 1991 (Termination Year: 1997)

Heath Mathematics Connections: Grades K-8
Heath, 192 (Termination Year: 1997)

*Math Every Day: An Activity-Based Mathemat-
ics Program*: Grades 1-2
Heath, 1992 (Termination Year: 1997)

Basic Learning System/Math: Grades K-8
Jostens Learning, 1990 (Termination
Year: 1997)

Prealgebra: An Accelerated Course: Grade 7
Houghton, 1992 (Termination Year: 1997)

Mathematics: Structure & Method: Grades 7-8
Houghton, 1992 (Termination Year: 1997)

Practice & Assessment Skill Pad: Grade 7
Heath, 1992 (Termination Year: 1997)

Prealgebra

Addison-Wesley Prealgebra: Grades 9-12
Addison-Wesley, 1992 (Termination
Year: 1997)

Prentice Hall Prealgebra: Grades 9-12
Prentice Hall, 192 (Termination Year: 1997)

Holt Prealgebra: Grades 9-12
Holt, 1992 (Termination Year: 1997)

*Mathematical Connections, A Bridge to Algebra
& Geometry*: Grades 9-12
Houghton, 192 (Termination Year: 1997)

*University of Chicago School Mathematics
Project: Transition Mathematics*: Grades 9-12
ScottForesman, 1992 (Termination Year: 1997)

Heath Prealgebra: Grades 9-12
Heath, 1992 (Termination Year: 1997)

Merrill Prealgebra: A Transition to Algebra:
Grades 9-12
Glencoe, 1992 (Termination Year: 1997)

Prealgebra: Grades 9-12
South-Western, 1992 (Termination Year: 1997)

Algebra I

Addison-Wesley Algebra: Grades 9-12
Addison-Wesley, 192 (Termination
Year: 1997)

*Algebra 1: Expressions, Equations & Applica-
tions*: Grades 9-12
Addison-Wesley, 1990 (Termination
Year: 1997)

Prentice Hall Algebra 1: Grades 9-12
Prentice Hall, 1990 (Termination Year: 1997)

Holt Algebra 1: Grades 9-12
Holt, 1992 (Termination Year: 1997)

Algebra: Structure & Method: Grades 9-12
Houghton, 192 (Termination Year: 1997)

*University of Chicago School Mathematics
Project: Algebra*: Grades 9-12
ScottForesman, 1990 (Termination Year: 1997)

Merrill Algebra 1: Applications & Connections:
Grades 9-12
Glencoe, 1992 (Termination Year: 1997)

Merrill Algebra Essentials: Grades 9-12
Glencoe, 1988 (Termination Year: 1997)

Algebra 1, An Integrated Approach:
Grades 9-12
McDougal, 1991 (Termination Year: 1997)

Algebra II

Addison-Wesley Algebra & Trigonometry:
Grades 9-12
Addison-Wesley, 1992 (Termination
Year: 1997)

Algebra & Trigonometry: Grades 9-12
Addison-Wesley, 1990 (Termination Year: 1997)

Prentice Hall Algebra II with Trigonometry: Grades 9-12
Prentice Hall, 1990 (Termination Year: 1997)

Holt Algebra with Trigonometry: Grades 9-12
Holt, 1990 (Termination Year: 1997)

Algebra & Trigonometry: Structure & Method: Grades 9-12
Houghton, 1992 (Termination Year: 1997)

University of Chicago School Mathematics Project: Advanced Algebra: Grades 9-12
ScottForesman, 1990 (Termination Year: 1997)

Merrill Algebra 2 with Trigonometry: Grades 9-12
Glencoe, 1992 (Termination Year: 1997)

Algebra 2 & Trigonometry: Grades 9-12
McDougal, 1991 (Termination Year: 1997)

Precalculus

Precalculus Mathematics: A Graphing Approach: Grades 9-12
Addison-Wesley, 1990 (Termination Year: 1997)

Precalculus Mathematics: Grades 9-12
Prentice Hall, 1991(Termination Year: 1997)

HBJ Advanced Mathematics: A Preparation for Calculus: Grades 9-12
Holt, 1988 (Termination Year: 1997)

Contemporary Precalculus through Applications: Functions, Data Analysis & Matrices: Grades 9-12
Janson, 1991 (Termination Year: 1997)

Advanced Mathematics: Precalculus with Discrete Mathematics & Data Analysis: Grades 9-12
Houghton, 1992 (Termination Year: 1997)

University of Chicago School Mathematics Project: Mathematics, Precalculus & Discrete Mathematics: Grades 9-12
ScottForesman, 1992 (Termination Year: 1997)

Merrill Precalculus Mathematics: Grades 9-12
Glencoe, 1988 (Termination Year: 1997)

Precalculus with Unit Circle Trigonometry: Grades 9-12
West, 1990 (Termination Year: 1997)

Calculus

Calculus of a Single Variable: Grades 9-12
Wadsworth, 1991 (Termination Year: 1997)

Single Variable Calculus: Grades 9-12
Prentice Hall, 1992 (Termination Year: 1997)

Elements of Calculus & Analytic Geometry: Grades 9-12
Addison-Wesley, 1989 (Termination Year: 1997)

Calculus with Analytic Geometry: Grades 9-12
Prentice Hall, 1992 (Termination Year: 1997)

HBJ Calculus with Analytic Geometry: Grades 9-12
Holt, 1989 (Termination Year: 1997)

The Calculus of a Single Variable with Analytic Geometry: Grades 9-12
ScottForesman, 1990 (Termination Year: 1997)

Geometry, Combined

Addison-Wesley Geometry: Grades 9-12
Addison-Wesley, 1992 (Termination Year: 1997)

Addison-Wesley Informal Geometry: Grades 9-12
Addison-Wesley, 1992 (Termination Year: 1997)

Prentice Hall Geometry: Grades 9-12
Prentice Hall, 1990 (Termination Year: 1997)

Informal Geometry: Grades 9-12
Prentice Hall, 1992 (Termination Year: 1997)

Discovering Geometry: An Inductive Approach: Grades 9-12
Key Curriculum, 1989 (Termination Year: 1997)

Geometry: Grades 9-12
Holt, 1991 (Termination Year: 1997)

Geometry: Grades 9–12
Houghton, 1992 (Termination Year: 1997)

University of Chicago School Mathematics Project: Geometry: Grades 9–12
ScottForesman, 1991 (Termination Year: 1997)

Geometry: Grades 9–12
Glencoe, 1990 (Termination Year: 1997)

Mathematical Analysis

Introductory Analysis: Grades 9–12
Houghton, 1991 (Termination Year: 1997)

Merrill Advance Mathematical Concepts:
Grades 9–12
Glencoe, 1991 (Termination Year: 1997)

Mathematics

Merrill Integrated Mathematics: Grades 9–12
Glencoe, 1991 (Termination Year: 1997)

Mathematics of Money: Grades 9–12
South-Western, 1992 (Termination Year: 1997)

Understanding Math Using Appleworks Spreadsheets: Grades 9–12
West, 1990 (Termination Year: 1997)

Mathematics, Consumer

Addison-Wesley Consumer Mathematics:
Grades 9–12
Addison-Wesley, 1992 (Termination Year: 1997)

Prentice Hall Consumer Mathematics:
Grades 9–12
Prentice Hall, 1989 (Termination Year: 1997)

Practical Mathematics: Consumer Applications:
Grades 9–12
Holt, 1989 (Termination Year: 1997)

Consumer & Career Mathematics:
Grades 9–12
ScottForesman, 1989 (Termination Year: 1997)

Merrill Applications of Mathematics:
Grades 9–12
Glencoe, 1989 (Termination Year: 1997)

Consumer Mathematics: Grades 9–12
Glencoe, 1989 (Termination Year: 1997)

Dollars & Sense: Problem Solving Strategies in Consumer Mathematics: Grades 9–12
South-Western, 1989 (Termination Year: 1997)

Mathematics of Money: Grades 9–12
West, 1992 (Termination Year: 1997)

Mathematics, General

Essentials of Mathematics: Grades 9–12
Addison-Wesley, 1992 (Termination Year: 1997)

Prentice Hall Refresher Mathematics:
Grades 9–12
Prentice Hall, 1989 (Termination Year: 1997)

Practical Mathematics: Skills & Concepts:
Grades 9–12
Holt, 1989 (Termination Year: 1997)

Applications of High School Mathematics:
Grades 9–12
Houghton, 1992 (Termination Year: 1997)

Mathematics in Life: Grades 9–12
ScottForesman, 1989 (Termination Year: 1997)

Merrill Mathematics Connections: Essentials & Applications: Grades 9–12
Glencoe, 1992 (Termination Year: 1997)

Geometry for Decision Making: Grades 9–12
South-Western, 1992 (Termination Year: 1997)

Mathematics Skill Builder: Grades 9–12
South-Western, 1985 (Termination Year: 1997)

Mathematics, Remedial

Spectrum Mathematics: Grades 9–12
Glencoe, 1990 (Termination Year: 1997)

Sound Foundations: A Practical Mathematics Simulation: Grades 9–12
South-Western, 1987 (Termination Year: 1997)

Study Guides for Solving Algebraic Word Problems: Grades 9–12
South Western, 1987 (Termination Year: 1997)

Statistics & Probabilities

Introduction to the Practice of Statistics:
Grades 9–12
Freeman, 1989 (Termination Year: 1997)

Probability & Statistics: Grades 9-12
South-Western, 1989 (Termination Year: 1997)

Trigonometry

Trigonometry: Functions & Applications:
Grades 9-12
Addison-Wesley, 1990 (Termination
Year: 1997)

Prentice Hall Trigonometry: Grades 9-12
Prentice Hall, 1990 (Termination Year: 1997)

Trigonometry with Applications: Grades 9-12
Houghton, 1990 (Termination Year: 1997)

*University of Chicago School Mathematics
Project: Functions, Statistics, & Trigonometry*:
Grades 9-12
ScottForesman, 1992 (Termination Year: 1997)

ScottForesman Trigonometry: Grades 9-12
ScottForesman, 1989 (Termination Year: 1997)

Trigonometry & Its Applications: Grades 9-12
Glencoe, 1990 (Termination Year: 1997)

Oregon

Mathematics—Grades K-3

Explorations: Grades K-2
Addison-Wesley, 1988 (Termination
Year: 1995)

Addison-Wesley Mathematics: Grades K-3
Addison-Wesley, 1989 (Termination
Year: 1995)

Holt Mathematics Unlimited: Grades K-3
Harcourt, 1988 (Termination Year: 1995)

Houghton Mifflin Mathematics: Grades K-3
Houghton, 1989 (Termination Year: 1995)

Merrill Mathematics: Grades K-3
Glencoe, 1987 (Termination Year: 1995)

The Real Math Program: Grades K-3
Open Court, 1991 (Termination Year: 1995)

Invitation to Mathematics: Grades K-3
ScottForesman, 1988 (Termination Year: 1995)

Silver Burdett & Ginn Mathematics:
Grades K-3
Silver, 1988 (Termination Year: 1995)

Mathematics—Grades 4-5

Addison-Wesley Mathematics: Grades 4-5
Addison-Wesley, 1989 (Termination
Year: 1995)

Mathematics Today: Grades 4-5
Harcourt, 1987 (Termination Year: 1995)

Holt Mathematics Unlimited: Grades 4-5
Harcourt, 1988 (Termination Year: 1995)

Houghton Mifflin Mathematics: Grades 4-5
Houghton, 1989 (Termination Year: 1995)

Merrill Mathematics: Grades 4-5
Glencoe, 1987 (Termination Year: 1995)

The Real Math Program: Grades 4-5
Open Court, 1991 (Termination Year: 1995)

Invitation to Mathematics: Grades 4-5
ScottForesman, 1988 (Termination Year: 1995)

Silver Burdett & Ginn Mathematics:
Grades 4-5
Silver, 1988 (Termination Year: 1995)

Mathematics—Grades 6-8

Addison-Wesley Mathematics: Grades 6-8
Addison-Wesley, 1989 (Termination
Year: 1995)

Mathematics Today: Grades 6-8
Harcourt, 1987 (Termination Year: 1995)

Holt Mathematics Unlimited: Grades 6-8
Harcourt, 1988 (Termination Year: 1995)

Houghton Mifflin Mathematics: Grades 6-8
Houghton, 1989 (Termination Year: 1995)

Merrill Mathematics: Grades 6-8
Glencoe, 1987 (Termination Year: 1995)

The Real Math Program: Grades 6-8
Open Court, 1987 (Termination Year: 1995)

Invitation to Mathematics: Grades 6-8
ScottForesman, 1988 (Termination Year: 1995)

Silver Burdett & Ginn Mathematics:
Grades 6-8
Silver, 1988 (Termination Year: 1995)

General Mathematics

Essentials of Mathematics: 9-12
Addison-Wesley, 1989 (Termination
Year: 1995)

Globe High School Mathematics: 9-12
Globe, 1989 (Termination Year: 1995)

Heath General Mathematics: 9-12
Heath, 1989 (Termination Year: 1995)

Merrill General Mathematics: 9-12
Glencoe, 1987 (Termination Year: 1995)

Algebra 1 and 2

Algebra I: Grades 9-12
Addison-Wesley, 1990 (Termination
Year: 1995)

Algebra & Trigonometry: Grades 9-12
Addison-Wesley, 1990 (Termination
Year: 1995)

HBJ Algebra 1: Grades 9-12
Holt, 1990 (Termination Year: 1995)

HBJ Algebra 2 with Trigonometry:
Grades 9-12
Holt, 1990 (Termination Year: 1995)

Heath Algebra 1: Grades 9-12
Heath, 1987 (Termination Year: 1995)

Heath Algebra 2 with Trigonometry:
Grades 9-12
Heath, 1987 (Termination Year: 1995)

Algebra: Structure & Method: Grades 9-12
Houghton, 1990 (Termination Year: 1995)

Algebra & Trigonometry: Structure & Method:
Grades 9-12 Houghton, 1990 (Termination
Year: 1995)

Merrill Algebra One: Grades 9-12
Glencoe, 1990 (Termination Year: 1995)

Merrill Algebra Two with Trigonometry:
Grades 9-12
Glencoe, 1990 (Termination Year: 1995)

Prentice hall Algebra 1: Grades 9-12
Prentice Hall, 1990 (Termination Year: 1995)

Prentice Hall Algebra 2 with Trigonometry:
Grades 9-12
Prentice Hall, 1990 (Termination Year: 1995)

Geometry

Addison-Wesley Geometry: Grades 9-12
Addison-Wesley, 1990 (Termination
Year: 1995)

Geometry: Grades 9-12
Freeman, 1987 (Termination Year: 1995)

Geometry: Grades 9-12
Houghton, 1990 (Termination Year: 1995)

Discovering Geometry: An Inductive Approach:
Grades 9-12
Key Curriculum, 1990 (Termination
Year: 1995)

Geometry for Enjoyment & Challenge:
Grades 9-12
McDougal, 1988 (Termination Year: 1995)

Geometry: Grades 9-12
Glencoe, 1990 (Termination Year: 1995)

Prentice Hall Geometry: Grades 9-12
Prentice Hall, 1990 (Termination Year: 1995)

ScottForesman Geometry: Grades 9-12
ScottForesman, 1990 (Termination Year: 1995)

Tennessee

Mathematics

Addison-Wesley Mathematics: Grades 1-8
Addison-Wesley, 1987 (Termination
Year: 1993)

Mathematics Today: Grades 1-8
Harcourt, 1987 (Termination Year: 1993)

Heath Mathematics Program: Grades 1-3
Heath, 1987 (Termination Year: 1993)

Mathematics Unlimited: Grades 1-8
Harcourt, 1987 (Termination Year: 1993)

Houghton Mifflin Mathematics: Grades 1-8
Houghton, 1987 (Termination Year: 1993)

Mathematics: Structure & Method: Grades 7-8
Houghton, 1985 (Termination Year: 1993)

Laidlaw Mathematics: Grades 1-8
Macmillan, 1987 (Termination Year: 1993)

Macmillan Mathematics: Grades 1-8
Macmillan, 1987 (Termination Year: 1993)

McGraw-Hill Mathematics: Grades 1-8
Macmillan, 1987 (Termination Year: 1993)

Merrill Mathematics: Grades 1-8
Macmillan, 1987 (Termination Year: 1993)

Invitation to Mathematics: Grades 6-8
ScottForesman, 1987 (Termination Year: 1993)

Silver Burdett & Ginn Mathematics:
Grades 1-8
Silver, 1987 (Termination Year: 1993)

Arithmetic (9)

General Mathematics: Grade 9
Addison-Wesley, 1986 (Termination
Year: 1993)

Refresher Mathematics: Grade 9
Prentice Hall, 1986 (Termination Year: 1993)

General Mathematics: Skills/Problem Solving/ Applications: Grade 9
Harcourt, 1987 (Termination Year: 1993)

Heath General Mathematics: Grade 9
Heath, 1985 (Termination Year: 1993)

Holt Essential Mathematics: Grade 9
Holt, 1987 (Termination Year: 1993)

Mathematics Skills for Daily Living: Grade 9
Glencoe, 1986 (Termination Year: 1993)

Basic Mathematics: Grade 9
Media Materials, 1982 (Termination
Year: 1993)

Merrill General Mathematics: Grade 9
Glencoe, 1987 (Termination Year: 1993)

Mathematics in Life: Grade 9
ScottForesman, 1987 (Termination Year: 1993)

Applied Mathematics

Basic Vocational–Technical Mathematics:
Grades 9-12
Delmar, 1985 (Termination Year: 1993)

Applying Mathematics in Daily Living:
Grades 9-12
Glencoe, 1986 (Termination Year: 1993)

Life Skills Mathematics: Grades 9-12
Media Materials, 1983 (Termination
Year: 1993)

Mathematics for Consumers: Grades 9-12
Media Materials, 1983 (Termination
Year: 1993)

Prealgebra

Addison-Wesley Prealgebra: Grades 8-11
Addison-Wesley, 1991 (Termination
Year: 1993)

Prealgebra: Skills/Problem Solving/Applications: Grades 7-9 Harcourt, 1996 (Termination
Year: 1993)

Introductory Algebra 1: Grades 9-12
Harcourt, 1982 (Termination Year: 1993)

Heath Prealgebra: Grades 7-12
Heath, 1986 (Termination Year: 1993)

Essentials for Algebra: Concepts & Skills:
Grades 9
Houghton, 1986 (Termination Year: 1993)

Prealgebra: An Accelerated Course: Grades 7
Houghton, 1985 (Termination Year: 1993)

Elementary Algebra, Part 1: Grades 9-12
Houghton, 1985 (Termination Year: 1993)

Merrill Prealgebra: a Problem-Solving Approach: Grades 7-9
Glencoe, 1986 (Termination Year: 1993)

ScottForesman Prealgebra: Grades 9–11
ScottForesman, 1986 (Termination Year: 1993)

Algebra

Algebra: Grades 9–12
Addison-Wesley, 1986 (Termination Year: 1993)

Algebra: Expressions, Equations & Applications: Grades 9–12
Addison-Wesley, 1984 (Termination Year: 1993)

Introductory Algebra 2: Grades 9–12
Harcourt, 1982 (Termination Year: 1993)

HBJ Algebra 1: Grades 9–12
Harcourt, 1987 (Termination Year: 1993)

Heath Algebra 1: Grades 9–12
Heath, 1987 (Termination Year: 1993)

Holt Algebra 1: Grades 9–12
Holt, 1987 (Termination Year: 1993)

Elementary Algebra, Part 2: Grades 9–12
Houghton, 1985 (Termination Year: 1993)

Algebra I: Grades 9–12
Houghton, 1986 (Termination Year: 1993)

Algebra: Structure & Method: Grades 9–12
Houghton, 1986 (Termination Year: 1993)

Laidlaw Algebra I: Grades 8–10
Glencoe, 1987 (Termination Year: 1993)

Merrill Algebra One: Grades 8–10
Glencoe, 1986 (Termination Year: 1993)

ScottForesman Algebra: Grades 9–12
ScottForesman, 1984 (Termination Year: 1993)

Harper & Row Algebra One: Grades 9–12
Glencoe, 1984 (Termination Year: 1993)

Algebra II

Algebra & Trigonometry: Grades 9–12
Addison-Wesley, 1986 (Termination Year: 1993)

HBJ Algebra 2 with Trigonometry:
Grades 9–12
Harcourt, 1987 (Termination Year: 1993)

Heath Algebra 2 with Trigonometry:
Grades 9–12
Heath, 1987 (Termination Year: 1993)

Holt Algebra 2 with Trigonometry:
Grades 9–12
Holt, 1986 (Termination Year: 1993)

Algebra & Trigonometry: Structure & Method:
Grades 9–12 Houghton, 1986 (Termination Year: 1993)

Algebra 2 & Trigonometry: Grades 9–12
Houghton, 1986 (Termination Year: 1993)

Laidlaw Algebra 2 with Trigonometry:
Grades 9–11
Glencoe, 1987 (Termination Year: 1993)

Merrill Algebra Two with Trigonometry:
Grades 10–12
Glencoe, 1986 (Termination Year: 1993)

Harper & Row Algebra Two with Trigonometry: Grades 9–12
Glencoe, 1985 (Termination Year: 1993)

Geometry

Geometry with Applications & Problem Solving: Grades 9–12
Addison-Wesley, 1984 (Termination Year: 1993)

HBJ Geometry: Grades 9–12
Harcourt, 1987 (Termination Year: 1993)

Geometry: Grades 10–12
Houghton, 1985 (Termination Year: 1993)

Holt Geometry: Grades 10–12
Holt, 1985 (Termination Year: 1993)

Laidlaw Geometry: Grades 9–11
Glencoe, 1987 (Termination Year: 1993)

Merrill Geometry: Grades 10–11
Glencoe, 1987 (Termination Year: 1993)

ScottForesman Geometry: Grades 10-12
ScottForesman, 1987 (Termination Year: 1993)

Harper & Row Geometry: Grades 9-12
Glencoe, 1986 (Termination Year: 1993)

Trigonometry

Trigonometry: Functions & Applications:
Grades 9-12
Addison-Wesley, 1984 (Termination
Year: 1993)

Trigonometry: Grades 10-12
Harcourt, 1987 (Termination Year: 1993)

Trigonometry with Applications: Grades 12
Houghton, 1987 (Termination Year: 1993)

Plane Trigonometry: Grades 11-12
Prentice Hall, 1984 (Termination Year: 1993)

ScottForesman Trigonometry: Grades 11-12
ScottForesman, 1983 (Termination Year: 1993)

Trigonometry: Grades 10-12
West, 1984 (Termination Year: 1993)

Advanced Mathematics

*Precalculus with Trigonometry: Functions &
Applications*: Grades 11-12
Addison-Wesley, 1987 (Termination
Year: 1993)

*HBJ Advanced Mathematics: A Preparation for
Calculus*: Grades 10-12
Harcourt, 1984 (Termination Year: 1993)

Precalculus: Grades 9-12
Heath, 1985 (Termination Year: 1993)

Advanced Mathematics: A Precalculus Course:
Grade 12
Houghton, 1987 (Termination Year: 1993)

Modern Introductory Analysis: Grade 12
Houghton, 1984 (Termination Year: 1993)

Merrill Advanced Mathematical Concepts:
Grades 11-12
Glencoe, 1986 (Termination Year: 1993)

Merrill Precalculus Mathematics: Grades
11-12
Glencoe, 1983 (Termination Year: 1993)

ScottForesman Precalculus: Grades 9-12
ScottForesman, 1987 (Termination Year: 1993)

Calculus

Elements of Calculus & Analytic Geometry:
Grades 10-12
Addison-Wesley, 1981 (Termination
Year: 1993)

Calculus with Analytic Geometry: Grades 9-12
Heath, 1986 (Termination Year: 1993)

Calculus with Analytic Geometry: Grades 9-12
Heath, 1986 (Termination Year: 1993)

Analytical Geometry

Analytic Geometry: Grades 11-12
Addison-Wesley, 1986 (Termination
Year: 1993)

Modern Analytic Geometry: Grade 12
Houghton, 1984 (Termination Year: 1993)

Advanced Algebra

*Algebra & Trigonometry: Functions & Applica-
tions*: Grades 9-12
Addison-Wesley, 1984 (Termination
Year: 1993)

College Algebra: Grades 9-12
West, 1984 (Termination Year: 1993)

College Algebra: Grades 9-12
West, 1986 (Termination Year: 1993)

College Algebra & Trigonometry: Grades 9-12
West, 1984 (Termination Year: 1993)

College Algebra & Trigonometry: Grades 9-12
West, 1986 (Termination Year: 1993)

Probability and Statistics

Using Statistics: Grades 11-12
Addison-Wesley, 1985 (Termination
Year: 1993)

Introductory Statistics & Probability: A Basis for Decision Making: Grade 12
Houghton, 1984 (Termination Year: 1993)

Introductory Statistics: Grades 9–12
West, 1985 (Termination Year: 1993)

Statistics: Concepts & Applications: Grades 9–12
West, 1986 (Termination Year: 1993)

Statistics: The Exploration of Data: Grades 9–12
West, 1986 (Termination Year: 1993)

Personal Computing

Addison-Wesley Computer Literacy Awareness, Applications & Programming: Grades 7–11
Addison-Wesley, 1986 (Termination Year: 1993)

Computers & Information Systems: Grades 9–12
Holt, 1986 (Termination Year: 1993)

Introduction to Computer Applications: Grades 9–12 Glencoe, 1986 (Termination Year: 1993)

Computer Literacy: A Hands on Approach: Grades 9–12
Glencoe, 1986 (Termination Year: 1993)

Macmillan Computer Literacy: Grades 9–12
Macmillan, 1986 (Termination Year: 1993)

Information Processing: Concepts, Principles, & Procedures: Grades 9–12
South-Western, 1985 (Termination Year: 1993)

Computer Fundamentals with Basic Programming: Grades 7–9
West, 1985 (Termination Year: 1993)

Introduction to Computers Using the Apple: Grades 9–12
West, 1985 (Termination Year: 1993)

Introduction to Computers Using the IBM PC: Grades 9–12
West, 1985 (Termination Year: 1993)

Introduction to Computers Using the TRS 80: Grades 9–1286
West, 1985 (Termination Year: 1993)

Mind Tool: Grades 9–12
West, 1986 (Termination Year: 1993)

West's Microcomputing Series: Grades 9–12
West, 1986 (Termination Year: 1993)

Basic Programming

Computer Programming in the Basic Language: Grades 9–12
Harcourt, 1981 (Termination Year: 1993)

Computer Mathematics: Basic Programming with Applications: Grades 9–12
Heath, 1985 (Termination Year: 1993)

Computer Programming in Basic: Grades 9–12
Houghton, 1986 (Termination Year: 1993)

Basic Steps to BASIC: Grades 7–12
Media Material, 1985 (Termination Year: 1993)

Programming in Basic: Grades 8–12
Glencoe, 1983 (Termination Year: 1993)

Business Basic: Grades 9–12
Macmillan, 1986 (Termination Year: 1993)

Introduction to Programming in Basic for the Apple II/IIe/IIc: Grades 9–12
Macmillan, 1985 (Termination Year: 1993)

Structured Basic: Grades 9–12
South-Western, 1989 (Termination Year: 1993)

Advanced Structured Basic: Grades 9–12
South-Western, 1985 (Termination Year: 1993)

Basic Programming Today: A structured Approach: Grades 9–12
West, 1986 (Termination Year: 1993)

Complete Basic Programming: Grades 9–12
West, 1984 (Termination Year: 1993)

Fundamentals of Basic Programming: A Structured Approach: Grades 7–9
West, 1985 (Termination Year: 1993)

Understanding Basic: A Structured Approach:
Grades 9-12
West, 1985 (Termination Year: 1993)

Strictly Structured Basic: Grades 10-12
West, 1985 (Termination Year: 1993)

*Basic: Concepts with Structured Problem
Solving*: Grades 9-12
West, 1984 (Termination Year: 1993)

Basic: Grades 9-12
West, 1986 (Termination Year: 1993)

Basic Programming for the IBM PC:
Grades 9-12
West, 1986 (Termination Year: 1993)

Advanced Basic: A Structured Approach:
Grades 9-12
West, 1986 (Termination Year: 1993)

Pascal Programming

*Problem Solving & Structured Programming in
Pascal*: Grades 11-12
Addison-Wesley, 1985 (Termination
Year: 1993)

Introduction to Pascal & Structured Design:
Grades 9-12
Heath, 1987 (Termination Year: 1993)

*Pascal Plus Data Structures, Algorithms, &
Advanced Programming*: Grades 9-12
Heath, 1985 (Termination Year: 1993)

Problem Solving with Pascal: Grades 10-12
Holt, 1986 (Termination Year: 1993)

*Pascal: An Introduction to Methodical Pro-
gramming*: Grades 10-12
Houghton, 1985 (Termination Year: 1993)

Introduction to Pascal: Grades 9-12
West, 1983 (Termination Year: 1993)

Understanding Pascal: A Structural Approach:
Grades 9-12
West, 1985 (Termination Year: 1993)

Apple Pascal: A Problem Solving Approach:
Grades 9-12
West, 1985 (Termination Year: 1993)

*Fundamentals of Pascal: Understanding Pro-
gramming & Problem Solving*: Grades 9-12
West, 1986 (Termination Year: 1993)

*Computer Science with Pascal for Advanced
Placement*: Grades 9-12
West, 1986 (Termination Year: 1993)

Business Mathematics

Business Mathematics: Grades 9-12
Glencoe, 1988 (Termination Year: 1993)

Business Mathematics: Grades 9-12
Houghton, 1986 (Termination Year: 1993)

Business Mathematics: Grades 10-12
Glencoe, 1988 (Termination Year: 1993)

Mathematics for Business: Grades 7-12
Media Materials, 1986 (Termination
Year: 1993)

Applied Business Mathematics: Grades 9-12
South-Western, 1985 (Termination Year: 1993)

Texas

Mathematics

Addison-Wesley Mathematics: Grades 1-8
(also in Spanish for Grades 1-5)
Addison-Wesley, 1991 (Termination
Year: 1997)

Mathematics in Action: Grades 1-8 (also in
Spanish for Grades 1-5)
Macmillan/McGraw-Hill, 1992 (Termination
Year: 1997)

Exploring Mathematics: Grades 1-8 (also in
Spanish for Grades 1-5)
ScottForesman, 1991 (Termination Year: 1997)

Mathematics: Exploring Your World: Grades 1-
8 (also in Spanish for Grades 1-5)
Silver, 1991 (Termination Year: 1997)

Fundamentals of Mathematics

Addison-Wesley Essentials of Mathematics:
Grades 9-12
Addison-Wesley, 1989 (Termination
Year: 1995)

HBJ Fundamentals of Mathematics:
Grades 9-12
Holt, 1989 (Termination Year: 1995)

Heath General Mathematics: Grades 9-12
Heath, 1989 (Termination Year: 1995)

Practical Math: Skills & Concepts:
Grades 9-12
Holt, 1989 (Termination Year: 1995)

Essentials for High School Mathematics:
Grades 9-12
Houghton, 1989 (Termination Year: 1995)

Refresher Mathematics: Grades 9-12
Prentice Hall, 1989 (Termination Year: 1995)

Mathematics in Life: Grades 9-12
ScottForesman, 1989 (Termination Year: 1995)

Consumer Mathematics

Addison-Wesley Consumer Mathematics:
Grades 9-12
Addison-Wesley, 1989 (Termination
Year: 1995)

HBJ Consumer Mathematics: Grades 9-12
Holt, 1989 (Termination Year: 1995)

Practical Math: Consumer Applications:
Grades 9-12
Holt, 1989 (Termination Year: 1995)

Consumer Mathematics: Grades 9-12
Glencoe, 1989 (Termination Year: 1995)

Consumer & Career Mathematics:
Grades 9-12
ScottForesman, 1989 (Termination Year: 1995)

*Dollars & Sense: Problem Solving in Consumer
Math*: Grades 9-12
South-Western, 1989 (Termination Year: 1995)

Prealgebra

*Addison-Wesley Pre Algebra: A Transition to
Algebra*: Grades 9-12
Addison-Wesley, 1992 (Termination
Year: 1998)

Prealgebra: A Transition to Algebra:
Grades 9-12
Glencoe, 1992 (Termination Year: 1998)

*Math Connections, a Bridge to Algebra &
Geometry*: Grades 9-12
Houghton, 1992 (Termination Year: 1998)

Prentice-Hall Prealgebra: Grades 9-12
Prentice Hall, 1992 (Termination Year: 1998)

*University of Chicago School Math Project:
Transition Math*: Grades 9-12
ScottForesman, 1992 (Termination Year: 1998)

Prealgebra: Grades 9-12
South-Western, 1992 (Termination Year: 1998)

Algebra I

Addison-Wesley Algebra: Grades 9-12
Addison-Wesley, 1990 (Termination
Year: 1996)

Algebra: Grades 9-12
Addison-Wesley, 1990 (Termination
Year: 1996)

HBJ Algebra: Grades 9-12
Holt, 1990 (Termination Year: 1996)

Heath Algebra: Grades 9-12
Heath, 1990 (Termination Year: 1996)

Algebra: Structure & Method: Grades 9-12
Houghton, 1990 (Termination Year: 1996)

Merrill Algebra One: Grades 9-12
Glencoe, 1990 (Termination Year: 1996)

Prentice Hall Algebra: Grades 9-12
Prentice Hall, 1990 (Termination Year: 1996)

*University of Chicago School Mathematics
Project: Algebra:* Grades 9-12
ScottForesman, 1990 (Termination Year: 1996)

Algebra II

Addison-Wesley Algebra & Trigonometry:
Grades 9-12
Addison-Wesley, 1990 (Termination
Year: 1996)

Algebra & Trigonometry: Functions & Applications: Grades 9–12
Addison-Wesley, 1990 (Termination Year: 1996)

HBJ Algebra 2 with Trigonometry: Grades 9–12
Holt, 1990 (Termination Year: 1996)

Heath Algebra 2 with Trigonometry: Grades 9–12
Heath, 1990 (Termination Year: 1996)

Algebra & Trigonometry: Structure & Method: Grades 9–12
Houghton, 1990 (Termination Year: 1996)

Merrill Algebra Two with Trigonometry: Grades 9–12
Glencoe, 1990 (Termination Year: 1996)

Prentice Hall Algebra 2 with Trigonometry: Grades 9–12
Prentice Hall, 1990 (Termination Year: 1996)

University of Chicago School Mathematics Project: Advanced Algebra: Grades 9–12
ScottForesman, 1990 (Termination Year: 1996)

Informal Geometry

Addison-Wesley Informal Geometry: Grades 9–12
Addison-Wesley, 1992 (Termination Year: 1998)

Informal Geometry: Grades 9–12
Prentice Hall, 1990 (Termination Year: 1998)

Geometry for Decision Making: Grades 9–12
South-Western, 1992 (Termination Year: 1998)

Geometry

Addison-Wesley Geometry: Grades 9–12
Addison-Wesley, 1990 (Termination Year: 1996)

Geometry: Grades 9–12
Houghton, 1990 (Termination Year: 1996)

Prentice Hall Geometry: Grades 9–12
Prentice Hall, 1990 (Termination Year: 1996)

ScottForesman Geometry: Grades 9–12
ScottForesman, 1990 (Termination Year: 1996)

Analytic Geometry

Analytic Geometry: Grades 9–12
Addison-Wesley, 1986 (Termination Year: 1994)

Trigonometry

Trigonometry: Functions & Applications: Grades 9–12
Addison-Wesley, 1990 (Termination Year: 1996)

Trigonometry & Its Applications: Grades 9–12
Glencoe, 1990 (Termination Year: 1996)

Trigonometry: Grades 9–12
Holt, 1990 (Termination Year: 1996)

Trigonometry with Applications: Grades 9–12
Houghton, 1990 (Termination Year: 1996)

Prentice Hall Trigonometry: Grades 9–12
Prentice Hall, 1990 (Termination Year: 1996)

Precalculus

Precalculus: Grades 9–12
Addison-Wesley, 1987 (Termination Year: 1993)

Precalculus: Grades 9–12
Glencoe, 1983 (Termination Year: 1993)

Precalculus: Grades 9–12
ScottForesman, 1987 (Termination Year: 1993)

Advanced Math: Grades 9–12
Houghton, 1987 (Termination Year: 1993)

Calculus

Elements of Calculus & Analytic Geometry: Grades 9–12
Addison-Wesley, 1989 (Termination Year: 1995)

HBJ Calculus with Analytic Geometry: Grades 9–12
Holt, 1989 (Termination Year: 1995)

Calculus: Grades 9–12
Heath, 1986 (Termination Year: 1995)

Elementary Analysis

HBJ Advanced Math: A Preparation for Calculus: Grades 9–12
Holt, 1988 (Termination Year: 1994)

Introductory Analysis: Grades 9–12
Houghton, 1988 (Termination Year: 1994)

Merrill Advanced Mathematical Concepts: Grades 9–12
Glencoe, 1986 (Termination Year: 1994)

Probability and Statistics

Using Statistics: Grades 9–12
Addison-Wesley, 1985 (Termination Year: 1994)

Elementary Statistics: Grades 9–12
Addison-Wesley, 1986 (Termination Year: 1994)

Computer Mathematics

Computer Math: BASIC Programming with Applications: Grades 9–12
Heath, 1985 (Termination Year: 1996)

Introduction to Computer Mathematics: Grades 9–12
Freeman, 1985 (Termination Year: 1996)

Using Computers in Mathematics: Grades 9–12
Addison-Wesley, 1986 (Termination Year: 1996)

Computer Programming in the BASIC Language: Grades 9–12
Holt, 1981 (Termination Year: 1996)

Virginia

Algebra I

Algebra I, Expressions, Equations, & Applications: Grades 9–12
Addison-Wesley, 1984 (Termination Year: 1993)

Merrill Algebra One: Grades 9–12
Glencoe, 1986 (Termination Year: 1993)

Harper & Row Algebra One: Grades 9–12
Glencoe, 1985 (Termination Year: 1993)

Laidlaw Algebra 1: Grades 9–12
Glencoe, 1987 (Termination Year: 1993)

HBJ Algebra I: Grades 9–12
Holt, 1987 (Termination Year: 1993)

Holt Algebra I: Grades 9–12
Holt, 1986 (Termination Year: 1993)

Algebra I: Structure & Method: Grades 9–12
Houghton, 1986 (Termination Year: 1993)

Algebra 1/Two-Year Sequence

Elementary Algebra, Part 1: Grades 9–12
Glencoe, 1985 (Termination Year: 1993)

Elementary Algebra, Part 2: Grades 9–12
Glencoe, 1985 (Termination Year: 1993)

Introductory Algebra 1 & 2: Grades 9–12
Holt, 1982 (Termination Year: 1993)

Algebra II

Algebra & Trigonometry, Functions & Applications: Grades 9–12
Addison-Wesley, 1984 (Termination Year: 1993)

Merrill Algebra Two with Trigonometry: Grades 9–12
Glencoe, 1986 (Termination Year: 1993)

Harper & Row Algebra Two with Trigonometry: Grades 9–12
Glencoe, 1986 (Termination Year: 1993)

Laidlaw Algebra 2 with Trigonometry: Grades 9–12
Glencoe, 1987 (Termination Year: 1993)

HBJ Algebra 2 with Trigonometry: Grades 9–12
Holt, 1986 (Termination Year: 1993)

Algebra & Trigonometry, Structure & Method: Grades 9–12
Houghton, 1986 (Termination Year: 1993)

Algebra 2 & Trigonometry: Grades 9–12
Houghton, 1986 (Termination Year: 1993)

Applications Mathematics

Basic Vocational-Technical Mathematics: Grades 9-12
Delmar, 1985 (Termination Year: 1993)

Applying Mathematics in Daily Living: Grades 9-12
Glencoe, 1986 (Termination Year: 1993)

Life Skills Mathematics: Grades 9-12
Media Materials, 1983 (Termination Year: 1993)

Occupational Mathematics: Grades 9-12
South-Western, 1986 (Termination Year: 1993)

Computer Mathematics

Using Computers in Mathematics: Grades 9-12
Addison-Wesley, 1986 (Termination Year: 1993)

Computer Mathematics: Basic Programming with Applications: Grades 9-12
Heath, 1985 (Termination Year: 1993)

Computer Programming in the Basic Language: Grades 9-12
Holt, 1981 (Termination Year: 1993)

Computer Programming in Basic: Grades 9-12
Houghton, 1986 (Termination Year: 1993)

Fundamentals of Pascal: Grades 9-12
West, 1986 (Termination Year: 1993)

Basic: Grades 9-12
West, 1986 (Termination Year: 1993)

Complete Basic Programming: Grades 9-12
West, 1984 (Termination Year: 1993)

Basic Programming Today: A Structured Approach: Grades 9-12
West, 1986 (Termination Year: 1993)

Consumer Mathematics

Applying Mathematics, A Consumer/Career Approach: Grades 9-12
Addison-Wesley, 1989 (Termination Year: 1993)

Consumer Mathematics: Grades 9-12
Glencoe, 1989 (Termination Year: 1993)

Essentials of Mathematics: Consumer & Career: Grades 9-12
Holt, 1987 (Termination Year: 1993)

Consumer & Career Mathematics: Grades 9-12
ScottForesman, 1987 (Termination Year: 1993)

Consumer Math: Grades 9-12
South-Western, 1983 (Termination Year: 1993)

Elementary Mathematics

Addison-Wesley Mathematics: Grades 3-8
Addison-Wesley, 1987 (Termination Year: 1993)

Mathematics Unlimited: Grades 3-7
Harcourt, 1987 (Termination Year: 1993)

Houghton Mifflin Mathematics: Grades 3-8
Houghton, 1987, (Termination Year: 1993)

Mathematics: Structure & Method: Grade 7-8
Houghton, 1985 (Termination Year: 1993)

Merrill Mathematics: Grades 3-8
Macmillan/McGraw-Hill, 1987 (Termination Year: 1993)

Laidlaw Mathematics, Series 2000: Grades 3-8
Macmillan/McGraw-Hill, 1987 (Termination Year: 1993)

McGraw-Hill Mathematics: Grades 3-8
Macmillan/McGraw Hill, 1987 (Termination Year: 1993)

Macmillan Mathematics: Grades 3-8
Macmillan/McGraw Hill, 1987 (Termination Year: 1993)

Open Court Real Math Series: Grades 3-7
Open Court, 1987 (Termination Year: 1993)

ScottForesman Invitation to Mathematics: Grades 3-7
ScottForesman, 1987 (Termination Year: 1993)

Silver Burdett & Ginn Mathematics:
Grades 3-8
Silver, 1987 (Termination Year: 1993)

General Mathematics

Pre Algebra: Skills/Program/Problem Solving Applications: Grade 8
Harcourt, 1986 (Termination Year: 1993)

Mathematics Today: Grade 8
Harcourt, 1987 (Termination Year: 1993)

Mathematics Unlimited: Grade 8
Holt, 1985 (Termination Year: 1993)

Essential Mathematics: Grade 8
Holt, 1987 (Termination Year: 1993)

Holt Prealgebra: Grade 8
Holt, 1986 (Termination Year: 1993)

Merrill Prealgebra: A Problem Solving Approach: Grade 8
Macmillan/McGraw-Hill, 1986 (Termination Year: 1993)

General Mathematics I

General Mathematics, A Fundamental Approach: Grade 9
Addison-Wesley, 1986 (Termination Year: 1993)

Heath General Mathematics: Grade 9
Heath, 1985 (Termination Year: 1993)

Merrill General Mathematics: Grade 9
Glencoe, 1987 (Termination Year: 1993)

General Mathematics: Skills/Problem Solving Applications: Grade 9
Holt, 1987 (Termination Year: 1993)

Essentials For Algebra: Concepts & Skills:
Grade 9
Houghton, 1986 (Termination Year: 1993)

Refresher Mathematics: Grade 9
Prentice-Hall, 1986 (Termination Year: 1993)

Geometry

Geometry: With Applications & Problems:
Grades 9-12
Addison-Wesley, 1984 (Termination Year: 1993)

Merrill Geometry: Grades 9-12
Glencoe, 1987 (Termination Year: 1993)

Harper & Row Geometry: Grades 9-12
Glencoe, 1986 (Termination Year: 1993)

Holt Geometry: Grades 9-12
Holt, 1986 (Termination Year: 1993)

Geometry: Grades 9-12
Houghton, 1985 (Termination Year: 1993)

ScottForesman Geometry: Grades 9-12
ScottForesman, 1987 (Termination: 1993)

Informal Geometry

Addison-Wesley Informal Geometry:
Grades 9-12
Addison-Wesley, 1986 (Termination Year: 1993)

Basic Geometry: Grades 9-12
Houghton, 1984 (Termination Year: 1993)

Intermediate Algebra and Trigonometry

Algebra & Trigonometry, Functions & Applications: Grades 9-12
Addison-Wesley, 1984 (Termination Year: 1993)

Holt Algebra 2 with Trigonometry:
Grades 9-12
Holt, 1986 (Termination Year: 1993)

Trigonometry

Trigonometry: Functions & Applications:
Grades 9-12
Addison-Wesley, 1984 (Termination Year: 1993)

HBJ Trigonometry: Grades 9-12
Holt, 1987 (Termination Year: 1993)

Trigonometry with Applications: Grades 9-12
Houghton, 1987 (Termination Year: 1993)

West Virginia

Mathematics K-8

Addison-Wesley Mathematics: Grades K–8
Addison-Wesley, 1991 (Termination Year: 1998)

Heath Mathematics Connections: Grades K–8
Heath, 1992 (Termination Year: 1998)

Mathematics Plus: Grades K–8
Harcourt, 1992 (Termination Year: 1998)

MathWorks: Grades K–2
Houghton, 1992 (Termination Year: 1998)

Houghton Mifflin: The Mathematics Experience: Grades K–8
Houghton, 1992 (Termination Year: 1998)

Mathematics in Action: Grades K–8
Macmillan/McGraw-Hill, K-8 (Termination Year: 1998)

Great Beginnings: Grades K–2
ScottForesman, 1992 (Termination Year: 1998)

Exploring Mathematics: Grades K–8
ScottForesman, 1991 (Termination Year: 1998)

Silver Burdett & Ginn Mathematics: Exploring Your World: Grades K–8
Silver, 1992 (Termination Year: 1998)

Prealgebra

Addison-Wesley Prealgebra: Grades 7–12
Addison-Wesley, 1992 (Termination Year: 1998)

Heath Prealgebra: Grades 7–12
Heath, 1992 (Termination Year: 1998)

Merrill Prealgebra: A Transition to Algebra: Grades 7–12
Glencoe, 1992 (Termination Year: 1998)

Holt Prealgebra: Grades 7–12
Holt, 1992 (Termination Year: 1998)

Mathematics: Structure & Method: Grades 7–12
Houghton, 1992 (Termination Year: 1998)

Mathematical Connections: A Bridge to Algebra & Geometry: Grades 7–12
Houghton, 1992 (Termination Year: 1998)

Prealgebra: An Accelerated Course: Grades 7–12
Houghton, 1992 (Termination Year: 1998)

Prentice Hall Prealgebra: Grades 7–12
Prentice Hall, 1992 (Termination Year: 1998)

Algebra 1/2, An Incremental Development: Grades 7–12
Saxon, 1990 (Termination Year: 1998)

The University of Chicago School Mathematics Project Transition Mathematics: Grades 7–12
ScottForesman, 1992 (Termination Year: 1998)

Prealgebra: Grades 7–12
South-Western, 1992 (Termination Year: 1998)

Transitional Math I

Essentials of Mathematics: Grades 9–12
Addison-Wesley, 1992 (Termination Year: 1998)

Merrill Mathematics Connections: Essentials & Applications: Grades 9–12
Glencoe, 1992 (Termination Year: 1998)

Practical Math: Skills & Concepts: Grades 9–12
Holt, 1989 (Termination Year: 1998)

Essentials for High School Mathematics: Grades 9–12
Houghton, 1989 (Termination Year: 1998)

Mathematical Connections: A Bridge to Algebra & Geometry: Grades 9–12
Houghton, 1992 (Termination Year: 1998)

Refresher Mathematics: Grades 9–12
Prentice Hall, 1989 (Termination Year: 1998)

Transitional Math II

Applications of High School Mathematics: Grades 9–12
Houghton, 1992 (Termination Year: 1998)

Mathematical Connections: A Bridge to Algebra & Geometry: Grades 9–12
Houghton, 1992 (Termination Year: 1998)

Refresher Mathematics: Grades 9–12
Prentice Hall, 1989 (Termination Year: 1998)

Algebra I

Addison-Wesley Algebra: Grades 7–12
Addison-Wesley, 1992 (Termination Year: 1998)

Merrill Algebra 1: Applications & Connections: Grades 7–12
Macmillan/McGraw Hill, 1992 (Termination Year: 1998)

Merrill Algebra Essentials: Grades 7–12
Macmillan/McGraw Hill, 1988 (Termination Year: 1998)

Holt Algebra I: Grades 7–12
Holt, 1992 (Termination Year: 1998)

Algebra: Structure & Method: Grades 7–12
Houghton, 1992 (Termination Year: 1998)

Algebra 1: Grades 7–12
Houghton, 1992 (Termination Year: 1998)

Basic Algebra: Grades 7–12
Houghton, 1990 (Termination Year: 1998)

Algebra 1 An Integrated Approach: Grades 7–12
McDougal, 1991 (Termination Year: 1998)

Prentice Hall Algebra 1: Grades 7–12
Prentice Hall, 1990 (Termination Year: 1998)

Algebra 1, An Incremental Development: Grades 7–12
Saxon, 1990 (Termination Year: 1998)

The University of Chicago School Mathematics Project Algebra: Grades 7–12
ScottForesman, 1990 (Termination Year: 1998)

Algebra II

Addison-Wesley Algebra & Trigonometry: Grades 9–12
Addison-Wesley, 1992 (Termination Year: 1998)

Merrill Algebra 2 with Trigonometry: Applications & Connections: Grades 9–12
Glencoe, 1992 (Termination Year: 1998)

Holt Algebra with Trigonometry: Grades 9–12
Holt, 1992 (Termination Year: 1998)

Algebra & Trigonometry: Structure & Method: Grades 9–12
Houghton, 1992 (Termination Year: 1998)

Algebra 2 & Trigonometry: Grades 9–12
Houghton, 1992 (Termination Year: 1998)

Algebra 2 & Trigonometry: Grades 9–12
McDougal, 1991 (Termination Year: 1998)

Prentice Hall Algebra 2 with Trigonometry: Grades 9–12
Prentice Hall, 1990 (Termination Year: 1998)

Algebra 2, An Incremental Development: Grades 9–12
Saxon, 1991 (Termination Year: 1998)

University of Chicago School Math Project: Advanced Algebra: Grades 9–12
ScottForesman, 1990 (Termination Year: 1998)

Geometry

Addison-Wesley Geometry: Grades 9–12
Addison-Wesley, 1992 (Termination Year: 1998)

Addison-Wesley Informal Geometry: Grades 9–12
Addison-Wesley, 1992 (Termination Year: 1998)

Geometry: Grades 9–12
Glencoe, 1990 (Termination Year: 1998)

Merrill Informal Geometry: Grades 9–12
Glencoe, 1988 (Termination Year: 1998)

Geometry: Grades 9–12
Holt, 1991 (Termination Year: 1998)

Geometry: Grades 9–12
Houghton, 1992 (Termination Year: 1998)

Basic Geometry: Grades 9–12
Houghton, 1990 (Termination Year: 1998)

Geometry for Enjoyment & Challenge:
Grades 9–12
McDougal, 1991 (Termination Year: 1998)

Prentice Hall Geometry: Grades 9–12
Prentice Hall, 1990 (Termination Year: 1998)

Informal Geometry: Grades 9–12
Prentice Hall, 1992 (Termination Year: 1998)

*University of Chicago School Math Project:
Geometry*: Grades 9–12
ScottForesman, 1991 (Termination Year: 1998)

ScottForesman Geometry: Grades 9–12
ScottForesman, 1990 (Termination Year: 1998)

Geometry for Decision Making: Grades 9–12
South-Western, 1992 (Termination Year: 1998)

Trigonometry

Trigonometry & Its Applications: Grades 9–12
Glencoe, 1990 (Termination Year: 1998)

Holt Algebra with Trigonometry: Grades 9–12
Holt, 1992 (Termination Year: 1998)

Trigonometry with Applications: Grades 9–12
Houghton, 1990 (Termination Year: 1998)

Prentice Hall Trigonometry: Grades 9–12
Prentice Hall, 1990 (Termination Year: 1998)

*University of Chicago School Math Project:
Functions, Statistics & Trigonometry*:
Grades 9–12
ScottForesman, 1992 (Termination Year: 1998)

Probability and Statistics

*Introductory Statistics & Probability: A Basis
for Decision Making*: Grades 9–12
Houghton, 1988 (Termination Year: 1998)

*University of Chicago School Math Project:
Functions, Statistics, & Trigonometry*:
Grades 9–12
ScottForesman, 1992 (Termination Year: 1998)

Probability & Statistics: Grades 9–12
South-Western, 1989 (Termination Year: 1998)

Precalculus

Precalculus Mathematics: Grades 9–12
Addison-Wesley, 1990 (Termination
Year: 1998)

Precalculus: Grades 9–12
Heath, 1989 (Termination Year: 1998)

Merrill Precalculus Mathematics: Grades 9–12
Glencoe, 1988 (Termination Year: 1998)

Merrill Advance Mathematical Concepts:
Grades 9–12
Glencoe, 1988 (Termination Year: 1998)

*Advanced Mathematics: Precalculus with Dis-
crete Mathematics & Data Analysis*:
Grades 9–12
Houghton, 1992 (Termination Year: 1998)

Introductory Analysis: Grades 9–12
Houghton, 1991 (Termination Year: 1998)

Precalculus Mathematics: Grades 9–12
Prentice Hall, 1991 (Termination Year: 1998)

*Advanced Mathematics, An Incremental Devel-
opment*: Grades 9–12
Saxon, 1989 (Termination Year: 1998)

*University of Chicago School Math Project:
Precalculus & Discrete Mathematics*:
Grades 9–12
ScottForesman, 1992 (Termination Year: 1998)

Precalculus with Unit Circle Trigonometry:
Grades 9–12
West, 1990 (Termination Year: 1998)

AB Calculus

Elements of Calculus & Analytic Geometry:
Grades 9–12
Addison-Wesley, 1989 (Termination
Year: 1998)

Calculus with Analytic Geometry: Grades 9–12
Heath, 1990 (Termination Year: 1998)

Calculus with Analytic Geometry, 5/E:
Grades 9–12
Prentice Hall, 1987 (Termination Year: 1998)

Calculus with Trigonometry: Grades 9–12
Saxon, 1988 (Termination Year: 1998)

The Calculus of a Single Variable with Analytic Geometry: Grades 9–12
ScottForesman, 1990 (Termination Year: 1998)

BC Calculus

Elements of Calculus & Analytic Geometry:
Grades 9–12
Addison-Wesley, 1989 (Termination
Year: 1998)

The Calculus with Analytic Geometry:
Grades 9–12
ScottForesman, 1990 (Termination Year: 1998)

Business Mathematics

Applied Business Mathematics: Grades 9–12
South-Western, 19909 (Termination
Year: 1998)

Mathematics of Money: Grades 9–12
West, 1992 (Termination Year: 1998)

Computer Mathematics

Computer Mathematics: Structured BASIC with Math Applications: Grades 9–12
South-Western, 1989 (Termination Year: 1998)

Understanding Structured Programming in BASIC Apple: Grades 9–12
West, 1991 (Termination Year: 1998)

Understanding Structured Programming in Basic MS DOS: Grades 9–12
West, 1991 (Termination Year: 1998)

Fundamentals of PASCAL: Grades 9–12
West, 1990 (Termination Year: 1998)

Consumer Mathematics

Consumer Mathematics: Grades 9–12
Addison-Wesley, 1992 (Termination
Year: 1998)

Practical Math: Consumer Applications:
Grades 9–12
Holt, 1989 (Termination Year: 1998)

Consumer Mathematics: Grades 9–12
Glencoe, 1989 (Termination Year: 1998)

Consumer Mathematics: Grades 9–12
Prentice Hall, 1989 (Termination Year: 1998)

Consumer & Career Mathematics:
Grades 9–12
ScottForesman, 1989 (Termination Year: 1998)

Dollars & Sense: Problem Solving Strategies in Consumer Mathematics: Grades 9–12
South-Western, 1989 (Termination Year: 1998)

Mathematics of Money: Grades 9–12
South-Western, 1992 (Termination Year: 1998)

Mathematics of Money: Grades 9–12
West, 1992 (Termination Year: 1998)

Integrated Mathematics

Merrill Integrated Mathematics: Grades 9–12
Glencoe, 1991 (Termination Year: 1998)

Houghton Mifflin Unified Mathematics:
Grades 9–12
Houghton, 1991 (Termination Year: 1998)

McDougal, Littell Integrated Mathematics:
Grades 9–12
McDougal, 1991 (Termination Year: 1998)

Technical Mathematics

Addison-Wesley Informal Geometry:
Grades 9–12
Addison-Wesley, 1992 (Termination
Year: 1998)

Merrill Informal Geometry: Grades 9–12
Glencoe, 1988 (Termination Year: 1998)

Algebra & Trigonometry: Structure & Method:
Grades 9–12
Houghton, 1992 (Termination Year: 1998)

Basic Geometry: Grades 9–12
Houghton, 1990 (Termination Year: 1998)

Informal Geometry: Grades 9–12
Prentice Hall, 1992 (Termination Year: 1998)

INDEX TO REVIEWS

THIS index cites reviews of recently published materials for use in early childhood classrooms, including curriculum guides, lesson plans, project books, software programs, videos, and filmstrips. The citations cover reviews from the past two years (up to August 1992), and they reflect a search of educational journals, magazines, and newsletters that would include reviews of early childhood materials. The journals chosen are those that are available in teacher college libraries, in other college and university collections, and in many public libraries. They also include the major publications sent to members of the appropriate educational organizations. The review for each item can be found under the following listings:

- the title of the item
- the author(s)
- the publisher or producer/distributor
- subject (a broad subject arrangement is used)
- special medium (for Software packages, Audiotape, Manipulatives, etc.)

Aaberg, Nancy
Math Stories for Problem Solving Success: Ready-To-Use Activities for Grades 7-12, by Jim Overholt, Nancy Aaberg, and Jim Lindsey (Old Tappen, NJ: Prentice-Hall, 1990). Reviewed in: *Mathematics Teacher* 84, no.7 (Oct. 1991): 578.

Add-Ventures for Girls: Building Math Confidence (Elementary)
by Margaret Franklin (Newton, MA: Educational Development Center, 1990). Reviewed in: *Arithmetic Teacher* 40, no.1 (Sep. 1992): 58.

Add-Ventures for Girls: Building Math Confidence (Junior High)
by Margaret Franklin (Newton, MA: Educational Development Center, 1990). Reviewed in: *Arithmetic Teacher* 40, no.1 (Sep. 1992): 58.

Addison-Wesley
Algebra (Teacher's ed.), by Stanley A. Smith, Randall I. Charles, and John A. Dossey (Menlo Park, CA: Addison-Wesley, 1992). Reviewed in: *Mathematics Teacher* 85, no.7 (Oct. 1992): 589.

Algebra and Trigonometry (Teacher's ed.), by Stanley A. Smith, Randall I. Charles, and John A. Dossey (Menlo Park, CA: Addison-Wesley, 1992). Reviewed in: *Mathematics Teacher* 85, no.7 (Oct. 1992): 589-90.

Calculus, by Ross L. Finney and George B. Thomas, Jr. (Reading, MA: Addison-Wesley, 1990). Reviewed in: *Mathematics Teacher* 84, no.1 (Jan. 1991): 66.

Addison-Wesley *(cont'd)*
Calculus with Analytic Geometry, 3d ed., by John B. Fraleigh (Reading, MA: Addison-Wesley). Reviewed in *Mathematics and Computer Education*: 24, no.1 (Winter 1991): 93.

Chaos, Fractals, and Dynamics: Computer Experiments in Math, by Robert L. Devaney (Reading, MA: Addison-Wesley). Reviewed in: *Mathematics Teacher* 84, no.4 (Apr. 1991): 236.

Concepts and Applications of Intermediate Algebra, 2d ed., by Marvin L. Bittinger, Marvin L. Keedy, and David Ellenbogen (Reading, MA: Addison-Wesley, 1991). Reviewed in: *Mathematics and Computer Education* 26, no.2 (Spring 1992): 205-07.

Elementary Algebra: Concepts and Applications, 3d ed., by Marvin L. Bittinger and Marvin L. Keedy (Reading, MA: Addison-Wesley, n.d.). Reviewed in: *Mathematics and Computer Education* 26, no.1 (Winter 1992): 95-96.

Geometry (Teacher's ed.), by Stanley R. Clemens, Phares G. O'Daffer, and Thomas J. Cooney (Menlo Park, CA: Addison-Wesley, 1992). Reviewed in: *Mathematics Teacher* 85, no.7 (Oct. 1992): 590.

Graphic Calculator and Computer Graphing Laboratory Manual (2nd ed.), by Franklin Demana and Bert K. Waits (Menlo Park, CA: Addison-Wesley, 1992). Reviewed in: *Mathematics Teacher* 85, no.7 (Oct. 1992): 590-91.

Intermediate Algebra: Concepts and Applications, 3d ed., by Marvin L. Bittinger, Mervin L. Keedy, and David Ellenbogen (Reading, MA: Addison-Wesley, 1990). Reviewed in: *Mathematics Teacher* 84, no.3 (Mar. 1991): 238.

Pre-Algebra: A Transition to Algebra (Teacher's ed.), by Stanley R. Clemens, Phares G. O'Daffer, and Randall I. Charles (Menlo Park, CA: Addison-Wesley, 1992). Reviewed in: *Mathematics Teacher* 85, no.7 (Oct. 1992): 592.

Precalculus Functions and Graphs, by Jimmie Gilbert and Linda Gilbert (Reading, MA: Addison-Wesley, 1990). Reviewed in: *Mathematics Teacher* 84, no.1 (Jan. 1991): 67.

Precalculus Mathematics: A Graphing Approach (2d ed.), by Franklin Demana, Bert K. Waits, and Stanley E. Clemens (Menlo Park, CA: Addison-Wesley, 1992). Reviewed in: *Mathematics Teacher* 85, no.7 (Oct. 1992): 592.

Addition Rap: For Grades 1 to 2 audiocassette (Boise, ID: Star Trax, 1990). Reviewed in: *Arithmetic Teacher* 40, no.1 (Sep. 1992): 62.

Advanced Algebra by Sharon L. Senk, Denisse R. Thompson and Steven S. Vitatora (Glenview, IL: Scott Foresman, 1990). Reviewed in: *Mathematics Teacher* 84, no.3 (Mar. 1991): 235-36.

Africk, Henry
Interactive Algebra, software, by Henry Africk, Ely Stern, and Harvey Broverman (Hicksville, NY: Technical Educational Consultants, 1990). Reviewed in: *Mathematics Teacher* 84, no.3 (Mar. 1991): 235.

Alchemy Mindworks
Graphic Workshop 6.1f, software (Ontario, Canada: Alchemy Mindworks, n.d.). Reviewed in: *Mathematics and Computer Education* 26, no.1 (Winter 1992): 106-08.

Alexander, Bob
Integrated Mathematics Course 1, by Brendan Kelly, Paul Atkinson, and Bob Alexander (St. Charles, IL: McDougal Littell, 1991). Reviewed in: *Mathematics Teacher* 84, no.9 (Dec. 1991): 768.

Integrated Mathematics Course 2, by Brenden Kelly, Bob Alexander, and Paul Atkinson (St. Charles, IL: McDougal Littell, 1991). Reviewed in: *Mathematics Teacher* 84, no.9 (Dec. 1991): 768; *Mathematics Teacher* 85, no.2 (Feb. 1992): 148.

Alfors, Douglas
Analyzer, software, by Douglas Alfors and Beverly West (Ithaca, NY: Delta-Epsilon Software, 1990). Reviewed in: *Mathematics Teacher* 85, no.3 (Mar. 1992): 240.

Algebra
Advanced Algebra, by Sharon L. Senk, Denisse R. Thompson, and Steven S. Vitatora (Glenview, IL: Scott Foresman, 1990). Reviewed in: *Mathematics Teacher* 84, no.3 (Mar. 1991): 235-36.

Algebra, software (Canton, MT: National Appleworks Users Group, 1991). Reviewed in: *Mathematics Teacher* 85, no.5 (May 1991): 391.

Algebra (Teacher's ed.), by Stanley A. Smith, Randall I. Charles, and John A. Dossey (Menlo Park, CA: Addison-Wesley, 1992). Reviewed in: *Mathematics Teacher* 85, no.7 (Oct. 1992): 589.

Algebra: A First Course, 3d ed., by John Baley and Martin Holstege (Belmont, CA: Wadsworth, 1990). Reviewed in: *Mathematics Teacher* 84, no.7 (Oct. 1991): 571.

Algebra and Trigonometry (2d ed.), by Thomas W. Hungerford and Richard Mercer (Orlando, FL: Saunders College, 1991). Reviewed in: *Mathematics Teacher* 85, no.9 (Dec. 1992): 763.

Algebra and Trigonometry (Teacher's ed.), by Stanley A. Smith, Randall I. Charles, and John A. Dossey (Menlo Park, CA: Addison-Wesley, 1992). Reviewed in: *Mathematics Teacher* 85, no.7 (Oct. 1992): 589-90.

Algebra from 0 to 3: Constants to Cubics, (Pleasantville, NY: Sunburst). Reviewed in: *Mathematics and Computer Education* 26, no.2 (Spring 1992): 222-23.

The Algebra Lab: High School: A Comprehensive Manipulative Program for Algebra I, by Henri Piccotti (Sunnyvale, CA: Creative, 1990). Reviewed in: *Mathematics Teacher* 38, no.5 (May 1991): 406.

The Algebra Lab: Middle School: Exploring Algebra Concepts with Manipulatives, by Henri Piccotti (Sunnyvale, CA: Creative, 1990). Reviewed in: *Mathematics Teacher* 38, no.4 (Apr. 1991): 332.

Algebra Made Easy, software (San Francisco: Britannica Software, 1991). Reviewed in: *Mathematics Teacher* 85, no.7 (Oct. 1992): 587.

Algebra I: An Intergrated Approach (Evanston, IL: McDougal Littell, 1991). Reviewed in: *Mathematics Teacher* 38, no.4 (Apr. 1991): 324.

Algebra II and Trigonometry (Evanston, IL: McDougal Littell, 1991). Reviewed in: *Mathematics Teacher* 38, no.4 (Apr. 1991): 325.

Beginning Algebra, by Alfonse Gobran (Belmont, CA: Wadsworth, 1991). Reviewed in: *Mathematics Teacher* 84, no.8 (Nov. 1991): 674.

Beginning Algebra, by James W. Hall (Boston: PWS-KENT, 1992). Reviewed in: *Mathematics Teacher* 85, no.5 (May 1992): 393.

Concepts and Applications of Intermediate Algebra, 2d ed., by Marvin L. Bittinger, Marvin L. Keedy, and David Ellenbogen (Reading, MA: Addison-Wesley, 1991). Reviewed in: *Mathematics and Computer Education* 26, no.2 (Spring 1992): 205-07.

Concepts in Mathematics: Conic Sections, software (Cary, NC: TV Ontario Video, 1991). Reviewed in: *Mathematics Teacher* 85, no.9 (Dec. 1992): 767-68.

Contemporary Abstract Algebra (2d ed.), by Joseph A. Gallian (Lexington, MA: D. C. Heath, 1990). Reviewed in: *Mathematics and Computer Education* 26, no.3 (Fall 1992): 334-35.

Developmental Mathematics (3d ed.), by C. L. Johnston, Alden T. Willis, and Gale M. Hughes (Belmont, CA: Wadsworth, 1991). Reviewed in: *Mathematics Teacher* 85, no.6 (Sep. 1992): 491.

Algebra *(cont'd)*

Elementary Algebra (2d ed.), by Jack Barker, James Rogers, and James Van Dyke (Orlando, FL: Saunders College, 1992). Reviewed in: *Mathematics Teacher* 85, no.9 (Dec. 1992): 764.

Elementary Algebra: Concepts and Applications, 3d ed., by Marvin L. Bittinger and Marvin L. Keedy (Reading, MA: Addison-Wesley, n.d.). Reviewed in: *Mathematics and Computer Education* 26, no.1 (Winter 1992): 95-96.

Essential Algebra (6th ed.), by C. L. Johnston, Alden T. Willis, and Jeanne Lazaris (Belmont, CA: Wadsworth, 1991). Reviewed in: *Mathematics Teacher* 85, no.6 (Sep. 1992): 491-92.

GrafEq, software (Terrace, British Columbia: Pedagoguery, 1990). Reviewed in: *Mathematics Teacher* 85, no.2 (Mar. 1992): 241-42.

Interactive Algebra, software, by Henry Africk, Ely Stern, and Harvey Broverman (Hicksville, NY: Technical Educational Consultants, 1990). Reviewed in: *Mathematics Teacher* 84, no.3 (Mar. 1991): 235.

Intermediate Algebra, by Dale E. Boye, Ed Kavanaugh, and Larry G. Williams (Boston: PWS-KENT, 1991). Reviewed in: *Mathematics Teacher* 84, no.8 (Nov. 1991): 678.

Intermediate Algebra, by James W. Hall (Boston, MA: PWS-KENT, 1992). Reviewed in: *Mathematics Teacher* 85, no.9 (Dec. 1992): 764-65.

Intermediate Algebra, by Ronald Hatton and Gene R. Sellers (San Diego: Harcourt Brace Jovanovich, 1991). Reviewed in: *Mathematics Teacher* 84, no.9 (Dec. 1991): 769.

Intermediate Algebra (2d ed.), by Jack Barker, James Rogers, and James Van Dyke (Orlando, FL: Saunders College, 1992). Reviewed in: *Mathematics Teacher* 85, no.9 (Dec. 1992): 765-66.

Intermediate Algebra: Concepts and Applications, 3d ed., by Marvin L. Bittinger, Mervin L. Keedy, and David Ellenbogen (Reading, MA: Addison-Wesley, 1990). Reviewed in: *Mathematics Teacher* 84, no.3 (Mar. 1991): 238.

Introductory Algebra, by Ronald Hatton and Gene R. Sellers (San Diego: Harcourt Brace Jovanovich, 1991). Reviewed in: *Mathematics Teacher* 85, no.1 (Jan. 1992): 74.

Introductory Alegbra: An Interactive Approach, by Linda Pulsinelli and Patricia Hooper (New York: Macmillan, 1991). Reviewed in: *Mathematics Teacher* 84, no.8 (Nov. 1991): 678.

Investigations in Algebra, by Albert Cuoco (Cambridge, MA: MIT Press, 1990). Reviewed in: *Mathematics Teacher* 85, no.1 (Jan. 1992): 74.

Logo Math: Tools and Games, software, by Henri Piccotti (Portland, ME: Terrapin, 1990). Reviewed in: *Mathematics Teacher* 84, no.6 (Sep. 1991): 486-87.

Maple V: Student Edition, software (Pacific Grove, CA: Symbolic Computation Group, n.d.). Reviewed in: *Mathematics and Computer Education* 26, no.3 (Fall 1992): 337-39.

Math Connections: Algebra 1, software (Scotts Valley, CA: Wings for Learning). Reviewed in: *Technology & Learning* 12, no.6 (Mar. 1992): 10-11.

Math Connections: Algebra I, software by Jon Rosenberg (Scotts Valley, CA: Wings for Learning, 1991). Reviewed in: *Mathematics Teacher* 85, no.6 (Sep. 1992): 489.

Math Connections: Algebra II, software (Scotts Valley, CA: Wings for Learning, 1992). Reviewed in: *Technology & Learning* 13, no.3 (Nov./Dec. 1992): 18.

NCSMP Algebra, by John W. McConnell, Susan Brown, Susan Eddins, Margaret Hackworth, and Zalman Usiskin (Tucker, GA: Scott Foresman, 1990). Reviewed in: *Mathematics Teacher* 84, no.2 (Feb. 1991): 144-46.

Algebra
software (Canton, MT: National Appleworks Users Group, 1991). Reviewed in: *Mathematics Teacher* 85, no.5 (May 1991): 391.

Algebra (Teacher's ed.)
by Stanley A. Smith, Randall I. Charles, and John A. Dossey (Menlo Park, CA: Addison-Wesley, 1992). Reviewed in: *Mathematics Teacher* 85, no.7 (Oct. 1992): 589.

Algebra: A First Course, 3d ed.
by John Baley and Martin Holstege (Belmont, CA: Wadsworth, 1990). Reviewed in: *Mathematics Teacher* 84, no.7 (Oct. 1991): 571.

Algebra and Trigonometry (2d ed.)
by Thomas W. Hungerford and Richard Mercer (Orlando, FL: Saunders College, 1991). Reviewed in: *Mathematics Teacher* 85, no.9 (Dec. 1992): 763.

Algebra and Trigonometry (Teacher's ed.)
by Stanley A. Smith, Randall I. Charles, and John A. Dossey (Menlo Park, CA: Addison-Wesley, 1992). Reviewed in: *Mathematics Teacher* 85, no.7 (Oct. 1992): 589-90.

Algebra from 0 to 3: Constants to Cubics
(Pleasantville, NY: Sunburst). Reviewed in: *Mathematics and Computer Education* 26, no.2 (Spring 1992): 222-23.

The Algebra Lab: High School: A Comprehensive Manipulative Program for Algebra I
by Henri Piccotti (Sunnyvale, CA: Creative, 1990). Reviewed in: *Mathematics Teacher* 38, no.5 (May 1991): 406.

The Algebra Lab: Middle School: Exploring Algebra Concepts with Manipulatives
by Henri Piccotti (Sunnyvale, CA: Creative, 1990). Reviewed in: *Mathematics Teacher* 38, no.4 (Apr. 1991): 332.

Algebra Made Easy
software (San Francisco: Britannica Software, 1991). Reviewed in: *Mathematics Teacher* 85, no.7 (Oct. 1992): 587.

Algebra I: An Intergrated Approach
(Evanston, IL: McDougal Littell, 1991). Reviewed in: *Mathematics Teacher* 38, no.4 (Apr. 1991): 324.

Algebra II and Trigonometry
(Evanston, IL: McDougal Littell, 1991). Reviewed in: *Mathematics Teacher* 38, no.4 (Apr. 1991): 325.

Amazing Arithmos
manipulatives (Forest Hills, NY: Marc Preven, 1990). Reviewed in: *Arithmetic Teacher* 40, no.4 (Dec. 1992): 242.

Analytic Geometry, 2d ed.
by Karl J. Smith (Pacific Grove, CA: Brooks/Cole, 1991). Reviewed in: *Mathematics Teacher* 85, no.2 (Feb. 1992): 146.

Analyzer
software, by Douglas Alfors and Beverly West (Ithaca, NY: Delta-Epsilon Software, 1990). Reviewed in: *Mathematics Teacher* 85, no.3 (Mar. 1992): 240.

Anton, Howard
Calculus, 3d ed. (Late Trigonometry Version), by Howard Anton (Somerset, NJ: John Wiley). Reviewed in: 24, no.2 (Spring 1992): 212-13.

Apostol, Tom
Calculus, 2d ed., vol. 1, by Tom M. Apostol (Somerset, NJ: John Wiley, 1988). Reviewed in: *Mathematics Teacher* 84, no.3 (Mar. 1991): 236.

M! Project MATHEMATICS!—Similarity, video by Tom Apostol (Reston, VA: NCTM, 1990). Reviewed in: *Mathematics Teacher* 85, no.6 (Sep. 1992): 496.

Apple Family
Primary Geometry Workshop, K-2, software by Apple Family (Glenville, IL: Scott Foresman, 1991). Reviewed in: *Arithmetic Teacher* 39, no.8 (Apr. 1992): 58.

Archimedean and Archimedean Dual Polyhydra
video, by Lorrainne L. Foster (Northridge, CA: California State University). Reviewed in: *Mathematics Teacher* 84, no.7 (Oct. 1991): 576.

Association for the Advancement of Science
Math Power in the Community, by Gerald Kulm (Waldorf, MD: American Association for the Advancement of Science, 1990). Reviewed in: *Arithmetic Teacher* 40, no.1 (Sep. 1992): 64.

Math Power at Home, by Gerald Kulm (Waldorf, MD: American Association for the Advancement of Science, 1990). Reviewed in: *Arithmetic Teacher* 40, no.1 (Sep. 1992): 64.

Math Power in School, by Gerald Kulm (Waldorf, MD: American Association for the Advancement of Science, 1990). Reviewed in: *Arithmetic Teacher* 40, no.1 (Sep. 1992): 64.

AstroNUMBERS
software (Washington, DC: PC Gradeworks, 1990). Reviewed in: *Arithmetic Teacher* 39, no.1 (Sep. 1991): 47.

Atkinson, Paul
Integrated Mathematics Course 1, by Brendan Kelly, Paul Atkinson, and Bob Alexander (St. Charles, IL: McDougal Littell, 1991). Reviewed in: *Mathematics Teacher* 84, no.9 (Dec. 1991): 768.

Integrated Mathematics Course 2, by Brenden Kelly, Bob Alexander, and Paul Atkinson (St. Charles, IL: McDougal Littell, 1991). Reviewed in: *Mathematics Teacher* 84, no.9 (Dec. 1991): 768; *Mathematics Teacher* 85, no.2 (Feb. 1992): 148.

Audiocassette
Addition Rap: For Grades 1 to 2, audiocassette (Boise, ID: Star Trax, 1990). Reviewed in: *Arithmetic Teacher* 40, no.1 (Sep. 1992): 62.

Multiplication Rap: For Grades 3 to 6, audiocassette (Boise, ID: Star Trax, 1988). Reviewed in: *Arithmetic Teacher* 40, no.1 (Sep. 1992): 62.

Rap-ability, audiocassette, by Mike Ross and Suzanne Rossi (Lake Orion, MI: Aynn Visual, 1990). Reviewed in: *Arithmetic Teacher* 38, no.9 (May 1991): 49.

Subtraction Rap: For Grades 1 to 2, audiocassette (Boise, ID: Star Trax, 1990). Reviewed in: *Arithmetic Teacher* 40, no.1 (Sep. 1992): 62.

Aynn Visual
Rap-ability, audiocassette, by Mike Ross and Suzanne Rossi (Lake Orion, MI: Aynn Visual, 1990). Reviewed in: *Arithmetic Teacher* 38, no.9 (May 1991): 49.

Baker, Ann
Raps and Rhymes in Maths, by Ann Baker and Johnny Baker (Portsmouth, NH: Heinemann, 1991). Reviewed in: *Arithmetic Teacher* 40, no.2 (Oct. 1992): 128.

Baker, Jim
Fastmath, software, by Jim Baker and Dale Huline (Minneapolis: New Directions, 1990). Reviewed in: *Arithmetic Teacher* 39, no.1 (Sep. 1991): 47.

Baker, Johnny
Raps and Rhymes in Maths, by Ann Baker and Johnny Baker (Portsmouth, NH: Heinemann, 1991). Reviewed in: *Arithmetic Teacher* 40, no.2 (Oct. 1992): 128.

Bruce Bakke (publisher)
Facts Master, manipulatives by Jivan Patel (Carrollton, TX: Bruce Bakke, n.d.). Reviewed in: *Arithmetic Teacher* 40, no.2 (Oct. 1992): 128.

Baley, John
Algebra: A First Course, 3d ed., by John Baley and Martin Holstege (Belmont, CA: Wadsworth, 1990). Reviewed in: *Mathematics Teacher* 84, no.7 (Oct. 1991): 571.

Balka, Don
Exploring Fractions and Decimals with Manipulatives, manipulatives, by Don Balka (Peabody, MA: Didax, 1991). Reviewed in: *Arithmetic Teacher* 39, no.9 (May 1992): 53.

Balloons
Number Squares, software (Washington, DC: Balloons, 1988). Reviewed in: *Arithmetic Teacher* 39, no.1 (Sep. 1991): 48.

Barclay, Tim
Koetke's Challenge, software by Tim Barclay and Jonathan Choate (Acton, MA: William K. Bradford, 1991). Reviewed in: *Mathematics Teacher* 85, no.9 (Dec. 1992): 761.

Barker, Jack
Elementary Algebra (2d ed.), by Jack Barker, James Rogers, and James Van Dyke (Orlando, FL: Saunders College, 1992). Reviewed in: *Mathematics Teacher* 85, no.9 (Dec. 1992): 764.

Intermediate Algebra (2d ed.), by Jack Barker, James Rogers, and James Van Dyke (Orlando, FL: Saunders College, 1992). Reviewed in: *Mathematics Teacher* 85, no.9 (Dec. 1992): 765-66.

Basic Mathematics
by David Novak (Lexington, MA: D. C. Heath, 1991). Reviewed in: *Mathematics Teacher* 85, no.6 (Sep. 1992): 491.

Basic Mathematics (3d ed.)
by Charles P. McKeague (Florence, KY: Brooks/Cole, 1992). Reviewed in: *Mathematics Teacher* 85, no.6 (Sep. 1992): 490-91.

Basic Technical Mathematics with Calculus, 2d ed.
by Peter Kuhfittig (Pacific Grove, CA: Brooks/Cole, 1989). Reviewed in: *Mathematics Teacher* 84, no.2 (Feb. 1991): 140.

Battista, Michael
Logo Geometry, software by Michael Battista and Douglas H. Clements (Morristown, NJ: Simon & Schuster, 1991). Reviewed in: *Arithmetic Teacher* 40, no.1 (Sep. 1992): 56.

Beckmann, Jerry
Euclid's Toolbox—Triangles, software, by Jerry Beckmann and Charles Friesen (n.p.: Heartland, 1990). Reviewed in: *Mathematics Teacher* 84, no.6 (Sep. 1991): 484.

Beginning Algebra
by Alfonse Gobran (Belmont, CA: Wadsworth, 1991). Reviewed in: *Mathematics Teacher* 84, no.8 (Nov. 1991): 674.

Beginning Algebra
by James W. Hall (Boston: PWS-KENT, 1992). Reviewed in: *Mathematics Teacher* 85, no.5 (May 1992): 393.

Bell, Eric T.
Magic of Numbers, by Eric T. Bell (Mineola, NY: Dover, 1991). Reviewed in: *Mathematics Teacher* 85, no.6 (Sep. 1992): 493.

Benjamin Banneker
poster (Burlington, NC: Cabisco Mathematics, 1991). Reviewed in: *Mathematics Teacher* 85, no.6 (Sep. 1992): 496.

Berenbom, Joshua
Prealgebra, by Joshua Berenbom and Umesh Nagarkatte (San Diego: Harcourt Brace Jovanovich, 1991). Reviewed in: *Arithmetic Teacher* 39, no.3 (Nov. 1991): 60.

Berstein, Bob
Mathemactivities, by Bob Berstein and illustrated by Bron Smith (Carthage, IL: Good Apple, 1991). Reviewed in: *Arithmetic Teacher* 40, no.1 (Sep. 1992): 62-63.

Bilingual games (English/Spanish)
Amazing Arithmos, manipulatives (Forest Hills, NY: Marc Preven, 1990). Reviewed in: *Arithmetic Teacher* 40, no.4 (Dec. 1992): 242.

Bitter, Gary G.
Mental Math and Estimation, by Gary G. Bitter (Allen, TX: DLM, 1989). Reviewed in: *Arithmetic Teacher* 39, no.1 (Sep. 1991): 53.

Bittinger, Marvin L.
Concepts and Applications of Intermediate Algebra, 2d ed., by Marvin L. Bittinger, Marvin L. Keedy, and David Ellenbogen (Reading, MA: Addison-Wesley, 1991). Reviewed in: *Mathematics and Computer Education* 26, no.2 (Spring 1992): 205-07.

Elementary Algebra: Concepts and Applications, 3d ed., by Marvin L. Bittinger and Marvin L. Keedy (Reading, MA: Addison-Wesley, n.d.). Reviewed in: *Mathematics and Computer Education* 26, no.1 (Winter 1992): 95-96.

Bittinger, Marvin L. *(cont'd)*
Intermediate Algebra: Concepts and Applications, 3d ed., by Marvin L. Bittinger, Mervin L. Keedy, and David Ellenbogen (Reading, MA: Addison-Wesley, 1990). Reviewed in: *Mathematics Teacher* 84, no.3 (Mar. 1991): 238.

Bloom, Marjorie W.
Estimate! Calculate! Evaluate! Calculator Activities for the Middle Grades, by Marjorie W. Bloom and Grace K. Galton (New Rochelle, NY: Cuisenaire, 1990). Reviewed in: *Arithmetic Teacher* 38, no.8 (Apr. 1991): 56.

Bobo, Betty
Math around the World: Math/Geography Enrichment Activities for the Middle Grades, by Lynn Embry and Betty Bobo (Carthage, IL: Good Apple, 1991). Reviewed in: *Arithmetic Teacher* 39, no.9 (May 1992): 53-60.

Bolt, Brian
Math Meets Technology, by Brian Bolt (Port Chester, NY: Cambridge University Press, 1991). Reviewed in: *Mathematics Teacher* 85, no.8 (Nov. 1992): 692.

Boyd, Linda Hawkins
Precalculus Algebra and Trigonometry, by Sharon Cutler Ross and Linda Hawkins Boyd (Pacific Grove, CA: Brooks/Cole, 1991). Reviewed in: *Mathematics Teacher* 85, no.1 (Jan. 1992): 75.

Boye, Dale E.
Intermediate Algebra, by Dale E. Boye, Ed Kavanaugh, and Larry G. Williams (Boston: PWS-KENT, 1991). Reviewed in: *Mathematics Teacher* 84, no.8 (Nov. 1991): 678.

William K. Bradford (publisher)
Bradford Graphmaker, software (Concord, MA: William K. Bradford, 1990). Reviewed in: *Mathematics Teacher* 84, no.5 (May 1991): 391-92; *Electronic Learning* 11, no.3 (Nov./Dec. 1991): 34-35.

Graph Wiz 1.0 or 1.2, software, by Alan Hoffer and Tom Lippincott (Concord, MA: William K. Bradford, 1989). Reviewed in: *Mathematics Teacher* 84, no.6 (Sep. 1991): 485-86.

Koetke's Challenge, software by Tim Barclay and Jonathan Choate (Acton, MA: William K. Bradford, 1991). Reviewed in: *Mathematics Teacher* 85, no.9 (Dec. 1992): 761.

Mystery Castle: Strategies for Problem Solving, Level I and II, software (Acton, MA: William K. Bradford, 1990). Reviewed in: *Arithmetic Teacher* 39, no.3 (Nov. 1991): 57.

Bradford Graphmaker
software (Concord, MA: William K. Bradford, 1990). Reviewed in: *Mathematics Teacher* 84, no.5 (May 1991): 391-92; *Electronic Learning* 11, no.3 (Nov./Dec. 1991): 34-35.

Bradford School
3D Images, software (Acton, MA: Bradford School, 1991). Reviewed in: *Mathematics Teacher* 85, no.6 (May 1992): 391; *Electronic Learning* 11, no.3 (Nov./Dec. 1991): 34-35.

Brisby, Linda Sue
Patterns and Functions, by Linda Sue Brisby (Solvang, CA: Hands On, 1990). Reviewed in: *Arithmetic Teacher* 40, no.1 (Sep. 1992): 64.

Britannica Software
Algebra Made Easy, software (San Francisco: Britannica Software, 1991). Reviewed in: *Mathematics Teacher* 85, no.7 (Oct. 1992): 587.

Britten, Jill
Introduction to Tessellations, by Dale Seymour and Jill Britten (Palo Alto, CA: Dale Seymour, 1989). Reviewed in: *Mathematics Teacher* 83, no.6 (Sep. 1992): 482.

Broderbund
Geometry 1.2, software (San Rafael, CA: Broderbund, n.d.). Reviewed in: *Journal of Geographical Education* 39, no.4 (Sep. 1991): n.p.

Mental Math Games, Series I, software (San Rafael, CA: Broderbund, 1992). Reviewed in: *Technology & Learning* 13, no.3 (Nov./Dec. 1992): 18.

Brooks/Cole
Analytic Geometry, 2d ed., by Karl J. Smith (Pacific Grove, CA: Brooks/Cole, 1991). Reviewed in: *Mathematics Teacher* 85, no.2 (Feb. 1992): 146.

Basic Mathematics (3d ed.), by Charles P. McKeague (Florence, KY: Brooks/Cole, 1992). Reviewed in: *Mathematics Teacher* 85, no.6 (Sep. 1992): 490-91.

Basic Technical Mathematics with Calculus, 2d ed., by Peter Kuhfittig (Pacific Grove, CA: Brooks/Cole, 1989). Reviewed in: *Mathematics Teacher* 84, no.2 (Feb. 1991): 140.

Calculus T/L: A Program for Doing, Teaching, and Learning Calculus, software, by J. Douglas Child (Pacific Grove, CA: Brooks/Cole, 1990). Reviewed in: *Mathematics Teacher* 84, no.3 (Mar. 1991): 234-35.

Geometry, by Karl J. Smith (Pacific Grove, CA: Brooks/Cole, 1991). Reviewed in: *Mathematics Teacher* 85, no.1 (Jan. 1992): 73-74.

Introduction to Statistics with Data Analysis, by Shelley Rasmussen (Pacific Grove, CA: Brooks/Cole, 1992). Reviewed in: *Mathematics Teacher* 85, no.6 (Sep. 1992): 492.

Math Writer, software (Pacific Grove, CA: Brooks/Cole). Reviewed in: *Mathematics and Computer Education* 26, no.1 (Winter, 1992): 109-10.

Prealgebra for Problem Solvers: A New Beginning, by Loyd V. Wilcox (Pacific Grove, CA: Brooks/Cole, 1991). Reviewed in: *Mathematics Teacher* 84, no.6 (Sep. 1991): 492.

Problem Solving (S,C,R), by Karl J. Smith (Pacific Grove, CA: Brooks/Cole, 1991). Reviewed in: *Mathematics Teacher* 85, no.1 (Jan. 1992): 75.

Single Variable Calculus, 2d ed., by James Stewart (Pacific Grove, CA: Brooks/Cole, 1991). Reviewed in: *Mathematics Teacher* 85, no.1 (Jan. 1992): 76.

TI-81 Graphing Calculator Activities for Finite Mathematics, by Wayne L. Miller, Donald Perry, and Gloria A. Tveten (Pacific Grove, CA: Brooks/Cole, 1992). Reviewed in: *Mathematics Teacher* 85, no.8 (Nov. 1992): 693.

TrigPak: Software & Tutorials for Trigonometry, version 2.0, software, by John R. Monbrag (Pacific Grove, CA: Brooks/Cole, 1989). Reviewed in: *Mathematics Teacher* 38, no.5 (May 1991): 406.

Broverman, Harvey
Interactive Algebra, software, by Henry Africk, Ely Stern, and Harvey Broverman (Hicksville, NY: Technical Educational Consultants, 1990). Reviewed in: *Mathematics Teacher* 84, no.3 (Mar. 1991): 235.

Brown, Susan
NCSMP Algebra, by John W. McConnell, Susan Brown, Susan Eddins, Margaret Hackworth, and Zalman Usiskin (Tucker, GA: Scott Foresman, 1990). Reviewed in: *Mathematics Teacher* 84, no.2 (Feb. 1991): 144-46.

Buck, Donna Kay
Math-O-Graphs: Critical Thinking through Graphing, by Donna Kay Buck and Francis Hildebrand (Pacific Grove, CA: Midwest, 1990). Reviewed in: *Arithmetic Teacher* 38, no.9 (May 1991): 47.

Building Self-Confidence in Math: A Student Workbook (2d ed.)
by Sally Wilding and Elizabeth Shearn (Dubuque, IA: Kendall/Hunt, 1991). Reviewed in: *Mathematics Teacher* 85, no.6 (Sep. 1992): 496.

Burns, Marilyn
A Collection of MATH Lessons, Grades 6-8, by Cathy McLaughlin (New Rochelle, NY: Cuisenaire, 1990). Reviewed in: *Arithmetic Teacher* 39, no.1 (Sep. 1991): 49.

The $1.00 Word Riddle Book, by Marilyn Burns and illustrated by Martha Weston (New Rochelle, NY: Cuisenaire, 1990). Reviewed in: *Arithmetic Teacher* 39, no.2 (Oct. 1991): 51-52.

Cabisco Mathematics
  Benjamin Banneker, poster (Burlington, NC: Cabisco Mathematics, 1991). Reviewed in: *Mathematics Teacher* 85, no.6 (Sep. 1992): 496.

  Colorful Characters of Mathematics, poster (Burlington, NC: Cabisco Mathematics, 1992). Reviewed in: *Mathematics Teacher* 85, no.8 (Nov. 1992): 694.

CAE Software
  Mathematics Achievement Project, software (Washington, DC: CAE Software, 1992). Reviewed in: *Mathematics Teacher* 85, no.7 (Oct. 1992): 587-88.

Cain, Michael
  Making the Most of Twenty Minutes: Warm Up Math Activities, by Michael Cain (Victoria, Australia: Dellasta, 1989). Reviewed in: *Arithmetic Teacher* 39, no.1 (Sep. 1991): 50.

The Calculator, Grades 4-8
  video (Columbus, OH: Silver Burdett & Ginn, 1991). Reviewed in: *Arithmetic Teacher* 37, no.7 (Mar. 1992): 42.

Calculator Conundrums
  by Thomas Camilli (Pacific Grove, CA: Thinking Press, 1991). Reviewed in: *Arithmetic Teacher* 40, no.4 (Dec. 1992): 242.

Calculators
  The Calculator, Grades 4-8, video (Columbus, OH: Silver Burdett & Ginn, 1991). Reviewed in: *Arithmetic Teacher* 37, no.7 (Mar. 1992): 42.

  Estimate! Calculate! Evaluate! Calculator Activities for the Middle Grades, by Marjorie W. Bloom and Grace K. Galton (New Rochelle, NY: Cuisenaire, 1990). Reviewed in: *Arithmetic Teacher* 38, no.8 (Apr. 1991): 56.

  The Problem Solver with Calculators, by Terence G. Coburn, Shirley Hoogeboom, and Judy Goodnon (Sunnyvale, CA: Creative, 1989). Reviewed in: *Arithmetic Teacher* 39, no.1 (Sep. 1991): 50-51.

What Do You Do with a Broken Calculator?, software (Pleasantville, NY: Sunburst, 1989). Reviewed in: *Arithmetic Teacher* 39, no.3 (Nov. 1991): 58.

Calculators in the Classroom, K-12
  by Katherine Pedersen and Jack Mummert (Carbondale, IL: ICTM, 1990). Reviewed in: *Arithmetic Teacher* 40, no.1 (Sep. 1992): 62.

Calculus
  Analytic Geometry, 2d ed., by Karl J. Smith (Pacific Grove, CA: Brooks/Cole, 1991). Reviewed in: *Mathematics Teacher* 85, no.2 (Feb. 1992): 146.

  Analyzer, software, by Douglas Alfors and Beverly West (Ithaca, NY: Delta-Epsilon Software, 1990). Reviewed in: *Mathematics Teacher* 85, no.3 (Mar. 1992): 240.

  Basic Technical Mathematics with Calculus, 2d ed., by Peter Kuhfittig (Pacific Grove, CA: Brooks/Cole, 1989). Reviewed in: *Mathematics Teacher* 84, no.2 (Feb. 1991): 140.

  Calculus, by Ross L. Finney and George B. Thomas, Jr. (Reading, MA: Addison-Wesley, 1990). Reviewed in: *Mathematics Teacher* 84, no.1 (Jan. 1991): 66.

  Calculus, by Earl W. Swokowski (Florence, KY: PWS-KENT, 1991). Reviewed in: *Mathematics Teacher* 85, no.4 (Apr. 1992): 314.

  Calculus, 2d ed., vol. 1, by Tom M. Apostol (Somerset, NJ: John Wiley, 1988). Reviewed in: *Mathematics Teacher* 84, no.3 (Mar. 1991): 236.

  Calculus, 3d ed. (Late Trigonometry Version), by Howard Anton (Somerset, NJ: John Wiley). Reviewed in: 24, no.2 (Spring 1992): 212-13.

  Calculus: Applications of Differentiations, Parts 1-3, video (Brooklyn: Video Tutorial Service, 1990). Reviewed in: *Mathematics Teacher* 84, no.6 (Sep. 1991): 496-97.

Calculus *(cont'd)*

Calculus: Late Trigonometry Version (5th ed.), by Earl W. Swokowski (Boston, MA: PWS-KENT, 1992). Reviewed in: *Mathematics Teacher* 85, no.8 (Nov. 1992): 690.

Calculus T/L: A Program for Doing, Teaching, and Learning Calculus, software, by J. Douglas Child (Pacific Grove, CA: Brooks/Cole, 1990). Reviewed in: *Mathematics Teacher* 84, no.3 (Mar. 1991): 234-35.

Calculus with Analytic Geometry, 3d. ed., by John B. Fraleigh (Reading, MA: Addison-Wesley). Reviewed in *Mathematics and Computer Education*: 24, no.1 (Winter 1991): 93.

Chaos, Fractals, and Dynamics: Computer Experiments in Math, by Robert L. Devaney (Reading, MA: Addison-Wesley). Reviewed in: *Mathematics Teacher* 84, no.4 (Apr. 1991): 236.

Finite Mathematics with Calculus: An Applied Approach (2d ed.), by David E. Zitarelli and Raymond F. Coughlin (Orlando, FL: Saunders College, 1992). Reviewed in: *Mathematics Teacher* 85, no.9 (Dec. 1992): 764.

GyroGraphics, software, by Nancy Anne Johnson (Stillwater, OK: Cipher Systems, 1990). Reviewed in: *Mathematics Teacher* 85, no.3 (Mar. 1992): 242.

Logo Math: Tools and Games, software, by Henri Piccotti (Portland, ME: Terrapin, 1990). Reviewed in: *Mathematics Teacher* 84, no.6 (Sep. 1991): 486-87.

Single Variable Calculus, 2d ed., by James Stewart (Pacific Grove, CA: Brooks/Cole, 1991). Reviewed in: *Mathematics Teacher* 85, no.1 (Jan. 1992): 76.

Calculus

by Earl W. Swokowski (Florence, KY: PWS-KENT, 1991). Reviewed in: *Mathematics Teacher* 85, no.4 (Apr. 1992): 314.

Calculus

by Ross L. Finney and George B. Thomas, Jr. (Reading, MA: Addison-Wesley, 1990). Reviewed in: *Mathematics Teacher* 84, no.1 (Jan. 1991): 66.

Calculus, 2d ed., vol. 1

by Tom M. Apostol (Somerset, NJ: John Wiley, 1988). Reviewed in: *Mathematics Teacher* 84, no.3 (Mar. 1991): 236.

Calculus, 3d ed. (Late Trigonometry Version)

by Howard Anton (Somerset, NJ: John Wiley). Reviewed in: 24, no.2 (Spring 1992): 212-13.

Calculus: Applications of Differentiations, Parts 1-3

video (Brooklyn: Video Tutorial Service, 1990). Reviewed in: *Mathematics Teacher* 84, no.6 (Sep. 1991): 496-97.

Calculus by and for Young People (Ages 7, Yes 7 and Up)

by Don Cohen (n.p.: Jonathan Press, 1991). Reviewed in: *Mathematics Teaching* 140 (Sep. 1992): 41.

Calculus Gems

by George F. Simmons (New York: McGraw-Hill, 1992). Reviewed in: *Mathematics and Computer Education* 26, no.3 (Fall 1992): 330.

Calculus: Late Trigonometry Version (5th ed.)

by Earl W. Swokowski (Boston, MA: PWS-KENT, 1992). Reviewed in: *Mathematics Teacher* 85, no.8 (Nov. 1992): 690.

Calculus T/L: A Program for Doing, Teaching, and Learning Calculus

software, by J. Douglas Child (Pacific Grove, CA: Brooks/Cole, 1990). Reviewed in: *Mathematics Teacher* 84, no.3 (Mar. 1991): 234-35.

Calculus with Analytic Geometry, 3d. ed.

by John B. Fraleigh (Reading, MA: Addison-Wesley). Reviewed in *Mathematics and Computer Education*: 24, no.1 (Winter 1991): 93.

California State Fullerton Press
Camp-LA, Books 1-4 (Grades K-8)
(Orange, CA: California State Fullerton
Press, 1991). Reviewed in: *Arithmetic
Teacher* 40, no.4 (Dec. 1992): 242-43.

California State University, Northridge
*Archimedean and Archimedean Dual
Polyhydra*, video, by Lorrainne L. Foster
(Northridge, CA: California State Univer-
sity). Reviewed in: *Mathematics Teacher* 84,
no.7 (Oct. 1991): 576.

Cambridge University Press
Math Meets Technology, by Brian Bolt (Port
Chester, NY: Cambridge University Press,
1991). Reviewed in: *Mathematics Teacher*
85, no.8 (Nov. 1992): 692.

*What's Your Game? A Resource Book for
Mathematical Activities*, by Michael Cornelius and Alan Parr (New York: Cambridge
University Press, 1991). Reviewed in: *Arithmetic Teacher* 40, no.4 (Dec. 1992): 241.

Camilli, Thomas
Calculator Conundrums, by Thomas Camilli
(Pacific Grove, CA: Thinking Press, 1991).
Reviewed in: *Arithmetic Teacher* 40, no.4
(Dec. 1992): 242.

Camp-LA, Books 1-4 (Grades K-8)
(Orange, CA: California State Fullerton
Press, 1991). Reviewed in: *Arithmetic
Teacher* 40, no.4 (Dec. 1992): 242-43.

Cavanagh, Mary
Math in Brief, game, by Mary Cavanagh
(Fort Collins, CO: Scott Resources, 1978).
Reviewed in: *Arithmetic Teacher* 39, no.1
(Sep. 1991): 53.

*Chaos, Fractals, and Dynamics: Computer
Experiments in Math*
by Robert L. Devaney (Reading, MA:
Addison-Wesley). Reviewed in: *Mathematics Teacher* 84, no.4 (Apr. 1991): 236.

Charles, Randall I.
Algebra (Teacher's ed.), by Stanley A.
Smith, Randall I. Charles, and John A.
Dossey (Menlo Park, CA: Addison-Wesley,
1992). Reviewed in: *Mathematics Teacher*
85, no.7 (Oct. 1992): 589.

Algebra and Trigonometry (Teacher's ed.),
by Stanley A. Smith, Randall I. Charles, and
John A. Dossey (Menlo Park, CA: Addison-
Wesley, 1992). Reviewed in: *Mathematics
Teacher* 85, no.7 (Oct. 1992): 589-90.

*Pre-Algebra: A Transition to Algebra
(Teacher's ed.)*, by Stanley R. Clemens,
Phares G. O'Daffer, and Randall I. Charles
(Menlo Park, CA: Addison-Wesley, 1992).
Reviewed in: *Mathematics Teacher* 85, no.7
(Oct. 1992): 592.

Child, J. Douglas
*Calculus T/L: A Program for Doing,
Teaching, and Learning Calculus*, software,
by J. Douglas Child (Pacific Grove, CA:
Fraleigh, John B.Brooks/Cole, 1990). Re-
viewed in: *Mathematics Teacher* 84, no.3
(Mar. 1991): 234-35.

Chip Publications
Fundamentally Math, Ages 8 and Up,
software by Howard Diamond (Chapel Hill,
NC: Chip Publications, 1990). Reviewed in:
Arithmetic Teacher 40, no.3 (Nov.
1992): 194.

Choate, Jonathan
Koetke's Challenge, software by Tim
Barclay and Jonathan Choate (Acton, MA:
William K. Bradford, 1991). Reviewed in:
Mathematics Teacher 85, no.9 (Dec.
1992): 761.

Choate, Laura Duncan
Graphing Primer, Grades K-2, by Laura
Duncan Choate and JoAnn King Okey (Palo
Alto, CA: Seymour Publications, 1989).
Reviewed in: *Arithmetic Teacher* 40, no.2
(Oct. 1992): 129.

*Pictograms: Graphing Pictures for a Reus-
able Classroom Grid*, manipulatives by
Laura Duncan Choate and JoAnn King Okey
(Palo Alto, CA: Dale Seymour, 1989).
Reviewed in: *Arithmetic Teacher* 40, no.2
(Oct. 1992): 130.

Chomsky, Carol
in COMMON Arithmetic, software, by Carol
Chomsky, Harry Chomsky, and Judah L.
Schwartz (Pleasantville, NY: Sunburst,
1990). Reviewed in: *Arithmetic Teacher* 39,
no.1 (Sep. 1991): 47.

Chomsky, Harry
in COMMON Arithmetic, software, by Carol Chomsky, Harry Chomsky, and Judah L. Schwartz (Pleasantville, NY: Sunburst, 1990). Reviewed in: *Arithmetic Teacher* 39, no.1 (Sep. 1991): 47.

Chronicle Books
Ten Little Rabbits, by Virginia Grossman and illustrated by Sylvia Long (San Francisco: Chronicle Books, 1991). Reviewed in: *Arithmetic Teacher* 40, no.1 (Sep. 1992): 56-57.

Cipher Systems
GyroGraphics, software, by Nancy Anne Johnson (Stillwater, OK: Cipher Systems, 1990). Reviewed in: *Mathematics Teacher* 85, no.3 (Mar. 1992): 242.

GyroGraphics, version 2.2, software (Stillwater, OK: Cipher Systems, 1990). Reviewed in: *Mathematics Teacher* 84, no.3 (Mar. 1991): 235.

The Circular Geoboard
by Judy Kevin (San Leandro, CA: Teaching Resource Center, 1990). Reviewed in: *Mathematics Teacher* 84, no.8 (Nov. 1991): 680.

Clark, John
Making Mathematics 8, by Gary Flewelling, Joan Routledge, and John Clark (Agincourt, Ontario: Gage Educational, 1991). Reviewed in: *Mathematics Teacher* 85, no.8 (Nov. 1992): 692.

Making Mathematics 7, by Gary Flewelling, Joan Routledge, and John Clark (Agincourt, Ontario: Gage Educational, 1991). Reviewed in: *Mathematics Teacher* 85, no.9 (Dec. 1992): 766.

Clemens, Stanley E.
Precalculus Mathematics: A Graphing Approach (2d ed.), by Franklin Demana, Bert K. Waits, and Stanley E. Clemens (Menlo Park, CA: Addison-Wesley, 1992). Reviewed in: *Mathematics Teacher* 85, no.7 (Oct. 1992): 592.

Clemens, Stanley R.
Geometry (Teacher's ed.), by Stanley R. Clemens, Phares G. O'Daffer, and Thomas J. Cooney (Menlo Park, CA: Addison-Wesley, 1992). Reviewed in: *Mathematics Teacher* 85, no.7 (Oct. 1992): 590.

Pre-Algebra: A Transition to Algebra (Teacher's ed.), by Stanley R. Clemens, Phares G. O'Daffer, and Randall I. Charles (Menlo Park, CA: Addison-Wesley, 1992). Reviewed in: *Mathematics Teacher* 85, no.7 (Oct. 1992): 592.

Clements, Douglas H.
Logo Geometry, software by Michael Battista and Douglas H. Clements (Morristown, NJ: Simon & Schuster, 1991). Reviewed in: *Arithmetic Teacher* 40, no.1 (Sep. 1992): 56.

Coburn, Terence G.
The Problem Solver with Calculators, by Terence G. Coburn, Shirley Hoogeboom, and Judy Goodnon (Sunnyvale, CA: Creative, 1989). Reviewed in: *Arithmetic Teacher* 39, no.1 (Sep. 1991): 50-51.

Cohen, David
Precalculus: A Problems-Oriented Approach, by David Cohen (St. Paul: West, 1990). Reviewed in: *Mathematics Teacher* 87 no. 7 (Oct. 1991): 574.

Cohen, Don
Calculus by and for Young People (Ages 7, Yes 7 and Up), by Don Cohen (n.p.: Jonathan Press, 1991). Reviewed in: *Mathematics Teaching* 140 (Sep. 1992): 41.

A Collection of MATH Lessons, Grades 6-8
by Cathy McLaughlin (New Rochelle, NY: Cuisenaire, 1990). Reviewed in: *Arithmetic Teacher* 39, no.1 (Sep. 1991): 49.

Colorful Characters of Mathematics
poster (Burlington, NC: Cabisco Mathematics, 1992). Reviewed in: *Mathematics Teacher* 85, no.8 (Nov. 1992): 694.

Computation
AstroNUMBERS, software (Washington, DC: PC Gradeworks, 1990). Reviewed in: *Arithmetic Teacher* 39, no.1 (Sep. 1991): 47.

Computation *(cont'd)*

Connections! Math in Action, software (SVE, 1992). Reviewed in: *Teaching Pre-K-8* 21, no.4 (Jan. 1992): 14.

Counters: An Action Approach to Counting and Arithmetic, software (Scotts Valley, CA: Wings for Learning, 1990). Reviewed in: *Arithmetic Teacher* 39, no.3 (Nov. 1991): 56.

Estimation Tutor, software by Richard E. Rand (Portland, ME: J. Weston Walch, 1989). Reviewed in: *Arithmetic Teacher* 38, no.8 (Apr. 1991): 54.

The Factor Pack, supplementary materials (Holywell, UK: Magic Mathworks, n.d.). Reviewed in: *Mathematics Teacher* 84, no.7 (Oct. 1991): 576.

Fastmath, software, by Jim Baker and Dale Huline (Minneapolis: New Directions, 1990). Reviewed in: *Arithmetic Teacher* 39, no.1 (Sep. 1991): 47.

Flashcards, supplementary materials, by Ellen Hechler (Farmington Hills, MI: Ellen Hechler, 1991). Reviewed in: *Mathematics Teacher* 85, no.1 (Jan. 1992): 78.

Fractions and Some Cool Distractions, video (Wynnewood, PA: Rahlic, 1990). Reviewed in: *Arithmetic Teacher* 37, no.7 (Mar. 1992): 42-43.

From the File Treasury, ed. by Jean M. Shaw (Reston, VA: NCTM, 1991). Reviewed in: *Arithmetic Teacher* 39, no.9 (May 1992): 52.

Fun with Money: A Problem-Solving Activity Book, by Carol A. Thornton and Judith K. Wells (Deerfield, IL: Learning Resources, 1991). Reviewed in: *Arithmetic Teacher* 40, no.2 (Oct. 1992): 128-29.

in COMMON Arithmetic, software, by Carol Chomsky, Harry Chomsky, and Judah L. Schwartz (Pleasantville, NY: Sunburst, 1990). Reviewed in: *Arithmetic Teacher* 39, no.1 (Sep. 1991): 47.

Junior Genius' Common Denominator, supplementary materials (Boca Raton, FL: Junior Genius, 1986). Reviewed in: *Arithmetic Teacher* 39, no.8 (Apr. 1992): 59-60.

Keep Your Balance, software (Pleasantville, NY: Sunburst, 1989). Reviewed in: *Arithmetic Teacher* 38, no.7 (Mar. 1991): 56.

Making the Most of Twenty Minutes: Warm Up Math Activities, by Michael Cain (Victoria, Australia: Dellasta, 1989). Reviewed in: *Arithmetic Teacher* 39, no.1 (Sep. 1991): 50.

Math around the World: Math/Geography Enrichment Activities for the Middle Grades, by Lynn Embry and Betty Bobo (Carthage, IL: Good Apple, 1991). Reviewed in: *Arithmetic Teacher* 39, no.9 (May 1992): 53-60.

Mathcad 3.0, software (Cambridge, MA: Mathsoft, 1991). Reviewed in: *Electronic Learning* 11, no.3 (Nov./Dec. 1991): 35.

Mathematical Olympiad Contest Problems for Children, by George Lenchner (Oceanside, NY: Glenwood, 1990). Reviewed in: *Arithmetic Teacher* 39, no.9 (May 1992): 52.

Mathematics Exploration Toolkit, software (Atlanta, GA: IBM Education). Reviewed in *Mathematics and Computer Education*: 25, no.3 (Fall 1991): 86.

Mathematics: Its Power and Utility, 3d ed., by Karl J. Smith (Belmont, CA: Wadsworth, 1990). Reviewed in: *Mathematics and Computer Education* 26, no.1 (Winter 1992): 92.

Math Extra, Early Bird ed., by Carole Greenes, Linda Schulman, and Rika Spungin (Allen, TX: DLM, 1991). Reviewed in: *Arithmetic Teacher* 39, no.7 (Mar. 1992): 43-44.

Math in Brief, game, by Mary Cavanagh (Fort Collins, CO: Scott Resources, 1978). Reviewed in: *Arithmetic Teacher* 39, no.1 (Sep. 1991): 53.

Math Intersections: A Look at Key Mathmatical Concepts, Grades 6-9, by David J. Glatzer and Joyce Glatzer (Palo Alto, CA: Dale Seymour, 1990). Reviewed in: *Arithmetic Teacher* 38, no.9 (May 1991): 46-47.

Computation *(cont'd)*

Math Mats: Hands-On Activities for Young Children, set 2, manipulatives, by Carole A. Thornton and Judith K. Wells (Allen, TX: Teaching Resources, 1990). Reviewed in: *Arithmetic Teacher* 37, no.7 (Mar. 1992): 44.

Maths at Play: Fun Ideas for 5-8 Year Olds, by Linn Maskell (Victoria, Australia: Dellasta, 1990). Reviewed in: *Arithmetic Teacher* 39, no.9 (May 1992): 51-52.

Math Shop Spotlight: Weights and Measures, software (Jefferson City, MO: Scholastic). Reviewed in: *Electronic Learning* 11, no.2 (Oct. 1991): 37.

The Maths Workshop: A Program of Maths Activities for Upper Primary, by Rob Vingerhoets (Victoria, Australia: Dellasta, 1990). Reviewed in: *Arithmetic Teacher* 38, no.9 (May 1991): 48.

Math Writer, software (Pacific Grove, CA: Brooks/Cole). Reviewed in: *Mathematics and Computer Education* 26, no.1 (Winter, 1992): 109-10.

Math-O-Graphs: Critical Thinking through Graphing, by Donna Kay Buck and Francis Hildebrand (Pacific Grove, CA: Midwest, 1990). Reviewed in: *Arithmetic Teacher* 38, no.9 (May 1991): 47.

Measure Up: The Visual Aid Ruler, by Kenneth L. Johannsen (Arlington Heights, IL: NEK, 1990). Reviewed in: *Arithmetic Teacher* 40, no.4 (Dec. 1992): 244.

Measuring: From Paces to Feet, Grades 3-4, by Rebecca B. Corwin and Susan Jo Russell (Palo Alto, CA: Dale Seymour, 1990). Reviewed in: *Arithmetic Teacher* 38, no.9 (May 1991): 48.

Measuring: From Used Numbers: Real Data in the Classroom, by Rebecca B. Corwin and Susan Jo Russell (Palo Alto, CA: Dale Seymour, 1990). Reviewed in: *Arithmetic Teacher* 38, no.9 (May 1991): 48.

Mental Math, supplementary materials, by Ellen Hechler (Farmington Hills, MI: Ellen Hechler, 1991). Reviewed in: *Mathematics Teacher* 85, no.1 (Jan. 1992): 78.

Mental Math and Estimation, by Gary G. Bitter (Allen, TX: DLM, 1989). Reviewed in: *Arithmetic Teacher* 39, no.1 (Sep. 1991): 53.

Minitab (State College, PA: Minitab Inc.). Reviewed in: *Mathematics and Computer Education* 25, no.3 (Fall 1991): 334-35.

Multiplication and Division Made Easy, by Catherine F. Debie (Artesia, CA: Scott Foresman, 1990). Reviewed in: *Arithmetic Teacher* 39, no.8 (Apr. 1992): 59.

Multiply with Balancing Bear, software (Pleasantville, NY: Sunburst, 1990). Reviewed in: *Arithmetic Teacher* 39, no.1 (Sep. 1991): 48.

Number Maze Decimals and Fractions, software (n.p.: Great Wave, n.d.). Reviewed in: *Media and Methods* 27, no.5 (May/June 1991): 60.

The $1.00 Word Riddle Book, by Marilyn Burns and illustrated by Martha Weston (New Rochelle, NY: Cuisenaire, 1990). Reviewed in: *Arithmetic Teacher* 39, no.2 (Oct. 1991): 51-52.

On the Button in Math: Activities for Young Children, by Carol A. Thornton and Judith K. Wells (Allen, TX: DLM, 1990). Reviewed in: *Arithmetic Teacher* 38, no.9 (May 1991): 46.

Overhead Math: Manipulative Lessons on the Overhead Projector, manipulatives, by Shirley Hoogeborm and Judy Goodnow (Sunnyvale, CA: Creative, 1990). Reviewed in: *Arithmetic Teacher* 39, no.8 (Apr. 1992): 60.

The Penguin Dictionary of Curious and Interesting Numbers, by David Wells (New York: Penguin, 1987). Reviewed in: *Mathematics and Computer Education* 24, no.1 (Winter 1991): 92.

PKZIP, software (Glendale, WI: PKWARE). Reviewed in: *Mathematics and Computer Education* 25, no.3 (Fall 1991): 332-33.

Computation *(cont'd)*
Prime Rollers, supplementary materials (Burlington, NC: Cabisco Mathematics, 1990). Reviewed in: *Arithmetic Teacher* 38, no.9 (May 1991): 48-49; *Mathematics Teacher* 85, no.7 (Oct. 1992): 594.

Projects To Enrich School Mathematics, Level 1, ed. by Judith Trowel (Reston, VA: NCTM, 1990). Reviewed in: *Mathematics Teacher* 83, no.8 (Nov. 1990): 677-80; *Arithmetic Teacher* 38, no.8 (Apr. 1991): 55.

Rap-ability, audiocassette, by Mike Ross and Suzanne Rossi (Lake Orion, MI: Aynn Visual, 1990). Reviewed in: *Arithmetic Teacher* 38, no.9 (May 1991): 49.

Taking Chances, software (Pleasantville, NY: Sunburst, 1991). Reviewed in: *Arithmetic Teacher* 39, no.3 (Nov. 1991): 57.

Talking Multiplication and Division: Grades 3-6, software (Pound Ridge, NY: Orange Cherry, 1990). Reviewed in: *Arithmetic Teacher* 39, no.3 (Nov. 1991): 57.

Teaching Primary Math with Music, cassette and songbook, by Esther Mardlesohn (Palo Alto, CA: Dale Seymour, 1990). Reviewed in: *Arithmetic Teacher* 39, no.1 (Sep. 1991): 53.

We Love MATHS: 4 Imaginative Themes for Early Primary Students, by Anne Marell and Susan Stajnko (Mount Waterly, Australia: Dellasta, 1990). Reviewed in: *Arithmetic Teacher* 39, no.1 (Sep. 1991): 53-54.

Zero in on Zero: Addition and Subtraction, software (Allen, TX: DLM, 1990). Reviewed in: *Arithmetic Teacher* 38, no.7 (Mar. 1991): 57-58.

Computer programming
Mathematical Topics for Computer Instructions: Grades 9-12, by Harris S. Shultz and William A. Leonard (Palo Alto, CA: Dale Seymour, 1990). Reviewed in: *Mathematics Teacher* 38, no.4 (Apr. 1991): 326-28.

Concepts and Applications of Intermediate Algebra, 2d ed.
by Marvin L. Bittinger, Marvin L. Keedy, and David Ellenbogen (Reading, MA: Addison-Wesley, 1991). Reviewed in: *Mathematics and Computer Education* 26, no.2 (Spring 1992): 205-07.

Concepts in Mathematics: Conic Sections
software (Cary, NC: TV Ontario Video, 1991). Reviewed in: *Mathematics Teacher* 85, no.9 (Dec. 1992): 767-68.

Concepts in Mathematics: Trigonometric Functions I--Solving Triangles
software (Cary, NC: TV Ontario Video, 1991). Reviewed in: *Mathematics Teacher* 85, no.9 (Dec. 1992): 768.

Connections! Math in Action
software (SVE, 1992). Reviewed in: *Teaching Pre-K-8* 21, no.4 (Jan. 1992): 14.

Contemporary Abstract Algebra (2d ed.)
by Joseph A. Gallian (Lexington, MA: D. C. Heath, 1990). Reviewed in: *Mathematics and Computer Education* 26, no.3 (Fall 1992): 334-35.

Cook, Wanda D.
Problem Solving through Critical Thinking, Grades 5-8, by Ronald R. Edwards and Wanda D. Cook (New Rochelle, NY: Cuisenaire, 1990). Reviewed in: *Arithmetic Teacher* 38, no.8 (Apr. 1991): 57-58.

Cooney, Thomas J.
Geometry (Teacher's ed.), by Stanley R. Clemens, Phares G. O'Daffer, and Thomas J. Cooney (Menlo Park, CA: Addison-Wesley, 1992). Reviewed in: *Mathematics Teacher* 85, no.7 (Oct. 1992): 590.

Cornelius, Michael
What's Your Game? A Resource Book for Mathematical Activities, by Michael Cornelius and Alan Parr (New York: Cambridge University Press, 1991). Reviewed in: *Arithmetic Teacher* 40, no.4 (Dec. 1992): 241.

Corwin, Rebecca B.
Measuring: From Paces to Feet, Grades 3-4, by Rebecca B. Corwin and Susan Jo Russell (Palo Alto, CA: Dale Seymour, 1990). Reviewed in: *Arithmetic Teacher* 38, no.9 (May 1991): 48.

Measuring: From Used Numbers: Real Data in the Classroom, by Rebecca B. Corwin and Susan Jo Russell (Palo Alto, CA: Dale Seymour, 1990). Reviewed in: *Arithmetic Teacher* 38, no.9 (May 1991): 48.

Statistics: The Shape of Data, Grades 4-6, by Susan Jo Russell and Rebecca B. Corwin (Palo Alto, CA: Dale Seymour, 1989). Reviewed in: *Arithmetic Teacher* 38, no.9 (May 1991): 49.

Coughlin, Raymond F.
Finite Mathematics with Calculus: An Applied Approach (2d ed.), by David E. Zitarelli and Raymond F. Coughlin (Orlando, FL: Saunders College, 1992). Reviewed in: *Mathematics Teacher* 85, no.9 (Dec. 1992): 764.

Counters: An Action Approach to Counting and Arithmetic
software (Scotts Valley, CA: Wings for Learning, 1990). Reviewed in: *Arithmetic Teacher* 39, no.3 (Nov. 1991): 56.

Counting books
Numbers at Play: A Counting Book, by Charles Sullivan (New York: Rizzoli, 1992). Reviewed in: *Childhood Education* 69, no.1 (Fall, 1992): 49.

Sea Squares, by Joy N. Hulme and illustrated by Carol Schwartz (Waltham, MA: Hyperion Books, 1991). Reviewed in: *Arithmetic Teacher* 40, no.1 (Sep. 1992): 56.

Ten Little Rabbits, by Virginia Grossman and illustrated by Sylvia Long (San Francisco: Chronicle Books, 1991). Reviewed in: *Arithmetic Teacher* 40, no.1 (Sep. 1992): 56-57.

What Comes in Two's, Three's, and Four's (New York: Simon & Schuster, 1990). Reviewed in: *Instructor* 102, no.2 (Sep. 1992): 80.

Creative (publisher)
The Algebra Lab: High School: A Comprehensive Manipulative Program for Algebra I, by Henri Piccotti (Sunnyvale, CA: Creative, 1990). Reviewed in: *Mathematics Teacher* 38, no.5 (May 1991): 406.

The Algebra Lab: Middle School: Exploring Algebra Concepts with Manipulatives, by Henri Piccotti (Sunnyvale, CA: Creative, 1990). Reviewed in: *Mathematics Teacher* 38, no.4 (Apr. 1991): 332.

Kaleidoscope Math, by Joe Kennedy and Diana Thomas (Sunnyvale, CA: Creative, 1989). Reviewed in: *Arithmetic Teacher* 38, no.6 (Feb. 1991): 63.

Overhead Math: Manipulative Lessons on the Overhead Projector, manipulatives, by Shirley Hoogeborm and Judy Goodnow (Sunnyvale, CA: Creative, 1990). Reviewed in: *Arithmetic Teacher* 39, no.8 (Apr. 1992): 60.

The Problem Solver with Calculators, by Terence G. Coburn, Shirley Hoogeboom, and Judy Goodnon (Sunnyvale, CA: Creative, 1989). Reviewed in: *Arithmetic Teacher* 39, no.1 (Sep. 1991): 50-51.

Crossmatics: A Challenging Collection of Cross-Number Puzzles: Grades 7-12
game, by Allan Dudley (Palo Alto, CA: Dale Seymour, 1990). Reviewed in: *Mathematics Teacher* 84, no.5 (May 1991): 406.

Cuisenaire
A Collection of MATH Lessons, Grades 6-8, by Cathy McLaughlin (New Rochelle, NY: Cuisenaire, 1990). Reviewed in: *Arithmetic Teacher* 39, no.1 (Sep. 1991): 49.

Estimate! Calculate! Evaluate! Calculator Activities for the Middle Grades, by Marjorie W. Bloom and Grace K. Galton (New Rochelle, NY: Cuisenaire, 1990). Reviewed in: *Arithmetic Teacher* 38, no.8 (Apr. 1991): 56.

The $1.00 Word Riddle Book, by Marilyn Burns and illustrated by Martha Weston (New Rochelle, NY: Cuisenaire, 1990). Reviewed in: *Arithmetic Teacher* 39, no.2 (Oct. 1991): 51-52.

Cuisenaire *(cont'd)*
Problem Solving through Critical Thinking, Grades 5-8, by Ronald R. Edwards and Wanda D. Cook (New Rochelle, NY: Cuisenaire, 1990). Reviewed in: *Arithmetic Teacher* 38, no.8 (Apr. 1991): 57-58.

Symmystries with Cuisenaire Rods, Sets, card set, by Patricia S. Davidson and Robert E. Willcutt (New Rochelle, NY: Cuisenaire, 1990). Reviewed in: *Arithmetic Teacher* 39, no.1 (Sep. 1991): 53.

Cumberland Educational Consultants
A New Look at Math, by Mary Jane Sweet (Providence, RI: Cumberland Educational Group, 1990). Reviewed in: *Arithmetic Teacher* 40, no.2 (Oct. 1992): 127-28.

Cuoco, Albert
Investigations in Algebra, by Albert Cuoco (Cambridge, MA: MIT Press, 1990). Reviewed in: *Mathematics Teacher* 85, no.1 (Jan. 1992): 74.

Curriculum guides
Discovering Meanings in Elementary School Mathematics, by Foster E. Grossnickle (Orlando, FL: Holt, Rinehart and Winston, 1990). Reviewed in: *Arithmetic Teacher* 40, no.1 (Sep. 1992): 58.

Mathematics for the Young Child, by Joseph N. Payne (Reston, VA: NCTM, 1991). Reviewed in: *Arithmetic Teacher* 40, no.1 (Sep. 1992): 57-58.

A New Look at Math, by Mary Jane Sweet (Providence, RI: Cumberland Educational Group, 1990). Reviewed in: *Arithmetic Teacher* 40, no.2 (Oct. 1992): 127-28.

Cushman, Jean
Do You Wanna Bet? Your Chance To Find Out about Probability, by Jean Cushman and illustrated by Martha Weston (New York: Clarion Books, 1991). Reviewed in: *Arithmetic Teacher* 40, no.4 (Dec. 1992): 240.

Daily Mathmatics: Critical Thinking and Problem Solving
(Evanston, IL: McDougal Littell, 1992). Reviewed in: *Arithmetic Teacher* 39, no.9 (May 1992): 52-53; *Mathematics Teacher* 85, no.6 (Sep. 1992): 496-97.

Data Collection and Analysis Activities (K-8)
by Katherine Pedersen and Judith Olson (Carbondale, IL: ICTM, 1990). Reviewed in: *Arithmetic Teacher* 40, no.1 (Sep. 1992): 62.

Data Insights
software, by Lois Edwards and K. M. Keogh (Pleasantville, NY: Sunburst, 1989). Reviewed in: *Mathematics Teacher* 84, no.4 (Apr. 1991): 320.

Davidson (publisher)
What's My Angle, software (Torrance, CA: Davidson, 1991). Reviewed in: *Media & Methods* 28, no.3 (Jan./Feb. 1992): 60.

Davidson, Patricia S.
Symmystries with Cuisenaire Rods, Sets, card set, by Patricia S. Davidson and Robert E. Willcutt (New Rochelle, NY: Cuisenaire, 1990). Reviewed in: *Arithmetic Teacher* 39, no.1 (Sep. 1991): 53.

Debie, Catherine F.
Multiplication and Division Made Easy, by Catherine F. Debie (Artesia, CA: Scott Foresman, 1990). Reviewed in: *Arithmetic Teacher* 39, no.8 (Apr. 1992): 59.

Decimals
Exploring Fractions and Decimals with Manipulatives, manipulatives, by Don Balka (Peabody, MA: Didax, 1991). Reviewed in: *Arithmetic Teacher* 39, no.9 (May 1992): 53.

Marcel Dekker (publisher)
A Primer in Probability, by Kathleen Subrahmaniam (New York: Marcel Dekker, 1990). Reviewed in: *Mathematics Teacher* 84, no.7 (Oct. 1991): 574.

Dellasta
Making the Most of Twenty Minutes: Warm Up Math Activities, by Michael Cain (Victoria, Australia: Dellasta, 1989). Reviewed in: *Arithmetic Teacher* 39, no.1 (Sep. 1991): 50.

Maths at Play: Fun Ideas for 5-8 Year Olds, by Linn Maskell (Victoria, Australia: Dellasta, 1990). Reviewed in: *Arithmetic Teacher* 39, no.9 (May 1992): 51-52.

Dellasta *(cont'd)*
The Maths Workshop: A Program of Maths Activities for Upper Primary, by Rob Vingerhoets (Victoria, Australia: Dellasta, 1990). Reviewed in: *Arithmetic Teacher* 38, no.9 (May 1991): 48.

Shapes Alive! Exploring Shapes with Primary Pupils, by Neville Leeson (Victoria, Australia: Dellasta, 1990). Reviewed in: *Arithmetic Teacher* 39, no.9 (May 1992): 60.

Solve It: Fun Folders for Fast Finishers, by Michael Richards and Wendy Lohse (Victoria, Australia: Dellasta, 1990). Reviewed in: *Arithmetic Teacher* 38, no.7 (Mar. 1991): 58.

Spy: A Teaching Resource of Fascinating Math Investigations, by Kevin Lees (Victoria, Australia: Dellasta, 1989). Reviewed in: *Arithmetic Teacher* 38, no.8 (Apr. 1991): 58.

We Love MATHS: 4 Imaginative Themes for Early Primary Students, by Anne Marell and Susan Stajnko (Mount Waterly, Australia: Dellasta, 1990). Reviewed in: *Arithmetic Teacher* 39, no.1 (Sep. 1991): 53-54.

Delta-Epsilon Software
Analyzer, software, by Douglas Alfors and Beverly West (Ithaca, NY: Delta-Epsilon Software, 1990). Reviewed in: *Mathematics Teacher* 85, no.3 (Mar. 1992): 240.

Demana, Franklin
Graphic Calculator and Computer Graphing Laboratory Manual (2nd ed.), by Franklin Demana and Bert K. Waits (Menlo Park, CA: Addison-Wesley, 1992). Reviewed in: *Mathematics Teacher* 85, no.7 (Oct. 1992): 590-91.

Precalculus Mathematics: A Graphing Approach (2d ed.), by Franklin Demana, Bert K. Waits, and Stanley E. Clemens (Menlo Park, CA: Addison-Wesley, 1992). Reviewed in: *Mathematics Teacher* 85, no.7 (Oct. 1992): 592.

Detective Stories for Math Problem Solving
software (Pleasantville, NY: Human Relations). Reviewed in: *Electronic Learning* 11, no.5 (Feb. 1992): 35.

Devaney, Robert L.
Chaos, Fractals, and Dynamics: Computer Experiments in Math, by Robert L. Devaney (Reading, MA: Addison-Wesley). Reviewed in: *Mathematics Teacher* 84, no.4 (Apr. 1991): 236.

Developing a Feel for Number (KS2)
(Shrewsbury, UK: Shropshire Mathematics Center, n.d.). Reviewed in: *Mathematics Teaching* 140 (Sep. 1992): 39.

Developing Number Awareness and Skills in the Early Years
(Shrewsbury, UK: Shropshire Mathematics Center, n.d.). Reviewed in: *Mathematics Teaching* 140 (Sep. 1992): 39.

Developmental Mathematics (3d ed.)
by C. L. Johnston, Alden T. Willis, and Gale M. Hughes (Belmont, CA: Wadsworth, 1991). Reviewed in: *Mathematics Teacher* 85, no.6 (Sep. 1992): 491.

Diamond, Howard
Fundamentally Math, Ages 8 and Up, software by Howard Diamond (Chapel Hill, NC: Chip Publications, 1990). Reviewed in: *Arithmetic Teacher* 40, no.3 (Nov. 1992): 194.

Didax
Exploring Fractions and Decimals with Manipulatives, manipulatives, by Don Balka (Peabody, MA: Didax, 1991). Reviewed in: *Arithmetic Teacher* 39, no.9 (May 1992): 53.

Discovering Meanings in Elementary School Mathematics
by Foster E. Grossnickle (Orlando, FL: Holt, Rinehart and Winston, 1990). Reviewed in: *Arithmetic Teacher* 40, no.1 (Sep. 1992): 58.

DLM (publisher)
Math Extra, Early Bird ed., by Carole Greenes, Linda Schulman, and Rika Spungin (Allen, TX: DLM, 1991). Reviewed in: *Arithmetic Teacher* 39, no.7 (Mar. 1992): 43-44.

Mental Math and Estimation, by Gary G. Bitter (Allen, TX: DLM, 1989). Reviewed in: *Arithmetic Teacher* 39, no.1 (Sep. 1991): 53.

DLM (publisher) *(cont'd)*
On the Button in Math: Activities for Young Children, by Carol A. Thornton and Judith K. Wells (Allen, TX: DLM, 1990). Reviewed in: *Arithmetic Teacher* 38, no.9 (May 1991): 46.

Zero in on Zero: Addition and Subtraction, software (Allen, TX: DLM, 1990). Reviewed in: *Arithmetic Teacher* 38, no.7 (Mar. 1991): 57-58.

Do You Wanna Bet? Your Chance To Find Out about Probability
by Jean Cushman and illustrated by Martha Weston (New York: Clarion Books, 1991). Reviewed in: *Arithmetic Teacher* 40, no.4 (Dec. 1992): 240.

Dossey, John A.
Algebra (Teacher's ed.), by Stanley A. Smith, Randall I. Charles, and John A. Dossey (Menlo Park, CA: Addison-Wesley, 1992). Reviewed in: *Mathematics Teacher* 85, no.7 (Oct. 1992): 589.

Algebra and Trigonometry (Teacher's ed.), by Stanley A. Smith, Randall I. Charles, and John A. Dossey (Menlo Park, CA: Addison-Wesley, 1992). Reviewed in: *Mathematics Teacher* 85, no.7 (Oct. 1992): 589-90.

Dover
Magic of Numbers, by Eric T. Bell (Mineola, NY: Dover, 1991). Reviewed in: *Mathematics Teacher* 85, no.6 (Sep. 1992): 493.

Shapes, Space and Symmetry, by Alan Holden (Mineola, NY: Dover, 1991). Reviewed in: *Mathematics Teacher* 85, no.6 (Sep. 1992): 494.

Dudley, Allan
Crossmatics: A Challenging Collection of Cross-Number Puzzles: Grades 7-12, game, by Allan Dudley (Palo Alto, CA: Dale Seymour, 1990). Reviewed in: *Mathematics Teacher* 84, no.5 (May 1991): 406.

Early Geometry: A Visual, Intuitive Introduction to Plane Geometry for Elementary School Children
by C. F. Navarro (Alexandria, VA: Start Smart Books, 1990). Reviewed in: *Arithmetic Teacher* 38, no.8 (Apr. 1991): 55-56.

Eddins, Susan
NCSMP Algebra, by John W. McConnell, Susan Brown, Susan Eddins, Margaret Hackworth, and Zalman Usiskin (Tucker, GA: Scott Foresman, 1990). Reviewed in: *Mathematics Teacher* 84, no.2 (Feb. 1991): 144-46.

Edmiston, Margaret C.
Merlin Book of Logic Puzzles, by Margaret C. Edmiston (New York: Sterling, 1991). Reviewed in: *Mathematics Teacher* 85, no.9 (Dec. 1992): 767.

Educational Design (publisher)
Strategies for Solving Math Word Problems (New York: Educational Design). Reviewed in: *Teaching Exceptional Children* 24, no.2 (Fall 1991): 73.

Educational Development Center
Add-Ventures for Girls: Building Math Confidence (Elementary), by Margaret Franklin (Newton, MA: Educational Development Center, 1990). Reviewed in: *Arithmetic Teacher* 40, no.1 (Sep. 1992): 58.

Add-Ventures for Girls: Building Math Confidence (Junior High), by Margaret Franklin (Newton, MA: Educational Development Center, 1990). Reviewed in: *Arithmetic Teacher* 40, no.1 (Sep. 1992): 58.

Edwards, Lois
Data Insights, software, by Lois Edwards and K. M. Keogh (Pleasantville, NY: Sunburst, 1989). Reviewed in: *Mathematics Teacher* 84, no.4 (Apr. 1991): 320.

Edwards, Ronald R.
Problem Solving through Critical Thinking, Grades 5-8, by Ronald R. Edwards and Wanda D. Cook (New Rochelle, NY: Cuisenaire, 1990). Reviewed in: *Arithmetic Teacher* 38, no.8 (Apr. 1991): 57-58.

Elementary
Addition Rap: For Grades 1 to 2, audiocassette (Boise, ID: Star Trax, 1990). Reviewed in: *Arithmetic Teacher* 40, no.1 (Sep. 1992): 62.

Elementary *(cont'd)*

Fractions and Some Cool Distractions, video (Wynnewood, PA: Rahlic, 1990). Reviewed in: *Arithmetic Teacher* 37, no.7 (Mar. 1992): 42-43.

From the File Treasury, ed. by Jean M. Shaw (Reston, VA: NCTM, 1991). Reviewed in: *Arithmetic Teacher* 39, no.9 (May 1992): 52.

Fun with Money: A Problem-Solving Activity Book, by Carol A. Thornton and Judith K. Wells (Deerfield, IL: Learning Resources, 1991). Reviewed in: *Arithmetic Teacher* 40, no.2 (Oct. 1992): 128-29.

Fundamentally Math, Ages 8 and Up, software by Howard Diamond (Chapel Hill, NC: Chip Publications, 1990). Reviewed in: *Arithmetic Teacher* 40, no.3 (Nov. 1992): 194.

Geometry Workshop, Grades 3-8, software (Glenville, IL: Scott Foresman, 1991). Reviewed in: *Arithmetic Teacher* 39, no.8 (Apr. 1992): 58.

Graphing and Probability Workshop, Grades 3-8, software (Glenview, IL: Scott Foresman, 1991). Reviewed in: *Arithmetic Teacher* 39, no.3 (Nov. 1991): 56.

Graphing Primer, Grades K-2, by Laura Duncan Choate and JoAnn King Okey (Palo Alto, CA: Seymour Publications, 1989). Reviewed in: *Arithmetic Teacher* 40, no.2 (Oct. 1992): 129.

in COMMON Arithmetic, software, by Carol Chomsky, Harry Chomsky, and Judah L. Schwartz (Pleasantville, NY: Sunburst, 1990). Reviewed in: *Arithmetic Teacher* 39, no.1 (Sep. 1991): 47.

Junior Genius' Common Denominator, supplementary materials (Boca Raton, FL: Junior Genius, 1986). Reviewed in: *Arithmetic Teacher* 39, no.8 (Apr. 1992): 59-60.

Kaleidoscope Math, by Joe Kennedy and Diana Thomas (Sunnyvale, CA: Creative Publications, 1989). Reviewed in: *Arithmetic Teacher* 38, no.6 (Feb. 1991): 63.

Koetke's Challenge, software by Tim Barclay and Jonathan Choate (Acton, MA: William K. Bradford, 1991). Reviewed in: *Mathematics Teacher* 85, no.9 (Dec. 1992): 761.

Label Land: A World of Inference and Problem Solving, software (Scotts Valley, CA: Wings for Learning, 1991). Reviewed in: *Arithmetic Teacher* 39, no.3 (Nov. 1991): 56-57.

Learning To Reason: Some, All, or None, software (Kalamazoo, MI: MCE, 1988). Reviewed in: *Arithmetic Teacher* 39, no.1 (Sep. 1991): 47.

Link 'n' Learn Activity Book, by Carol A. Thornton and Judith K. Wells (Deerfield, IL: Learning Resources, 1990). Reviewed in: *Arithmetic Teacher* 40, no.2 (Oct. 1992): 130.

The Little Shoppers Kit, software (Cambridge, MA: Tom Snyder, 1989). Reviewed in: *Arithmetic Teacher* 39, no.1 (Sep. 1991): 47.

Logo Geometry, software by Michael Battista and Douglas H. Clements (Morristown, NJ: Simon & Schuster, 1991). Reviewed in: *Arithmetic Teacher* 40, no.1 (Sep. 1992): 56.

Making the Most of Twenty Minutes: Warm Up Math Activities, by Michael Cain (Victoria, Australia: Dellasta, 1989). Reviewed in: *Arithmetic Teacher* 39, no.1 (Sep. 1991): 50.

Math around the World: Math/Geography Enrichment Activities for the Middle Grades, by Lynn Embry and Betty Bobo (Carthage, IL: Good Apple, 1991). Reviewed in: *Arithmetic Teacher* 39, no.9 (May 1992): 53-60.

Mathemactivities, by Bob Berstein and illustrated by Bron Smith (Carthage, IL: Good Apple, 1991). Reviewed in: *Arithmetic Teacher* 40, no.1 (Sep. 1992): 62-63.

Mathematical Olympiad Contest Problems for Children, by George Lenchner (Oceanside, NY: Glenwood, 1990). Reviewed in: *Arithmetic Teacher* 39, no.9 (May 1992): 52.

Elementary *(cont'd)*
Mathematical Olympiad Contest Problems for Children, by George Lenchner (East Meadow, NY: Glenwood, 1990). Reviewed in: *Curriculum Review* 31, no.7 (Mar. 1992): 26.

Mathematics for the Young Child, by Joseph N. Payne (Reston, VA: NCTM, 1991). Reviewed in: *Arithmetic Teacher* 40, no.1 (Sep. 1992): 57-58.

Math Extra, Early Bird ed., by Carole Greenes, Linda Schulman, and Rika Spungin (Allen, TX: DLM, 1991). Reviewed in: *Arithmetic Teacher* 39, no.7 (Mar. 1992): 43-44.

Math in Brief, game, by Mary Cavanagh (Fort Collins, CO: Scott Resources, 1978). Reviewed in: *Arithmetic Teacher* 39, no.1 (Sep. 1991): 53.

Math Mats: Hands-On Activities for Young Children, set 2, manipulatives, by Carole A. Thornton and Judith K. Wells (Allen, TX: Teaching Resources, 1990). Reviewed in: *Arithmetic Teacher* 37, no.7 (Mar. 1992): 44.

Math-O-Graphs: Critical Thinking through Graphing, by Donna Kay Buck and Francis Hildebrand (Pacific Grove, CA: Midwest, 1990). Reviewed in: *Arithmetic Teacher* 38, no.9 (May 1991): 47.

Math Power at Home, by Gerald Kulm (Waldorf, MD: American Association for the Advancement of Science, 1990). Reviewed in: *Arithmetic Teacher* 40, no.1 (Sep. 1992): 64.

Math Power in School, by Gerald Kulm (Waldorf, MD: American Association for the Advancement of Science, 1990). Reviewed in: *Arithmetic Teacher* 40, no.1 (Sep. 1992): 64.

Math Power in the Community, by Gerald Kulm (Waldorf, MD: American Association for the Advancement of Science, 1990). Reviewed in: *Arithmetic Teacher* 40, no.1 (Sep. 1992): 64.

Math Shop Spotlight: Weights and Measures, software (Jefferson City, MO: Scholastic). Reviewed in: *Electronic Learning* 11, no.2 (Oct. 1991): 37.

The Maths Workshop: A Program of Maths Activities for Upper Primary, by Rob Vingerhoets (Victoria, Australia: Dellasta, 1990). Reviewed in: *Arithmetic Teacher* 38, no.9 (May 1991): 48.

Math Talk, (Cambridge, MA: Tom Snyder, 1991). Reviewed in: *Teaching Pre-K-8* 21, no.2 (Oct. 1991): 16-17.

Maths at Play: Fun Ideas for 5-8 Year Olds, by Linn Maskell (Victoria, Australia: Dellasta, 1990). Reviewed in: *Arithmetic Teacher* 39, no.9 (May 1992): 51-52.

Measure Up: The Visual Aid Ruler, by Kenneth L. Johannsen (Arlington Heights, IL: NEK, 1990). Reviewed in: *Arithmetic Teacher* 40, no.4 (Dec. 1992): 244.

Measuring: From Paces to Feet, Grades 3-4, by Rebecca B. Corwin and Susan Jo Russell (Palo Alto, CA: Dale Seymour, 1990). Reviewed in: *Arithmetic Teacher* 38, no.9 (May 1991): 48.

Measuring: From Used Numbers: Real Data in the Classroom, by Rebecca B. Corwin and Susan Jo Russell (Palo Alto, CA: Dale Seymour, 1990). Reviewed in: *Arithmetic Teacher* 38, no.9 (May 1991): 48.

Mental Math, supplementary materials, by Ellen Hechler (Farmington Hills, MI: Ellen Hechler, 1991). Reviewed in: *Mathematics Teacher* 85, no.1 (Jan. 1992): 78.

Mental Math Games, Series I, software (San Rafael, CA: Broderbund, 1992). Reviewed in: *Technology & Learning* 13, no.3 (Nov./Dec. 1992): 18.

Merlin Book of Logic Puzzles, by Margaret C. Edmiston (New York: Sterling, 1991). Reviewed in: *Mathematics Teacher* 85, no.9 (Dec. 1992): 767.

Elementary *(cont'd)*

Mind Benders—A1, software, by Anita Harnadek (Pacific Grove, CA: Midwest, 1988). Reviewed in: *Arithmetic Teacher* 39, no.1 (Sep. 1991): 47-48.

Mr. Marfil's Last Will and Testament, video (Pleasantville, NY: Human Relations, 1991). Reviewed in: *Mathematics Teacher* 84, no.7 (Oct. 1991): 578.

Money and Time Workshop, Grades K-2, software (Glenview, IL: Scott Foresman, 1991). Reviewed in: *Arithmetic Teacher* 39, no.1 (Sep. 1991): 48.

Mosaic Magic, software (Encinitas, CA: Kinder Magic, 1990). Reviewed in: *Arithmetic Teacher* 39, no.1 (Sep. 1991): 48.

Multiplication and Division Made Easy, by Catherine F. Debie (Artesia, CA: Scott Foresman, 1990). Reviewed in: *Arithmetic Teacher* 39, no.8 (Apr. 1992): 59.

Multiplication Rap: For Grades 3 to 6, audiocassette (Boise, ID: Star Trax, 1988). Reviewed in: *Arithmetic Teacher* 40, no.1 (Sep. 1992): 62.

Multiply with Balancing Bear, software (Pleasantville, NY: Sunburst, 1990). Reviewed in: *Arithmetic Teacher* 39, no.1 (Sep. 1991): 48.

Mystery Castle: Strategies for Problem Solving, Level I and II, software (Acton, MA: William K. Bradford, 1990). Reviewed in: *Arithmetic Teacher* 39, no.3 (Nov. 1991): 57.

A New Look at Math, by Mary Jane Sweet (Providence, RI: Cumberland Educational Group, 1990). Reviewed in: *Arithmetic Teacher* 40, no.2 (Oct. 1992): 127-28.

Number Maze Decimals and Fractions, software (n.p.: Great Wave, n.d.). Reviewed in: *Media and Methods* 27, no.5 (May/June 1991): 60.

Numbers at Play: A Counting Book, by Charles Sullivan (New York: Rizzoli, 1992). Reviewed in: *Childhood Education* 69, no.1 (Fall, 1992): 49.

Number Squares, software (Washington, DC: Balloons, 1988). Reviewed in: *Arithmetic Teacher* 39, no.1 (Sep. 1991): 48.

The $1.00 Word Riddle Book, by Marilyn Burns and illustrated by Martha Weston (New Rochelle, NY: Cuisenaire, 1990). Reviewed in: *Arithmetic Teacher* 39, no.2 (Oct. 1991): 51-52.

On the Button in Math: Activities for Young Children, by Carol A. Thornton and Judith K. Wells (Allen, TX: DLM, 1990). Reviewed in: *Arithmetic Teacher* 38, no.9 (May 1991): 46.

Patterns and Functions, by Linda Sue Brisby (Solvang, CA: Hands On, 1990). Reviewed in: *Arithmetic Teacher* 40, no.1 (Sep. 1992): 64.

Patterns and Puzzles: Math Challenges for Grades 3 and 4, by Neville Leeson (Mt. Waverly, Australia: n.p., 1990). Reviewed in: *Arithmetic Teacher* 40, no.1 (Sep. 1992): 64.

Pictograms: Graphing Pictures for a Reusable Classroom Grid, manipulatives by Laura Duncan Choate and JoAnn King Okey (Palo Alto, CA: Dale Seymour, 1989). Reviewed in: *Arithmetic Teacher* 40, no.2 (Oct. 1992): 130.

Place Value-Plus, manipulatives (Goshen, IN: New Vision, n.d.). Reviewed in: *Arithmetic Teacher* 40, no.2 (Oct. 1992): 130.

Plane Geometry, supplementary materials (Little Rock, AR: High Q, 1988). Reviewed in: *Arithmetic Teacher* 38, no.8 (Apr. 1991): 60.

Polyominoes: A Guide to Puzzles and Problems in Tiling, by George E. Martin (Washington, DC: Mathematical Association of America, 1991). Reviewed in: *Mathematics Teacher* 85, no.6 (Sep. 1992): 493-94.

Primary Geoboard Activity Book, by Carol A. Thornton and Judith K. Wells (Deerfield, IL: Learning Resources, 1990). Reviewed in: *Arithmetic Teacher* 40, no.2 (Oct. 1992): 130.

Elementary *(cont'd)*
Ten Little Rabbits, by Virginia Grossman and illustrated by Sylvia Long (San Francisco: Chronicle Books, 1991). Reviewed in: *Arithmetic Teacher* 40, no.1 (Sep. 1992): 56-57.

Thinking through Math Word Problems, by Authur Whimbey, Jack Lochhead, and Paula Potter (Hillsdale, NJ: Laurence Erlbaum, 1990). Reviewed in: *Arithmetic Teacher* 39, no.2 (Oct. 1991): 52.

Troll Sports Math: Math Word Problems for Grades 4-6, (Mahwah, NJ: Troll, 1991). Reviewed in: *Teaching Pre-K-8* 21, no.2 (Oct. 1991): 14-16.

What Comes in Two's, Three's, and Four's (New York: Simon & Schuster, 1990). Reviewed in: *Instructor* 102, no.2 (Sep. 1992): 80.

What Do You Do with a Broken Calculator?, software (Pleasantville, NY: Sunburst, 1989). Reviewed in: *Arithmetic Teacher* 39, no.3 (Nov. 1991): 58.

What's Your Game? A Resource Book for Mathematical Activities, by Michael Cornelius and Alan Parr (New York: Cambridge University Press, 1991). Reviewed in: *Arithmetic Teacher* 40, no.4 (Dec. 1992): 241.

What's Your Problem? Posing and Solving Mathematical Problems, by Penny Skinner (Portsmouth, NH: Heinemann, 1990). Reviewed in: *Arithmetic Teacher* 40, no.4 (Dec. 1992): 241-42.

Zero in on Zero: Addition and Subtraction, software (Allen, TX: DLM, 1990). Reviewed in: *Arithmetic Teacher* 38, no.7 (Mar. 1991): 57-58.

Elementary Algebra (2d ed.)
by Jack Barker, James Rogers, and James Van Dyke (Orlando, FL: Saunders College, 1992). Reviewed in: *Mathematics Teacher* 85, no.9 (Dec. 1992): 764.

Elementary Algebra: Concepts and Applications, 3d ed.
by Marvin L. Bittinger and Marvin L. Keedy (Reading, MA: Addison-Wesley, n.d.). Reviewed in: *Mathematics and Computer Education* 26, no.1 (Winter 1992): 95-96.

Ellenbogen, David
Concepts and Applications of Intermediate Algebra, 2d ed., by Marvin L. Bittinger, Marvin L. Keedy, and David Ellenbogen (Reading, MA: Addison-Wesley, 1991). Reviewed in: *Mathematics and Computer Education* 26, no.2 (Spring 1992): 205-07.

Intermediate Algebra: Concepts and Applications, 3d ed., by Marvin L. Bittinger, Mervin L. Keedy, and David Ellenbogen (Reading, MA: Addison-Wesley, 1990). Reviewed in: *Mathematics Teacher* 84, no.3 (Mar. 1991): 238.

Ellis, Robert
Precalculus (4th ed.), by Robert Ellis and Deny Gulick (Orlando, FL: Saunders College, 1992). Reviewed in: *Mathematics Teacher* 85, no.9 (Dec. 1992): 767.

Embry, Lynn
Math around the World: Math/Geography Enrichment Activities for the Middle Grades, by Lynn Embry and Betty Bobo (Carthage, IL: Good Apple, 1991). Reviewed in: *Arithmetic Teacher* 39, no.9 (May 1992): 53-60.

Essential Algebra (6th ed.)
by C. L. Johnston, Alden T. Willis, *and Jeanne Lazaris (Belmont, CA: Wadsworth, 1991). Reviewed in: Mathematics Teacher* 85, no.6 (Sep. 1992): 491-92.

Essential Arithmetic (6th ed.)
by C. L. Johnston, Alden T. Willis, and Jeanne Lazaris (Belmont, CA: Wadsworth, 1991). Reviewed in: *Mathematics Teacher* 85, no.6 (Sep. 1992): 492.

Estimate! Calculate! Evaluate! Calculator Activities for the Middle Grades
by Marjorie W. Bloom and Grace K. Galton (New Rochelle, NY: Cuisenaire, 1990). Reviewed in: *Arithmetic Teacher* 38, no.8 (Apr. 1991): 56.

Estimation: Quick Solve I
software (St. Paul: MECC). Reviewed in: *Teaching Pre-K-8* 21, no.3 (Nov./Dec. 1991): 15-16; *Arithmetic Teacher* 38, no.8 (Apr. 1991): 54.

Estimation Tutor
software by Richard E. Rand (Portland, ME: J. Weston Walch, 1989). Reviewed in: *Arithmetic Teacher* 38, no.8 (Apr. 1991): 54.

Euclid's Toolbox—Triangles
software, by Jerry Beckmann and Charles Friesen (n.p.: Heartland, 1990). Reviewed in: *Mathematics Teacher* 84, no.6 (Sep. 1991): 484.

Exploring Fractions and Decimals with Manipulatives
manipulatives, by Don Balka (Peabody, MA: Didax, 1991). Reviewed in: *Arithmetic Teacher* 39, no.9 (May 1992): 53.

Exploring Sequences and Series
software (Minneapolis: MECC, 1991). Reviewed in: *Mathematics Teacher* 85, no.6 (Sep. 1992): 489.

EZ Graph
manipulatives (Englewood, CO: EZ Graph, 1992). Reviewed in: *Mathematics Teacher* 85, no.6 (Sep. 1992): 497.

The Factor Pack
supplementary materials (Holywell, UK: Magic Mathworks, n.d.). Reviewed in: *Mathematics Teacher* 84, no.7 (Oct. 1991): 576.

Facts Master
manipulatives by Jivan Patel (Carrollton, TX: Bruce Bakke, n.d.). Reviewed in: *Arithmetic Teacher* 40, no.2 (Oct. 1992): 128.

Fastmath
software, by Jim Baker and Dale Huline (Minneapolis: New Directions, 1990). Reviewed in: *Arithmetic Teacher* 39, no.1 (Sep. 1991): 47.

Fiendishly Difficult
by Ivan Moscovich (New York: Sterling, 1991). Reviewed in: *Mathematics Teacher* 84, no.9 (Dec. 1991): 766.

Films/Videos
Archimedean and Archimedean Dual Polyhydra, video, by Lorrainne L. Foster (Northridge, CA: California State University). Reviewed in: *Mathematics Teacher* 84, no.7 (Oct. 1991): 576.

The Calculator, Grades 4-8, video (Columbus, OH: Silver Burdett & Ginn, 1991). Reviewed in: *Arithmetic Teacher* 37, no.7 (Mar. 1992): 42.

Calculus: Applications of Differentiations, Parts 1-3, video (Brooklyn: Video Tutorial Service, 1990). Reviewed in: *Mathematics Teacher* 84, no.6 (Sep. 1991): 496-97.

Fractals: An Animated Discussion, video (New York: W. H. Freeman, 1990). Reviewed in: *Mathematics Teacher* 84, no.7 (Oct. 1991): 576-78.

Fractions and Some Cool Distractions, video (Wynnewood, PA: Rahlic, 1990). Reviewed in: *Arithmetic Teacher* 37, no.7 (Mar. 1992): 42-43.

Mr. Marfil's Last Will and Testament, video (Pleasantville, NY: Human Relations, 1991). Reviewed in: *Mathematics Teacher* 84, no.7 (Oct. 1991): 578.

M! Project MATHEMATICS!—Similarity, video by Tom Apostol (Reston, VA: NCTM, 1990). Reviewed in: *Mathematics Teacher* 85, no.6 (Sep. 1992): 496.

Pre-Calculus, video (Brooklyn: Video Tutorial, 1990). Reviewed in: *Mathematics Teacher* 84. no.6 (Sep. 1991): 498.

Problem Solving, video (Atlanta, GA: Silver Burdett & Ginn, 1991). Reviewed in: *Arithmetic Teacher* 39, no.2 (Oct. 1991): 52-53.

Reading Higher: A Problem-Solving Approach to Elementary School Mathematics, video and supplementary materials (Reston, VA: NCTE, 1990). Reviewed in: *Curriculum Review* 31, no.7 (Mar. 1992): 26-27.

The Stella Octangula Activity Set, video (Berkeley: Key Curriculum, 1991). Reviewed in: *Mathematics Teacher* 85, no.2 (Feb. 1992): 156.

Films/Videos *(cont'd)*
The Wonderful Problems of Fizz and Martina, video (Cambridge, MA: Tom Snyder). Reviewed in: *Media and Methods* 28, no.2 (Nov./Dec. 1991): 55-56.

Finite Mathematics with Calculus: An Applied Approach (2d ed.)
by David E. Zitarelli and Raymond F. Coughlin (Orlando, FL: Saunders College, 1992). Reviewed in: *Mathematics Teacher* 85, no.9 (Dec. 1992): 764.

Finney, Ross L.
Calculus, by Ross L. Finney and George B. Thomas, Jr. (Reading, MA: Addison-Wesley, 1990). Reviewed in: *Mathematics Teacher* 84, no.1 (Jan. 1991): 66.

Flashcards
supplementary materials, by Ellen Hechler (Farmington Hills, MI: Ellen Hechler, 1991). Reviewed in: *Mathematics Teacher* 85, no.1 (Jan. 1992): 78.

Flewelling, Gary
Making Mathematics 8, by Gary Flewelling, Joan Routledge, and John Clark (Agincourt, Ontario: Gage Educational, 1991). Reviewed in: *Mathematics Teacher* 85, no.8 (Nov. 1992): 692.

Making Mathematics 7, by Gary Flewelling, Joan Routledge, and John Clark (Agincourt, Ontario: Gage Educational, 1991). Reviewed in: *Mathematics Teacher* 85, no.9 (Dec. 1992): 766.

Foster, Lorrainne L.
Archimedean and Archimedean Dual Polyhydra, video, by Lorrainne L. Foster (Northridge, CA: California State University). Reviewed in: *Mathematics Teacher* 84, no.7 (Oct. 1991): 576.

Fractals
Fractals: An Animated Discussion, video (New York: W. H. Freeman, 1990). Reviewed in: *Mathematics Teacher* 84, no.7 (Oct. 1991): 576-78.

Fractals for the Classroom: Strategic Activities, vol. 1, by Heinz-Otto Peritgen, Hartmut Jurgens, Dietmar Saupe, Evan Maletsky, Terry Perciante, and Lee Yunker (Reston, VA: NCTM, 1991). Reviewed in: *Mathematics Teacher* 85, no.2 (Feb. 1992): 146; *Arithmetic Teacher* 39, no.7 (Mar. 1992): 40.

IFS Explorer, software (St. Louis, MO: Koyn, 1990). Reviewed in: *Mathematics Teacher* 84, no.6 (Sep. 1991): 486.

Fractals: An Animated Discussion
video (New York: W. H. Freeman, 1990). Reviewed in: *Mathematics Teacher* 84, no.7 (Oct. 1991): 576-78.

Fractals for the Classroom: Strategic Activities, vol. 1
by Heinz-Otto Peritgen, Hartmut Jurgens, Dietmar Saupe, Evan Maletsky, Terry Perciante, and Lee Yunker (Reston, VA: NCTM, 1991). Reviewed in: *Mathematics Teacher* 85, no.2 (Feb. 1992): 146; *Arithmetic Teacher* 39, no.7 (Mar. 1992): 40.

Fraction-oids
software (Danvers, MA: MindPlay, Methods and Solutions, 1988). Reviewed in: *Arithmetic Teacher* 39, no.1 (Sep. 1991): 46-47.

Fractions
Exploring Fractions and Decimals with Manipulatives, manipulatives, by Don Balka (Peabody, MA: Didax, 1991). Reviewed in: *Arithmetic Teacher* 39, no.9 (May 1992): 53.

Fraction-oids, software (Danvers, MA: MindPlay, Methods and Solutions, 1988). Reviewed in: *Arithmetic Teacher* 39, no.1 (Sep. 1991): 46-47.

Fractions and Some Cool Distractions
video (Wynnewood, PA: Rahlic, 1990). Reviewed in: *Arithmetic Teacher* 37, no.7 (Mar. 1992): 42-43.

Fraleigh, John
Calculus with Analytic Geometry, 3d. ed., by John B. Fraleigh (Reading, MA: Addison-Wesley). Reviewed in *Mathematics and Computer Education:* 24, no.1 (Winter 1991): 93.

Geometric Connectors: Transformations software (Pleasantville, NY: Sunburst, 1990). Reviewed in: *Mathematics Teacher* 85, no.9 (Dec. 1992): 761.

Geometry
Analytic Geometry, 2d ed., by Karl J. Smith (Pacific Grove, CA: Brooks/Cole, 1991). Reviewed in: *Mathematics Teacher* 85, no.2 (Feb. 1992): 146.

Archimedean and Archimedean Dual Polyhydra, video, by Lorrainne L. Foster (Northridge, CA: California State University). Reviewed in: *Mathematics Teacher* 84, no.7 (Oct. 1991): 576.

The Circular Geoboard, by Judy Kevin (San Leandro, CA: Teaching Resource Center, 1990). Reviewed in: *Mathematics Teacher* 84, no.8 (Nov. 1991): 680.

Early Geometry: A Visual, Intuitive Introduction to Plane Geometry for Elementary School Children, by C. F. Navarro (Alexandria, VA: Start Smart Books, 1990). Reviewed in: *Arithmetic Teacher* 38, no.8 (Apr. 1991): 55-56.

Euclid's Toolbox—Triangles, software, by Jerry Beckmann and Charles Friesen (n.p.: Heartland, 1990). Reviewed in: *Mathematics Teacher* 84, no.6 (Sep. 1991): 484.

GeoExplorer, software (Glenview, IL: Scott Foresman, n.d.). Reviewed in: *Electronic Learning* 12, no.1 (Sep. 1992): 23-24; *Mathematics Teacher* 85, no.7 (Oct. 1992): 587.

The Geometer's Sketchpad, software (Berkeley: Key Curriculum, 1991). Reviewed in: *Mathematics Teacher* 85, no.5 (May 1992): 392-93; *Electronic Learning* 11, no.3 (Nov./Dec. 1991): 34-35.

Geometric Connectors: Transformations, software (Pleasantville, NY: Sunburst, 1990). Reviewed in: *Mathematics Teacher* 85, no.9 (Dec. 1992): 761.

Geometry, by Karl J. Smith (Pacific Grove, CA: Brooks/Cole, Geometry1991). Reviewed in: *Mathematics Teacher* 85, no.1 (Jan. 1992): 73-74.

Geometry (Teacher's ed.), by Stanley R. Clemens, Phares G. O'Daffer, and Thomas J. Cooney (Menlo Park, CA: Addison-Wesley, 1992). Reviewed in: *Mathematics Teacher* 85, no.7 (Oct. 1992): 590.

Geometry 1.2, software (San Rafael, CA: Broderbund, n.d.). Reviewed in: *Journal of Geographical Education* 39, no.4 (Sep. 1991): n.p.

Geometry for Enjoyment and Challenges, by Richard Rhoad, George Milauskas, and Robert Whipple (Evanston, IL: McDougal Littell, 1991). Reviewed in: *Mathematics Teacher* 84, no.6 (Sep. 1991): 488-89.

Geometry Inventor, software (Scotts Valley, CA: Wings for Learning, n.d.). Reviewed in: *Electronic Learning* 12, no.1 (Sep. 1992): 23-24.

Geometry Workshop, Grades 3-8, software (Glenville, IL: Scott Foresman, 1991). Reviewed in: *Arithmetic Teacher* 39, no.8 (Apr. 1992): 58.

GyroGraphics, version 2.2, software (Stillwater, OK: Cipher Systems, 1990). Reviewed in: *Mathematics Teacher* 84, no.3 (Mar. 1991): 235.

Introduction to Tessellations, by Dale Seymour and Jill Britten (Palo Alto, CA: Dale Seymour, 1989). Reviewed in: *Mathematics Teacher* 83, no.6 (Sep. 1992): 482.

Journey into Geometries, by Martha Sved (Washington, DC: Mathematical Association of America, 1991). Reviewed in: *Mathematics Teacher* 85, no.8 (Nov. 1992): 691.

Kaleidoscope Math, by Joe Kennedy and Diana Thomas (Sunnyvale, CA: Creative Publications, 1989). Reviewed in: *Arithmetic Teacher* 38, no.6 (Feb. 1991): 63.

Logo Geometry, software by Michael Battista and Douglas H. Clements (Morristown, NJ: Simon & Schuster, 1991). Reviewed in: *Arithmetic Teacher* 40, no.1 (Sep. 1992): 56.

Gilbert, Jimmie
Precalculus Functions and Graphs, by Jimmie Gilbert and Linda Gilbert (Reading, MA: Addison-Wesley, 1990). Reviewed in: *Mathematics Teacher* 84, no.1 (Jan. 1991): 67.

Gilbert, Linda
Precalculus Functions and Graphs, by Jimmie Gilbert and Linda Gilbert (Reading, MA: Addison-Wesley, 1990). Reviewed in: *Mathematics Teacher* 84, no.1 (Jan. 1991): 67.

Glatzer, David J.
Math Intersections: A Look at Key Mathmatical Concepts, Grades 6-9, by David J. Glatzer and Joyce Glatzer (Palo Alto, CA: Dale Seymour, 1990). Reviewed in: *Arithmetic Teacher* 38, no.9 (May 1991): 46-47.

Glatzer, Joyce
Math Intersections: A Look at Key Mathmatical Concepts, Grades 6-9, by David J. Glatzer and Joyce Glatzer (Palo Alto, CA: Dale Seymour, 1990). Reviewed in: *Arithmetic Teacher* 38, no.9 (May 1991): 46-47.

Glenwood (publisher)
Mathematical Olympiad Contest Problems for Children, by George Lenchner (East Meadow, NY: Glenwood, 1990). Reviewed in: *Curriculum Review* 31, no.7 (Mar. 1992): 26; *Arithmetic Teacher* 39, no.9 (May 1992): 52.

Gobran, Alfonse
Beginning Algebra, by Alfonse Gobran (Belmont, CA: Wadsworth, 1991). Reviewed in: *Mathematics Teacher* 84, no.8 (Nov. 1991): 674.

Good Apple
Math around the World: Math/Geography Enrichment Activities for the Middle Grades, by Lynn Embry and Betty Bobo (Carthage, IL: Good Apple, 1991). Reviewed in: *Arithmetic Teacher* 39, no.9 (May 1992): 53-60.

Mathemactivities, by Bob Berstein and illustrated by Bron Smith (Carthage, IL: Good Apple, 1991). Reviewed in: *Arithmetic Teacher* 40, no.1 (Sep. 1992): 62-63.

Goodnow, Judy
Overhead Math: Manipulative Lessons on the Overhead Projector, manipulatives, by Shirley Hoogeborm and Judy Goodnow (Sunnyvale, CA: Creative, 1990). Reviewed in: *Arithmetic Teacher* 39, no.8 (Apr. 1992): 60.

The Problem Solver with Calculators, by Terence G. Coburn, Shirley Hoogeboom, and Judy Goodnow (Sunnyvale, CA: Creative, 1989). Reviewed in: *Arithmetic Teacher* 39, no.1 (Sep. 1991): 50-51.

GrafEq
software (Terrace, British Columbia: Pedagoguery, 1990). Reviewed in: *Mathematics Teacher* 85, no.2 (Mar. 1992): 241-42.

Graphic Calculator and Computer Graphing Laboratory Manual (2nd ed.)
by Franklin Demana and Bert K. Waits (Menlo Park, CA: Addison-Wesley, 1992). Reviewed in: *Mathematics Teacher* 85, no.7 (Oct. 1992): 590-91.

Graphic Workshop 6.1f
software (Ontario, Canada: Alchemy Mindworks, n.d.). Reviewed in: *Mathematics and Computer Education* 26, no.1 (Winter 1992). 106-08.

Graphing
Bradford Graphmaker, software (Concord, MA: William K. Bradford, 1990). Reviewed in: *Mathematics Teacher* 84, no.5 (May 1991): 391-92; *Electronic Learning* 11, no.3 (Nov./Dec. 1991): 34-35.

EZ Graph, manipulatives (Englewood, CO: EZ Graph, 1992). Reviewed in: *Mathematics Teacher* 85, no.6 (Sep. 1992): 497.

Graphic Calculator and Computer Graphing Laboratory Manual (2nd ed.), by Franklin Demana and Bert K. Waits (Menlo Park, CA: Addison-Wesley, 1992). Reviewed in: *Mathematics Teacher* 85, no.7 (Oct. 1992): 590-91.

Graphing *(cont'd)*
Graphic Workshop 6.1f, software (Ontario, Canada: Alchemy Mindworks, n.d.). Reviewed in: *Mathematics and Computer Education* 26, no.1 (Winter 1992): 106-08.

Graphing and Probability Workshop, Grades 3-8, software (Glenview, IL: Scott Foresman, 1991). Reviewed in: *Arithmetic Teacher* 39, no.3 (Nov. 1991): 56.

Graphing Primer, Grades K-2, by Laura Duncan Choate and JoAnn King Okey (Palo Alto, CA: Seymour Publications, 1989). Reviewed in: *Arithmetic Teacher* 40, no.2 (Oct. 1992): 129.

Graph Wiz 1.0 or 1.2, software, by Alan Hoffer and Tom Lippincott (Concord, MA: William K. Bradford, 1989). Reviewed in: *Mathematics Teacher* 84, no.6 (Sep. 1991): 485-86.

Math-O-Graphs: Critical Thinking through Graphing, by Donna Kay Buck and Francis Hildebrand (Pacific Grove, CA: Midwest, 1990). Reviewed in: *Arithmetic Teacher* 38, no.9 (May 1991): 47.

Pictograms: Graphing Pictures for a Reusable Classroom Grid, manipulatives by Laura Duncan Choate and JoAnn King Okey (Palo Alto, CA: Dale Seymour, 1989). Reviewed in: *Arithmetic Teacher* 40, no.2 (Oct. 1992): 130.

Precalculus Mathematics: A Graphing Approach (2d ed.), by Franklin Demana, Bert K. Waits, and Stanley E. Clemens (Menlo Park, CA: Addison-Wesley, 1992). Reviewed in: *Mathematics Teacher* 85, no.7 (Oct. 1992): 592.

TI-81 Graphing Calculator Activities for Finite Mathematics, by Wayne L. Miller, Donald Perry, and Gloria A. Tveten (Pacific Grove, CA: Brooks/Cole, 1992). Reviewed in: *Mathematics Teacher* 85, no.8 (Nov. 1992): 693.

Graphing and Probability Workshop, Grades 3-8
software (Glenview, IL: Scott Foresman, 1991). Reviewed in: *Arithmetic Teacher* 39, no.3 (Nov. 1991): 56.

Graphing Primer, Grades K-2
by Laura Duncan Choate and JoAnn King Okey (Palo Alto, CA: Seymour Publications, 1989). Reviewed in: *Arithmetic Teacher* 40, no.2 (Oct. 1992): 129.

Graph Wiz 1.0 or 1.2
software, by Alan Hoffer and Tom Lippincott (Concord, MA: William K. Bradford, 1989). Reviewed in: *Mathematics Teacher* 84, no.6 (Sep. 1991): 485-86.

Great Wave
Number Maze Decimals and Fractions, software (n.p.: Great Wave, n.d.). Reviewed in: *Media and Methods* 27, no.5 (May/June 1991): 60.

Greenes, Carole
Math Extra, Early Bird ed., by Carole Greenes, Linda Schulman, and Rika Spungin (Allen, TX: DLM, 1991). Reviewed in: *Arithmetic Teacher* 39, no.7 (Mar. 1992): 43-44.

Grossman, Virginia
Ten Little Rabbits, by Virginia Grossman and illustrated by Sylvia Long (San Francisco: Chronicle Books, 1991). Reviewed in: *Arithmetic Teacher* 40, no.1 (Sep. 1992): 56-57.

Grossnickle, Foster E.
Discovering Meanings in Elementary School Mathematics, by Foster E. Grossnickle (Orlando, FL: Holt, Rinehart and Winston, 1990). Reviewed in: *Arithmetic Teacher* 40, no.1 (Sep. 1992): 58.

Gulick, Deny
Precalculus (4th ed.), by Robert Ellis and Deny Gulick (Orlando, FL: Saunders College, 1992). Reviewed in: *Mathematics Teacher* 85, no.9 (Dec. 1992): 767.

GyroGraphics
software, by Nancy Anne Johnson (Stillwater, OK: Cipher Systems, 1990). Reviewed in: *Mathematics Teacher* 85, no.3 (Mar. 1992): 242.

GyroGraphics, version 2.2
software (Stillwater, OK: Cipher Systems, 1990). Reviewed in: *Mathematics Teacher* 84, no.3 (Mar. 1991): 235.

Hackworth, Margaret
NCSMP Algebra, by John W. McConnell, Susan Brown, Susan Eddins, Margaret Hackworth, and Zalman Usiskin (Tucker, GA: Scott Foresman, 1990). Reviewed in: *Mathematics Teacher* 84, no.2 (Feb. 1991): 144-46.

Hale, Norman F.
Practical Math for Everyone, by Norman F. Hale (New York: Thinker's Books, 1990). Reviewed in: *Mathematics Teacher* 85, no.7 (Oct. 1992): 591.

Hall, James W.
Beginning Algebra, by James W. Hall (Boston: PWS-KENT, 1992). Reviewed in: *Mathematics Teacher* 85, no.5 (May 1992): 393.

Intermediate Algebra, by James W. Hall (Boston, MA: PWS-KENT, 1992). Reviewed in: *Mathematics Teacher* 85, no.9 (Dec. 1992): 764-65.

Halmos, Paul R.
Problems for Mathematicians Young and Old (vol. 12), by Paul R. Halmos (Washington, DC: Mathematical Association of America, 1991). Reviewed in: *Mathematics Teacher* 85, no.7 (Oct. 1992): 592.

Hands On
Patterns and Functions, by Linda Sue Brisby (Solvang, CA: Hands On, 1990). Reviewed in: *Arithmetic Teacher* 40, no.1 (Sep. 1992): 64.

Harcourt Brace Jovanovich
Intermediate Algebra, by Ronald Hatton and Gene R. Sellers (San Diego: Harcourt Brace Jovanovich, 1991). Reviewed in: *Mathematics Teacher* 84, no.9 (Dec. 1991): 769.

Introductory Algebra, by Ronald Hatton and Gene R. Sellers (San Diego: Harcourt Brace Jovanovich, 1991). Reviewed in: *Mathematics Teacher* 85, no.1 (Jan. 1992): 74.

Prealgebra, by Joshua Berenbom and Umesh Nagarkatte (San Diego: Harcourt Brace Jovanovich, 1991). Reviewed in: *Arithmetic Teacher* 39, no.3 (Nov. 1991): 60.

Harnadek, Anita
Mind Benders—A1, software, by Anita Harnadek (Pacific Grove, CA: Midwest, 1988). Reviewed in: *Arithmetic Teacher* 39, no.1 (Sep. 1991): 47-48.

Hatton, Ronald
Intermediate Algebra, by Ronald Hatton and Gene R. Sellers (San Diego: Harcourt Brace Jovanovich, 1991). Reviewed in: *Mathematics Teacher* 84, no.9 (Dec. 1991): 769.

Introductory Algebra, by Ronald Hatton and Gene R. Sellers (San Diego: Harcourt Brace Jovanovich, 1991). Reviewed in: *Mathematics Teacher* 85, no.1 (Jan. 1992): 74.

Haylock, Derek
Teaching Mathematics to Low Attainers (8-12), by Derek Haylock (Bristol, PA: Paul Chapman, 1991). Reviewed in: *Mathematics Teacher* 85, no.8 (Nov. 1992): 693.

Heartland
Euclid's Toolbox—Triangles, software, by Jerry Beckmann and Charles Friesen (n.p.: Heartland, 1990). Reviewed in: *Mathematics Teacher* 84, no.6 (Sep. 1991): 484.

D. C. Heath
Basic Mathematics, by David Novak (Lexington, MA: D. C. Heath, 1991). Reviewed in: *Mathematics Teacher* 85, no.6 (Sep. 1992): 491.

Contemporary Abstract Algebra (2d ed.), by Joseph A. Gallian (Lexington, MA: D. C. Heath, 1990). Reviewed in: *Mathematics and Computer Education* 26, no.3 (Fall 1992): 334-35.

Ellen Hechler (publisher)
Flashcards, supplementary materials, by Ellen Hechler (Farmington Hills, MI: Ellen Hechler, 1991). Reviewed in: *Mathematics Teacher* 85, no.1 (Jan. 1992): 78.

Mental Math, supplementary materials, by Ellen Hechler (Farmington Hills, MI: Ellen Hechler, 1991). Reviewed in: *Mathematics Teacher* 85, no.1 (Jan. 1992): 78.

High school *(cont'd)*

Analytic Geometry, 2d ed., by Karl J. Smith (Pacific Grove, CA: Brooks/Cole, 1991). Reviewed in: *Mathematics Teacher* 85, no.2 (Feb. 1992): 146.

Analyzer, software, by Douglas Alfors and Beverly West (Ithaca, NY: Delta-Epsilon Software, 1990). Reviewed in: *Mathematics Teacher* 85, no.3 (Mar. 1992): 240.

Archimedean and Archimedean Dual Polyhydra, video, by Lorrainne L. Foster (Northridge, CA: California State University). Reviewed in: *Mathematics Teacher* 84, no.7 (Oct. 1991): 576.

Basic Technical Mathematics with Calculus, 2d ed., by Peter Kuhfittig (Pacific Grove, CA: Brooks/Cole, 1989). Reviewed in: *Mathematics Teacher* 84, no.2 (Feb. 1991): 140.

Beginning Algebra, by Alfonse Gobran (Belmont, CA: Wadsworth, 1991). Reviewed in: *Mathematics Teacher* 84, no.8 (Nov. 1991): 674.

Beginning Algebra, by James W. Hall (Boston: PWS-KENT, 1992). Reviewed in: *Mathematics Teacher* 85, no.5 (May 1992): 393.

Benjamin Banneker, poster (Burlington, NC: Cabisco Mathematics, 1991). Reviewed in: *Mathematics Teacher* 85, no.6 (Sep. 1992): 496.

Bradford Graphmaker, software (Concord, MA: William K. Bradford, 1990). Reviewed in: *Mathematics Teacher* 84, no.5 (May 1991): 391-92; *Electronic Learning* 11, no.3 (Nov./Dec. 1991): 34-35.

Calculators in the Classroom, K-12, by Katherine Pedersen and Jack Mummert (Carbondale, IL: ICTM, 1990). Reviewed in: *Arithmetic Teacher* 40, no.1 (Sep. 1992): 62.

Calculus, by Earl W. Swokowski (Florence, KY: PWS-KENT, 1991). Reviewed in: *Mathematics Teacher* 85, no.4 (Apr. 1992): 314.

Calculus, by Ross L. Finney and George B. Thomas, Jr. (Reading, MA: Addison-Wesley, 1990). Reviewed in: *Mathematics Teacher* 84, no.1 (Jan. 1991): 66.

Calculus, 2d ed., vol. 1, by Tom M. Apostol (Somerset, NJ: John Wiley, 1988). Reviewed in: *Mathematics Teacher* 84, no.3 (Mar. 1991): 236.

Calculus, 3d ed. (Late Trigonometry Version), by Howard Anton (Somerset, NJ: John Wiley). Reviewed in: 24, no.2 (Spring 1992): 212-13.

Calculus: Applications of Differentiations, Parts 1-3, video (Brooklyn: Video Tutorial Service, 1990). Reviewed in: *Mathematics Teacher* 84, no.6 (Sep. 1991): 496-97.

Calculus Gems, by George F. Simmons (New York: McGraw-Hill, 1992). Reviewed in: *Mathematics and Computer Education* 26, no.3 (Fall 1992): 330.

Calculus: Late Trigonometry Version (5th ed.), by Earl W. Swokowski (Boston, MA: PWS-KENT, 1992). Reviewed in: *Mathematics Teacher* 85, no.8 (Nov. 1992): 690.

Calculus T/L: A Program for Doing, Teaching, and Learning Calculus, software, by J. Douglas Child (Pacific Grove, CA: Brooks/Cole, 1990). Reviewed in: *Mathematics Teacher* 84, no.3 (Mar. 1991): 234-35.

Chaos, Fractals, and Dynamics: Computer Experiments in Math, by Robert L. Devaney (Reading, MA: Addison-Wesley). Reviewed in: *Mathematics Teacher* 84, no.4 (Apr. 1991): 236.

Colorful Characters of Mathematics, poster (Burlington, NC: Cabisco Mathematics, 1992). Reviewed in: *Mathematics Teacher* 85, no.8 (Nov. 1992): 694.

Concepts in Mathematics: Conic Sections, software (Cary, NC: TV Ontario Video, 1991). Reviewed in: *Mathematics Teacher* 85, no.9 (Dec. 1992): 767-68.

High school *(cont'd)*

Concepts in Mathematics: Trigonometric Functions I--Solving Triangles, software (Cary, NC: TV Ontario Video, 1991). Reviewed in: *Mathematics Teacher* 85, no.9 (Dec. 1992): 768.

Contemporary Abstract Algebra (2d ed.), by Joseph A. Gallian (Lexington, MA: D. C. Heath, 1990). Reviewed in: *Mathematics and Computer Education* 26, no.3 (Fall 1992): 334-35.

Crossmatics: A Challenging Collection of Cross-Number Puzzles: Grades 7-12, game, by Allan Dudley (Palo Alto, CA: Dale Seymour, 1990). Reviewed in: *Mathematics Teacher* 84, no.5 (May 1991): 406.

Data Insights, software, by Lois Edwards and K. M. Keogh (Pleasantville, NY: Sunburst, 1989). Reviewed in: *Mathematics Teacher* 84, no.4 (Apr. 1991): 320.

Developmental Mathematics (3d ed.), by C. L. Johnston, Alden T. Willis, and Gale M. Hughes (Belmont, CA: Wadsworth, 1991). Reviewed in: *Mathematics Teacher* 85, no.6 (Sep. 1992): 491.

Elementary Algebra (2d ed.), by Jack Barker, James Rogers, and James Van Dyke (Orlando, FL: Saunders College, 1992). Reviewed in: *Mathematics Teacher* 85, no.9 (Dec. 1992): 764.

Elementary Algebra: Concepts and Applications, 3d ed., by Marvin L. Bittinger and Marvin L. Keedy (Reading, MA: Addison-Wesley, n.d.). Reviewed in: *Mathematics and Computer Education* 26, no.1 (Winter 1992): 95-96.

Essential Algebra (6th ed.), by C. L. Johnston, Alden T. Willis, and Jeanne Lazaris (Belmont, CA: Wadsworth, 1991). Reviewed in: *Mathematics Teacher* 85, no.6 (Sep. 1992): 491-92.

Euclid's Toolbox—Triangles, software, by Jerry Beckmann and Charles Friesen (n.p.: Heartland, 1990). Reviewed in: *Mathematics Teacher* 84, no.6 (Sep. 1991): 484.

Exploring Sequences and Series, software (Minneapolis: MECC, 1991). Reviewed in: *Mathematics Teacher* 85, no.6 (Sep. 1992): 489.

EZ Graph, manipulatives (Englewood, CO: EZ Graph, 1992). Reviewed in: *Mathematics Teacher* 85, no.6 (Sep. 1992): 497.

The Factor Pack, supplementary materials (Holywell, UK: Magic Mathworks, n.d.). Reviewed in: *Mathematics Teacher* 84, no.7 (Oct. 1991): 576.

Fiendishly Difficult, by Ivan Moscovich (New York: Sterling, 1991). Reviewed in: *Mathematics Teacher* 84, no.9 (Dec. 1991): 766.

Finite Mathematics with Calculus: An Applied Approach (2d ed.), by David E. Zitarelli and Raymond F. Coughlin (Orlando, FL: Saunders College, 1992). Reviewed in: *Mathematics Teacher* 85, no.9 (Dec. 1992): 764.

Flashcards, supplementary materials, by Ellen Hechler (Farmington Hills, MI: Ellen Hechler, 1991). Reviewed in: *Mathematics Teacher* 85, no.1 (Jan. 1992): 78.

Fractals: An Animated Discussion, video (New York: W. H. Freeman, 1990). Reviewed in: *Mathematics Teacher* 84, no.7 (Oct. 1991): 576-78.

GeoExplorer, software (Glenview, IL: Scott Foresman, n.d.). Reviewed in: *Electronic Learning* 12, no.1 (Sep. 1992): 23-24; *Mathematics Teacher* 85, no.7 (Oct. 1992): 587.

The Geometer's Sketchpad, software (Berkeley: Key Curriculum, 1991). Reviewed in: *Mathematics Teacher* 85, no.5 (May 1992): 392-93; *Electronic Learning* 11, no.3 (Nov./Dec. 1991): 34-35.

Geometric Connectors: Transformations, software (Pleasantville, NY: Sunburst, 1990). Reviewed in: *Mathematics Teacher* 85, no.9 (Dec. 1992): 761.

High school *(cont'd)*

Geometry, by Karl J. Smith (Pacific Grove, CA: Brooks/Cole, 1991). Reviewed in: *Mathematics Teacher* 85, no.1 (Jan. 1992): 73-74.

Geometry (Teacher's ed.), by Stanley R. Clemens, Phares G. O'Daffer, and Thomas J. Cooney (Menlo Park, CA: Addison-Wesley, 1992). Reviewed in: *Mathematics Teacher* 85, no.7 (Oct. 1992): 590.

Geometry 1.2, software (San Rafael, CA: Broderbund, n.d.). Reviewed in: *Journal of Geographical Education* 39, no.4 (Sep. 1991): n.p.

Geometry for Enjoyment and Challenges, by Richard Rhoad, George Milauskas, and Robert Whipple (Evanston, IL: McDougal Littell, 1991). Reviewed in: *Mathematics Teacher* 84, no.6 (Sep. 1991): 488-89.

Geometry Inventor, software (Scotts Valley, CA: Wings for Learning, n.d.). Reviewed in: *Electronic Learning* 12, no.1 (Sep. 1992): 23-24.

GrafEq, software (Terrace, British Columbia: Pedagoguery, 1990). Reviewed in: *Mathematics Teacher* 85, no.2 (Mar. 1992): 241-42.

Graphic Calculator and Computer Graphing Laboratory Manual (2nd ed.), by Franklin Demana and Bert K. Waits (Menlo Park, CA: Addison-Wesley, 1992). Reviewed in: *Mathematics Teacher* 85, no.7 (Oct. 1992): 590-91.

Graphic Workshop 6.1f, software (Ontario, Canada: Alchemy Mindworks, n.d.). Reviewed in: *Mathematics and Computer Education* 26, no.1 (Winter 1992): 106-08.

Graph Wiz 1.0 or 1.2, software, by Alan Hoffer and Tom Lippincott (Concord, MA: William K. Bradford, 1989). Reviewed in: *Mathematics Teacher* 84, no.6 (Sep. 1991): 485-86.

GyroGraphics, software, by Nancy Anne Johnson (Stillwater, OK: Cipher Systems, 1990). Reviewed in: *Mathematics Teacher* 85, no.3 (Mar. 1992): 242.

GyroGraphics, version 2.2, software (Stillwater, OK: Cipher Systems, 1990). Reviewed in: *Mathematics Teacher* 84, no.3 (Mar. 1991): 235.

IFS Explorer, software (St. Louis, MO: Koyn, 1990). Reviewed in: *Mathematics Teacher* 84, no.6 (Sep. 1991): 486.

Integrated Mathematics Course 2, by Brenden Kelly, Bob Alexander, and Paul Atkinson (St. Charles, IL: McDougal Littell, 1991). Reviewed in: *Mathematics Teacher* 84, no.9 (Dec. 1991): 768; *Mathematics Teacher* 85, no.2 (Feb. 1992): 148.

Interactive Algebra, software, by Henry Africk, Ely Stern, and Harvey Broverman (Hicksville, NY: Technical Educational Consultants, 1990). Reviewed in: *Mathematics Teacher* 84, no.3 (Mar. 1991): 235.

Intermediate Algebra, by Dale E. Boye, Ed Kavanaugh, and Larry G. Williams (Boston: PWS-KENT, 1991). Reviewed in: *Mathematics Teacher* 84, no.8 (Nov. 1991): 678.

Intermediate Algebra, by James W. Hall (Boston, MA: PWS-KENT, 1992). Reviewed in: *Mathematics Teacher* 85, no.9 (Dec. 1992): 764-65.

Intermediate Algebra, by Ronald Hatton and Gene R. Sellers (San Diego: Harcourt Brace Jovanovich, 1991). Reviewed in: *Mathematics Teacher* 84, no.9 (Dec. 1991): 769.

Intermediate Algebra (2d ed.), by Jack Barker, James Rogers, and James Van Dyke (Orlando, FL: Saunders College, 1992). Reviewed in: *Mathematics Teacher* 85, no.9 (Dec. 1992): 765-66.

Intermediate Algebra: Concepts and Applications, 3d ed., by Marvin L. Bittinger, Mervin L. Keedy, and David Ellenbogen (Reading, MA: Addison-Wesley, 1990). Reviewed in: *Mathematics Teacher* 84, no.3 (Mar. 1991): 238.

Introduction to Statistics with Data Analysis, by Shelley Rasmussen (Pacific Grove, CA: Brooks/Cole, 1992). Reviewed in: *Mathematics Teacher* 85, no.6 (Sep. 1992): 492.

High school *(cont'd)*

Introduction to Tessellations, by Dale Seymour and Jill Britten (Palo Alto, CA: Dale Seymour, 1989). Reviewed in: *Mathematics Teacher* 83, no.6 (Sep. 1992): 482.

Introductory Algebra, by Ronald Hatton and Gene R. Sellers (San Diego: Harcourt Brace Jovanovich, 1991). Reviewed in: *Mathematics Teacher* 85, no.1 (Jan. 1992): 74.

Introductory Alegbra: An Interactive Approach, by Linda Pulsinelli and Patricia Hooper (New York: Macmillan, 1991). Reviewed in: *Mathematics Teacher* 84, no.8 (Nov. 1991): 678.

Investigations in Algebra, by Albert Cuoco (Cambridge, MA: MIT Press, 1990). Reviewed in: *Mathematics Teacher* 85, no.1 (Jan. 1992): 74.

Journey into Geometries, by Martha Sved (Washington, DC: Mathematical Association of America, 1991). Reviewed in: *Mathematics Teacher* 85, no.8 (Nov. 1992): 691.

Kaleidoscope Math, by Joe Kennedy and Diana Thomas (Sunnyvale, CA: Creative Publications, 1989). Reviewed in: *Arithmetic Teacher* 38, no.6 (Feb. 1991): 63.

Logo Math: Tools and Games, software, by Henri Piccotti (Portland, ME: Terrapin, 1990). Reviewed in: *Mathematics Teacher* 84, no.6 (Sep. 1991): 486-87.

M! Project MATHEMATICS!—Similarity, video by Tom Apostol (Reston, VA: NCTM, 1990). Reviewed in: *Mathematics Teacher* 85, no.6 (Sep. 1992): 496.

Magic of Numbers, by Eric T. Bell (Mineola, NY: Dover, 1991). Reviewed in: *Mathematics Teacher* 85, no.6 (Sep. 1992): 493.

Making the Most of Twenty Minutes: Warm Up Math Activities, by Michael Cain (Victoria, Australia: Dellasta, 1989). Reviewed in: *Arithmetic Teacher* 39, no.1 (Sep. 1991): 50.

Maple V: Student Edition, software (Pacific Grove, CA: Symbolic Computation Group, n.d.). Reviewed in: *Mathematics and Computer Education* 26, no.3 (Fall 1992): 337-39.

Mathcad 3.0, software (Cambridge, MA: Mathsoft, 1991). Reviewed in: *Electronic Learning* 11, no.3 (Nov./Dec. 1991): 35.

Math Connections: Algebra 1, software (Scotts Valley, CA: Wings for Learning). Reviewed in: *Technology & Learning* 12, no.6 (Mar. 1992): 10-11.

Math Connections: Algebra I, software by Jon Rosenberg (Scotts Valley, CA: Wings for Learning, 1991). Reviewed in: *Mathematics Teacher* 85, no.6 (Sep. 1992): 489.

Math Connections: Algebra II, software (Scotts Valley, CA: Wings for Learning, 1992). Reviewed in: *Technology & Learning* 13, no.3 (Nov./Dec. 1992): 18.

Mathematica: A System for Doing Mathematics by Computer (2d ed.), software by Stephen Wolfram (Champaign, IL: Wolfram Research, 1991). Reviewed in: *Mathematics Teacher* 85, no.8 (Nov. 1992): 688.

Mathematical Topics for Computer Instructions: Grades 9-12, by Harris S. Shultz and William A. Leonard (Palo Alto, CA: Dale Seymour, 1990). Reviewed in: *Mathematics Teacher* 38, no.4 (Apr. 1991): 326-28.

Mathematics Achievement Project, software (Washington, DC: CAE Software, 1992). Reviewed in: *Mathematics Teacher* 85, no.7 (Oct. 1992): 587-88.

Mathematics Exploration Toolkit, software (Atlanta, GA: IBM Education). Reviewed in *Mathematics and Computer Education*: 25, no.3 (Fall 1991): 86.

Mathematics: Its Power and Utility, 3d ed., by Karl J. Smith (Belmont, CA: Wadsworth, 1990). Reviewed in: *Mathematics and Computer Education* 26, no.1 (Winter 1992): 92.

Math Meets Technology, by Brian Bolt (Port Chester, NY: Cambridge University Press, 1991). Reviewed in: *Mathematics Teacher* 85, no.8 (Nov. 1992): 692.

High school *(cont'd)*

Math Stories for Problem Solving Success: Ready-To-Use Activities for Grades 7-12, by Jim Overholt, Nancy Aaberg, and Jim Lindsey (Old Tappen, NJ: Prentice-Hall, 1990). Reviewed in: *Mathematics Teacher* 84, no.7 (Oct. 1991): 578.

Math Writer, software (Pacific Grove, CA: Brooks/Cole). Reviewed in: *Mathematics and Computer Education* 26, no.1 (Winter, 1992): 109-10.

Mind Benders—A1, software, by Anita Harnadek (Pacific Grove, CA: Midwest, 1988). Reviewed in: *Arithmetic Teacher* 39, no.1 (Sep. 1991): 47-48.

Minitab (State College, PA: Minitab Inc.). Reviewed in: *Mathematics and Computer Education* 25, no.3 (Fall 1991): 334-35.

More Mathematical Morsels, by Ross Honsberger (Washington, DC: Mathematical Association). Reviewed in: *Mathematics Teacher* 84, no.6 (Sep. 1991): 490; *Mathematics and Computer Education* 25, no.3 (Fall 1991): 316.

NCSMP Algebra, by John W. McConnell, Susan Brown, Susan Eddins, Margaret Hackworth, and Zalman Usiskin (Tucker, GA: Scott Foresman, 1990). Reviewed in: *Mathematics Teacher* 84, no.2 (Feb. 1991): 144-46.

Nonroutine Problems: Doing Mathematics, by Robert London (Providence, RI: Janson, 1989). Reviewed in: *Mathematics Teacher* 38, no.4 (Apr. 1991): 328.

Operation Neptune, software (Fremont, CA: Learning Company, 1992). Reviewed in: *Technology & Learning* 13, no.3 (Nov./Dec. 1992): 18.

The Penguin Dictionary of Curious and Interesting Numbers, by David Wells (New York: Penguin, 1987). Reviewed in: *Mathematics and Computer Education* 24, no.1 (Winter 1991): 92.

PKZIP, software (Glendale, WI: PKWARE). Reviewed in: *Mathematics and Computer Education* 25, no.3 (Fall 1991): 332-33.

Polyominoes: A Guide to Puzzles and Problems in Tiling, by George E. Martin (Washington, DC: Mathematical Association of America, 1991). Reviewed in: *Mathematics Teacher* 85, no.6 (Sep. 1992): 493-94.

Practical Math for Everyone, by Norman F. Hale (New York: Thinker's Books, 1990). Reviewed in: *Mathematics Teacher* 85, no.7 (Oct. 1992): 591.

Pre-Algebra: A Transition to Algebra (Teacher's ed.), by Stanley R. Clemens, Phares G. O'Daffer, and Randall I. Charles (Menlo Park, CA: Addison-Wesley, 1992). Reviewed in: *Mathematics Teacher* 85, no.7 (Oct. 1992): 592.

Prealgebra for Problem Solvers: A New Beginning, by Loyd V. Wilcox (Pacific Grove, CA: Brooks/Cole, 1991). Reviewed in: *Mathematics Teacher* 84, no.6 (Sep. 1991): 492.

Precalculus, by Jerome E. Kaufmann (Florence, KY: PWS-KENT, 1991). Reviewed in: *Mathematics Teacher* 84, no.9 (Dec. 1991): 770.

Pre-Calculus, video (Brooklyn: Video Tutorial, 1990). Reviewed in: *Mathematics Teacher* 84. no.6 (Sep. 1991): 498.

Precalculus (4th ed.), by Robert Ellis and Deny Gulick (Orlando, FL: Saunders College, 1992). Reviewed in: *Mathematics Teacher* 85, no.9 (Dec. 1992): 767.

Precalculus Algebra and Trigonometry, by Sharon Cutler Ross and Linda Hawkins Boyd (Pacific Grove, CA: Brooks/Cole, 1991). Reviewed in: *Mathematics Teacher* 85, no.1 (Jan. 1992): 75.

Precalculus: A Problems-Oriented Approach, by David Cohen (St. Paul: West, 1990). Reviewed in: *Mathematics Teacher* 87 no. 7 (Oct. 1991): 574.

Precalculus Functions and Graphs, by Jimmie Gilbert and Linda Gilbert (Reading, MA: Addison-Wesley, 1990). Reviewed in: *Mathematics Teacher* 84, no.1 (Jan. 1991): 67.

High school *(cont'd)*
Precalculus Functions and Graphs, 5th ed., by M. A. Munem and J. P. Yizze (New York: Worth, 1990). Reviewed in: *Mathematics Teacher* 84, no.1 (Jan. 1991): 68.

Precalculus Functions and Graphs, 6th ed., by Earl W. Swokowski (Boston: Kent, 1990). Reviewed in: *Mathematics Teacher* 84, no.1 (Jan. 1991): 68-70.

Precalculus Mathematics: A Graphing Approach (2d ed.), by Franklin Demana, Bert K. Waits, and Stanley E. Clemens (Menlo Park, CA: Addison-Wesley, 1992). Reviewed in: *Mathematics Teacher* 85, no.7 (Oct. 1992): 592.

A Primer in Probability, by Kathleen Subrahmaniam (New York: Marcel Dekker, 1990). Reviewed in: *Mathematics Teacher* 84, no.7 (Oct. 1991): 574.

Probability Lab No. A-262, software (St. Paul: MECC, 1990). Reviewed in: *Mathematics Teacher* 84, no.6 (Sep. 1991): 487; *The Computing Teacher* 18, no.8 (May 1991): 46-47.

Problems for Mathematicians Young and Old (vol. 12), by Paul R. Halmos (Washington, DC: Mathematical Association of America, 1991). Reviewed in: *Mathematics Teacher* 85, no.7 (Oct. 1992): 592.

Problem Solving (S,C,R), by Karl J. Smith (Pacific Grove, CA: Brooks/Cole, 1991). Reviewed in: *Mathematics Teacher* 85, no.1 (Jan. 1992): 75.

Problem Solving and Comprehension (5th ed.), by Arthur Whimbey and Jack Lochhead (Hillsdale, NJ: Lawrence Erlbaum Associates, 1991). Reviewed in: *Mathematics Teacher* 85, no.7 (Oct. 1992): 592.

Problem Solving Series, by Derek Holton (Leicester, UK: Mathematical Association, 1988). Reviewed in: *Mathematics Teacher* 84, no.6 (Sep. 1991): 492-93.

Shapes, Space and Symmetry, by Alan Holden (Mineola, NY: Dover, 1991). Reviewed in: *Mathematics Teacher* 85, no.6 (Sep. 1992): 494.

Simulated Real-Life Experiences Using Classified Ads in the Classroom, by Ellen Hechler (Farmington Hills, MI: Ellen Hechler, 1991). Reviewed in: *Arithmetic Teacher* 40, no.1 (Sep. 1992): 64.

Single Variable Calculus, 2d ed., by James Stewart (Pacific Grove, CA: Brooks/Cole, 1991). Reviewed in: *Mathematics Teacher* 85, no.1 (Jan. 1992): 76.

Solve It: Fun Folders for Fast Finishers, by Michael Richards and Wendy Lohse (Victoria, Australia: Dellasta, 1990). Reviewed in: *Arithmetic Teacher* 38, no.7 (Mar. 1991): 58.

Spy: A Teaching Resource of Fascinating Math Investigations, by Kevin Lees (Victoria, Australia: Dellasta, 1989). Reviewed in: *Arithmetic Teacher* 38, no.8 (Apr. 1991): 58.

Statistics and Probability in Modern Life (5th ed.) (Orlando, FL: Saunders College, 1992). Reviewed in: *Mathematics Teacher* 85, no.8 (Nov. 1992): 692-93.

Symmystries with Cuisenaire Rods, Sets, card set, by Patricia S. Davidson and Robert E. Willcutt (New Rochelle, NY: Cuisenaire, 1990). Reviewed in: *Arithmetic Teacher* 39, no.1 (Sep. 1991): 53.

Teaching Mathematics to Low Attainers (8-12), by Derek Haylock (Bristol, PA: Paul Chapman, 1991). Reviewed in: *Mathematics Teacher* 85, no.8 (Nov. 1992): 693.

3D Images, software (Acton, MA: Bradford School, 1991). Reviewed in: *Mathematics Teacher* 85, no.6 (May 1992): 391; *Electronic Learning* 11, no.3 (Nov./Dec. 1991): 34-35.

TI-81 Graphing Calculator Activities for Finite Mathematics, by Wayne L. Miller, Donald Perry, and Gloria A. Tveten (Pacific Grove, CA: Brooks/Cole, 1992). Reviewed in: *Mathematics Teacher* 85, no.8 (Nov. 1992): 693.

Trigonometry, by Charles N. Gantner and Thomas E. Gantner (Belmont, CA: Wadsworth, 1990). Reviewed in: *Mathematics Teacher* 84, no.1 (Jan. 1991): 70.

High school *(cont'd)*
Trigonometry, by Thomas W. Hungerford and Richard Mercer (Orlando, FL: Saunders College, 1992). Reviewed in: *Mathematics Teacher* 85, no.8 (Nov. 1992): 693-94.

TrigPak: Software & Tutorials for Trigonometry, version 2.0, software, by John R. Monbrag (Pacific Grove, CA: Brooks/Cole, 1989). Reviewed in: *Mathematics Teacher* 38, no.5 (May 1991): 406.

The Unexpected Hanging and Other Mathematical Diversions, by Martin Gardner (Chicago, IL: University of Chicago Press, 1991). Reviewed in: *Mathematics Teacher* 85, no.6 (Sep. 1992): 494-95.

What Do You Do with a Broken Calculator?, software (Pleasantville, NY: Sunburst, 1989). Reviewed in: *Arithmetic Teacher* 39, no.3 (Nov. 1991): 58.

What To Solve? Problems and Suggestions for Young Mathematicians, by Judita Lofman (Cary, NC: Oxford University Press). Reviewed in: *Mathematics Teacher* 84, no.6 (Sep. 1991): 496.

What's My Angle, software (Torrance, CA: Davidson, 1991). Reviewed in: *Media & Methods* 28, no.3 (Jan./Feb. 1992): 60.

The Wonderful Problems of Fizz and Martina, video (Cambridge, MA: Tom Snyder). Reviewed in: *Media and Methods* 28, no.2 (Nov./Dec. 1991): 55-56.

World's Most Baffling Puzzles, by Charles Barry Townsend (New York: Sterling, 1991). Reviewed in: *Mathematics Teacher* 85, no.9 (Dec. 1992): 767.

High School (Remedial)
Basic Mathematics, by David Novak (Lexington, MA: D. C. Heath, 1991). Reviewed in: *Mathematics Teacher* 85, no.6 (Sep. 1992): 491.

Basic Mathematics (3d ed.), by Charles P. McKeague (Florence, KY: Brooks/Cole, 1992). Reviewed in: *Mathematics Teacher* 85, no.6 (Sep. 1992): 490-91.

Math Facts: Survival Guide to Basic Mathematics, by Theodore J. Szymanski and Anne Scanlan-Rohrer (Belmont, CA: Wadsworth, 1992). Reviewed in: *Mathematics Teacher* 85, no.6 (Sep. 1992): 493.

Prealgebra, by Joshua Berenbom and Umesh Nagarkatte (San Diego: Harcourt Brace Jovanovich, 1991). Reviewed in: *Arithmetic Teacher* 39, no.3 (Nov. 1991): 60.

Hildebrand, Francis
Math-O-Graphs: Critical Thinking through Graphing, by Donna Kay Buck and Francis Hildebrand (Pacific Grove, CA: Midwest, 1990). Reviewed in: *Arithmetic Teacher* 38, no.9 (May 1991): 47.

History of Mathematics
Benjamin Banneker, poster (Burlington, NC: Cabisco Mathematics, 1991). Reviewed in: *Mathematics Teacher* 85, no.6 (Sep. 1992): 496.

Calculus Gems, by George F. Simmons (New York: McGraw-Hill, 1992). Reviewed in: *Mathematics and Computer Education* 26, no.3 (Fall 1992): 330.

Colorful Characters of Mathematics, poster (Burlington, NC: Cabisco Mathematics, 1992). Reviewed in: *Mathematics Teacher* 85, no.8 (Nov. 1992): 694.

Magic of Numbers, by Eric T. Bell (Mineola, NY: Dover, 1991). Reviewed in: *Mathematics Teacher* 85, no.6 (Sep. 1992): 493.

Hoffer, Alan
Graph Wiz 1.0 or 1.2, software, by Alan Hoffer and Tom Lippincott (Concord, MA: William K. Bradford, 1989). Reviewed in: *Mathematics Teacher* 84, no.6 (Sep. 1991): 485-86.

Holden, Alan
Shapes, Space and Symmetry, by Alan Holden (Mineola, NY: Dover, 1991). Reviewed in: *Mathematics Teacher* 85, no.6 (Sep. 1992): 494.

Holstege, Martin
Algebra: A First Course, 3d ed., by John Baley and Martin Holstege (Belmont, CA: Wadsworth, 1990). Reviewed in: *Mathematics Teacher* 84, no.7 (Oct. 1991): 571.

Holt, Rinehart and Winston
Discovering Meanings in Elementary School Mathematics, by Foster E. Grossnickle (Orlando, FL: Holt, Rinehart and Winston, 1990). Reviewed in: *Arithmetic Teacher* 40, no.1 (Sep. 1992): 58.

Holton, Derek
Problem Solving Series, by Derek Holton (Leicester, UK: Mathematical Association, 1988). Reviewed in: *Mathematics Teacher* 84, no.6 (Sep. 1991): 492-93.

Honsberger, Ross
More Mathematical Morsels, by Ross Honsberger (Washington, DC: Mathematical Association). Reviewed in: *Mathematics Teacher* 84, no.6 (Sep. 1991): 490; *Mathematics and Computer Education* 25, no.3 (Fall 1991): 316.

Hoogeboom, Shirley
Overhead Math: Manipulative Lessons on the Overhead Projector, manipulatives, by Shirley Hoogeboom and Judy Goodnow (Sunnyvale, CA: Creative, 1990). Reviewed in: *Arithmetic Teacher* 39, no.8 (Apr. 1992): 60.

The Problem Solver with Calculators, by Terence G. Coburn, Shirley Hoogeboom, and Judy Goodnon (Sunnyvale, CA: Creative, 1989). Reviewed in: *Arithmetic Teacher* 39, no.1 (Sep. 1991): 50-51.

Hooper, Patricia
Introductory Alegbra: An Interactive Approach, by Linda Pulsinelli and Patricia Hooper (New York: Macmillan, 1991). Reviewed in: *Mathematics Teacher* 84, no.8 (Nov. 1991): 678.

Hughes, Gale M.
Developmental Mathematics (3d ed.), by C. L. Johnston, Alden T. Willis, and Gale M. Hughes (Belmont, CA: Wadsworth, 1991). Reviewed in: *Mathematics Teacher* 85, no.6 (Sep. 1992): 491.

Huline, Dale
Fastmath, software, by Jim Baker and Dale Huline (Minneapolis: New Directions, 1990). Reviewed in: *Arithmetic Teacher* 39, no.1 (Sep. 1991): 47.

Hulme, Joy N.
Sea Squares, by Joy N. Hulme and illustrated by Carol Schwartz (Waltham, MA: Hyperion Books, 1991). Reviewed in: *Arithmetic Teacher* 40, no.1 (Sep. 1992): 56.

Human Relations (publisher)
Detective Stories for Math Problem Solving, software (Pleasantville, NY: Human Relations). Reviewed in: *Electronic Learning* 11, no.5 (Feb. 1992): 35.

Mr. Marfil's Last Will and Testament, video (Pleasantville, NY: Human Relations, 1991). Reviewed in: *Mathematics Teacher* 84, no.7 (Oct. 1991): 578.

Hungerford, Thomas W.
Algebra and Trigonometry (2d ed.), by Thomas W. Hungerford and Richard Mercer (Orlando, FL: Saunders College, 1991). Reviewed in: *Mathematics Teacher* 85, no.9 (Dec. 1992): 763.

Trigonometry, by Thomas W. Hungerford and Richard Mercer (Orlando, FL: Saunders College, 1992). Reviewed in: *Mathematics Teacher* 85, no.8 (Nov. 1992): 693-94.

Hyperion Books
Sea Squares, by Joy N. Hulme and illustrated by Carol Schwartz (Waltham, MA: Hyperion Books, 1991). Reviewed in: *Arithmetic Teacher* 40, no.1 (Sep. 1992): 56.

IBM Education
Mathematics Exploration Toolkit, software (Atlanta, GA: IBM Education). Reviewed in *Mathematics and Computer Education*: 25, no.3 (Fall 1991): 86.

ICTM
Calculators in the Classroom, K-12, by Katherine Pedersen and Jack Mummert (Carbondale, IL: ICTM, 1990). Reviewed in: *Arithmetic Teacher* 40, no.1 (Sep. 1992): 62.

ICTM *(cont'd)*
Data Collection and Analysis Activities (K-8), by Katherine Pedersen and Judith Olson (Carbondale, IL: ICTM, 1990). Reviewed in: *Arithmetic Teacher* 40, no.1 (Sep. 1992): 62.

IFS Explorer
software (St. Louis, MO: Koyn, 1990). Reviewed in: *Mathematics Teacher* 84, no.6 (Sep. 1991): 486.

in COMMON Arithmetic
software, by Carol Chomsky, Harry Chomsky, and Judah L. Schwartz (Pleasantville, NY: Sunburst, 1990). Reviewed in: *Arithmetic Teacher* 39, no.1 (Sep. 1991): 47.

Integrated mathematics
Integrated Mathematics Course 1, by Brendan Kelly, Paul Atkinson, and Bob Alexander (St. Charles, IL: McDougal Littell, 1991). Reviewed in: *Mathematics Teacher* 84, no.9 (Dec. 1991): 768.

Integrated Mathematics Course 2, by Brenden Kelly, Bob Alexander, and Paul Atkinson (St. Charles, IL: McDougal Littell, 1991). Reviewed in: *Mathematics Teacher* 84, no.9 (Dec. 1991): 768; *Mathematics Teacher* 85, no.2 (Feb. 1992): 148.

Integrated Mathematics Course 1
by Brendan Kelly, Paul Atkinson, and Bob Alexander (St. Charles, IL: McDougal Littell, 1991). Reviewed in: *Mathematics Teacher* 84, no.9 (Dec. 1991): 768.

Integrated Mathematics Course 2
by Brenden Kelly, Bob Alexander, and Paul Atkinson (St. Charles, IL: McDougal Littell, 1991). Reviewed in: *Mathematics Teacher* 84, no.9 (Dec. 1991): 768; *Mathematics Teacher* 85, no.2 (Feb. 1992): 148.

Interactive Algebra
software, by Henry Africk, Ely Stern, and Harvey Broverman (Hicksville, NY: Technical Educational *Consultants, 1990). Reviewed in: Mathematics Teacher* 84, no.3 (Mar. 1991): 235.

Intermediate Algebra
by Ronald Hatton and Gene R. Sellers (San Diego: Harcourt Brace Jovanovich, 1991). Reviewed in: *Mathematics Teacher* 84, no.9 (Dec. 1991): 769.

Intermediate Algebra
by Dale E. Boye, Ed Kavanaugh, and Larry G. Williams (Boston: PWS-KENT, 1991). Reviewed in: *Mathematics Teacher* 84, no.8 (Nov. 1991): 678.

Intermediate Algebra
by James W. Hall (Boston, MA: PWS-KENT, 1992). Reviewed in: *Mathematics Teacher* 85, no.9 (Dec. 1992): 764-65.

Intermediate Algebra (2d ed.)
by Jack Barker, James Rogers, and James Van Dyke (Orlando, FL: Saunders College, 1992). Reviewed in: *Mathematics Teacher* 85, no.9 (Dec. 1992): 765-66.

Intermediate Algebra: Concepts and Applications, 3d ed.
by Marvin L. Bittinger, Mervin L. Keedy, and David Ellenbogen (Reading, MA: Addison-Wesley, 1990). Reviewed in: *Mathematics Teacher* 84, no.3 (Mar. 1991): 238.

Introduction to Statistics with Data Analysis
by Shelley Rasmussen (Pacific Grove, CA: Brooks/Cole, 1992). Reviewed in: *Mathematics Teacher* 85, no.6 (Sep. 1992): 492.

Introduction to Tessellations
by Dale Seymour and Jill Britten (Palo Alto, CA: Dale Seymour, 1989). Reviewed in: *Mathematics Teacher* 83, no.6 (Sep. 1992): 482.

Introductory Algebra
by Ronald Hatton and Gene R. Sellers (San Diego: Harcourt Brace Jovanovich, 1991). Reviewed in: *Mathematics Teacher* 85, no.1 (Jan. 1992): 74.

Introductory Alegbra: An Interactive Approach
by Linda Pulsinelli and Patricia Hooper (New York: Macmillan, 1991). Reviewed in: *Mathematics Teacher* 84, no.8 (Nov. 1991): 678.

Investigations in Algebra
by Albert Cuoco (Cambridge, MA: MIT Press, 1990). Reviewed in: *Mathematics Teacher* 85, no.1 (Jan. 1992): 74.

Keedy, Marvin L.
Elementary Algebra: Concepts and Applications, 3d ed., by Marvin L. Bittinger and Marvin L. Keedy (Reading, MA: Addison-Wesley, n.d.). Reviewed in: *Mathematics and Computer Education* 26, no.1 (Winter 1992): 95-96.

Keedy, Marvin L.
Concepts and Applications of Intermediate Algebra, 2d ed., by Marvin L. Bittinger, Marvin L. Keedy, and David Ellenbogen (Reading, MA: Addison-Wesley, 1991). Reviewed in: *Mathematics and Computer Education* 26, no.2 (Spring 1992): 205-07.

Intermediate Algebra: Concepts and Applications, 3d ed., by Marvin L. Bittinger, Marvin L. Keedy, and David Ellenbogen (Reading, MA: Addison-Wesley, 1990). Reviewed in: *Mathematics Teacher* 84, no.3 (Mar. 1991): 238.

Keep Your Balance
software (Pleasantville, NY: Sunburst, 1989). Reviewed in: *Arithmetic Teacher* 38, no.7 (Mar. 1991): 56.

Kelly, Brendan
Integrated Mathematics Course 1, by Brendan Kelly, Paul Atkinson, and Bob Alexander (St. Charles, IL: McDougal Littell, 1991). Reviewed in: *Mathematics Teacher* 84, no.9 (Dec. 1991): 768.

Integrated Mathematics Course 2, by Brendan Kelly, Bob Alexander, and Paul Atkinson (St. Charles, IL: McDougal Littell, 1991). Reviewed in: *Mathematics Teacher* 84, no.9 (Dec. 1991): 768; *Mathematics Teacher* 85, no.2 (Feb. 1992): 148.

Kendall/Hunt
Building Self-Confidence in Math: A Student Workbook (2d ed.), by Sally Wilding and Elizabeth Shearn (Dubuque, IA: Kendall/Hunt, 1991). Reviewed in: *Mathematics Teacher* 85, no.6 (Sep. 1992): 496.

Kennedy, Joe
Kaleidoscope Math, by Joe Kennedy and Diana Thomas (Sunnyvale, CA: Creative Publications, 1989). Reviewed in: *Arithmetic Teacher* 38, no.6 (Feb. 1991): 63.

Kent
Precalculus Functions and Graphs, 6th ed., by Earl W. Swokowski (Boston: Kent, 1990). Reviewed in: *Mathematics Teacher* 84, no.1 (Jan. 1991): 68-70.

Keogh, K. M.
Data Insights, software, by Lois Edwards and K. M. Keogh (Pleasantville, NY: Sunburst, 1989). Reviewed in: *Mathematics Teacher* 84, no.4 (Apr. 1991): 320.

Kevin, Judy
The Circular Geoboard, by Judy Kevin (San Leandro, CA: Teaching Resource Center, 1990). Reviewed in: *Mathematics Teacher* 84, no.8 (Nov. 1991): 680.

Key Curriculum (publisher)
The Geometer's Sketchpad, software (Berkeley: Key Curriculum, 1991). Reviewed in: *Mathematics Teacher* 85, no.5 (May 1992): 392-93; *Electronic Learning* 11, no.3 (Nov./Dec. 1991): 34-35.

The Stella Octangula Activity Set, video (Berkeley: Key Curriculum, 1991). Reviewed in: *Mathematics Teacher* 85, no.2 (Feb. 1992): 156.

Kinder Magic
Mosaic Magic, software (Encinitas, CA: Kinder Magic, 1990). Reviewed in: *Arithmetic Teacher* 39, no.1 (Sep. 1991): 48.

Koetke's Challenge
software by Tim Barclay and Jonathan Choate (Acton, MA: William K. Bradford, 1991). Reviewed in: *Mathematics Teacher* 85, no.9 (Dec. 1992): 761.

Koyn
IFS Explorer, software (St. Louis, MO: Koyn, 1990). Reviewed in: *Mathematics Teacher* 84, no.6 (Sep. 1991): 486.

Kuhfittig, Peter
Basic Technical Mathematics with Calculus, 2d ed., by Peter Kuhfittig (Pacific Grove, CA: Brooks/Cole, 1989). Reviewed in: *Mathematics Teacher* 84, no.2 (Feb. 1991): 140.

Leonard, William A.
Mathematical Topics for Computer Instructions: Grades 9-12, by Harris S. Shultz and William A. Leonard (Palo Alto, CA: Dale Seymour, 1990). Reviewed in: *Mathematics Teacher* 38, no.4 (Apr. 1991): 326-28.

Lindsey, Jim
Math Stories for Problem Solving Success: Ready-To-Use Activities for Grades 7-12, by Jim Overholt, Nancy Aaberg, and Jim Lindsey (Old Tappen, NJ: Prentice-Hall, 1990). Reviewed in: *Mathematics Teacher* 84, no.7 (Oct. 1991): 578.

Link 'n' Learn Activity Book
by Carol A. Thornton and Judith K. Wells (Deerfield, IL: Learning Resources, 1990). Reviewed in: *Arithmetic Teacher* 40, no.2 (Oct. 1992): 130.

Lippincott, Tom
Graph Wiz 1.0 or 1.2, software, by Alan Hoffer and Tom Lippincott (Concord, MA: William K. Bradford, 1989). Reviewed in: *Mathematics Teacher* 84, no.6 (Sep. 1991): 485-86.

The Little Shoppers Kit
software (Cambridge, MA: Tom Snyder, 1989). Reviewed in: *Arithmetic Teacher* 39, no.1 (Sep. 1991): 47.

Lochhead, Jack
Problem Solving and Comprehension (5th ed.), by Arthur Whimbey and Jack Lochhead (Hillsdale, NJ: Lawrence Erlbaum Associates, 1991). Reviewed in: *Mathematics Teacher* 85, no.7 (Oct. 1992): 592.

Thinking through Math Word Problems, by Authur Whimbey, Jack Lochhead, and Paula Potter (Hillsdale, NJ: Laurence Erlbaum, 1990). Reviewed in: *Arithmetic Teacher* 39, no.2 (Oct. 1991): 52.

Lofman, Judita
What To Solve? Problems and Suggestions for Young Mathematicians, by Judita Lofman (Cary, NC: Oxford University Press). Reviewed in: *Mathematics Teacher* 84, no.6 (Sep. 1991): 496.

Logic
Learning To Reason: Some, All, or None, software (Kalamazoo, MI: MCE, 1988). Reviewed in: *Arithmetic Teacher* 39, no.1 (Sep. 1991): 47.

Mathematical Challenges for the Middle Grades: From The Mathematics Teacher Calendar Problems, by William D. Jamski (Reston, VA: NCTM, 1991). Reviewed in: *Mathematics Teacher* 85, no.8 (Nov. 1992): 688-89.

Merlin Book of Logic Puzzles, by Margaret C. Edmiston (New York: Sterling, 1991). Reviewed in: *Mathematics Teacher* 85, no.9 (Dec. 1992): 767.

Problems for Mathematicians Young and Old (vol. 12), by Paul R. Halmos (Washington, DC: Mathematical Association of America, 1991). Reviewed in: *Mathematics Teacher* 85, no.7 (Oct. 1992): 592.

World's Most Baffling Puzzles, by Charles Barry Townsend (New York: Sterling, 1991). Reviewed in: *Mathematics Teacher* 85, no.9 (Dec. 1992): 767.

Logo Geometry
software by Michael Battista and Douglas H. Clements (Morristown, NJ: Simon & Schuster, 1991). Reviewed in: *Arithmetic Teacher* 40, no.1 (Sep. 1992): 56.

Logo Math: Tools and Games
software, by Henri Piccotti (Portland, ME: Terrapin, 1990). Reviewed in: *Mathematics Teacher* 84, no.6 (Sep. 1991): 486-87.

Lohse, Wendy
Solve It: Fun Folders for Fast Finishers, by Michael Richards and Wendy Lohse (Victoria, Australia: Dellasta, 1990). Reviewed in: *Arithmetic Teacher* 38, no.7 (Mar. 1991): 58.

London, Robert
Nonroutine Problems: Doing Mathematics, by Robert London (Providence, RI: Janson, 1989). Reviewed in: *Mathematics Teacher* 38, no.4 (Apr. 1991): 328.

Making Mathematics 8
by Gary Flewelling, Joan Routledge, and John Clark (Agincourt, Ontario: Gage Educational, 1991). Reviewed in: *Mathematics Teacher* 85, no.8 (Nov. 1992): 692.

Making the Most of Twenty Minutes: Warm Up Math Activities
by Michael Cain (Victoria, Australia: Dellasta, 1989). Reviewed in: *Arithmetic Teacher* 39, no.1 (Sep. 1991): 50.

Maletsky, Evan
Fractals for the Classroom: Strategic Activities, vol. 1, by Heinz-Otto Peritgen, Hartmut Jurgens, Dietmar Saupe, Evan Maletsky, Terry Perciante, and Lee Yunker (Reston, VA: NCTM, 1991). Reviewed in: *Mathematics Teacher* 85, no.2 (Feb. 1992): 146; *Arithmetic Teacher* 39, no.7 (Mar. 1992): 40.

Manipulatives
The Algebra Lab: Middle School: Exploring Algebra Concepts with Manipulatives, by Henri Piccotti (Sunnyvale, CA: Creative, 1990). Reviewed in: *Mathematics Teacher* 38, no.4 (Apr. 1991): 332.

Amazing Arithmos, manipulatives (Forest Hills, NY: Marc Preven, 1990). Reviewed in: *Arithmetic Teacher* 40, no.4 (Dec. 1992): 242.

Exploring Fractions and Decimals with Manipulatives, manipulatives, by Don Balka (Peabody, MA: Didax, 1991). Reviewed in: *Arithmetic Teacher* 39, no.9 (May 1992): 53.

EZ Graph, manipulatives (Englewood, CO: EZ Graph, 1992). Reviewed in: *Mathematics Teacher* 85, no.6 (Sep. 1992): 497.

Facts Master, manipulatives by Jivan Patel (Carrollton, TX: Bruce Bakke, n.d.). Reviewed in: *Arithmetic Teacher* 40, no.2 (Oct. 1992): 128.

Math Mats: Hands-On Activities for Young Children, set 2, manipulatives, by Carole A. Thornton and Judith K. Wells (Allen, TX: Teaching Resources, 1990). Reviewed in: *Arithmetic Teacher* 37, no.7 (Mar. 1992): 44.

Overhead Math: Manipulative Lessons on the Overhead Projector, manipulatives, by Shirley Hoogeborm and Judy Goodnow (Sunnyvale, CA: Creative, 1990). Reviewed in: *Arithmetic Teacher* 39, no.8 (Apr. 1992): 60.

Pictograms: Graphing Pictures for a Reusable Classroom Grid, manipulatives by Laura Duncan Choate and JoAnn King Okey (Palo Alto, CA: Dale Seymour, 1989). Reviewed in: *Arithmetic Teacher* 40, no.2 (Oct. 1992): 130.

Place Value-Plus, manipulatives (Goshen, IN: New Vision, n.d.). Reviewed in: *Arithmetic Teacher* 40, no.2 (Oct. 1992): 130.

Maple V: Student Edition
software (Pacific Grove, CA: Symbolic Computation Group, n.d.). Reviewed in: *Mathematics and Computer Education* 26, no.3 (Fall 1992): 337-39.

Mardlesohn, Esther
Teaching Primary Math with Music, cassette and songbook, by Esther Mardlesohn (Palo Alto, CA: Dale Seymour, 1990). Reviewed in: *Arithmetic Teacher* 39, no.1 (Sep. 1991): 53.

Marell, Anne
We Love MATHS: 4 Imaginative Themes for Early Primary Students, by Anne Marell and Susan Stajnko (Mount Waterly, Australia: Dellasta, 1990). Reviewed in: *Arithmetic Teacher* 39, no.1 (Sep. 1991): 53-54.

Marsh, Carole
Math for Girls: The Book with the Number To Girls To Love and Excel in Math, by Carole Marsh (Historic Bath, NC: Gallopade, 1989). Reviewed in: *Arithmetic Teacher* 39, no.1 (Sep. 1991): 50.

Martin, George E.
Polyominoes: A Guide to Puzzles and Problems in Tiling, by George E. Martin (Washington, DC: Mathematical Association of America, 1991). Reviewed in: *Mathematics Teacher* 85, no.6 (Sep. 1992): 493-94.

Mathematical Association of America *(cont'd)*
Problems for Mathematicians Young and Old (vol. 12), by Paul R. Halmos (Washington, DC: Mathematical Association of America, 1991). Reviewed in: *Mathematics Teacher* 85, no.7 (Oct. 1992): 592.

Mathematical Challenges for the Middle Grades: From The Mathematics Teacher Calendar Problems
by William D. Jamski (Reston, VA: NCTM, 1991). Reviewed in: *Mathematics Teacher* 85, no.8 (Nov. 1992): 688-89.

Mathematical concepts
Basic Mathematics, by David Novak (Lexington, MA: D. C. Heath, 1991). Reviewed in: *Mathematics Teacher* 85, no.6 (Sep. 1992): 491.

Basic Mathematics (3d ed.), by Charles P. McKeague (Florence, KY: Brooks/Cole, 1992). Reviewed in: *Mathematics Teacher* 85, no.6 (Sep. 1992): 490-91.

Building Self-Confidence in Math: A Student Workbook (2d ed.), by Sally Wilding and Elizabeth Shearn (Dubuque, IA: Kendall/Hunt, 1991). Reviewed in: *Mathematics Teacher* 85, no.6 (Sep. 1992): 496.

Essential Arithmetic (6th ed.), by C. L. Johnston, Alden T. Willis, and Jeanne Lazaris (Belmont, CA: Wadsworth, 1991). Reviewed in: *Mathematics Teacher* 85, no.6 (Sep. 1992): 492.

Exploring Sequences and Series, software (Minneapolis: MECC, 1991). Reviewed in: *Mathematics Teacher* 85, no.6 (Sep. 1992): 489.

Math Facts: Survival Guide to Basic Mathematics, by Theodore J. Szymanski and Anne Scanlan-Rohrer (Belmont, CA: Wadsworth, 1992). Reviewed in: *Mathematics Teacher* 85, no.6 (Sep. 1992): 493.

Mathematical concepts (Basic)
Facts Master, manipulatives by Jivan Patel (Carrollton, TX: Bruce Bakke, n.d.). Reviewed in: *Arithmetic Teacher* 40, no.2 (Oct. 1992): 128.

Mathematical concepts (Remedial)
Teaching Mathematics to Low Attainers (8-12), by Derek Haylock (Bristol, PA: Paul Chapman, 1991). Reviewed in: *Mathematics Teacher* 85, no.8 (Nov. 1992): 693.

Mathematical games
Amazing Arithmos, manipulatives (Forest Hills, NY: Marc Preven, 1990). Reviewed in: *Arithmetic Teacher* 40, no.4 (Dec. 1992): 242.

Raps and Rhymes in Maths, by Ann Baker and Johnny Baker (Portsmouth, NH: Heinemann, 1991). Reviewed in: *Arithmetic Teacher* 40, no.2 (Oct. 1992): 128.

The Unexpected Hanging and Other Mathematical Diversions, by Martin Gardner (Chicago, IL: University of Chicago Press, 1991). Reviewed in: *Mathematics Teacher* 85, no.6 (Sep. 1992): 494-95.

Mathematical Olympiad Contest Problems for Children
by George Lenchner (East Meadow, NY: Glenwood, 1990). Reviewed in: *Curriculum Review* 31, no.7 (Mar. 1992): 26; *Arithmetic Teacher* 39, no.9 (May 1992): 52.

Mathematical Topics for Computer Instructions: Grades 9-12
by Harris S. Shultz and William A. Leonard (Palo Alto, CA: Dale Seymour, 1990). Reviewed in: *Mathematics Teacher* 38, no.4 (Apr. 1991): 326-28.

Mathematics Achievement Project
software (Washington, DC: CAE Software, 1992). Reviewed in: *Mathematics Teacher* 85, no.7 (Oct. 1992): 587-88.

Mathematics Exploration Toolkit
software (Atlanta, GA: IBM Education). Reviewed in *Mathematics and Computer Education*: 25, no.3 (Fall 1991): 86.

Mathematics for the Young Child
by Joseph N. Payne (Reston, VA: NCTM, 1991). Reviewed in: *Arithmetic Teacher* 40, no.1 (Sep. 1992): 57-58.

Mathematics: Its Power and Utility, 3d ed.
by Karl J. Smith (Belmont, CA: Wadsworth, 1990). Reviewed in: *Mathematics and Computer Education* 26, no.1 (Winter 1992): 92.

Math Extra, Early Bird ed.
by Carole Greenes, Linda Schulman, and Rika Spungin (Allen, TX: DLM, 1991). Reviewed in: *Arithmetic Teacher* 39, no.7 (Mar. 1992): 43-44.

Math Facts: Survival Guide to Basic Mathematics
by Theodore J. Szymanski and Anne Scanlan-Rohrer (Belmont, CA: Wadsworth, 1992). Reviewed in: *Mathematics Teacher* 85, no.6 (Sep. 1992): 493.

Math for Girls: The Book with the Number To Girls To Love and Excel in Math
by Carole Marsh (Historic Bath, NC: Gallopade, 1989). Reviewed in: *Arithmetic Teacher* 39, no.1 (Sep. 1991): 50.

Math in Brief
game, by Mary Cavanagh (Fort Collins, CO: Scott Resources, 1978). Reviewed in: *Arithmetic Teacher* 39, no.1 (Sep. 1991): 53.

Math Intersections: A Look at Key Mathmatical Concepts, Grades 6-9
by David J. Glatzer and Joyce Glatzer (Palo Alto, CA: Dale Seymour, 1990). Reviewed in: *Arithmetic Teacher* 38, no.9 (May 1991): 46-47.

Math Mats: Hands-On Activities for Young Children, set 2
manipulatives, by Carole A. Thornton and Judith K. Wells (Allen, TX: Teaching Resources, 1990). Reviewed in: *Arithmetic Teacher* 37, no.7 (Mar. 1992): 44.

Math Meets Technology
by Brian Bolt (Port Chester, NY: Cambridge University Press, 1991). Reviewed in: *Mathematics Teacher* 85, no.8 (Nov. 1992): 692.

Math-O-Graphs: Critical Thinking through Graphing
by Donna Kay Buck and Francis Hildebrand (Pacific Grove, CA: Midwest, 1990). Reviewed in: *Arithmetic Teacher* 38, no.9 (May 1991): 47.

Math Power at Home
by Gerald Kulm (Waldorf, MD: American Association for the Advancement of Science, 1990). Reviewed in: *Arithmetic Teacher* 40, no.1 (Sep. 1992): 64.

Math Power in School
by Gerald Kulm (Waldorf, MD: American Association for the Advancement of Science, 1990). Reviewed in: *Arithmetic Teacher* 40, no.1 (Sep. 1992): 64.

Math Power in the Community
by Gerald Kulm (Waldorf, MD: American Association for the Advancement of Science, 1990). Reviewed in: *Arithmetic Teacher* 40, no.1 (Sep. 1992): 64.

Maths at Play: Fun Ideas for 5-8 Year Olds
by Linn Maskell (Victoria, Australia: Dellasta, 1990). Reviewed in: *Arithmetic Teacher* 39, no.9 (May 1992): 51-52.

Math Shop Spotlight: Weights and Measures
software (Jefferson City, MO: Scholastic). Reviewed in: *Electronic Learning* 11, no.2 (Oct. 1991): 37.

Mathsoft
Mathcad 3.0, software (Cambridge, MA: Mathsoft, 1991). Reviewed in: *Electronic Learning* 11, no.3 (Nov./Dec. 1991): 35.

Math Stories for Problem Solving Success: Ready-To-Use Activities for Grades 7-12
by Jim Overholt, Nancy Aaberg, and Jim Lindsey (Old Tappen, NJ: Prentice-Hall, 1990). Reviewed in: *Mathematics Teacher* 84, no.7 (Oct. 1991): 578.

The Maths Workshop: A Program of Maths Activities for Upper Primary
by Rob Vingerhoets (Victoria, Australia: Dellasta, 1990). Reviewed in: *Arithmetic Teacher* 38, no.9 (May 1991): 48.

Math Talk
(Cambridge, MA: Tom Snyder, 1991). Reviewed in: *Teaching Pre-K-8* 21, no.2 (Oct. 1991): 16-17.

Math Writer
software (Pacific Grove, CA: Brooks/Cole). Reviewed in: *Mathematics and Computer Education* 26, no.1 (Winter, 1992): 109-10.

MCE
Learning To Reason: Some, All, or None,
software (Kalamazoo, MI: MCE, 1988).
Reviewed in: *Arithmetic Teacher* 39, no.1
(Sep. 1991): 47.

Measure Up: The Visual Aid Ruler
by Kenneth L. Johannsen (Arlington Heights,
IL: NEK, 1990). Reviewed in: *Arithmetic
Teacher* 40, no.4 (Dec. 1992): 244.

Measuring: From Paces to Feet, Grades 3-4
by Rebecca B. Corwin and Susan Jo Russell
(Palo Alto, CA: Dale Seymour, 1990).
Reviewed in: *Arithmetic Teacher* 38, no.9
(May 1991): 48.

*Measuring: From Used Numbers: Real Data
in the Classroom*
by Rebecca B. Corwin and Susan Jo Russell
(Palo Alto, CA: Dale Seymour, 1990).
Reviewed in: *Arithmetic Teacher* 38, no.9
(May 1991): 48.

MECC
Estimation: Quick Solve I, software (St. Paul:
MECC). Reviewed in: *Teaching Pre-K-8*
21, no.3 (Nov./Dec. 1991): 15-16; *Arithmetic Teacher* 38, no.8 (Apr. 1991): 54.

Exploring Sequences and Series, software
(Minneapolis: MECC, 1991). Reviewed in:
Mathematics Teacher 85, no.6 (Sep.
1992): 489.

Probability Lab No. A-262, software (St.
Paul: MECC, 1990). Reviewed in: *Mathematics Teacher* 84, no.6 (Sep. 1991): 487;
The Computing Teacher 18, no.8 (May
1991): 46-47.

Mental Math
supplementary materials, by Ellen Hechler
(Farmington Hills, MI: Ellen Hechler, 1991).
Reviewed in: *Mathematics Teacher* 85, no.1
(Jan. 1992): 78.

Mental Math and Estimation
by Gary G. Bitter (Allen, TX: DLM, 1989).
Reviewed in: *Arithmetic Teacher* 39, no.1
(Sep. 1991): 53.

Mental Math Games, Series I
software (San Rafael, CA: Broderbund,
1992). Reviewed in: *Technology & Learning*
13, no.3 (Nov./Dec. 1992): 18.

Mercer, Richard
Algebra and Trigonometry (2d ed.), by
Thomas W. Hungerford and Richard Mercer
(Orlando, FL: Saunders College, 1991).
Reviewed in: *Mathematics Teacher* 85, no.9
(Dec. 1992): 763.

Trigonometry, by Thomas W. Hungerford
and Richard Mercer (Orlando, FL: Saunders
College, 1992). Reviewed in: *Mathematics
Teacher* 85, no.8 (Nov. 1992): 693-94.

Merlin Book of Logic Puzzles
by Margaret C. Edmiston (New York:
Sterling, 1991). Reviewed in: *Mathematics
Teacher* 85, no.9 (Dec. 1992): 767.

Middle school
*The Algebra Lab: Middle School: Exploring
Algebra Concepts with Manipulatives*, by
Henri Piccotti (Sunnyvale, CA: Creative,
1990). Reviewed in: *Mathematics Teacher*
38, no.4 (Apr. 1991): 332.

Beginning Algebra, by Alfonse Gobran
(Belmont, CA: Wadsworth, 1991). Reviewed
in: *Mathematics Teacher* 84, no.8 (Nov.
1991): 674.

Benjamin Banneker, poster (Burlington, NC:
Cabisco Mathematics, 1991). Reviewed in:
Mathematics Teacher 85, no.6 (Sep.
1992): 496.

*Building Self-Confidence in Math: A Student
Workbook (2d ed.)*, by Sally Wilding and
Elizabeth Shearn (Dubuque, IA: Kendall/
Hunt, 1991). Reviewed in: *Mathematics
Teacher* 85, no.6 (Sep. 1992): 496.

The Calculator, Grades 4-8, video (Columbus, OH: Silver Burdett & Ginn, 1991).
Reviewed in: *Arithmetic Teacher* 37, no.7
(Mar. 1992): 42.

Calculator Conundrums, by Thomas Camilli
(Pacific Grove, CA: Thinking Press, 1991).
Reviewed in: *Arithmetic Teacher* 40, no.4
(Dec. 1992): 242.

Middle school *(cont'd)*

Facts Master, manipulatives by Jivan Patel (Carrollton, TX: Bruce Bakke, n.d.). Reviewed in: *Arithmetic Teacher* 40, no.2 (Oct. 1992): 128.

Fastmath, software, by Jim Baker and Dale Huline (Minneapolis: New Directions, 1990). Reviewed in: *Arithmetic Teacher* 39, no.1 (Sep. 1991): 47.

Flashcards, supplementary materials, by Ellen Hechler (Farmington Hills, MI: Ellen Hechler, 1991). Reviewed in: *Mathematics Teacher* 85, no.1 (Jan. 1992): 78.

Fractals for the Classroom: Strategic Activities, vol. 1, by Heinz-Otto Peritgen, Hartmut Jurgens, Dietmar Saupe, Evan Maletsky, Terry Perciante, and Lee Yunker (Reston, VA: NCTM, 1991). Reviewed in: *Mathematics Teacher* 85, no.2 (Feb. 1992): 146; *Arithmetic Teacher* 39, no.7 (Mar. 1992): 40.

Fraction-oids, software (Danvers, MA: MindPlay, Methods and Solutions, 1988). Reviewed in: *Arithmetic Teacher* 39, no.1 (Sep. 1991): 46-47.

Fractions and Some Cool Distractions, video (Wynnewood, PA: Rahlic, 1990). Reviewed in: *Arithmetic Teacher* 37, no.7 (Mar. 1992): 42-43.

Fundamentally Math, Ages 8 and Up, software by Howard Diamond (Chapel Hill, NC: Chip Publications, 1990). Reviewed in: *Arithmetic Teacher* 40, no.3 (Nov. 1992): 194.

GeoExplorer, software (Glenview, IL: Scott Foresman, n.d.). Reviewed in: *Electronic Learning* 12, no.1 (Sep. 1992): 23-24; *Mathematics Teacher* 85, no.7 (Oct. 1992): 587.

The Geometer's Sketchpad, software (Berkeley: Key Curriculum, 1991). Reviewed in: *Mathematics Teacher* 85, no.5 (May 1992): 392-93; *Electronic Learning* 11, no.3 (Nov./Dec. 1991): 34-35.

Geometry Inventor, software (Scotts Valley, CA: Wings for Learning, n.d.). Reviewed in: *Electronic Learning* 12, no.1 (Sep. 1992): 23-24.

Geometry Workshop, Grades 3-8, software (Glenville, IL: Scott Foresman, 1991). Reviewed in: *Arithmetic Teacher* 39, no.8 (Apr. 1992): 58.

Graphing and Probability Workshop, Grades 3-8, software (Glenview, IL: Scott Foresman, 1991). Reviewed in: *Arithmetic Teacher* 39, no.3 (Nov. 1991): 56.

Integrated Mathematics Course 1, by Brendan Kelly, Paul Atkinson, and Bob Alexander (St. Charles, IL: McDougal Littell, 1991). Reviewed in: *Mathematics Teacher* 84, no.9 (Dec. 1991): 768.

Introductory Alegbra: An Interactive Approach, by Linda Pulsinelli and Patricia Hooper (New York: Macmillan, 1991). Reviewed in: *Mathematics Teacher* 84, no.8 (Nov. 1991): 678.

Junior Genius' Common Denominator, supplementary materials (Boca Raton, FL: Junior Genius, 1986). Reviewed in: *Arithmetic Teacher* 39, no.8 (Apr. 1992): 59-60.

Kaleidoscope Math, by Joe Kennedy and Diana Thomas (Sunnyvale, CA: Creative Publications, 1989). Reviewed in: *Arithmetic Teacher* 38, no.6 (Feb. 1991): 63.

Keep Your Balance, software (Pleasantville, NY: Sunburst, 1989). Reviewed in: *Arithmetic Teacher* 38, no.7 (Mar. 1991): 56.

Koetke's Challenge, software by Tim Barclay and Jonathan Choate (Acton, MA: William K. Bradford, 1991). Reviewed in: *Mathematics Teacher* 85, no.9 (Dec. 1992): 761.

Learning Mathematics: A Program for Classroom Teachers, by Ross McKeown (Portsmouth, NH: Heinemann, 1990). Reviewed in: *Arithmetic Teacher* 40, no.1 (Sep. 1992): 59.

Making Mathematics 7, by Gary Flewelling, Joan Routledge, and John Clark (Agincourt, Ontario: Gage Educational, 1991). Reviewed in: *Mathematics Teacher* 85, no.9 (Dec. 1992): 766.

Middle school *(cont'd)*

Math Shop Spotlight: Weights and Measures, software (Jefferson City, MO: Scholastic). Reviewed in: *Electronic Learning* 11, no.2 (Oct. 1991): 37.

Math Stories for Problem Solving Success: Ready-To-Use Activities for Grades 7-12, by Jim Overholt, Nancy Aaberg, and Jim Lindsey (Old Tappen, NJ: Prentice-Hall, 1990). Reviewed in: *Mathematics Teacher* 84, no.7 (Oct. 1991): 578.

The Maths Workshop: A Program of Maths Activities for Upper Primary, by Rob Vingerhoets (Victoria, Australia: Dellasta, 1990). Reviewed in: *Arithmetic Teacher* 38, no.9 (May 1991): 48.

Measure Up: The Visual Aid Ruler, by Kenneth L. Johannsen (Arlington Heights, IL: NEK, 1990). Reviewed in: *Arithmetic Teacher* 40, no.4 (Dec. 1992): 244.

Mental Math, supplementary materials, by Ellen Hechler (Farmington Hills, MI: Ellen Hechler, 1991). Reviewed in: *Mathematics Teacher* 85, no.1 (Jan. 1992): 78.

Mental Math and Estimation, by Gary G. Bitter (Allen, TX: DLM, 1989). Reviewed in: *Arithmetic Teacher* 39, no.1 (Sep. 1991): 53.

Mental Math Games, Series I, software (San Rafael, CA: Broderbund, 1992). Reviewed in: *Technology & Learning* 13, no.3 (Nov./ Dec. 1992): 18.

Merlin Book of Logic Puzzles, by Margaret C. Edmiston (New York: Sterling, 1991). Reviewed in: *Mathematics Teacher* 85, no.9 (Dec. 1992): 767.

Mind Benders—A1, software, by Anita Harnadek (Pacific Grove, CA: Midwest, 1988). Reviewed in: *Arithmetic Teacher* 39, no.1 (Sep. 1991): 47-48.

Mr. Marfil's Last Will and Testament, video (Pleasantville, NY: Human Relations, 1991). Reviewed in: *Mathematics Teacher* 84, no.7 (Oct. 1991): 578.

Number Maze Decimals and Fractions, software (n.p.: Great Wave, n.d.). Reviewed in: *Media and Methods* 27, no.5 (May/June 1991): 60.

Operation Neptune, software (Fremont, CA: Learning Company, 1992). Reviewed in: *Technology & Learning* 13, no.3 (Nov./Dec. 1992): 18.

Overhead Math: Manipulative Lessons on the Overhead Projector, manipulatives, by Shirley Hoogeborm and Judy Goodnow (Sunnyvale, CA: Creative, 1990). Reviewed in: *Arithmetic Teacher* 39, no.8 (Apr. 1992): 60.

Polyominoes: A Guide to Puzzles and Problems in Tiling, by George E. Martin (Washington, DC: Mathematical Association of America, 1991). Reviewed in: *Mathematics Teacher* 85, no.6 (Sep. 1992): 493-94.

Practical Math for Everyone, by Norman F. Hale (New York: Thinker's Books, 1990). Reviewed in: *Mathematics Teacher* 85, no.7 (Oct. 1992): 591.

Prealgebra, by Joshua Berenbom and Umesh Nagarkatte (San Diego: Harcourt Brace Jovanovich, 1991). Reviewed in: *Arithmetic Teacher* 39, no.3 (Nov. 1991): 60.

Prime Rollers, supplementary materials (Burlington, NC: Cabisco Mathematics, 1990). Reviewed in: *Arithmetic Teacher* 38, no.9 (May 1991): 48-49; *Mathematics Teacher* 85, no.7 (Oct. 1992): 594.

The Problem Solver with Calculators, by Terence G. Coburn, Shirley Hoogeboom, and Judy Goodnon (Sunnyvale, CA: Creative, 1989). Reviewed in: *Arithmetic Teacher* 39, no.1 (Sep. 1991): 50-51.

Problem Solving (S,C,R), by Karl J. Smith (Pacific Grove, CA: Brooks/Cole, 1991). Reviewed in: *Mathematics Teacher* 85, no.1 (Jan. 1992): 75.

Problem Solving Series, by Derek Holton (Leicester, UK: Mathematical Association, 1988). Reviewed in: *Mathematics Teacher* 84, no.6 (Sep. 1991): 492-93.

Middle school *(cont'd)*

Problem Solving through Critical Thinking, Grades 5-8, by Ronald R. Edwards and Wanda D. Cook (New Rochelle, NY: Cuisenaire, 1990). Reviewed in: *Arithmetic Teacher* 38, no.8 (Apr. 1991): 57-58.

Projects To Enrich School Mathematics, Level 1, ed. by Judith Trowel (Reston, VA: NCTM, 1990). Reviewed in: *Mathematics Teacher* 83, no.8 (Nov. 1990): 677-80; *Arithmetic Teacher* 38, no.8 (Apr. 1991): 55.

Simulated Real-Life Experiences Using Classified Ads in the Classroom, by Ellen Hechler (Farmington Hills, MI: Ellen Hechler, 1991). Reviewed in: *Arithmetic Teacher* 40, no.1 (Sep. 1992): 64.

Solve It: Fun Folders for Fast Finishers, by Michael Richards and Wendy Lohse (Victoria, Australia: Dellasta, 1990). Reviewed in: *Arithmetic Teacher* 38, no.7 (Mar. 1991): 58.

Space Shuttle Word Problems, software (Pound Ridge, NY: Orange Cherry, 1991). Reviewed in: *Teaching Pre-K-8* 21, no.4 (Jan. 1992): 14-15.

Spy: A Teaching Resource of Fascinating Math Investigations, by Kevin Lees (Victoria, Australia: Dellasta, 1989). Reviewed in: *Arithmetic Teacher* 38, no.8 (Apr. 1991): 58.

Statistics: The Shape of Data, Grades 4-6, by Susan Jo Russell and Rebecca B. Corwin (Palo Alto, CA: Dale Seymour, 1989). Reviewed in: *Arithmetic Teacher* 38, no.9 (May 1991): 49.

The Stella Octangula Activity Set, video (Berkeley: Key Curriculum, 1991). Reviewed in: *Mathematics Teacher* 85, no.2 (Feb. 1992): 156.

Strategies for Solving Math Word Problems (New York: Educational Design). Reviewed in: *Teaching Exceptional Children* 24, no.2 (Fall 1991): 73.

Symmystries with Cuisenaire Rods, Sets, card set, by Patricia S. Davidson and Robert E. Willcutt (New Rochelle, NY: Cuisenaire, 1990). Reviewed in: *Arithmetic Teacher* 39, no.1 (Sep. 1991): 53.

Taking Chances, software (Pleasantville, NY: Sunburst, 1991). Reviewed in: *Arithmetic Teacher* 39, no.3 (Nov. 1991): 57.

Teaching Mathematics to Low Attainers (8-12), by Derek Haylock (Bristol, PA: Paul Chapman, 1991). Reviewed in: *Mathematics Teacher* 85, no.8 (Nov. 1992): 693.

Troll Sports Math: Math Word Problems for Grades 4-6, (Mahwah, NJ: Troll, 1991). Reviewed in: *Teaching Pre-K-8* 21, no.2 (Oct. 1991): 14-16.

What Do You Do with a Broken Calculator?, software (Pleasantville, NY: Sunburst, 1989). Reviewed in: *Arithmetic Teacher* 39, no.3 (Nov. 1991): 58.

What's My Angle, software (Torrance, CA: Davidson, 1991). Reviewed in: *Media & Methods* 28, no.3 (Jan./Feb. 1992): 60.

What's Your Game? A Resource Book for Mathematical Activities, by Michael Cornelius and Alan Parr (New York: Cambridge University Press, 1991). Reviewed in: *Arithmetic Teacher* 40, no.4 (Dec. 1992): 241.

The Wonderful Problems of Fizz and Martina, video (Cambridge, MA: Tom Snyder). Reviewed in: *Media and Methods* 28, no.2 (Nov./Dec. 1991): 55-56.

World's Most Baffling Puzzles, by Charles Barry Townsend (New York: Sterling, 1991). Reviewed in: *Mathematics Teacher* 85, no.9 (Dec. 1992): 767.

Midwest (publisher)

Math-O-Graphs: Critical Thinking through Graphing, by Donna Kay Buck and Francis Hildebrand (Pacific Grove, CA: Midwest, 1990). Reviewed in: *Arithmetic Teacher* 38, no.9 (May 1991): 47.

Mind Benders—A1, software, by Anita Harnadek (Pacific Grove, CA: Midwest, 1988). Reviewed in: *Arithmetic Teacher* 39, no.1 (Sep. 1991): 47-48.

Milauskas, George
Geometry for Enjoyment and Challenges, by Richard Rhoad, George Milauskas, and Robert Whipple (Evanston, IL: McDougal Lippincott, TomLittell, 1991). Reviewed in: *Mathematics Teacher* 84, no.6 (Sep. 1991): 488-89.

Miller, Wayne L.
TI-81 Graphing Calculator Activities for Finite Mathematics, by Wayne L. Miller, Donald Perry, and Gloria A. Tveten (Pacific Grove, CA: Brooks/Cole, 1992). Reviewed in: *Mathematics Teacher* 85, no.8 (Nov. 1992): 693.

Mind Benders—A1
software, by Anita Harnadek (Pacific Grove, CA: Midwest, 1988). Reviewed in: *Arithmetic Teacher* 39, no.1 (Sep. 1991): 47-48.

MindPlay, Methods and Solutions
Fraction-oids, software (Danvers, MA: MindPlay, Methods and Solutions, 1988). Reviewed in: *Arithmetic Teacher* 39, no.1 (Sep. 1991): 46-47.

Minitab
(State College, PA: Minitab Inc.). Reviewed in: *Mathematics and Computer Education* 25, no.3 (Fall 1991): 334-35.

Mr. Marfil's Last Will and Testament
video (Pleasantville, NY: Human Relations, 1991). Reviewed in: *Mathematics Teacher* 84, no.7 (Oct. 1991): 578.

MIT Press
Investigations in Algebra, by Albert Cuoco (Cambridge, MA: MIT Press, 1990). Reviewed in: *Mathematics Teacher* 85, no.1 (Jan. 1992): 74.

Monbrag, John R.
TrigPak: Software & Tutorials for Trigonometry, version 2.0, software, by John R. Monbrag (Pacific Grove, CA: Brooks/Cole, 1989). Reviewed in: *Mathematics Teacher* 38, no.5 (May 1991): 406.

Money and Time Workshop, Grades K-2
software (Glenview, IL: Scott Foresman, 1991). Reviewed in: *Arithmetic Teacher* 39, no.1 (Sep. 1991): 48.

More Mathematical Morsels
by Ross Honsberger (Washington, DC: Mathematical Association). Reviewed in: *Mathematics Teacher* 84, no.6 (Sep. 1991): 490; *Mathematics and Computer Education* 25, no.3 (Fall 1991): 316.

Mosaic Magic
software (Encinitas, CA: Kinder Magic, 1990). Reviewed in: *Arithmetic Teacher* 39, no.1 (Sep. 1991): 48.

Moscovich, Ivan
Fiendishly Difficult, by Ivan Moscovich (New York: Sterling, 1991). Reviewed in: *Mathematics Teacher* 84, no.9 (Dec. 1991): 766.

Multiplication and Division Made Easy
by Catherine F. Debie (Artesia, CA: Scott Foresman, 1990). Reviewed in: *Arithmetic Teacher* 39, no.8 (Apr. 1992): 59.

Multiplication Rap: For Grades 3 to 6
audiocassette (Boise, ID: Star Trax, 1988). Reviewed in: *Arithmetic Teacher* 40, no.1 (Sep. 1992): 62.

Multiply with Balancing Bear
software (Pleasantville, NY: Sunburst, 1990). Reviewed in: *Arithmetic Teacher* 39, no.1 (Sep. 1991): 48.

Mummert, Jack
Calculators in the Classroom, K-12, by Katherine Pedersen and Jack Mummert (Carbondale, IL: ICTM, 1990). Reviewed in: *Arithmetic Teacher* 40, no.1 (Sep. 1992): 62.

Munem, M. A.
Precalculus Functions and Graphs, 5th ed., by M. A. Munem and J. P. Yizze (New York: Worth, 1990). Reviewed in: *Mathematics Teacher* 84, no.1 (Jan. 1991): 68.

Mystery Castle: Strategies for Problem Solving, Level I and II
software (Acton, MA: William K. Bradford, 1990). Reviewed in: *Arithmetic Teacher* 39, no.3 (Nov. 1991): 57.

Number Squares
software (Washington, DC: Balloons, 1988). Reviewed in: *Arithmetic Teacher* 39, no.1 (Sep. 1991): 48.

Numbers at Play: A Counting Book
by Charles Sullivan (New York: Rizzoli, 1992). Reviewed in: *Childhood Education* 69, no.1 (Fall, 1992): 49.

O'Daffer, Phares G.
Geometry (Teacher's ed.), by Stanley R. Clemens, Phares G. O'Daffer, and Thomas J. Cooney (Menlo Park, CA: Addison-Wesley, 1992). Reviewed in: *Mathematics Teacher* 85, no.7 (Oct. 1992): 590.

Pre-Algebra: A Transition to Algebra (Teacher's ed.), by Stanley R. Clemens, Phares G. O'Daffer, and Randall I. Charles (Menlo Park, CA: Addison-Wesley, 1992). Reviewed in: *Mathematics Teacher* 85, no.7 (Oct. 1992): 592.

Okey, JoAnn King
Graphing Primer, Grades K-2, by Laura Duncan Choate and JoAnn King Okey (Palo Alto, CA: Seymour Publications, 1989). Reviewed in: *Arithmetic Teacher* 40, no.2 (Oct. 1992): 129.

Pictograms: Graphing Pictures for a Reusable Classroom Grid, manipulatives by Laura Duncan Choate and JoAnn King Okey (Palo Alto, CA: Dale Seymour, 1989). Reviewed in: *Arithmetic Teacher* 40, no.2 (Oct. 1992): 130.

Olson, Judith
Data Collection and Analysis Activities (K-8), by Katherine Pedersen and Judith Olson (Carbondale, IL: ICTM, 1990). Reviewed in: *Arithmetic Teacher* 40, no.1 (Sep. 1992): 62.

The $1.00 Word Riddle Book
by Marilyn Burns and illustrated by Martha Weston (New Rochelle, NY: Cuisenaire, 1990). Reviewed in: *Arithmetic Teacher* 39, no.2 (Oct. 1991): 51-52.

On the Button in Math: Activities for Young Children
by Carol A. Thornton and Judith K. Wells (Allen, TX: DLM, 1990). Reviewed in: *Arithmetic Teacher* 38, no.9 (May 1991): 46.

Operation Neptune
software (Fremont, CA: Learning Company, 1992). Reviewed in: *Technology & Learning* 13, no.3 (Nov./Dec. 1992): 18.

Orange Cherry
Space Shuttle Word Problems, software (Pound Ridge, NY: Orange Cherry, 1991). Reviewed in: *Teaching Pre-K-8* 21, no.4 (Jan. 1992): 14-15.

Talking Multiplication and Division: Grades 3-6, software (Pound Ridge, NY: Orange Cherry, 1990). Reviewed in: *Arithmetic Teacher* 39, no.3 (Nov. 1991): 57.

Overhead Math: Manipulative Lessons on the Overhead Projector
manipulatives, by Shirley Hoogeborm and Judy Goodnow (Sunnyvale, CA: Creative, 1990). Reviewed in: *Arithmetic Teacher* 39, no.8 (Apr. 1992): 60.

Overholt, Jim
Math Stories for Problem Solving Success: Ready-To-Use Activities for Grades 7-12, by Jim Overholt, Nancy Aaberg, and Jim Lindsey (Old Tappen, NJ: Prentice-Hall, 1990). Reviewed in: *Mathematics Teacher* 84, no.7 (Oct. 1991): 578.

Oxford University Press
What To Solve? Problems and Suggestions for Young Mathematicians, by Judita Lofman (Cary, NC: Oxford University Press). Reviewed in: *Mathematics Teacher* 84, no.6 (Sep. 1991): 496.

Parr, Alan
What's Your Game? A Resource Book for Mathematical Activities, by Michael Cornelius and Alan Parr (New York: Cambridge University Press, 1991). Reviewed in: *Arithmetic Teacher* 40, no.4 (Dec. 1992): 241.

Patel, Jivan
Facts Master, manipulatives by Jivan Patel (Carrollton, TX: Bruce Bakke, n.d.). Reviewed in: *Arithmetic Teacher* 40, no.2 (Oct. 1992): 128.

Patterns and Functions
by Linda Sue Brisby (Solvang, CA: Hands On, 1990). Reviewed in: *Arithmetic Teacher* 40, no.1 (Sep. 1992): 64.

Patterns and Puzzles: Math Challenges for Grades 3 and 4
by Neville Leeson (Mt. Waverly, Australia: n.p., 1990). Reviewed in: *Arithmetic Teacher* 40, no.1 (Sep. 1992): 64.

Paul Chapman
Teaching Mathematics to Low Attainers (8-12), by Derek Haylock (Bristol, PA: Paul Chapman, 1991). Reviewed in: *Mathematics Teacher* 85, no.8 (Nov. 1992): 693.

Payne, Joseph N.
Mathematics for the Young Child, by Joseph N. Payne (Reston, VA: NCTM, 1991). Reviewed in: *Arithmetic Teacher* 40, no.1 (Sep. 1992): 57-58.

PC Gradeworks
AstroNUMBERS, software (Washington, DC: PC Gradeworks, 1990). Reviewed in: *Arithmetic Teacher* 39, no.1 (Sep. 1991): 47.

Pedagoguery (publisher)
GrafEq, software (Terrace, British Columbia: Pedagoguery, 1990). Reviewed in: *Mathematics Teacher* 85, no.2 (Mar. 1992): 241-42.

Pedersen, Katherine
Calculators in the Classroom, K-12, by Katherine Pedersen and Jack Mummert (Carbondale, IL: ICTM, 1990). Reviewed in: *Arithmetic Teacher* 40, no.1 (Sep. 1992): 62.

Data Collection and Analysis Activities (K-8), by Katherine Pedersen and Judith Olson (Carbondale, IL: ICTM, 1990). Reviewed in: *Arithmetic Teacher* 40, no.1 (Sep. 1992): 62.

Penguin
The Penguin Dictionary of Curious and Interesting Numbers, by David Wells (New York: Penguin, 1987). Reviewed in: *Mathematics and Computer Education* 24, no.1 (Winter 1991): 92.

The Penguin Dictionary of Curious and Interesting Numbers,
by David Wells (New York: Penguin, 1987). Reviewed in: *Mathematics and Computer Education* 24, no.1 (Winter 1991): 92.

Perciante, Terry
Fractals for the Classroom: Strategic Activities, vol. 1, by Heinz-Otto Peritgen, Hartmut Jurgens, Dietmar Saupe, Evan Maletsky, Terry Perciante, and Lee Yunker (Reston, VA: NCTM, 1991). Reviewed in: *Mathematics Teacher* 85, no.2 (Feb. 1992): 146; *Arithmetic Teacher* 39, no.7 (Mar. 1992): 40.

Peritgen, Heinz-Otto
Fractals for the Classroom: Strategic Activities, vol. 1, by Heinz-Otto Peritgen, Hartmut Jurgens, Dietmar Saupe, Evan Maletsky, Terry Perciante, and Lee Yunker (Reston, VA: NCTM, 1991). Reviewed in: *Mathematics Teacher* 85, no.2 (Feb. 1992): 146; *Arithmetic Teacher* 39, no.7 (Mar. 1992): 40.

Perry, Donald
TI-81 Graphing Calculator Activities for Finite Mathematics, by Wayne L. Miller, Donald Perry, and Gloria A. Tveten (Pacific Grove, CA: Brooks/Cole, 1992). Reviewed in: *Mathematics Teacher* 85, no.8 (Nov. 1992): 693.

Piccotti, Henri
The Algebra Lab: High School: A Comprehensive Manipulative Program for Algebra I, by Henri Piccotti (Sunnyvale, CA: Creative, 1990). Reviewed in: *Mathematics Teacher* 38, no.5 (May 1991): 406.

The Algebra Lab: Middle School: Exploring Algebra Concepts with Manipulatives, by Henri Piccotti (Sunnyvale, CA: Creative, 1990). Reviewed in: *Mathematics Teacher* 38, no.4 (Apr. 1991): 332.

Piccotti, Henri *(cont'd)*
  Logo Math: Tools and Games, software, by
  Henri Piccotti (Portland, ME: Terrapin,
  1990). Reviewed in: *Mathematics Teacher*
  84, no.6 (Sep. 1991): 486-87.

*Pictograms: Graphing Pictures for a Reusable
Classroom Grid*
  manipulatives by Laura Duncan Choate and
  JoAnn King Okey (Palo Alto, CA: Dale
  Seymour, 1989). Reviewed in: *Arithmetic
  Teacher* 40, no.2 (Oct. 1992): 130.

PKWARE
  PKZIP, software (Glendale, WI: PKWARE).
  Reviewed in: *Mathematics and Computer
  Education* 25, no.3 (Fall 1991): 332-33.

PKZIP
  software (Glendale, WI: PKWARE). Reviewed in: *Mathematics and Computer Education* 25, no.3 (Fall 1991): 332-33.

Place Value-Plus
  manipulatives (Goshen, IN: New Vision,
  n.d.). Reviewed in: *Arithmetic Teacher* 40,
  no.2 (Oct. 1992): 130.

Plane Geometry
  supplementary materials (Little Rock, AR:
  High Q, 1988). Reviewed in: *Arithmetic
  Teacher* 38, no.8 (Apr. 1991): 60.

*Polyominoes: A Guide to Puzzles and Problems
in Tiling*
  by George E. Martin (Washington, DC:
  Mathematical Association of America, 1991).
  Reviewed in: *Mathematics Teacher* 85, no.6
  (Sep. 1992): 493-94.

Posters
  Benjamin Banneker, poster (Burlington, NC:
  Cabisco Mathematics, 1991). Reviewed in:
  Mathematics Teacher 85, no.6 (Sep.
  1992): 496.

  Colorful Characters of Mathematics, poster
  (Burlington, NC: Cabisco Mathematics,
  1992). Reviewed in: *Mathematics Teacher*
  85, no.8 (Nov. 1992): 694.

Potter, Paula
  Thinking through Math Word Problems, by
  Authur Whimbey, Jack Lochhead, and Paula
  Potter (Hillsdale, NJ: Laurence Erlbaum,
  1990). Reviewed in: *Arithmetic Teacher* 39,
  no.2 (Oct. 1991): 52.

Practical Math for Everyone
  by Norman F. Hale (New York: Thinker's
  Books, 1990). Reviewed in: *Mathematics
  Teacher* 85, no.7 (Oct. 1992): 591.

Prealgebra
  *Math Intersections: A Look at Key
  Mathmatical Concepts, Grades 6-9*, by
  David J. Glatzer and Joyce Glatzer (Palo
  Alto, CA: Dale Seymour, 1990). Reviewed
  in: *Arithmetic Teacher* 38, no.9 (May 1991):
  46-47.

  Prealgebra, by Joshua Berenbom and Umesh
  Nagarkatte (San Diego: Harcourt Brace
  Jovanovich, 1991). Reviewed in: *Arithmetic
  Teacher* 39, no.3 (Nov. 1991): 60.

  *Pre-Algebra: A Transition to Algebra
  (Teacher's ed.)*, by Stanley R. Clemens,
  Phares G. O'Daffer, and Randall I. Charles
  (Menlo Park, CA: Addison-Wesley, 1992).
  Reviewed in: *Mathematics Teacher* 85, no.7
  (Oct. 1992): 592.

  *Prealgebra for Problem Solvers: A New
  Beginning*, by Loyd V. Wilcox (Pacific
  Grove, CA: Brooks/Cole, 1991). Reviewed
  in: *Mathematics Teacher* 84, no.6 (Sep.
  1991): 492.

Prealgebra
  by Joshua Berenbom and Umesh Nagarkatte
  (San Diego: Harcourt Brace Jovanovich,
  1991). Reviewed in: *Arithmetic Teacher* 39,
  no.3 (Nov. 1991): 60.

*Pre-Algebra: A Transition to Algebra
(Teacher's ed.)*
  by Stanley R. Clemens, Phares G. O'Daffer,
  and Randall I. Charles (Menlo Park, CA:
  Addison-Wesley, 1992). Reviewed in:
  Mathematics Teacher 85, no.7 (Oct.
  1992): 592.

Precalculus Mathematics: A Graphing Approach (2d ed.)
by Franklin Demana, Bert K. Waits, and Stanley E. Clemens (Menlo Park, CA: Addison-Wesley, 1992). Reviewed in: *Mathematics Teacher* 85, no.7 (Oct. 1992): 592.

Prentice-Hall
Math Stories for Problem Solving Success: Ready-To-Use Activities for Grades 7-12, by Jim Overholt, Nancy Aaberg, and Jim Lindsey (Old Tappen, NJ: Prentice-Hall, 1990). Reviewed in: *Mathematics Teacher* 84, no.7 (Oct. 1991): 578.

Marc Preven (publisher)
Amazing Arithmos, manipulatives (Forest Hills, NY: Marc Preven, 1990). Reviewed in: *Arithmetic Teacher* 40, no.4 (Dec. 1992): 242.

Primary Geoboard Activity Book
by Carol A. Thornton and Judith K. Wells (Deerfield, IL: Learning Resources, 1990). Reviewed in: *Arithmetic Teacher* 40, no.2 (Oct. 1992): 130.

Primary Geometry Workshop, K-2
software by Apple Family (Glenville, IL: Scott Foresman, 1991). Reviewed in: *Arithmetic Teacher* 39, no.8 (Apr. 1992): 58.

Prime Rollers
supplementary materials (Burlington, NC: Cabisco Mathematics, 1990). Reviewed in: *Arithmetic Teacher* 38, no.9 (May 1991): 48-49; *Mathematics Teacher* 85, no.7 (Oct. 1992): 594.

A Primer in Probability
by Kathleen Subrahmaniam (New York: Marcel Dekker, 1990). Reviewed in: *Mathematics Teacher* 84, no.7 (Oct. 1991): 574.

Probability
Do You Wanna Bet? Your Chance To Find Out about Probability, by Jean Cushman and illustrated by Martha Weston (New York: Clarion Books, 1991). Reviewed in: *Arithmetic Teacher* 40, no.4 (Dec. 1992): 240.

Graphing and Probability Workshop, Grades 3-8, software (Glenview, IL: Scott Foresman, 1991). Reviewed in: *Arithmetic Teacher* 39, no.3 (Nov. 1991): 56.

A Primer in Probability, by Kathleen Subrahmaniam (New York: Marcel Dekker, 1990). Reviewed in: *Mathematics Teacher* 84, no.7 (Oct. 1991): 574.

Probability Lab No. A-262, software (St. Paul: MECC, 1990). Reviewed in: *Mathematics Teacher* 84, no.6 (Sep. 1991): 487; *The Computing Teacher* 18, no.8 (May 1991): 46-47.

Statistics and Probability in Modern Life (5th ed.) (Orlando, FL: Saunders College, 1992). Reviewed in: *Mathematics Teacher* 85, no.8 (Nov. 1992): 692-93.

Taking Chances, software (Pleasantville, NY: Sunburst, 1991). Reviewed in: *Arithmetic Teacher* 39, no.3 (Nov. 1991): 57.

Probability Lab No. A-262
software (St. Paul: MECC, 1990). Reviewed in: *Mathematics Teacher* 84, no.6 (Sep. 1991): 487; *The Computing Teacher* 18, no.8 (May 1991): 46-47.

The Problem Solver with Calculators
by Terence G. Coburn, Shirley Hoogeboom, and Judy Goodnon (Sunnyvale, CA: Creative, 1989). Reviewed in: *Arithmetic Teacher* 39, no.1 (Sep. 1991): 50-51.

Problem solving
Add-Ventures for Girls: Building Math Confidence (Elementary), by Margaret Franklin (Newton, MA: Educational Development Center, 1990). Reviewed in: *Arithmetic Teacher* 40, no.1 (Sep. 1992): 58.

Add-Ventures for Girls: Building Math Confidence (Junior High), by Margaret Franklin (Newton, MA: Educational Development Center, 1990). Reviewed in: *Arithmetic Teacher* 40, no.1 (Sep. 1992): 58.

Basic Mathematics, by David Novak (Lexington, MA: D. C. Heath, 1991). Reviewed in: *Mathematics Teacher* 85, no.6 (Sep. 1992): 491.

Basic Mathematics (3d ed.), by Charles P. McKeague (Florence, KY: Brooks/Cole, 1992). Reviewed in: *Mathematics Teacher* 85, no.6 (Sep. 1992): 490-91.

Problem solving *(cont'd)*

Building Self-Confidence in Math: A Student Workbook (2d ed.), by Sally Wilding and Elizabeth Shearn (Dubuque, IA: Kendall/Hunt, 1991). Reviewed in: *Mathematics Teacher* 85, no.6 (Sep. 1992): 496.

Calculator Conundrums, by Thomas Camilli (Pacific Grove, CA: Thinking Press, 1991). Reviewed in: *Arithmetic Teacher* 40, no.4 (Dec. 1992): 242.

Calculators in the Classroom, K-12, by Katherine Pedersen and Jack Mummert (Carbondale, IL: ICTM, 1990). Reviewed in: *Arithmetic Teacher* 40, no.1 (Sep. 1992): 62.

Calculus by and for Young People (Ages 7, Yes 7 and Up), by Don Cohen (n.p.: Jonathan Press, 1991). Reviewed in: *Mathematics Teaching* 140 (Sep. 1992): 41.

Camp-LA, Books 1-4 (Grades K-8) (Orange, CA: California State Fullerton Press, 1991). Reviewed in: *Arithmetic Teacher* 40, no.4 (Dec. 1992): 242-43.

A Collection of MATH Lessons, Grades 6-8, by Cathy McLaughlin (New Rochelle, NY: Cuisenaire, 1990). Reviewed in: *Arithmetic Teacher* 39, no.1 (Sep. 1991): 49.

Crossmatics: A Challenging Collection of Cross-Number Puzzles: Grades 7-12, game, by Allan Dudley (Palo Alto, CA: Dale Seymour, 1990). Reviewed in: *Mathematics Teacher* 84, no.5 (May 1991): 406.

Daily Mathmatics: Critical Thinking and Problem Solving, (Evanston, IL: McDougal Littell, 1992). Reviewed in: *Arithmetic Teacher* 39, no.9 (May 1992): 52-53; *Mathematics Teacher* 85, no.6 (Sep. 1992): 496-97.

Data Collection and Analysis Activities (K-8), by Katherine Pedersen and Judith Olson (Carbondale, IL: ICTM, 1990). Reviewed in: *Arithmetic Teacher* 40, no.1 (Sep. 1992): 62.

Detective Stories for Math Problem Solving, software (Pleasantville, NY: Human Relations). Reviewed in: *Electronic Learning* 11, no.5 (Feb. 1992): 35.

Do You Wanna Bet? Your Chance To Find Out about Probability, by Jean Cushman and illustrated by Martha Weston (New York: Clarion Books, 1991). Reviewed in: *Arithmetic Teacher* 40, no.4 (Dec. 1992): 240.

Estimation: Quick Solve I, software (St. Paul: MECC). Reviewed in: *Teaching Pre-K-8* 21, no.3 (Nov./Dec. 1991): 15-16; *Arithmetic Teacher* 38, no.8 (Apr. 1991): 54.

Fiendishly Difficult, by Ivan Moscovich (New York: Sterling, 1991). Reviewed in: *Mathematics Teacher* 84, no.9 (Dec. 1991): 766.

Fundamentally Math, Ages 8 and Up, software by Howard Diamond (Chapel Hill, NC: Chip Publications, 1990). Reviewed in: *Arithmetic Teacher* 40, no.3 (Nov. 1992): 194.

Journey into Geometries, by Martha Sved (Washington, DC: Mathematical Association of America, 1991). Reviewed in: *Mathematics Teacher* 85, no.8 (Nov. 1992): 691.

Koetke's Challenge, software by Tim Barclay and Jonathan Choate (Acton, MA: William K. Bradford, 1991). Reviewed in: *Mathematics Teacher* 85, no.9 (Dec. 1992): 761.

Label Land: A World of Inference and Problem Solving, software (Scotts Valley, CA: Wings for Learning, 1991). Reviewed in: *Arithmetic Teacher* 39, no.3 (Nov. 1991): 56-57.

Learning Mathematics: A Program for Classroom Teachers, by Ross McKeown (Portsmouth, NH: Heinemann, 1990). Reviewed in: *Arithmetic Teacher* 40, no.1 (Sep. 1992): 59.

Learning To Reason: Some, All, or None, software (Kalamazoo, MI: MCE, 1988). Reviewed in: *Arithmetic Teacher* 39, no.1 (Sep. 1991): 47.

Link 'n' Learn Activity Book, by Carol A. Thornton and Judith K. Wells (Deerfield, IL: Learning Resources, 1990). Reviewed in: *Arithmetic Teacher* 40, no.2 (Oct. 1992): 130.

Problem solving *(cont'd)*

The Little Shoppers Kit, software (Cambridge, MA: Tom Snyder, 1989). Reviewed in: *Arithmetic Teacher* 39, no.1 (Sep. 1991): 47.

M! Project MATHEMATICS!—Similarity, video by Tom Apostol (Reston, VA: NCTM, 1990). Reviewed in: *Mathematics Teacher* 85, no.6 (Sep. 1992): 496.

Making Mathematics 7, by Gary Flewelling, Joan Routledge, and John Clark (Agincourt, Ontario: Gage Educational, 1991). Reviewed in: *Mathematics Teacher* 85, no.9 (Dec. 1992): 766.

Making Mathematics 8, by Gary Flewelling, Joan Routledge, and John Clark (Agincourt, Ontario: Gage Educational, 1991). Reviewed in: *Mathematics Teacher* 85, no.8 (Nov. 1992): 692.

Math Challenges for the Middle Grades from the Arithmetic Teacher, by William D. Jamski (Reston, VA: NCTM, 1990). Reviewed in: *Arithmetic Teacher* 38, no.7 (Mar. 1991): 58.

Mathemactivities, by Bob Berstein and illustrated by Bron Smith (Carthage, IL: Good Apple, 1991). Reviewed in: *Arithmetic Teacher* 40, no.1 (Sep. 1992): 62-63.

Mathematica: A System for Doing Mathematics by Computer (2d ed.), software by Stephen Wolfram (Champaign, IL: Wolfram Research, 1991). Reviewed in: *Mathematics Teacher* 85, no.8 (Nov. 1992): 688.

Mathematical Challenges for the Middle Grades: From The Mathematics Teacher Calendar Problems, by William D. Jamski (Reston, VA: NCTM, 1991). Reviewed in: *Mathematics Teacher* 85, no.8 (Nov. 1992): 688-89.

Mathematical Olympiad Contest Problems for Children, by George Lenchner (East Meadow, NY: Glenwood, 1990). Reviewed in: *Curriculum Review* 31, no.7 (Mar. 1992): 26.

Mathematics Achievement Project, software (Washington, DC: CAE Software, 1992). Reviewed in: *Mathematics Teacher* 85, no.7 (Oct. 1992): 587-88.

Math Facts: Survival Guide to Basic Mathematics, by Theodore J. Szymanski and Anne Scanlan-Rohrer (Belmont, CA: Wadsworth, 1992). Reviewed in: *Mathematics Teacher* 85, no.6 (Sep. 1992): 493.

Math for Girls: The Book with the Number To Girls To Love and Excel in Math, by Carole Marsh (Historic Bath, NC: Gallopade, 1989). Reviewed in: *Arithmetic Teacher* 39, no.1 (Sep. 1991): 50.

Math Meets Technology, by Brian Bolt (Port Chester, NY: Cambridge University Press, 1991). Reviewed in: *Mathematics Teacher* 85, no.8 (Nov. 1992): 692.

Math-O-Graphs: Critical Thinking through Graphing, by Donna Kay Buck and Francis Hildebrand (Pacific Grove, CA: Midwest, 1990). Reviewed in: *Arithmetic Teacher* 38, no.9 (May 1991): 47.

Math Power at Home, by Gerald Kulm (Waldorf, MD: American Association for the Advancement of Science, 1990). Reviewed in: *Arithmetic Teacher* 40, no.1 (Sep. 1992): 64.

Math Power in School, by Gerald Kulm (Waldorf, MD: American Association for the Advancement of Science, 1990). Reviewed in: *Arithmetic Teacher* 40, no.1 (Sep. 1992): 64.

Math Power in the Community, by Gerald Kulm (Waldorf, MD: American Association for the Advancement of Science, 1990). Reviewed in: *Arithmetic Teacher* 40, no.1 (Sep. 1992): 64.

Math Stories for Problem Solving Success: Ready-To-Use Activities for Grades 7-12, by Jim Overholt, Nancy Aaberg, and Jim Lindsey (Old Tappen, NJ: Prentice-Hall, 1990). Reviewed in: *Mathematics Teacher* 84, no.7 (Oct. 1991): 578.

Problem solving *(cont'd)*

Math Talk, (Cambridge, MA: Tom Snyder, 1991). Reviewed in: *Teaching Pre-K-8* 21, no.2 (Oct. 1991): 16-17.

Mental Math Games, Series I, software (San Rafael, CA: Broderbund, 1992). Reviewed in: *Technology & Learning* 13, no.3 (Nov./Dec. 1992): 18.

Merlin Book of Logic Puzzles, by Margaret C. Edmiston (New York: Sterling, 1991). Reviewed in: *Mathematics Teacher* 85, no.9 (Dec. 1992): 767.

Mind Benders—A1, software, by Anita Harnadek (Pacific Grove, CA: Midwest, 1988). Reviewed in: *Arithmetic Teacher* 39, no.1 (Sep. 1991): 47-48.

Mr. Marfil's Last Will and Testament, video (Pleasantville, NY: Human Relations, 1991). Reviewed in: *Mathematics Teacher* 84, no.7 (Oct. 1991): 578.

Money and Time Workshop, Grades K-2, software (Glenview, IL: Scott Foresman, 1991). Reviewed in: *Arithmetic Teacher* 39, no.1 (Sep. 1991): 48.

More Mathematical Morsels, by Ross Honsberger (Washington, DC: Mathematical Association). Reviewed in: *Mathematics Teacher* 84, no.6 (Sep. 1991): 490; *Mathematics and Computer Education* 25, no.3 (Fall 1991): 316.

Mystery Castle: Strategies for Problem Solving, Level I and II, software (Acton, MA: William K. Bradford, 1990). Reviewed in: *Arithmetic Teacher* 39, no.3 (Nov. 1991): 57.

Nonroutine Problems: Doing Mathematics, by Robert London (Providence, RI: Janson, 1989). Reviewed in: *Mathematics Teacher* 38, no.4 (Apr. 1991): 328.

Number Squares, software (Washington, DC: Balloons, 1988). Reviewed in: *Arithmetic Teacher* 39, no.1 (Sep. 1991): 48.

Operation Neptune, software (Fremont, CA: Learning Company, 1992). Reviewed in: *Technology & Learning* 13, no.3 (Nov./Dec. 1992): 18.

Patterns and Functions, by Linda Sue Brisby (Solvang, CA: Hands On, 1990). Reviewed in: *Arithmetic Teacher* 40, no.1 (Sep. 1992): 64.

Patterns and Puzzles: Math Challenges for Grades 3 and 4, by Neville Leeson (Mt. Waverly, Australia: n.p., 1990). Reviewed in: *Arithmetic Teacher* 40, no.1 (Sep. 1992): 64.

Place Value-Plus, manipulatives (Goshen, IN: New Vision, n.d.). Reviewed in: *Arithmetic Teacher* 40, no.2 (Oct. 1992): 130.

Polyominoes: A Guide to Puzzles and Problems in Tiling, by George E. Martin (Washington, DC: Mathematical Association of America, 1991). Reviewed in: *Mathematics Teacher* 85, no.6 (Sep. 1992): 493-94.

Practical Math for Everyone, by Norman F. Hale (New York: Thinker's Books, 1990). Reviewed in: *Mathematics Teacher* 85, no.7 (Oct. 1992): 591.

Prealgebra for Problem Solvers: A New Beginning, by Loyd V. Wilcox (Pacific Grove, CA: Brooks/Cole, 1991). Reviewed in: *Mathematics Teacher* 84, no.6 (Sep. 1991): 492.

Primary Geoboard Activity Book, by Carol A. Thornton and Judith K. Wells (Deerfield, IL: Learning Resources, 1990). Reviewed in: *Arithmetic Teacher* 40, no.2 (Oct. 1992): 130.

Problems for Mathematicians Young and Old (vol. 12), by Paul R. Halmos (Washington, DC: Mathematical Association of America, 1991). Reviewed in: *Mathematics Teacher* 85, no.7 (Oct. 1992): 592.

The Problem Solver with Calculators, by Terence G. Coburn, Shirley Hoogeboom, and Judy Goodnon (Sunnyvale, CA: Creative, 1989). Reviewed in: *Arithmetic Teacher* 39, no.1 (Sep. 1991): 50-51.

Problem Solving, video (Atlanta, GA: Silver Burdett & Ginn, 1991). Reviewed in: *Arithmetic Teacher* 39, no.2 (Oct. 1991): 52-53.

Problem solving *(cont'd)*

Problem Solving (S,C,R), by Karl J. Smith (Pacific Grove, CA: Brooks/Cole, 1991). Reviewed in: *Mathematics Teacher* 85, no.1 (Jan. 1992): 75.

Problem Solving and Comprehension (5th ed.), by Arthur Whimbey and Jack Lochhead (Hillsdale, NJ: Lawrence Erlbaum Associates, 1991). Reviewed in: *Mathematics Teacher* 85, no.7 (Oct. 1992): 592.

Problem Solving Series, by Derek Holton (Leicester, UK: Mathematical Association, 1988). Reviewed in: *Mathematics Teacher* 84, no.6 (Sep. 1991): 492-93.

Problem Solving through Critical Thinking, Grades 5-8, by Ronald R. Edwards and Wanda D. Cook (New Rochelle, NY: Cuisenaire, 1990). Reviewed in: *Arithmetic Teacher* 38, no.8 (Apr. 1991): 57-58.

Reading Higher: A Problem-Solving Approach to Elementary School Mathematics, video and supplementary materials (Reston, VA: NCTE, 1990). Reviewed in: *Curriculum Review* 31, no.7 (Mar. 1992): 26-27.

Simulated Real-Life Experiences Using Classified Ads in the Classroom, by Ellen Hechler (Farmington Hills, MI: Ellen Hechler, 1991). Reviewed in: *Arithmetic Teacher* 40, no.1 (Sep. 1992): 64.

Solve It: Fun Folders for Fast Finishers, by Michael Richards and Wendy Lohse (Victoria, Australia: Dellasta, 1990). Reviewed in: *Arithmetic Teacher* 38, no.7 (Mar. 1991): 58.

Space Shuttle Word Problems, software (Pound Ridge, NY: Orange Cherry, 1991). Reviewed in: *Teaching Pre-K-8* 21, no.4 (Jan. 1992): 14-15.

Strategies for Solving Math Word Problems (New York: Educational Design). Reviewed in: *Teaching Exceptional Children* 24, no.2 (Fall 1991): 73.

Symmystries with Cuisenaire Rods, Sets, card set, by Patricia S. Davidson and Robert E. Willcutt (New Rochelle, NY: Cuisenaire, 1990). Reviewed in: *Arithmetic Teacher* 39, no.1 (Sep. 1991): 53.

Teaching Mathematics to Low Attainers (8-12), by Derek Haylock (Bristol, PA: Paul Chapman, 1991). Reviewed in: *Mathematics Teacher* 85, no.8 (Nov. 1992): 693.

Thinking through Math Word Problems, by Authur Whimbey, Jack Lochhead, and Paula Potter (Hillsdale, NJ: Laurence Erlbaum, 1990). Reviewed in: *Arithmetic Teacher* 39, no.2 (Oct. 1991): 52.

TI-81 Graphing Calculator Activities for Finite Mathematics, by Wayne L. Miller, Donald Perry, and Gloria A. Tveten (Pacific Grove, CA: Brooks/Cole, 1992). Reviewed in: *Mathematics Teacher* 85, no.8 (Nov. 1992): 693.

Troll Sports Math: Math Word Problems for Grades 4-6, (Mahwah, NJ: Troll, 1991). Reviewed in: *Teaching Pre-K-8* 21, no.2 (Oct. 1991): 14-16.

What Do You Do with a Broken Calculator?, software (Pleasantville, NY: Sunburst, 1989). Reviewed in: *Arithmetic Teacher* 39, no.3 (Nov. 1991): 58.

What To Solve? Problems and Suggestions for Young Mathematicians, by Judita Lofman (Cary, NC: Oxford University Press). Reviewed in: *Mathematics Teacher* 84, no.6 (Sep. 1991): 496.

What's Your Game? A Resource Book for Mathematical Activities, by Michael Cornelius and Alan Parr (New York: Cambridge University Press, 1991). Reviewed in: *Arithmetic Teacher* 40, no.4 (Dec. 1992): 241.

What's Your Problem? Posing and Solving Mathematical Problems, by Penny Skinner (Portsmouth, NH: Heinemann, 1990). Reviewed in: *Arithmetic Teacher* 40, no.4 (Dec. 1992): 241-42.

The Wonderful Problems of Fizz and Martina, video (Cambridge, MA: Tom Snyder). Reviewed in: *Media and Methods* 28, no.2 (Nov./Dec. 1991): 55-56.

World's Most Baffling Puzzles, by Charles Barry Townsend (New York: Sterling, 1991). Reviewed in: *Mathematics Teacher* 85, no.9 (Dec. 1992): 767.

Reading Higher: A Problem-Solving Approach to Elementary School Mathematics video and supplementary materials (Reston, VA: NCTE, 1990). Reviewed in: *Curriculum Review* 31, no.7 (Mar. 1992): 26-27.

Rhoad, Richard
Geometry for Enjoyment and Challenges, by Richard Rhoad, George Milauskas, and Robert Whipple (Evanston, IL: McDougal Littell, 1991). Reviewed in: *Mathematics Teacher* 84, no.6 (Sep. 1991): 488-89.

Richards, Michael
Solve It: Fun Folders for Fast Finishers, by Michael Richards and Wendy Lohse (Victoria, Australia: Dellasta, 1990). Reviewed in: *Arithmetic Teacher* 38, no.7 (Mar. 1991): 58.

Rizzoli (publisher)
Numbers at Play: A Counting Book, by Charles Sullivan (New York: Rizzoli, 1992). Reviewed in: *Childhood Education* 69, no.1 (Fall, 1992): 49.

Rogers, James
Elementary Algebra (2d ed.), by Jack Barker, James Rogers, and James Van Dyke (Orlando, FL: Saunders College, 1992). Reviewed in: *Mathematics Teacher* 85, no.9 (Dec. 1992): 764.

Intermediate Algebra (2d ed.), by Jack Barker, James Rogers, and James Van Dyke (Orlando, FL: Saunders College, 1992). Reviewed in: *Mathematics Teacher* 85, no.9 (Dec. 1992): 765-66.

Rosenberg, Jon
Math Connections: Algebra I, software by Jon Rosenberg (Scotts Valley, CA: Wings for Learning, 1991). Reviewed in: *Mathematics Teacher* 85, no.6 (Sep. 1992): 489.

Ross, Mike
Rap-ability, audiocassette, by Mike Ross and Suzanne Rossi (Lake Orion, MI: Aynn Visual, 1990). Reviewed in: *Arithmetic Teacher* 38, no.9 (May 1991): 49.

Ross, Sharon Cutler
Precalculus Algebra and Trigonometry, by Sharon Cutler Ross and Linda Hawkins Boyd (Pacific Grove, CA: Brooks/Cole, 1991). Reviewed in: *Mathematics Teacher* 85, no.1 (Jan. 1992): 75.

Rossi, Suzanne
Rap-ability, audiocassette, by Mike Ross and Suzanne Rossi (Lake Orion, MI: Aynn Visual, 1990). Reviewed in: *Arithmetic Teacher* 38, no.9 (May 1991): 49.

Routledge, Joan
Making Mathematics 7, by Gary Flewelling, Joan Routledge, and John Clark (Agincourt, Ontario: Gage Educational, 1991). Reviewed in: *Mathematics Teacher* 85, no.9 (Dec. 1992): 766.

Making Mathematics 8, by Gary Flewelling, Joan Routledge, and John Clark (Agincourt, Ontario: Gage Educational, 1991). Reviewed in: *Mathematics Teacher* 85, no.8 (Nov. 1992): 692.

Russell, Susan Jo
Measuring: From Paces to Feet, Grades 3-4, by Rebecca B. Corwin and Susan Jo Russell (Palo Alto, CA: Dale Seymour, 1990). Reviewed in: *Arithmetic Teacher* 38, no.9 (May 1991): 48.

Measuring: From Used Numbers: Real Data in the Classroom, by Rebecca B. Corwin and Susan Jo Russell (Palo Alto, CA: Dale Seymour, 1990). Reviewed in: *Arithmetic Teacher* 38, no.9 (May 1991): 48.

Statistics: The Shape of Data, Grades 4-6, by Susan Jo Russell and Rebecca B. Corwin (Palo Alto, CA: Dale Seymour, 1989). Reviewed in: *Arithmetic Teacher* 38, no.9 (May 1991): 49.

Saunders College
Algebra and Trigonometry (2d ed.), by Thomas W. Hungerford and Richard Mercer (Orlando, FL: Saunders College, 1991). Reviewed in: *Mathematics Teacher* 85, no.9 (Dec. 1992): 763.

ScottForesman *(cont'd)*
NCSMP Algebra, by John W. McConnell, Susan Brown, Susan Eddins, Margaret Hackworth, and Zalman Usiskin (Tucker, GA: Scott Foresman, 1990). Reviewed in: *Mathematics Teacher* 84, no.2 (Feb. 1991): 144-46.

Primary Geometry Workshop, K-2, software by Apple Family (Glenville, IL: Scott Foresman, 1991). Reviewed in: *Arithmetic Teacher* 39, no.8 (Apr. 1992): 58.

Scott Resources
Math in Brief, game, by Mary Cavanagh (Fort Collins, CO: Scott Resources, 1978). Reviewed in: *Arithmetic Teacher* 39, no.1 (Sep. 1991): 53.

Sea Squares
by Joy N. Hulme and illustrated by Carol Schwartz (Waltham, MA: Hyperion Books, 1991). Reviewed in: *Arithmetic Teacher* 40, no.1 (Sep. 1992): 56.

Sellers, Gene R.
Intermediate Algebra, by Ronald Hatton and Gene R. Sellers (San Diego: Harcourt Brace Jovanovich, 1991). Reviewed in: *Mathematics Teacher* 84, no.9 (Dec. 1991): 769.

Introductory Algebra, by Ronald Hatton and Gene R. Sellers (San Diego: Harcourt Brace Jovanovich, 1991). Reviewed in: *Mathematics Teacher* 85, no.1 (Jan. 1992): 74.

Senk, Sharon L.
Advanced Algebra, by Sharon L. Senk, Denisse R. Thompson and Steven S. Vitatora (Glenview, IL: Scott Foresman, 1990). Reviewed in: *Mathematics Teacher* 84, no.3 (Mar. 1991): 235-36.

Sequences
Exploring Sequences and Series, software (Minneapolis: MECC, 1991). Reviewed in: *Mathematics Teacher* 85, no.6 (Sep. 1992): 489.

Dale Seymour (publisher)
Graphing Primer, Grades K-2, by Laura Duncan Choate and JoAnn King Okey (Palo Alto, CA: Seymour Publications, 1989). Reviewed in: *Arithmetic Teacher* 40, no.2 (Oct. 1992): 129.

Introduction to Tessellations, by Dale Seymour and Jill Britten (Palo Alto, CA: Dale Seymour, 1989). Reviewed in: *Mathematics Teacher* 83, no.6 (Sep. 1992): 482.

Mathematical Topics for Computer Instructions: Grades 9-12, by Harris S. Shultz and William A. Leonard (Palo Alto, CA: Dale Seymour, 1990). Reviewed in: *Mathematics Teacher* 38, no.4 (Apr. 1991): 326-28.

Math Intersections: A Look at Key Mathmatical Concepts, Grades 6-9, by David J. Glatzer and Joyce Glatzer (Palo Alto, CA: Dale Seymour, 1990). Reviewed in: *Arithmetic Teacher* 38, no.9 (May 1991): 46-47.

Measuring: From Paces to Feet, Grades 3-4, by Rebecca B. Corwin and Susan Jo Russell (Palo Alto, CA: Dale Seymour, 1990). Reviewed in: *Arithmetic Teacher* 38, no.9 (May 1991): 48.

Measuring: From Used Numbers: Real Data in the Classroom, by Rebecca B. Corwin and Susan Jo Russell (Palo Alto, CA: Dale Seymour, 1990). Reviewed in: *Arithmetic Teacher* 38, no.9 (May 1991): 48.

Pictograms: Graphing Pictures for a Reusable Classroom Grid, manipulatives by Laura Duncan Choate and JoAnn King Okey (Palo Alto, CA: Dale Seymour, 1989). Reviewed in: *Arithmetic Teacher* 40, no.2 (Oct. 1992): 130.

Statistics: The Shape of Data, Grades 4-6, by Susan Jo Russell and Rebecca B. Corwin (Palo Alto, CA: Dale Seymour, 1989). Reviewed in: *Arithmetic Teacher* 38, no.9 (May 1991): 49.

Teaching Primary Math with Music, cassette and songbook, by Esther Mardlesohn (Palo Alto, CA: Dale Seymour, 1990). Reviewed in: *Arithmetic Teacher* 39, no.1 (Sep. 1991): 53.

Seymour, Dale
Introduction to Tessellations, by Dale Seymour and Jill Britten (Palo Alto, CA: Dale Seymour, 1989). Reviewed in: *Mathematics Teacher* 83, no.6 (Sep. 1992): 482.

Smith, Stanley A.
Algebra (Teacher's ed.), by Stanley A. Smith, Randall I. Charles, and John A. Dossey (Menlo Park, CA: Addison-Wesley, 1992). Reviewed in: *Mathematics Teacher* 85, no.7 (Oct. 1992): 589.

Algebra and Trigonometry (Teacher's ed.), by Stanley A. Smith, Randall I. Charles, and John A. Dossey (Menlo Park, CA: Addison-Wesley, 1992). Reviewed in: *Mathematics Teacher* 85, no.7 (Oct. 1992): 589-90.

Software packages
Algebra, software (Canton, MT: National Appleworks Users Group, 1991). Reviewed in: *Mathematics Teacher* 85, no.5 (May 1991): 391.

Algebra from 0 to 3: Constants to Cubics, (Pleasantville, NY: Sunburst). Reviewed in: *Mathematics and Computer Education* 26, no.2 (Spring 1992): 222-23.

Algebra Made Easy, software (San Francisco: Britannica Software, 1991). Reviewed in: *Mathematics Teacher* 85, no.7 (Oct. 1992): 587.

Analyzer, software, by Douglas Alfors and Beverly West (Ithaca, NY: Delta-Epsilon Software, 1990). Reviewed in: *Mathematics Teacher* 85, no.3 (Mar. 1992): 240.

AstroNUMBERS, software (Washington, DC: PC Gradeworks, 1990). Reviewed in: *Arithmetic Teacher* 39, no.1 (Sep. 1991): 47.

Bradford Graphmaker, software (Concord, MA: William K. Bradford, 1990). Reviewed in: *Mathematics Teacher* 84, no.5 (May 1991): 391-92; *Electronic Learning* 11, no.3 (Nov./Dec. 1991): 34-35.

Calculus T/L: A Program for Doing, Teaching, and Learning Calculus, software, by J. Douglas Child (Pacific Grove, CA: Brooks/Cole, 1990). Reviewed in: *Mathematics Teacher* 84, no.3 (Mar. 1991): 234-35.

Concepts in Mathematics: Conic Sections, software (Cary, NC: TV Ontario Video, 1991). Reviewed in: *Mathematics Teacher* 85, no.9 (Dec. 1992): 767-68.

Concepts in Mathematics: Trigonometric Functions I--Solving Triangles, software (Cary, NC: TV Ontario Video, 1991). Reviewed in: *Mathematics Teacher* 85, no.9 (Dec. 1992): 768.

Connections! Math in Action, software (SVE, 1992). Reviewed in: *Teaching Pre-K-8* 21, no.4 (Jan. 1992): 14.

Counters: An Action Approach to Counting and Arithmetic, software (Scotts Valley, CA: Wings for Learning, 1990). Reviewed in: *Arithmetic Teacher* 39, no.3 (Nov. 1991): 56.

Data Insights, software, by Lois Edwards and K. M. Keogh (Pleasantville, NY: Sunburst, 1989). Reviewed in: *Mathematics Teacher* 84, no.4 (Apr. 1991): 320.

Detective Stories for Math Problem Solving, software (Pleasantville, NY: Human Relations). Reviewed in: *Electronic Learning* 11, no.5 (Feb. 1992): 35.

Estimation: Quick Solve I, software (St. Paul: MECC). Reviewed in: *Teaching Pre-K-8* 21, no.3 (Nov./Dec. 1991): 15-16; *Arithmetic Teacher* 38, no.8 (Apr. 1991): 54.

Estimation Tutor, software by Richard E. Rand (Portland, ME: J. Weston Walch, 1989). Reviewed in: *Arithmetic Teacher* 38, no.8 (Apr. 1991): 54.

Euclid's Toolbox—Triangles, software, by Jerry Beckmann and Charles Friesen (n.p.: Heartland, 1990). Reviewed in: *Mathematics Teacher* 84, no.6 (Sep. 1991): 484.

Exploring Sequences and Series, software (Minneapolis: MECC, 1991). Reviewed in: *Mathematics Teacher* 85, no.6 (Sep. 1992): 489.

Fastmath, software, by Jim Baker and Dale Huline (Minneapolis: New Directions, 1990). Reviewed in: *Arithmetic Teacher* 39, no.1 (Sep. 1991): 47.

Fraction-oids, software (Danvers, MA: MindPlay, Methods and Solutions, 1988). Reviewed in: *Arithmetic Teacher* 39, no.1 (Sep. 1991): 46-47.

Software packages *(cont'd)*

The Little Shoppers Kit, software (Cambridge, MA: Tom Snyder, 1989). Reviewed in: *Arithmetic Teacher* 39, no.1 (Sep. 1991): 47.

Logo Geometry, software by Michael Battista and Douglas H. Clements (Morristown, NJ: Simon & Schuster, 1991). Reviewed in: *Arithmetic Teacher* 40, no.1 (Sep. 1992): 56.

Logo Math: Tools and Games, software, by Henri Piccotti (Portland, ME: Terrapin, 1990). Reviewed in: *Mathematics Teacher* 84, no.6 (Sep. 1991): 486-87.

Maple V: Student Edition, software (Pacific Grove, CA: Symbolic Computation Group, n.d.). Reviewed in: *Mathematics and Computer Education* 26, no.3 (Fall 1992): 337-39.

Mathcad 3.0, software (Cambridge, MA: Mathsoft, 1991). Reviewed in: *Electronic Learning* 11, no.3 (Nov./Dec. 1991): 35.

Math Connections: Algebra 1, software (Scotts Valley, CA: Wings for Learning). Reviewed in: *Technology & Learning* 12, no.6 (Mar. 1992): 10-11.

Math Connections: Algebra I, software by Jon Rosenberg (Scotts Valley, CA: Wings for Learning, 1991). Reviewed in: *Mathematics Teacher* 85, no.6 (Sep. 1992): 489.

Math Connections: Algebra II, software (Scotts Valley, CA: Wings for Learning, 1992). Reviewed in: *Technology & Learning* 13, no.3 (Nov./Dec. 1992): 18.

Mathematica: A System for Doing Mathematics by Computer (2d ed.), software by Stephen Wolfram (Champaign, IL: Wolfram Research, 1991). Reviewed in: *Mathematics Teacher* 85, no.8 (Nov. 1992): 688.

Mathematics Achievement Project, software (Washington, DC: CAE Software, 1992). Reviewed in: *Mathematics Teacher* 85, no.7 (Oct. 1992): 587-88.

Mathematics Exploration Toolkit, software (Atlanta, GA: IBM Education). Reviewed in: *Mathematics and Computer Education* 25, no.3 (Fall 1991): 86.

Math Shop Spotlight: Weights and Measures, software (Jefferson City, MO: Scholastic). Reviewed in: *Electronic Learning* 11, no.2 (Oct. 1991): 37.

Math Writer, software (Pacific Grove, CA: Brooks/Cole). Reviewed in: *Mathematics and Computer Education* 26, no.1 (Winter, 1992): 109-10.

Mental Math Games, Series I, software (San Rafael, CA: Broderbund, 1992). Reviewed in: *Technology & Learning* 13, no.3 (Nov./Dec. 1992): 18.

Mind Benders—A1, software, by Anita Harnadek (Pacific Grove, CA: Midwest, 1988). Reviewed in: *Arithmetic Teacher* 39, no.1 (Sep. 1991): 47-48.

Money and Time Workshop, Grades K-2, software (Glenview, IL: Scott Foresman, 1991). Reviewed in: *Arithmetic Teacher* 39, no.1 (Sep. 1991): 48.

Mosaic Magic, software (Encinitas, CA: Kinder Magic, 1990). Reviewed in: *Arithmetic Teacher* 39, no.1 (Sep. 1991): 48.

Multiply with Balancing Bear, software (Pleasantville, NY: Sunburst, 1990). Reviewed in: *Arithmetic Teacher* 39, no.1 (Sep. 1991): 48.

Mystery Castle: Strategies for Problem Solving, Level I and II, software (Acton, MA: William K. Bradford, 1990). Reviewed in: *Arithmetic Teacher* 39, no.3 (Nov. 1991): 57.

Number Maze Decimals and Fractions, software (n.p.: Great Wave, n.d.). Reviewed in: *Media and Methods* 27, no.5 (May/June 1991): 60.

Number Squares, software (Washington, DC: Balloons, 1988). Reviewed in: *Arithmetic Teacher* 39, no.1 (Sep. 1991): 48.

Operation Neptune, software (Fremont, CA: Learning Company, 1992). Reviewed in: *Technology & Learning* 13, no.3 (Nov./Dec. 1992): 18.

Statistics
Data Insights, software, by Lois Edwards and K. M. Keogh (Pleasantville, NY: Sunburst, 1989). Reviewed in: *Mathematics Teacher* 84, no.4 (Apr. 1991): 320.

Introduction to Statistics with Data Analysis, by Shelley Rasmussen (Pacific Grove, CA: Brooks/Cole, 1992). Reviewed in: *Mathematics Teacher* 85, no.6 (Sep. 1992): 492.

Logo Math: Tools and Games, software, by Henri Piccotti (Portland, ME: Terrapin, 1990). Reviewed in: *Mathematics Teacher* 84, no.6 (Sep. 1991): 486-87.

Probability Lab No. A-262, software (St. Paul: MECC, 1990). Reviewed in: *Mathematics Teacher* 84, no.6 (Sep. 1991): 487; *The Computing Teacher* 18, no.8 (May 1991): 46-47.

Statistics and Probability in Modern Life (5th ed.) (Orlando, FL: Saunders College, 1992). Reviewed in: *Mathematics Teacher* 85, no.8 (Nov. 1992): 692-93.

Statistics: The Shape of Data, Grades 4-6, by Susan Jo Russell and Rebecca B. Corwin (Palo Alto, CA: Dale Seymour, 1989). Reviewed in: *Arithmetic Teacher* 38, no.9 (May 1991): 49.

Statistics and Probability in Modern Life (5th ed.)
(Orlando, FL: Saunders College, 1992). Reviewed in: *Mathematics Teacher* 85, no.8 (Nov. 1992): 692-93.

Statistics: The Shape of Data, Grades 4-6
by Susan Jo Russell and Rebecca B. Corwin (Palo Alto, CA: Dale Seymour, 1989). Reviewed in: *Arithmetic Teacher* 38, no.9 (May 1991): 49.

The Stella Octangula Activity Set
video (Berkeley: Key Curriculum, 1991). Reviewed in: *Mathematics Teacher* 85, no.2 (Feb. 1992): 156.

Sterling
Fiendishly Difficult, by Ivan Moscovich (New York: Sterling, 1991). Reviewed in: *Mathematics Teacher* 84, no.9 (Dec. 1991): 766.

Merlin Book of Logic Puzzles, by Margaret C. Edmiston (New York: Sterling, 1991). Reviewed in: *Mathematics Teacher* 85, no.9 (Dec. 1992): 767.

World's Most Baffling Puzzles, by Charles Barry Townsend (New York: Sterling, 1991). Reviewed in: *Mathematics Teacher* 85, no.9 (Dec. 1992): 767.

Stern, Ely
Interactive Algebra, software, by Henry Africk, Ely Stern, and Harvey Broverman (Hicksville, NY: Technical Educational Consultants, 1990). Reviewed in: *Mathematics Teacher* 84, no.3 (Mar. 1991): 235.

Stewart, James
Single Variable Calculus, 2d ed., by James Stewart (Pacific Grove, CA: Brooks/Cole, 1991). Reviewed in: *Mathematics Teacher* 85, no.1 (Jan. 1992): 76.

Strategies for Solving Math Word Problems (New York: Educational Design). Reviewed in: *Teaching Exceptional Children* 24, no.2 (Fall 1991): 73.

Subrahmaniam, Kathleen
A Primer in Probability, by Kathleen Subrahmaniam (New York: Marcel Dekker, 1990). Reviewed in: *Mathematics Teacher* 84, no.7 (Oct. 1991): 574.

Subtraction Rap: For Grades 1 to 2 audiocassette (Boise, ID: Star Trax, 1990). Reviewed in: *Arithmetic Teacher* 40, no.1 (Sep. 1992): 62.

Sullivan, Charles
Numbers at Play: A Counting Book, by Charles Sullivan (New York: Rizzoli, 1992). Reviewed in: *Childhood Education* 69, no.1 (Fall, 1992): 49.

Sunburst
Algebra from 0 to 3: Constants to Cubics, (Pleasantville, NY: Sunburst). Reviewed in: *Mathematics and Computer Education* 26, no.2 (Spring 1992): 222-23.

Data Insights, software, by Lois Edwards and K. M. Keogh (Pleasantville, NY: Sunburst, 1989). Reviewed in: *Mathematics Teacher* 84, no.4 (Apr. 1991): 320.

Swokowski, Earl W.
Calculus, by Earl W. Swokowski (Florence, KY: PWS-KENT, 1991). Reviewed in: *Mathematics Teacher* 85, no.4 (Apr. 1992): 314.

Calculus: Late Trigonometry Version (5th ed.), by Earl W. Swokowski (Boston, MA: PWS-KENT, 1992). Reviewed in: *Mathematics Teacher* 85, no.8 (Nov. 1992): 690.

Precalculus Functions and Graphs, 6th ed., by Earl W. Swokowski (Boston: Kent, 1990). Reviewed in: *Mathematics Teacher* 84, no.1 (Jan. 1991): 68-70.

Symbolic Computation Group
Maple V: Student Edition, software (Pacific Grove, CA: Symbolic Computation Group, n.d.). Reviewed in: *Mathematics and Computer Education* 26, no.3 (Fall 1992): 337-39.

Symmystries with Cuisenaire Rods, Sets
card set, by Patricia S. Davidson and Robert E. Willcutt (New Rochelle, NY: Cuisenaire, 1990). Reviewed in: *Arithmetic Teacher* 39, no.1 (Sep. 1991): 53.

Szymanski, Theodore J.
Math Facts: Survival Guide to Basic Mathematics, by Theodore J. Szymanski and Anne Scanlan-Rohrer (Belmont, CA: Wadsworth, 1992). Reviewed in: *Mathematics Teacher* 85, no.6 (Sep. 1992): 493.

Taking Chances
software (Pleasantville, NY: Sunburst, 1991). Reviewed in: *Arithmetic Teacher* 39, no.3 (Nov. 1991): 57.

Talking Multiplication and Division: Grades 3-6
software (Pound Ridge, NY: Orange Cherry, 1990). Reviewed in: *Arithmetic Teacher* 39, no.3 (Nov. 1991): 57.

Teaching Mathematics to Low Attainers (8-12)
by Derek Haylock (Bristol, PA: Paul Chapman, 1991). Reviewed in: *Mathematics Teacher* 85, no.8 (Nov. 1992): 693.

Teaching Primary Math with Music
cassette and songbook, by Esther Mardlesohn (Palo Alto, CA: Dale Seymour, 1990). Reviewed in: *Arithmetic Teacher* 39, no.1 (Sep. 1991): 53.

Teaching Resource Center
The Circular Geoboard, by Judy Kevin (San Leandro, CA: Teaching Resource Center, 1990). Reviewed in: *Mathematics Teacher* 84, no.8 (Nov. 1991): 680.

Teaching Resources
Math Mats: Hands-On Activities for Young Children, set 2, manipulatives, by Carole A. Thornton and Judith K. Wells (Allen, TX: Teaching Resources, 1990). Reviewed in: *Arithmetic Teacher* 37, no.7 (Mar. 1992): 44.

Ten Little Rabbits
by Virginia Grossman and illustrated by Sylvia Long (San Francisco: Chronicle Books, 1991). Reviewed in: *Arithmetic Teacher* 40, no.1 (Sep. 1992): 56-57.

Terrapin
Logo Math: Tools and Games, software, by Henri Piccotti (Portland, ME: Terrapin, 1990). Reviewed in: *Mathematics Teacher* 84, no.6 (Sep. 1991): 486-87.

Thinker's Books
Practical Math for Everyone, by Norman F. Hale (New York: Thinker's Books, 1990). Reviewed in: *Mathematics Teacher* 85, no.7 (Oct. 1992): 591.

Thinking Press
Calculator Conundrums, by Thomas Camilli (Pacific Grove, CA: Thinking Press, 1991). Reviewed in: *Arithmetic Teacher* 40, no.4 (Dec. 1992): 242.

Thinking through Math Word Problems
by Authur Whimbey, Jack Lochhead, and Paula Potter (Hillsdale, NJ: Laurence Erlbaum, 1990). Reviewed in: *Arithmetic Teacher* 39, no.2 (Oct. 1991): 52.

Thomas, Diana
Kaleidoscope Math, by Joe Kennedy and Diana Thomas (Sunnyvale, CA: Creative Publications, 1989). Reviewed in: *Arithmetic Teacher* 38, no.6 (Feb. 1991): 63.

Thomas, George B.
Calculus, by Ross L. Finney and George B. Thomas, Jr. (Reading, MA: Addison-Wesley, 1990). Reviewed in: *Mathematics Teacher* 84, no.1 (Jan. 1991): 66.

Thompson, Denisse R.
Advanced Algebra, by Sharon L. Senk, Denisse R. Thompson and Steven S. Vitatora (Glenview, IL: Scott Foresman, 1990). Reviewed in: *Mathematics Teacher* 84, no.3 (Mar. 1991): 235-36.

Thornton, Carol A.
Fun with Money: A Problem-Solving Activity Book, by Carol A. Thornton and Judith K. Wells (Deerfield, IL: Learning Resources, 1991). Reviewed in: *Arithmetic Teacher* 40, no.2 (Oct. 1992): 128-29.

Link 'n' Learn Activity Book, by Carol A. Thornton and Judith K. Wells (Deerfield, IL: Learning Resources, 1990). Reviewed in: *Arithmetic Teacher* 40, no.2 (Oct. 1992): 130.

Math Mats: Hands-On Activities for Young Children, set 2, manipulatives, by Carole A. Thornton and Judith K. Wells (Allen, TX: Teaching Resources, 1990). Reviewed in: *Arithmetic Teacher* 37, no.7 (Mar. 1992): 44.

On the Button in Math: Activities for Young Children, by Carol A. Thornton and Judith K. Wells (Allen, TX: DLM, 1990). Reviewed in: *Arithmetic Teacher* 38, no.9 (May 1991): 46.

Primary Geoboard Activity Book, by Carol A. Thornton and Judith K. Wells (Deerfield, IL: Learning Resources, 1990). Reviewed in: *Arithmetic Teacher* 40, no.2 (Oct. 1992): 130.

3D Images
software (Acton, MA: Bradford School, 1991). Reviewed in: *Mathematics Teacher* 85, no.6 (May 1992): 391; *Electronic Learning* 11, no.3 (Nov./Dec. 1991): 34-35.

TI-81 Graphing Calculator Activities for Finite Mathematics
by Wayne L. Miller, Donald Perry, and Gloria A. Tveten (Pacific Grove, CA: Brooks/Cole, 1992). Reviewed in: *Mathematics Teacher* 85, no.8 (Nov. 1992): 693.

Tom Snyder
The Little Shoppers Kit, software (Cambridge, MA: Tom Snyder, 1989). Reviewed in: *Arithmetic Teacher* 39, no.1 (Sep. 1991): 47.

Math Talk, (Cambridge, MA: Tom Snyder, 1991). Reviewed in: *Teaching Pre-K-8* 21, no.2 (Oct. 1991): 16-17.

The Wonderful Problems of Fizz and Martina, video (Cambridge, MA: Tom Snyder). Reviewed in: *Media and Methods* 28, no.2 (Nov./Dec. 1991): 55-56.

Townsend, Charles Barry
World's Most Baffling Puzzles, by Charles Barry Townsend (New York: Sterling, 1991). Reviewed in: *Mathematics Teacher* 85, no.9 (Dec. 1992): 767.

Trigonometry
Algebra and Trigonometry (2d ed.), by Thomas W. Hungerford and Richard Mercer (Orlando, FL: Saunders College, 1991). Reviewed in: *Mathematics Teacher* 85, no.9 (Dec. 1992): 763.

Algebra II and Trigonometry (Evanston, IL: McDougal Littell, 1991). Reviewed in: *Mathematics Teacher* 38, no.4 (Apr. 1991): 325.

Calculus: Late Trigonometry Version (5th ed.), by Earl W. Swokowski (Boston, MA: PWS-KENT, 1992). Reviewed in: *Mathematics Teacher* 85, no.8 (Nov. 1992): 690.

Concepts in Mathematics: Trigonometric Functions I--Solving Triangles, software (Cary, NC: TV Ontario Video, 1991). Reviewed in: *Mathematics Teacher* 85, no.9 (Dec. 1992): 768.

GrafEq, software (Terrace, British Columbia: Pedagoguery, 1990). Reviewed in: *Mathematics Teacher* 85, no.2 (Mar. 1992): 241-42.

Precalculus Algebra and Trigonometry, by Sharon Cutler Ross and Linda Hawkins Boyd (Pacific Grove, CA: Brooks/Cole, 1991). Reviewed in: *Mathematics Teacher* 85, no.1 (Jan. 1992): 75.

Trigonometry *(cont'd)*
Trigonometry, by Thomas W. Hungerford and Richard Mercer (Orlando, FL: Saunders College, 1992). Reviewed in: *Mathematics Teacher* 85, no.8 (Nov. 1992): 693-94.

Trigonometry, by Charles N. Gantner and Thomas E. Gantner (Belmont, CA: Wadsworth, 1990). Reviewed in: *Mathematics Teacher* 84, no.1 (Jan. 1991): 70.

TrigPak: Software & Tutorials for Trigonometry, version 2.0, software, by John R. Monbrag (Pacific Grove, CA: Brooks/Cole, 1989). Reviewed in: *Mathematics Teacher* 38, no.5 (May 1991): 406.

Trigonometry
by Thomas W. Hungerford and Richard Mercer (Orlando, FL: Saunders College, 1992). Reviewed in: *Mathematics Teacher* 85, no.8 (Nov. 1992): 693-94.

Trigonometry
by Charles N. Gantner and Thomas E. Gantner (Belmont, CA: Wadsworth, 1990). Reviewed in: *Mathematics Teacher* 84, no.1 (Jan. 1991): 70.

TrigPak: Software & Tutorials for Trigonometry, version 2.0
software, by John R. Monbrag (Pacific Grove, CA: Brooks/Cole, 1989). Reviewed in: *Mathematics Teacher* 38, no.5 (May 1991): 406.

Troll
Troll Sports Math: Math Word Problems for Grades 4-6, (Mahwah, NJ: Troll, 1991). Reviewed in: *Teaching Pre-K-8* 21, no.2 (Oct. 1991): 14-16.

Troll Sports Math: Math Word Problems for Grades 4-6
(Mahwah, NJ: Troll, 1991). Reviewed in: *Teaching Pre-K-8* 21, no.2 (Oct. 1991): 14-16.

Tveten, Gloria A.
TI-81 Graphing Calculator Activities for Finite Mathematics, by Wayne L. Miller, Donald Perry, and Gloria A. Tveten (Pacific Grove, CA: Brooks/Cole, 1992). Reviewed in: *Mathematics Teacher* 85, no.8 (Nov. 1992): 693.

TV Ontario Video
Concepts in Mathematics: Trigonometric Functions I--Solving Triangles, software (Cary, NC: TV Ontario Video, 1991). Reviewed in: *Mathematics Teacher* 85, no.9 (Dec. 1992): 768.

The Unexpected Hanging and Other Mathematical Diversions
by Martin Gardner (Chicago, IL: University of Chicago Press, 1991). Reviewed in: *Mathematics Teacher* 85, no.6 (Sep. 1992): 494-95.

University of Chicago Press
The Unexpected Hanging and Other Mathematical Diversions, by Martin Gardner (Chicago, IL: University of Chicago Press, 1991). Reviewed in: *Mathematics Teacher* 85, no.6 (Sep. 1992): 494-95.

Usiskin, Zalman
NCSMP Algebra, by John W. McConnell, Susan Brown, Susan Eddins, Margaret Hackworth, and Zalman Usiskin (Tucker, GA: Scott Foresman, 1990). Reviewed in: *Mathematics Teacher* 84, no.2 (Feb. 1991): 144-46.

Van Dyke, James
Elementary Algebra (2d ed.), by Jack Barker, James Rogers, and James Van Dyke (Orlando, FL: Saunders College, 1992). Reviewed in: *Mathematics Teacher* 85, no.9 (Dec. 1992): 764.

Intermediate Algebra (2d ed.), by Jack Barker, James Rogers, and James Van Dyke (Orlando, FL: Saunders College, 1992). Reviewed in: *Mathematics Teacher* 85, no.9 (Dec. 1992): 765-66.

Video Tutorial Service (publisher)
Calculus: Applications of Differentiations, Parts 1-3, video (Brooklyn: Video Tutorial Service, 1990). Reviewed in: *Mathematics Teacher* 84, no.6 (Sep. 1991): 496-97.

Pre-Calculus, video (Brooklyn: Video Tutorial, 1990). Reviewed in: *Mathematics Teacher* 84. no.6 (Sep. 1991): 498.

Vingerhoets, Rob
The Maths Workshop: A Program of Maths Activities for Upper Primary, by Rob Vingerhoets (Victoria, Australia: Dellasta, 1990). Reviewed in: *Arithmetic Teacher* 38, no.9 (May 1991): 48.

Vitatora, Steven S.
Advanced Algebra, by Sharon L. Senk, Denisse R. Thompson and Steven S. Vitatora (Glenview, IL: Scott Foresman, 1990). Reviewed in: *Mathematics Teacher* 84, no.3 (Mar. 1991): 235-36.

Wadsworth
Algebra: A First Course, 3d ed., by John Baley and Martin Holstege (Belmont, CA: Wadsworth, 1990). Reviewed in: *Mathematics Teacher* 84, no.7 (Oct. 1991): 571.

Beginning Algebra, by Alfonse Gobran (Belmont, CA: Wadsworth, 1991). Reviewed in: *Mathematics Teacher* 84, no.8 (Nov. 1991): 674.

Developmental Mathematics (3d ed.), by C. L. Johnston, Alden T. Willis, and Gale M. Hughes (Belmont, CA: Wadsworth, 1991). Reviewed in: *Mathematics Teacher* 85, no.6 (Sep. 1992): 491.

Essential Arithmetic (6th ed.), by C. L. Johnston, Alden T. Willis, and Jeanne Lazaris (Belmont, CA: Wadsworth, 1991). Reviewed in: *Mathematics Teacher* 85, no.6 (Sep. 1992): 492.

Mathematics: Its Power and Utility, 3d ed., by Karl J. Smith (Belmont, CA: Wadsworth, 1990). Reviewed in: *Mathematics and Computer Education* 26, no.1 (Winter 1992): 92.

Math Facts: Survival Guide to Basic Mathematics, by Theodore J. Szymanski and Anne Scanlan-Rohrer (Belmont, CA: Wadsworth, 1992). Reviewed in: *Mathematics Teacher* 85, no.6 (Sep. 1992): 493.

Trigonometry, by Charles N. Gantner and Thomas E. Gantner (Belmont, CA: Wadsworth, 1990). Reviewed in: *Mathematics Teacher* 84, no.1 (Jan. 1991): 70.

Waits, Bert K.
Graphic Calculator and Computer Graphing Laboratory Manual (2nd ed.), by Franklin Demana and Bert K. Waits (Menlo Park, CA: Addison-Wesley, 1992). Reviewed in: *Mathematics Teacher* 85, no.7 (Oct. 1992): 590-91.

Precalculus Mathematics: A Graphing Approach (2d ed.), by Franklin Demana, Bert K. Waits, and Stanley E. Clemens (Menlo Park, CA: Addison-Wesley, 1992). Reviewed in: *Mathematics Teacher* 85, no.7 (Oct. 1992): 592.

J. Weston Walch (publisher)
Estimation Tutor, software by Richard E. Rand (Portland, ME: J. Weston Walch, 1989). Reviewed in: *Arithmetic Teacher* 38, no.8 (Apr. 1991): 54.

We Love MATHS: 4 Imaginative Themes for Early Primary Students
by Anne Marell and Susan Stajnko (Mount Waterly, Australia: Dellasta, 1990). Reviewed in: *Arithmetic Teacher* 39, no.1 (Sep. 1991): 53-54.

Wells, David
The Penguin Dictionary of Curious and Interesting Numbers, by David Wells (New York: Penguin, 1987). Reviewed in: *Mathematics and Computer Education* 24, no.1 (Winter 1991): 92.

Wells, Judith K.
Fun with Money: A Problem-Solving Activity Book, by Carol A. Thornton and Judith K. Wells (Deerfield, IL: Learning Resources, 1991). Reviewed in: *Arithmetic Teacher* 40, no.2 (Oct. 1992): 128-29.

Link 'n' Learn Activity Book, by Carol A. Thornton and Judith K. Wells (Deerfield, IL: Learning Resources, 1990). Reviewed in: *Arithmetic Teacher* 40, no.2 (Oct. 1992): 130.

Math Mats: Hands-On Activities for Young Children, set 2, manipulatives, by Carole A. Thornton and Judith K. Wells (Allen, TX: Teaching Resources, 1990). Reviewed in: *Arithmetic Teacher* 37, no.7 (Mar. 1992): 44.

Wells, Judith K. *(cont'd)*
On the Button in Math: Activities for Young Children, by Carol A. Thornton and Judith K. Wells (Allen, TX: DLM, 1990). Reviewed in: *Arithmetic Teacher* 38, no.9 (May 1991): 46.

Primary Geoboard Activity Book, by Carol A. Thornton and Judith K. Wells (Deerfield, IL: Learning Resources, 1990). Reviewed in: *Arithmetic Teacher* 40, no.2 (Oct. 1992): 130.

West (publisher)
Precalculus: A Problems-Oriented Approach, by David Cohen (St. Paul: West, 1990). Reviewed in: *Mathematics Teacher* 87 no. 7 (Oct. 1991): 574.

West, Beverly
Analyzer, software, by Douglas Alfors and Beverly West (Ithaca, NY: Delta-Epsilon Software, 1990). Reviewed in: *Mathematics Teacher* 85, no.3 (Mar. 1992): 240.

Weston, Martha
Do You Wanna Bet? Your Chance To Find Out about Probability, by Jean Cushman and illustrated by Martha Weston (New York: Clarion Books, 1991). Reviewed in: *Arithmetic Teacher* 40, no.4 (Dec. 1992): 240.

The $1.00 Word Riddle Book, by Marilyn Burns and illustrated by Martha Weston (New Rochelle, NY: Cuisenaire, 1990). Reviewed in: *Arithmetic Teacher* 39, no.2 (Oct. 1991): 51-52.

What Comes in Two's, Three's, and Four's (New York: Simon & Schuster, 1990). Reviewed in: *Instructor* 102, no.2 (Sep. 1992): 80.

What Do You Do with a Broken Calculator? software (Pleasantville, NY: Sunburst, 1989). Reviewed in: *Arithmetic Teacher* 39, no.3 (Nov. 1991): 58.

What's My Angle software (Torrance, CA: Davidson, 1991). Reviewed in: *Media & Methods* 28, no.3 (Jan./Feb. 1992): 60.

What's Your Game? A Resource Book for Mathematical Activities by Michael Cornelius and Alan Parr (New York: Cambridge University Press, 1991). Reviewed in: *Arithmetic Teacher* 40, no.4 (Dec. 1992): 241.

What's Your Problem? Posing and Solving Mathematical Problems by Penny Skinner (Portsmouth, NH: Heinemann, 1990). Reviewed in: *Arithmetic Teacher* 40, no.4 (Dec. 1992): 241-42.

What To Solve? Problems and Suggestions for Young Mathematicians by Judita Lofman (Cary, NC: Oxford University Press). Reviewed in: *Mathematics Teacher* 84, no.6 (Sep. 1991): 496.

Whimbey, Arthur
Problem Solving and Comprehension (5th ed.), by Arthur Whimbey and Jack Lochhead (Hillsdale, NJ: Lawrence Erlbaum Associates, 1991). Reviewed in: *Mathematics Teacher* 85, no.7 (Oct. 1992): 592.

Thinking through Math Word Problems, by Authur Whimbey, Jack Lochhead, and Paula Potter (Hillsdale, NJ: Laurence Erlbaum, 1990). Reviewed in: *Arithmetic Teacher* 39, no.2 (Oct. 1991): 52.

Whipple, Robert
Geometry for Enjoyment and Challenges, by Richard Rhoad, George Milauskas, and Robert Whipple (Evanston, IL: McDougal Schwartz, Judah L.Littell, 1991). Reviewed in: *Mathematics Teacher* 84, no.6 (Sep. 1991): 488-89.

Wilcox, Loyd V.
Prealgebra for Problem Solvers: A New Beginning, by Loyd V. Wilcox (Pacific Grove, CA: Brooks/Cole, 1991). Reviewed in: *Mathematics Teacher* 84, no.6 (Sep. 1991): 492.

Wilding, Sally
Building Self-Confidence in Math: A Student Workbook (2d ed.), by Sally Wilding and Elizabeth Shearn (Dubuque, IA: Kendall/ Hunt, 1991). Reviewed in: *Mathematics Teacher* 85, no.6 (Sep. 1992): 496.

Zitarelli, David E.
Finite Mathematics with Calculus: An Applied Approach (2d ed.), by David E. Zitarelli and Raymond F. Coughlin (Orlando, FL: Saunders College, 1992). Reviewed in: *Mathematics Teacher* 85, no.9 (Dec. 1992): 764.

KRAUS CURRICULUM DEVELOPMENT LIBRARY CUSTOMERS

THE following list shows the current subscribers to the Kraus Curriculum Development Library (KCDL), Kraus's annual program of curriculum guides on microfiche. Customers marked with an asterisk (*) do not currently have standing orders to KCDL, but do have recent editions of the program. This information is provided for readers who want to use KCDL for models of curriculum in particular subject areas or grade levels.

Alabama

Auburn University
Ralph Brown Draughton Library/Serials
Mell Street
Auburn University, AL 36849

Jacksonville State University
Houston Cole Library/Serials
Jacksonville, AL 36265

University of Alabama at Birmingham
Mervyn H. Sterne Library
University Station
Birmingham, AL 35294

*University of Alabama at Tuscaloosa
University Libraries
204 Capstone Drive
Tuscaloosa, AL 35487-0266

Alaska

*University of Alaska—Anchorage
Library
3211 Providence Drive
Anchorage, AK 99508

Arizona

Arizona State University, Phoenix
Fletcher Library/Journals
West Campus
4701 West Thunderbird Road
Phoenix, AZ 85069-7100

Arizona State University, Tempe
Library/Serials
Tempe, AZ 85287-0106

Northern Arizona University
University Library
Flagstaff, AZ 86011

University of Arizona
Library/Serials
Tuson, AZ 85721

Arkansas

Arkansas State University
Dean B. Ellis Library
State University, AR 72467

Southern Arkansas University
The Curriculum Center
SAU Box 1389
Magnolia, AR 71753

*University of Arkansas at Monticello
Library
Highway 425 South
Monticello, AR 71656

University of Central Arkansas
The Center for Teaching & Human
    Development
Box H, Room 104
Conway, AR 72032

California

California Polytechnic State University
Library/Serials
San Luis Obispo, CA 93407

California State Polytechnic University
Library/Serials
3801 West Temple Avenue
Pomona, CA 91768

California State University at Chico
Meriam Library
Chico, CA 95929-0295

*California State University, Dominguez Hills
Library
800 East Victoria Street
Carson, CA 90747

California State University at Fresno
Henry Madden Library/Curriculum Department
Fresno, CA 93740

California State University at Fresno
College of the Sequoia Center
5241 North Maple, Mail Stop 106
Fresno, CA 93740

California State University at Fullerton
Library Serials BIC
Fullerton, CA 92634

California State University at Long Beach
Library/Serials Department
1250 Bellflower Boulevard
Long Beach, CA 90840

*California State University at Sacramento
Library
2000 Jed Smith Drive
Sacramento, CA 95819

California State University, Stanislaus
Library
801 West Monte Vista Avenue
Turlock, CA 95380

*La Sierra University
Library
Riverside, CA 92515

Los Angeles County Education Center
Professional Reference Center
9300 East Imperial Highway
Downey, CA 90242

National University
Library
4007 Camino del Rio South
San Diego, CA 92108

San Diego County Office of Education
Research and Reference Center
6401 Linda Vista Road
San Diego, CA 92111-7399

San Diego State University
Library/Serials
San Diego, CA 92182-0511

*San Francisco State University
J. Paul Leonard Library
1630 Holloway Avenue
San Francisco, CA 94132

San Jose State University
Clark Library, Media Department
San Jose, CA 95192-0028

*Stanford University
Cubberly Library
School of Education
Stanford, CA 94305

*University of California at Santa Cruz
Library
Santa Cruz, CA 95064

Colorado

Adams State College
Library
Alamosa, CO 81102

University of Northern Colorado
Michener Library
Greeley, CO 80639

Connecticut

*Central Connecticut State University
Burritt Library
1615 Stanley Street
New Britain, CT 06050

District of Columbia

The American University
Library
Washington, DC 20016-8046

*United States Department of Education/OERI
Room 101
555 New Jersey Avenue, N.W., C.P.
Washington, DC 20202-5731

*University of the District of Columbia
Learning Resource Center
11100 Harvard Street, N.W.
Washington, DC 20009

Florida

*Florida Atlantic University
Library/Serials
Boca Raton, FL 33431-0992

Florida International University
Library/Serials
Bay Vista Campus
North Miami, FL 33181

Florida International University
Library/Serials
University Park
Miami, FL 33199

Marion County School Board
Professional Library
406 S.E. Alvarez Avenue
Ocala, FL 32671-2285

*University of Central Florida
Library
Orlando, FL 32816-0666

University of Florida
Smathers Library/Serials
Gainesville, FL 32611-2047

*University of North Florida
Library
4567 St. Johns Bluff Road South
Jacksonville, FL 32216

*University of South Florida
Library/University Media Center
4202 Fowler Avenue
Tampa, FL 33620

University of West Florida
John C. Pace Library/Serials
11000 University Parkway
Pensacola, FL 32514

Georgia

*Albany State College
Margaret Rood Hazard Library
Albany, GA 31705

Atlanta University Center in Georgia
Robert W. Woodruff Library
111 James P. Brawley Drive
Atlanta, GA 30314

*Columbus College
Library
Algonquin Drive
Columbus, GA 31993

Kennesaw College
TRAC
3455 Frey Drive
Kennesaw, GA 30144

University of Georgia
Main Library
Athens, GA 30602

Guam

*University of Guam
Curriculum Resources Center
College of Education
UOG Station
Mangilao, GU 96923

Idaho

*Boise State University
Curriculum Resource Center
1910 University Drive
Boise, ID 83725

Illinois

Community Consolidated School District 15
Educational Service Center
505 South Quentin Road
Palatine, IL 60067

DePaul University
Library/Serials
2323 North Seminary
Chicago, IL 60614

Illinois State University
Milner Library/Periodicals
Normal, IL 61761

Loyola University
Instructional Materials Library
Lewis Towers Library
820 North Michigan Avenue
Chicago, Illinois 60611

National—Louis University
Library/Technical Services
2840 North Sheridan Road
Evanston, IL 60201

Northeastern Illinois University
Library/Serials
5500 North St. Louis Avenue
Chicago, IL 60625

*Northern Illinois University
Founders Memorial Library
DeKalb, IL 60115

Southern Illinois University
Lovejoy Library/Periodicals
Edwardsville, IL 62026

*University of Illinois at Chicago
Library/Serials
Box 8198
Chicago, IL 60680

University of Illinois at Urbana—Champaign
246 Library
1408 West Gregory Drive
Urbana, IL 61801

Indiana

Indiana State University
Cunningham Memorial Library
Terre Haute, IN 47809

Indiana University
Library/Serials
Bloomington, IN 47405-1801

Kentucky

Cumberland College
Instructional Media Library
Williamsburg, KY 40769

* Jefferson County Public Schools
  The Greens Professional Development
    Academy
  4425 Preston Highway
  Louisville, KY 40213

Maine

University of Maine
Raymond H. Fogler Library/Serials
Orono, ME 04469

Maryland

* Bowie State University
  Library
  Jericho Park Road
  Bowie, MD 20715

* University of Maryland
  Curriculum Laboratory
  Building 143, Room 0307
  College Park, MD 20742

Western Maryland College
Hoover Library
2 College Hill
Westminster, MD 21157

Massachusetts

* Barnstable Public Schools
  230 South Street
  Hyannis, MA 02601

Boston College
Educational Resource Center
Campion Hall G13
Chestnut Hill, MA 02167

Bridgewater State College
Library
3 Shaw Road
Bridgewater, MA 02325

Framingham State College
Curriculum Library
Henry Whittemore Library
Box 2000
Framingham, MA 01701

Harvard University
School of Education
Monroe C. Gutman Library
6 Appian Way
Cambridge, MA 02138

* Lesley College
  Library
  30 Mellen Street
  Cambridge, MA 02138

* Salem State College
  Professional Studies Resource Center
  Library
  Lafayette Street
  Salem, MA 01970

Tufts University
Wessell Library
Medford, MA 02155-5816

* University of Lowell
  OLeary Library
  Wilder Street
  Lowell, MA 01854

* Worcester State College
  Learning Resource Center
  486 Chandler Street
  Worcester, MA 01602

Michigan

Grand Valley State University
Library
Allendale, MI 49401

* Wayne County Regional Educational Services
    Agency
Technical Services
5454 Venoy
Wayne, MI 48184

Wayne State University
Purdy Library
Detroit, MI 48202

* Western Michigan University
Dwight B. Waldo Library
Kalamazoo, MI 49008

Minnesota

Mankato State University
Memorial Library
Educational Resource Center
Mankato, MN 56002-8400

Moorhead State University
Library
Moorhead, MN 56563

University of Minnesota
170 Wilson Library/Serials
309 19th Avenue South
Minneapolis, MN 55455

Winona State University
Maxwell Library/Curriculum Laboratory
Sanborn and Johnson Streets
Winona, MN 55987

Mississippi

Mississippi State University
Mitchell Memorial Library
Mississippi State, MS 39762

University of Southern Mississippi
Cook Memorial Library/Serials
Box 5053
Hattiesburg, MS 39406-5053

Missouri

Central Missouri State University
Ward Edwards Library
Warrensburg, MO 64093-5020

Missouri Southern State College
George A. Spiva Library
3950 Newman Road
Joplin, MO 64801-1595

Northeast Missouri State University
Pickler Library/Serials
Kirksville, MO 63501

Southwest Baptist University
ESTEP Library
Bolivar, MO 65613-2496

Southwest Missouri State University
#175 Library
Springfield, MO 65804-0095

* University of Missouri at Kansas City
Instructional Materials Center
School of Education
5100 Rockhill Road
Kansas City, MO 64110-2499

University of Missouri at St. Louis
Library
St. Louis, MO 63121

Webster University
Library
470 East Lockwood Avenue
St. Louis, MO 63119-3194

Nebraska

Chadron State College
Library
10th and Main Streets
Chadron, NE 69337

University of Nebraska
University Libraries
Lincoln, NE 68588

University of Nebraska at Kearney
Calvin T. Ryan Library/Serials
Kearney, NE 68849-0700

*University of Nebraska at Omaha
Education Technology Center/Instructional
  Material
Kayser Hall, Room 522
Omaha, NE 68182-0169

Nevada

*University of Nevada, Las Vegas
Materials Center—101 Education
Las Vegas, NV 89154

*University of Nevada, Reno
Library (322)
Reno, NV 89557-0044

New Hampshire

Plymouth State College
Herbert H. Lamson Library
Plymouth, NH 03264

New Jersey

Caldwell College
Library
9 Ryerson Avenue
Caldwell, NJ 07006

Georgian Court College
Farley Memorial Library
Lakewood, NJ 08701

Jersey City State College
Forrest A. Irwin Library
2039 Kennedy Boulevard
Jersey City, NJ 07305

*Kean College of New Jersey
Library
Union, NJ 07083

Paterson Board of Education
Media Center
823 East 28th Street
Paterson, NJ 07513

*Rutgers University
Alexander Library/Serials
New Brunswick, NJ 08903

St. Peters College
George F. Johnson Library
Kennedy Boulevard
Jersey City, NJ 07306

Trenton State College
West Library
Pennington Road CN4700
Trenton, NJ 08650-4700

William Paterson College
Library
300 Pompton Road
Wayne, NJ 07470

New Mexico

University of New Mexico
General Library/Serials
Albuquerque, NM 87131

New York

*BOCES—REPIC
Carle Place Center Concourse
234 Glen Cove Road
Carle Place, NY 11514

*Canisius College
Curriculum Materials Center
Library
2001 Main Street
Buffalo, NY 14208

Fordham University
Duane Library
Bronx, NY 10458

Hofstra University
Library
1000 Hempstead Turnpike
Hempstead, NY 11550

*Hunter College
Library
695 Park Avenue
New York, NY 10021

*Lehman College
Library/Serials
Bedford Park Boulevard West
Bronx, NY 10468

*New York University
Bobst Library
70 Washington Square South
New York, NY 10012

*Niagara University
Library/Serials
Niagara, NY 14109

Queens College
Benjamin Rosenthal Library
Flushing, NY 11367

St. Johns University
Library
Grand Central and Utopia Parkways
Jamaica, NY 11439

State University of New York at Albany
University Library/Serials
1400 Washington Avenue
Alany, NY 12222

State University of New York, College at
Buffalo
E. H. Butler Library
1300 Elmwood Avenue
Buffalo, NY 14222

State University of New York, College at
Cortland
Teaching Materials Center
Cortland, NY 13045

State University of New York, College at
Oneonta
James M. Milne Library
Oneonta, NY 13820

Teachers College of Columbia University
Millbank Memorial Library/Serials
525 West 120th Street
New York, NY 10027

North Carolina

*Appalachian State University
Instructional Materials Center
Belk Library
Boone, NC 28608

Charlotte—Mecklenburg Schools
Curriculum Resource Center
Staff Development Center
428 West Boulevard
Charlotte, NC 28203

*East Carolina University
Joyner Library
Greenville, NC 27858-4353

North Carolina A&T State University
F. D. Bluford Library
Greeensboro, NC 27411

North Carolina State University
D. H. Hill Library
Box 7111
Raleigh, NC 27695-7111

University of North Carolina at Chapel Hill
Davis Library/Serials
Campus Box 3938
Chapel Hill, NC 27599-3938

University of North Carolina at Charlotte
Atkins Library
UNCC Station
Charlotte, NC 28223

University of North Carolina at Wilmington
William M. Randall Library
601 South College Road
Wilmington, NC 28403-3297

*Western Carolina University
Hunter Library/Acquisitions
Cullowhee, NC 28723

Ohio

Bowling Green State University
Curriculum Center
Jerome Library
Bowling Green, OH 43403-0177

Miami University
Library
Oxford, OH 45056

*Ohio State University
2009 Millikin Road
Columbus, OH 43210

University of Akron
Bierce Library/Serials
Akron, OH 44325

*University of Rio Grande
Davis Library
Rio Grande, OH 45674

*Wright State University
Educational Resource Center
Dayton, OH 45435

Oklahoma

Southwestern Oklahoma State University
Al Harris Library
809 North Custer Street
Weatherford, OK 73096

*University of Tulsa
McFarlin Library
600 South College
Tulsa, OK 74104

Oregon

Oregon State University
Kerr Library/Serials
Corvallis, OR 97331-4503

Portland State University
Library/Serials
Portland, OR 97207

University of Oregon
Knight Library/Serials
Eugene, OR 97403

Pennsylvania

*Bucks County Intermediate Unit #22
705 Shady Retreat Road
Doylestown, PA 18901

*Cheyney University
Library
Cheyney, PA 19319

East Stroudsburg University of Pennsylvania
Library
East Stroudsburg, PA 18301

Holy Family College
Grant and Frankford Avenues
Philadelphia, PA 19114

*Indiana University of Pennsylvania
Media Resource Department
Stapleton Library
Indiana, PA 15705

Kutztown University
Curriculum Materials Center
Rohrbach Library
Kutztown, PA 19530

La Salle College
Instructional Materials Center
The Connelly Library
Olney Avenue at 20th Street
Philadelphia, PA 19141

Lock Haven University of Pennsylvania
Library
Lock Haven, PA 17745

*Millersville University
Ganser Library
Millersville, PA 17551-0302

*Pennsylvania State University
Pattee Library/Serials
University Park, PA 16802

*Shippensburg University of Pennsylvania
Ezra Lehman Library
Shippensburg, PA 17257-2299

*Slippery Rock University
Bailey Library
Instructional Materials Center
Slippery Rock, PA 16057

University of Pittsburgh
Hillman Library/Serials
Pittsburgh, PA 15260

West Chester University
Francis H. Green Library
West Chester, PA 19383

Rhode Island

Rhode Island College
Curriculum Resources Center
600 Mt. Pleasant Avenue
Providence, RI 02908

South Dakota

Northern State University
Williams Library
Aberdeen, SD 57401

University of South Dakota
I. D. Weeks Library
414 East Clark
Vermillion, SD 57069

Tennessee

Tennessee Technological University
Library
Cookeville, TN 38505

Trevecca Nazarene College
Curriculum Library
Mackey Library
33 Murfreesboro Road
Nashville, TN 37210-2877

*University of Tennessee at Chattanooga
Library/Serials
Chattanooga, TN 37403

*University of Tennessee at Martin
Instructional Improvement
Gooch HallRoom 217
Martin, TN 38238

*Vanderbilt University
Curriculum Laboratory
Peabody Library
Peabody Campus, Magnolia Circle
Nashville, TN 37203-5601

Texas

Baylor University
School of Education
Waco, TX 76798-7314

East Texas State University
Curriculum Library
Commerce, TX 75429

*East Texas State University
Library
Texarkana, TX 75501

*Houston Baptist University
Moody Library
7502 Fondren Road
Houston, TX 77074

*Incarnate Word College
Library
4301 Broadway
San Antonio, TX 78209

*Sam Houston State University
Library
Huntsville, TX 77341

*Southern Methodist University
Fondren Library
Dallas, TX 75275-0135

*Stephen F. Austin State University
Library/Serials
Box 13055 SFA Station
Nacogdoches, TX 75962

Texas A&M University
Library/Serials
College Station, TX 77843-5000

*Texas Tech University
Library
Lubbock, TX 79409

Texas Womans University
Library
Box 23715 TWU Station
Denton, TX 76204

University of Houston—University Park
University of Houston Library
Central Serial
4800 Calhoun
Houston, TX 77004

University of North Texas
Library
Denton, TX 76203

*University of Texas at Arlington
Library
702 College Street
Arlington, TX 76019-0497

University of Texas at Austin
General Libraries/Serials
Austin, TX 78713-7330

University of Texas at El Paso
Library
El Paso, TX 79968-0582

*University of Texas—Pan American
School of Education
1201 West University Drive
Edinburg, TX 78539

Utah

Utah State University
Educational Resources Center
College of Education
Logan, UT 84322-2845

Vermont

University of Vermont
Guy W. Bailey Library/Serials
Burlington, VT 05405

Virginia

Longwood College
Dabney Lancaster Library
Farmville, VA 23909-1897

*Regent University
Library
Virginia Beach, VA 23464-9877

University of Virginia
Alderman Library
Serials/Periodicals
Charlottesville, VA 22901

*Virginia Beach Public Schools
Instruction and Curriculum
School Administration Building
2512 George Mason Drive
Virginia Beach, VA 23456

Washington

Central Washington University
Library/Serials
Ellensburg, WA 98926

University of Puget Sound
Collins Library
Tacoma, WA 98416

University of Washington
Library/Serials
Seattle, WA 98195

Washington State University
Library
Pullman, WA 99164-5610

Western Washington University
Wilson Library
Bellingham, WA 98225

Wisconsin

University of Wisconsin—Eau Claire
Instructional Media Center
Eau Claire, WI 54702-4004

University of Wisconsin—Madison
Instructional Materials Center
225 North Mills
Madison, Wisconsin 53706

University of Wisconsin—Oshkosh
F. R. Polk Library
Oshkosh, WI 54901

University of Wisconsin—Platteville
Library
One University Plaza
Platteville, WI 53818-3099

University of Wisconsin-River Falls
Chalmer Davee Library
River Falls, WI 54022

University of Wisconsin—Whitewater
Learning Resources
Whitewater, WI 53190

Wyoming

*University of Wyoming
Coe Library
15th and Lewis
Laramie, WY 82071

AUSTRALIA

Griffith University
Library
Mount Gravatt Campus
Nathan, Queensland 4111

CANADA

The Ontario Institute for Studies in Education
Library
252 Bloor Street West
Toronto, Ontario M5S 1V6

Queens University
Education Library
McArthur Hall
Kingston, Ontario K7L 3H6

*University of British Columbia
Education Library
2125 Main Mall
Vancouver, British Columbia V6T 1Z5

*University of New Brunswick
Harriet Irving Library/Serials
Fredericton, New Brunswick E3B 5H5

University of Regina
Library/Serials
Regina, Saskatchewan S4S 0A2

University of Saskatchewan
Library
Saskatoon, Saskatchewan S7N 0W0

University of Windsor
Leddy Library/Serials
Windsor, Ontario N9B 3P4

*Vancouver School Board
Teachers Professional Library
123 East 6th Avenue
Vancouver, British Columbia V5T 1J6

HONG KONG

*The Chinese University of Hong Kong
University Library
Shatin, N.T.

KOREA

*Kyungpook National University
Department of Education
Taegu 702-701, Korea

THE NETHERLANDS

National Institute for Curriculum Development
(Stichting voor de Leerplanontwikkeling)
7500 CA Enschede

APPENDIX: CURRICULUM GUIDE REPRINT

T EACHERS and other people involved with curriculum can often benefit from seeing exemplary guides from other states, districts, or individual teachers. In this appendix we have reprinted large sections from a mathematics guide that we believe our readers will find helpful and interesting:

Add-Ventures for Girls: Building Math Confidence, by the Research and Educational Planning Center (Reno, NV: University of Nevada, 1990).

The research that produced this elementary teacher's guide arose out of a concern about the dearth of women in highly math-related fields. Girls seem to start out on a par with boys in math ability, but somewhere during the school years they lose confidence in their ability and tend to stay away from upper-level math classes and, consequently, math-intensive careers.

The sections reproduced are indicated with a checkmark in the guide's table of contents. The sections are structured similarly throughout the book: each starts out with sample research findings on the section's topic and goes on to offer strategies for combatting problems related to that topic. After a number of activities and worksheets, each section ends with an annotated list of resources.

Some pages have had to be omitted due to copyrighted material from another source appearing on those pages. The following pages have been left out of this reprint section: 43-44, 47, 148-56, 226-27, 234-35, and 244.

If you would like to order this guide in its entirety, contact the Women's Educational Equity Act Program's Publishing Center, Education Development Center, 55 Chapel Street, Newton, MA 02160.

Add-Ventures for Girls: Building Math Confidence

Elementary Teacher's Guide

Project Director
Margaret Franklin

Project Staff
Michelle Dotson
Roberta Evans
Eunice Foldesy
Barbara Gardner
Diane Rhea

Research and Educational Planning Center
University of Nevada
Reno, Nevada

Women's Educational Equity Act Program
U.S. Department of Education
Lauro F. Cavazos, Secretary

1990
WEEA Publishing Center
Education Development Center, Inc.
55 Chapel Street
Newton, Massachusetts 02160

Cover design by Annie Grear

Contents

Preface

The issue of girls and mathematics is important for all teachers. At the elementary level, girls often enjoy math and attain achievement levels equal to or higher than those of boys. However, by the time they reach high school, many bright girls become disinterested in mathematics, enroll in fewer advanced math classes, achieve lower math scores than boys on college placement tests, and are less likely to choose careers that are highly math-related. Mathematics is an important foundation for many rewarding occupations. In neglecting to develop their math skills, many girls are excluding themselves from a large number of potentially satisfying careers.

The reasons for high school girls' loss of interest and relatively lower achievement in math have been carefully studied by many researchers over the past ten years. No single cause has emerged. Instead, we find an interrelated set of attitudes, self-perceptions, and feelings, reinforced by society, parents, teachers, and peers that can combine to produce strong barriers toward girls excelling in math.

When girls enter elementary school, most of them enjoy math and do well in it. During the elementary grades, however, they begin to perceive mathematics as a subject more appropriate for boys and to lose confidence in their own mathematical abilities. It is particularly important that elementary teachers recognize the subtle and blatant messages that society gives to girls and take action to counteract these messages before girls drop out of math.

Although equitable mathematics instruction is very important, it is an area often neglected in teacher education. To fill this gap, many excellent supplementary publications on girls and math have been developed. This book describes the attitudes and the practices that cause girls to lose interest in math; it then provides strategies, activities, and resources that teachers can use to help girls overcome barriers and reach their full potential in mathematics. A second book has been prepared for junior high school math teachers.

To develop this book, we spent several months researching the subject of girls and math. We reviewed countless research studies and many resource books for parents and teachers. After delineating the problem and outlining the topics we wanted to cover in this guide, we invited local elementary, middle, and high school math teachers and administrators to brainstorm with us. These teachers and administrators discussed strategies, activities, and resources they had used in each of the selected topic areas. We incorporated their ideas, concepts from research materials, and our own thoughts into this guide. Two outstanding local elementary teachers, Darleen Azizi and Jackie Berrum, reviewed the book and added their comments and suggestions.

The final product represents the work of many persons. The authors wish to thank the members of our Advisory Committee: Kenneth Johns, Carol Olmstead, Jennifer Salls, Barbara Schlenker, Jeanne Reitz, Bob Huwe, Diane Barone, Jackie Berrum, Shane Templeton, Randy McClanahan, Jesse McClanahan, Terri Walsh, and Elaine Enarson. We would also like to thank the teachers and administrators who contributed ideas: Terry Terras, Joe Elcano, Marian Marks, Margaret Mason, Marge Sill, Pat Haller, Al Babb, Yvonne Shaw, Joan Mueller, Dan Carter and Shirley Williams. Also, thanks to the members of the Research and Educational Planning Center office staff who prepared the manuscript: Sandra Walsh, Janet Oxborrow, Pat Downey, Ted Muller, Claudia Eaker, and Tina Wilkinson.

Introduction

In the past ten years there have been many attempts to explain sex differences in persistence and achievement in mathematics. Cases have been made for differences in brain development and lateralization, in spatial ability, in hormonal balance—even for the presence of a (male) math gene. Such research has a tendency to excuse and preserve the status quo. Indeed, it implies that the differences are natural and necessary, universal, and therefore just. I, on the other hand, have been content to be more modest: I simply visited schools where these sex differences in achievement were minimal or absent and looked around. The same hormones, the same brain lobes, the same maturation patterns were at work as prevail elsewhere. But the young women were learning mathematics—principally, so far as I could see, because they had been given good reason to think they could and should.

—Patricia Casserly, "Encouraging young women
to persist and achieve in mathematics," p. 12

This book will help you give young women those "good reasons" to think they can and should learn mathematics. As an elementary teacher, you know that mathematics is an important subject. However, you may not be aware of how math acts as a "critical filter" when students enter postsecondary school. Without a sound advanced mathematics background, students are excluded from a large portion of college majors. A solid mathematics background is also crucial for most postsecondary vocational training programs and for most entry-level jobs. The purpose of this book is to help you prepare a foundation so that when they reach high school, students will want to take as many math courses as possible and will be able to achieve in mathematics.

During their early school years, students develop the skills and attitudes toward learning that form the basis of future academic growth. If students develop a negative learning pattern toward a subject, it is extremely difficult to change. We know that when girls reach adolescence, a number of factors may combine to produce strong internal (attitudinal) and external (societal) barriers to reaching their mathematics potential. Therefore, it is vitally important that elementary teachers do all they can to build girls' positive attitudes and skills in mathematics in the early years. This will help girls withstand later societal pressures, continue math studies, and feel free to select math-related careers.

This guide includes strategies, activities, and resources that deal with five major topics: Attitudes and Math, Math Relevance, the Learning Environment, Other Issues, and Mathematics Promotion. Within each of these major topic areas are several subtopics.

Each section of this guide begins with a discussion of research findings on the practices and/or attitudes that affect girls' math attitudes and performance. Strategies, activities, and resources that you can use to address each topic are described. Within each topic, activities for the primary grades are generally listed before those for the intermediate grades. An annotated resource list at the end of each chapter provides resources that contain valuable strategies, activities, and ideas.

Because many of the subtopics in the guide are interrelated, several of the strategies, activities, and resources are appropriate for more than one area. In those cases, the reader is referred to the appropriate section for additional information.

Most, but not all, of the activities in the book include math skill practice while covering the topics of concern. To use the guide, review it to see how the activities fit with what your students are currently learning. You may want to use the topic ideas, but modify the math portion of the activity to better fit your students' needs. The sections on attitudes and math relevancy are at the beginning of the book for two reasons. First, because negative and stereotypical attitudes about girls and math and lack of information about math usefulness and relevancy form the basis for many problems that surface later. And second, because we know that teachers usually devote little or no time to such topics. We encourage teachers to devise ways to spend more class time exploring and remediating negative attitudes and stereotypes and explaining why math is important; the time lost from math drill-and-practice will be more than returned when students develop positive attitudes and a high level of interest in math.

Another point that needs to be stressed is that although the activities and strategies suggested in this guide are particularly focused on providing equitable math instruction for girls, they can benefit all students. All of the suggestions are based on research findings, published resources, and practical ideas from math teachers. They represent sound educational practice and, when used as part of your mathematics curriculum, will provide a positive learning environment for both boys and girls. The book is based on the premise that we can encourage students' positive feelings as well as providing information about mathematical facts and processes. All students will benefit from this approach.

Part 1

Attitudes and Math

This section contains suggestions that will

1. help you build students' math confidence

2. raise students' aspirations and expectations

3. change detrimental attribution patterns

4. deal with sex-role stereotyping and stereotyping of mathematics as a "male" subject

Each of the following attitudes can deter a young woman from taking advanced math courses: having low self-confidence about her math abilities, having low expectations for success in math, attributing failure in math to lack of ability, and viewing math as "unfeminine." When these attitudinal factors are combined, they can form an almost impenetrable barrier to math for a young woman.

Many psychologists believe that changes in attitudes can follow, rather than precede, changes in behavior. The following section includes ways to encourage girls to exhibit positive behaviors toward mathematics.

Building Math Confidence

Researchers have consistently found that confidence in math is directly related to later math achievement and decisions to enroll in elective high school math courses. Girls' and boys' self-confidence in their mathematical abilities do not differ in the primary grades. However, by grade six, boys have more confidence, even though their math test scores and grades are not any higher than those of girls. This difference in confidence becomes more pronounced and more detrimental for girls as they reach high school. A sample of research findings on math confidence is listed below.

- At each grade level from six through eleven, boys are more confident of their math abilities. (Fennema and Sherman 1978)

- At all grade levels, girls are more likely to experience "math anxiety"—an extremely debilitating fear of mathematics. (Boswell and Katz 1980; Tobias and Weissbrod 1980)

- Even though boys performed no better than girls in math at age 13, boys were much more confident of their math abilities. (Lantz, cited in Chipman and Wilson 1985)

- Bright female students are the most likely of any student group to underestimate their chances of success in math. (Licht and Dweck 1983)

- Even female math graduate students who were achieving on a par with male students doubted their ability to complete their advanced degrees in mathematics. (Becker 1984)

- Students' confidence as math learners is strongly related to their perceptions of teacher encouragement. (Sherman, cited in Chipman and Wilson 1985)

- From elementary school to college age, girls consistently rate their intellectual abilities lower than do boys, despite the fact that girls generally get better grades and score higher on most aptitude tests. (Russo 1985)

Many research findings also indicate that girls have less confidence in their mathematical abilities, independent of any real difference in performance. Thus, it is essential that elementary teachers work to develop girls' confidence in their math abilities, so that as math becomes more difficult in junior high and high school, girls will continue to feel that they have the ability to learn it.

Skolnick et al. (1982) have suggested that teachers utilize tasks that offer success for each student, feature many approaches with many possible answers, and offer confidence-building opportunities for guessing, checking, and estimating. The strategies, activities, and resources described on the following pages are designed to help your students gain confidence in their math abilities.

Strategies

1. Build students' confidence by publicly and privately acknowledging their academic and intellectual accomplishments (not their effort); e.g., "Mary, you figured out that answer very well" or "Joanne, you're really learning to solve these problems. With a little more practice, you'll have no difficulty with decimals." Try to focus on the intellectual aspects of girls' performance rather than neatness, organizational skills, or "just trying."

2. Practice is extremely important in building confidence. Make sure that girls get enough practice so that they can feel confident with their math skills.

3. Another confidence-building method that encourages student involvement without individual risk is the use of slates. All the students can show you their answers at the same time, and you can determine how well you have done in teaching the concept.

4. Try to structure math learning activities so that all students will be able to achieve success at some level.

5. Incorporate some math problems that call for many approaches with several right answers. Stress the idea that, in most cases, there is more than one way to solve a problem.

6. Provide opportunities for estimating, guessing, and checking.

7. Recognize students' math achievement, and especially improvement, by creating a "Math Star" bulletin board.

8. Girls are often reluctant to recognize and acknowledge their own ability— especially in mathematics. Help them learn that it's okay to say, "I'm very good at math." It's not always easy for us to acknowledge our own abilities. If you are a female teacher, try setting an example for girls by saying something like, "I've always been good at math." If that statement is unrealistic, say, "I've always wanted to do well in math."

9. Create more opportunities for cooperative learning and minimize overt competition between classmates. Efforts to utilize groups where everyone must participate equally and will be given feedback collectively can be beneficial in building girls' math confidence. See the "Encouraging Cooperative Learning" section in this guide for further information.

10. Turn the tables in class, and let students take turns asking the teacher questions about math. This technique can generate some good discussion and promote the idea that "there's no such thing as a stupid question." In fact, questions become the learning environment norm with this practice.

11. Have children keep track of their own increases in computational speed and accuracy. Compare week-to-week results and reward students for bettering their performance records. In using this strategy, make sure students are competing with themselves, not with other students.

12. Use girls as peer tutors in math. Being asked to help others will definitely build the confidence of the tutor. Tutoring also helps both students gain increased understanding.

Activity

Guesses Galore

Objective

To provide students with an accepting environment in which they feel free to guess at math responses, minimizing the pressure for answer accuracy

Grade Level

Grades K–3

Time

10 minutes or longer

Materials

Chalkboard; jar of discrete items such as marbles, cookies, pennies, paper clips, etc.—size and number of items can vary. (For party days, use a related material, e.g., jar of candy corn for Halloween.)

Procedure

Ask students to guess how many items are in the jar. Encourage lots of guesses from each student and make sure all of the girls contribute their guesses. Depending on your class you may have to establish behavioral rules—no looking, etc.—during the guessing. Have students write down the guesses arranged from low to high or high to low to practice counting forward and counting back. Assign students to count the items and report the answer to the class. As they count, ask the students to count out the items into cups by 5s, 10s, or 2s, so that other counting can be practiced. Lay the objects out on a place value board, so the children can see a graphic representation of the number. For example, count 10 items into a cup; put 10 cups on the board.

Example

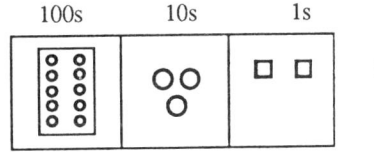

100s   10s   1s   132

Variation

You can vary the things students guess about—e.g., how many petals on a flower, how many students have birthdays this month, how far is it from your town or city to some other city.

This guessing activity can be spread over several days. Let the children look at the container (use an adequate number of objects so they cannot be counted readily). Also change the container shape and size, but keep the quantity of items the same, and see if the children's estimations change. (Let the children see you change the items from one container to the next, so they understand that the number did not change.)

At Halloween, the number of seeds in a pumpkin can be used for guessing and checking. Also, students can compare a large pumpkin and a smaller pumpkin. Have the children decide which they believe will have more seeds, then count and report to the class.

8   Part 1   Attitudes and Math

Activity

Many Answers

Objective

To demonstrate to students that many problems have more than one "correct" answer; to allow all students to experience success and thus build confidence

Grade Level

Grades 1–6, depending on complexity

Time

15 minutes or longer

Materials

Slates, blank 5" x 8" response cards or pieces of paper, "Sample Questions" worksheet on the following page

Procedure

Use appropriate sample questions from the list on the following page, or construct your own. Have students work the problems. After students have shown their answers, determine how many different correct answers were given for each question, and discuss the idea of multiple approaches and correct answers with students. If students are responding on paper rather than slates, you may also want to let them determine how many children chose each correct answer. Make a sample bar graph of frequency of correct answers for each question.

Variations

1. Let students prepare or submit ideas for sample questions along with all the correct answers they can think of.

2. Give students a number, say "12," and see how many ways they can think of to add or multiply whole numbers or fractions to get this result.

3. The game of cribbage uses the principle of adding card values to reach 15. Teach your elementary students to play cribbage in its regular form, or vary the target number—use 12 or 23, for example.

4. Use real-life problems that could have many answers. For example, if Mary's allowance is $5, how could she spend (or save) it?

Worksheet

Sample Questions

What two numbers add up to 4? _____ + _____ = 4

What two numbers add up to 12? _____ + _____ = 12

What three numbers add up to 10? _____ + _____ + _____ = 10

What three numbers add up to 18? _____ + _____ + _____ = 18

What two numbers add up to 40? _____ + _____ = 40

What two numbers can be multiplied to get 6? _____ x _____ = 6

What two numbers can be multiplied to get 12? _____ x _____ = 12

What two numbers can be multiplied to get 100? _____ x _____ = 100

What two mixed numbers can be added to get 8? _____ + _____ = 8

What two mixed numbers can be added to get 13? _____ + _____ = 13

What two decimal numbers can be added to get 1? _____ + _____ = 1

Activity

Fraction Grids

Objective	To build students' confidence in their ability to understand and use fractions by using manipulatives
Grade Level	Grades 1–6
Time	Variable
Materials	"Fraction Grids" on the following page (For grades 1–2, limit fractions to 1/2, 1/3, and 1/4.)
Procedure	Duplicate one copy of the fraction grids on the following page for each student. Grids may then be colored, laminated, and cut apart. They can be used in a variety of activities with fractions, and they are especially helpful for visualizing relationships between fractions. Ask a variety of questions that students can answer by looking at the grids. For example, which is larger, 1/3 or 2/5? How many eighths are in 1/4, and so on. After using the fraction grids, store them in individual coin envelopes.

Handout

Fraction Grids

1

1/2	1/2

1/3	1/3	1/3

1/4	1/4	1/4	1/4

1/5	1/5	1/5	1/5	1/5

1/6	1/6	1/6	1/6	1/6	1/6

1/7	1/7	1/7	1/7	1/7	1/7	1/7

1/8	1/8	1/8	1/8	1/8	1/8	1/8	1/8

$\frac{1}{9}$	$\frac{1}{9}$	$\frac{1}{9}$	$\frac{1}{9}$	$\frac{1}{9}$	$\frac{1}{9}$	$\frac{1}{9}$	$\frac{1}{9}$	$\frac{1}{9}$

$\frac{1}{10}$	$\frac{1}{10}$	$\frac{1}{10}$	$\frac{1}{10}$	$\frac{1}{10}$	$\frac{1}{10}$	$\frac{1}{10}$	$\frac{1}{10}$	$\frac{1}{10}$	$\frac{1}{10}$

12   Part 1   Attitudes and Math

Activity

More or Less Game

Objective	To build confidence by giving children practice recognizing relative sizes of numbers; to emphasize place value (see variation 5)
Grade Level	Grades 2–6 (You can vary the complexity, depending on game rules and how you number the cards.)
Time	5–15 minutes or longer per game
Materials	Deck of 50–100 numbered cards
Procedure	Construct the cards by cutting 3" x 5" cards in half and numbering each of the resulting 3" x 2 1/2" cards. The numbers do not have to be sequential. For example, you can use 10 two-digit numbers, 10 three-digit numbers, 10 four-digit numbers, and 10 five-digit numbers.

The game should be played in small groups of 3–6 students. Deal five cards to each student, and turn one card face up from the pack to start the game. Place the turned-up card in the center of the playing surface. Moving clockwise around the group, each student plays one of his or her cards by laying it face up to the right if it is *more* than the faced (played) card or to the left if it is *less* than the faced card. In the example below, card 51 was turned face up to start the game. The next player played 50 (less than 51) to the left. The next player played 73 (more than 51) to the right; 87 (more than 73) was then played to the right, and so forth.

A player who does not hold a playable card draws one from the pack and waits until her or his next turn to play. All players must monitor the game to be sure that no errors are made in placing the cards. If a player makes an error, the card must be placed back in his or her hand and another card drawn from the pack. Then the player must wait until his or her next turn to play a card. The object of the game is to play as many cards as possible and/or to be the person with the fewest total "points" left in his or her hands. The game ends when one player is out of cards or when no one else can play and all of the cards in the pack have been drawn. At this point, players with cards left in their hands must add the card values to determine the number of points they have accumulated.

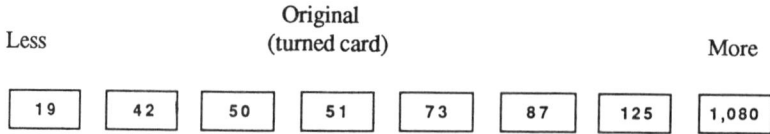

Less			Original (turned card)				More
19	42	50	51	73	87	125	1,080

Variations

1. To make the game very simple, use only numbers below 20 or 30 and play only left or right, i.e., *only more* or *only less*. For very young children, limit the number of cards to no more than 10, and include a template of the correct layout of those 10 cards so the children can self-check the game and reinforce the correct sequence.

2. To make the game more challenging, use larger numbers, deal more cards to each student, and/or allow students to play "up" (for less) and "down" (for more) from the already played cards as in Scrabble.

Example

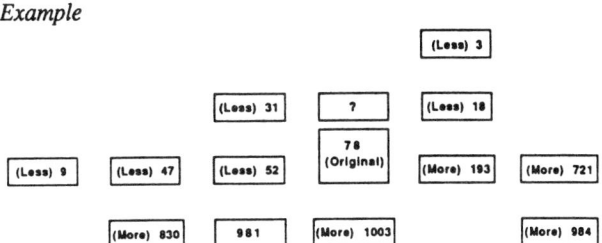

In the example above, the space above the original "78" would be unplayable because the card played would have to be more than 31, and less than 78, but also less than 18! Similarly, the space below the 193 would be unplayable, but card 981 can be played because it is more than 830, more than 52, and less than 1003.

3. To emphasize cooperative learning, make the object of the game be that the group play as many cards as possible. Older students can discuss strategies for maximizing the number of cards played by the group.

4. In the lower grades, make correct reading of all numbers a task that gains the team a point. Numbers can also be identified as odd or even to earn a point for the team.

5. These cards can be used to practice place value. Divide children into two teams, and let each child randomly choose a card. The child must count out the correct number on each card using unifix cubes, beans, or sticks.

Example   *31*   or   *31*

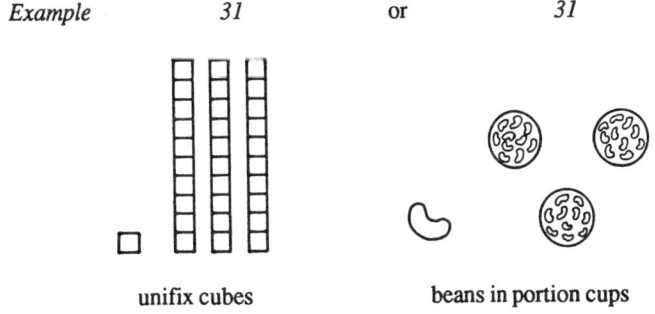

unifix cubes   beans in portion cups

Activity

I Am the Greatest! Game

Objective

To give students practice reading large numbers and assist them in learning place value

Grade Level

Grades 4–6

Time

15 minutes per game

Materials

Chalkboard, numbered cards or tickets, small jar

Procedure

Cut out 20 small (1" or 2" square) cards, and number them 0 through 9. (You will have two of each number.) Draw nine large squares with two large commas on the board.

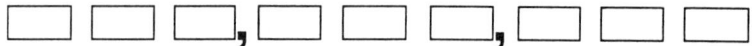

First, tell students that the object of the game is to create the largest number possible. Ask each student to draw the same squares on scratch paper. Put the numbered cards in the jar. Shake the jar to mix the numbers, draw out the first number, and announce it to the class. The students then write the number in one of their squares. Continue to draw and announce each number. Make sure students put their pencils down between numbers—no erasing or changing allowed. After you have drawn nine cards, all students should have constructed a nine-digit number.

Ask the students who has the greatest number. The student who thinks he or she has constructed the largest number reads it out loud—and has to read it correctly. Ask if any one has a larger number. Arrange the drawn numbers in order of magnitude to determine the largest possible nine-digit number that could have been constructed. Ask the students who came up with the greatest numbers to explain their strategies.

Variations

1. To make the game simpler and more appropriate for younger children use only three- or four-digit numbers.

2. This game can be excellent as a cooperative learning tool. Have small groups of students cut out squares or cards and number them. Each group tries to place their numbered cards or tiles in order to make the largest number. They then briefly discuss why they placed each card in the position they chose and work out a strategy for rearranging the numbers so that the largest number is obtained. Let small groups develop a strategy for playing the game when the actual numbers are unknown.

Note: Many teachers use Marcy Cook's tiling program (see resource list). Each student keeps a ziplock bag containing 10 one-inch square, light-colored ceramic tiles that are numbered 0–9 with a permanent marker. The tiles are great for games like this, as well as other active participation activities.

Activity

"Guesstimating" and Measuring

Objective

To allow students to practice estimating and then checking their answers through measurement; to build confidence in estimating and measuring skills

Grade Level

Grades 4–6 (For grades 1–3, see variation 2.)

Time

15–30 minutes

Materials

Chalkboard, scratch paper, rulers, and other appropriate measuring instruments

Procedure

Ask students, individually or in pairs, to estimate (or "guesstimate") a distance, weight, area, number of objects in a container, and so on (see examples below). Then, let students check their answers by actual measurement. Before the measurement process, discuss how students might find the answer.

It is important that you help students build estimating skills. Begin by working with small amounts or distances that they can actually count or measure, i.e., a small jar containing 15 crayons. The children may touch (handle) the jar.

As they progress, use larger numbers, distances, and so on, and provide reference points on which students can base their estimates. For example, "Here is a jar filled with marbles; write your estimate. Now, let me tell you that 10 marbles cover the bottom of the jar; do you want to change your estimate? Thirty marbles fill half the jar; do you want to adjust your estimate?" You can give the same type of reference points for distances, i.e., "It is 20 feet from the wall to this point."

Using this procedure, children will learn a method for determining their "guesses" or estimates, and not just respond with their "father's age" or a favorite number. For every guesstimate, help students develop a strategy for making a reasonable estimate.

Examples

How many feet wide is our classroom?

How far (in feet or yards) is it from our classroom to the cafeteria?

How long is a minute?

What is the area of our chalkboard?

How many boots are in the closet today (on a rainy or snowy day)?

Which of several flowers has more petals? How many petals does each have?

Which of these objects weighs more? What are two ways we could find the answer?

Which of these containers holds more water? What are some ways we could measure the amount of water each holds?

Use the suggestions above as ideas. Let students think up their own questions for the class.

Variations

1. Develop some guesstimating problems to use as take-home assignments—questions for which answers can be estimated at school and measured at home. For example, how many square feet are there in your kitchen? How far is it from your house to your neighbor's house?

2. For younger children, grades 1–2, this activity can be done using concrete objects for the measurement. For example, how many erasers long is your desk? How many new pencils will it take to go from the door to the wall? Next, compare two objects. How many more paper clips longer is this book than that book? (Subtract to find the answer.)

Resources

Burns, M. 1975. *The I hate mathematics book.* Boston: Little, Brown.

> For those students who "seemingly" hate mathematics, this book provides many relevant activities to boost confidence and aspirations. Positive attitudes toward mathematics develop as students experiment with and investigate the uses of mathematics in solving everyday problems. Suitable only for upper elementary students.

Cook, M. *Math materials.* Catalog. Balboa Island, California.

> These materials include tiling sets, task cards, and books designed to add variety to math. The materials emphasize problem solving and focus on active student involvement. Also included are several books on cooperative learning. The catalog is available from Marcy Cook, P. O. Box 5840, Balboa Island, CA 92662.

Downie, D.; Slesnick, T.; and Stenmark, J. K. 1981. *Math for girls and other problem solvers.* Berkeley: University of California, Math/Science Network.

> The activities in this book encourage independent thinking and creativity in mathematics. Students and teachers are encouraged to think about problem solving in versatile ways and forms. Although this book was originally designed for females, the activities are appropriate and interesting to both boys and girls, ages 7–14.

Fennell, F., and Williams, D. 1986. *Ideas from the arithmetic teacher: Grades 4–6 intermediate school.* Reston, VA: National Council of Teachers of Mathematics.

> This is a collection of classroom-tested activities from the journal *Arithmetic Teacher*. Perforated pages allow easy duplication of activities, which can be used for supplementing, extending, or reinforcing daily mathematics lessons in numeration, whole number computation, rational numbers, geometry, measurement, and problem solving. Objectives, grade levels, directions, and answers are given for each activity.

Holden, L. 1987. "Math: Even middle graders can learn with manipulatives." *Learning 87* 16, no. 3: 52–55.

> Learning fractions can be frustrating. This article includes many ways that manipulatives can be used to help upper elementary students understand fractions and learn geometric concepts.

Jensen, R., and Spector, D. 1984. *Teaching mathematics to young children.* Englewood Cliffs, NJ: Prentice-Hall.

> This teacher and parent resource suggests many activities that can be used to explore math concepts with young children. The suggested vehicles for math instruction—including manipulative and creative movements, art activities, and games—allow children to tackle new ideas while developing problem-solving skills. Chapters and activities are organized from concrete to abstract, so that young children may master mathematical concepts and gain confidence in learning.

Kaseberg, A.; Kreinberg, N.; and Downie, D. 1980. *Use EQUALS to promote the participation of women in mathematics.* Berkeley: University of California, Math/Science Network.

> This handbook assists educators in conducting teacher training to increase awareness of the problem of female math avoidance, enhance female interest and competence in mathematics, and provide information about

opportunities for women in nontraditional careers. Ultimately, the purpose of the program is to help teachers promote positive math attitudes and bring about changes in the occupational patterns of women. The book includes activities that increase girls' confidence in their math abilities and relate the usefulness of mathematics to future career choices. An excellent sampling of strategy games, spatial activities, and logic problems is also included, as well as bibliographies on problem solving in mathematics and sex-fair counseling and instruction. Materials are suitable for grades 4–12.

Skolnick, J.; Langbort, C.; and Day, L. 1982. *How to encourage girls in math and science: Strategies for parents and educators.* Palo Alto, CA: Dale Seymour Publications.

This excellent resource examines the effect of sex-role socialization on girls' math/science skills and confidence. It explains how attitudes, parenting and teaching practices, stereotypical play activities and books, peer pressure, and career and family expectations cause girls to question their abilities in math and science, and thus hinder their development in these areas.

In addition to the summary of the socialization process, the book contains a variety of compensatory educational strategies and activities that may be used to encourage females in mathematics. These particularly focus on increasing math confidence, spatial visualization skills, and problem solving. Both parents and educators can benefit from this book.

Souviney, R. J. 1981. *Solving problems kids care about.* Palo Alto, CA: Scott, Foresman.

Solving problems kids care about is divided into two parts. The first section includes notes and strategies for teaching mathematical problem solving. The second section contains thirty real-world problems that encourage divergent and logical thinking. Many of the problems have a range of acceptable solutions and multiple solution strategies, so students have the opportunity to be creative, independent thinkers. Activities are designed for elementary through junior high school students; teachers will enjoy them also.

Stenmark, J. K.; Thompson, V.; and Cossey, R. 1986. *Family math.* Berkeley: University of California, Lawrence Hall of Science.

If mathematics promotion is a goal of your teaching, *Family math* activities will help you introduce parents and children to ideas that improve their math skills and help them gain an appreciation for math. Hands-on mathematical experiences provide families an opportunity to develop problem-solving skills by using the following strategies: looking for patterns, drawing pictures, working backwards, working cooperatively with a partner, and eliminating possibilities. Spatial relationships (geometry), estimation, data interpretation (probability and statistics), and mathematical reasoning are the mathematical concepts learned from *Family math.* Materials suitable for ages 5–18.

Stereotyping and Mathematics

Stereotyping means making generalizations about people or things based on commonly held beliefs or societal expectations rather than on actual individual characteristics. For example, we are using gender stereotypes when we believe that boys are "supposed to be aggressive" or that girls are "supposed to be quiet." Another common gender stereotype is the belief that boys have greater mathematical ability than girls. Some people also stereotype mathematics as a "masculine" subject and mathematicians as cold, unfeeling, unfeminine persons.

We are taught stereotypes by our parents, the media, textbooks, peers, and teachers. As girls reach adolescence, the stereotype about femininity and mathematics is one of the important barriers that prevent many of them from forming positive attitudes toward math. One manifestation of this can be seen in high school-age girls who achieve well in math but have low self-concepts and perceive themselves as being very unpopular. According to Fennema and Ayer (1984), when young girls believe that mathematics is inappropriate for them or their sex roles (i.e., not feminine), they feel anxious about succeeding in math and have more negative attitudes toward it. Male peer pressure is also an extremely important factor for adolescent girls.

- Occupational and subject-related stereotypes are well developed in children by grade three. (Hughes et al. 1985)

- Early in their school years, children learn that mathematics is closely identified with the male role; these stereotypes increase with age and become particularly debilitating for females' math achievement (Sheridan and Fizdale 1981); there is a strong negative relationship between the degree of stereotyping and female math achievement. (Boswell and Katz 1980)

- There is a negative relationship between high math achievement and self-image in sixth and seventh grade girls (Roberts et al., 1987); high school girls who were high math achievers rated their popularity extremely low as compared with popularity ratings of high math-achieving boys and those of girls who were high verbal/low math achievers. (Franklin and Wong 1987)

- Girls who took four years of theoretical math exhibited more conflict between sex roles and achievement than did cognitively equated girls who enrolled in fewer math courses. (Sherman 1982)

- High school-age boys and their parents are significantly more likely than girls and their parents to feel that math is more appropriate for males and that males' math skills are superior to those of females (Visser 1986; Franklin and Wong 1987); high school students classify mathematics as a "male" achievement domain. (Stage et al. 1985)

- Northam (1986) studied a number of math books for ages 3–13 that were published in England between 1970 and 1978. In these books, mathematical and scientific skills became increasingly defined as masculine as pupils moved through middle and junior high school. Women and girls almost disappeared from books for ages 12 and 13. In problems, boys and men were typically described in active terms—they were solving problems, explaining to others, devising, planning, performing, and competing. Girls were typically shown repeating or elaborating on a process already learned, cooperating or helping, and correcting another's behavior.

- Female college students are much less likely than males to select math-related majors. (Boli et al. 1984)

- The majority of female Ph.D.'s in mathematics believe that their field is stereotyped by other persons as masculine. (Boswell 1985)

- When asked why more young women do not pursue mathematics-related careers, "fear of masculine disapproval" is often given by parents as a reason. (Franklin and Wong 1987)

- Mathematically gifted girls are very reluctant to skip a grade or to enroll early in college math courses for fear of male peer rejection. Girls who take advanced placement courses in math stress the importance of girl friends' support in helping them deal with the disapproval of boys. (Fox 1981)

Students' needs to establish their masculinity or femininity become extremely important during the adolescent years. If, at that time, girls see mathematics as a "masculine" subject, and if they perceive the world of mathematics as a male-dominated place in which they do not belong, girls will begin to make educational and career decisions that exclude math.

Since gender and math stereotypes appear to develop during the early elementary years, it is critical that we begin to intervene during those years to offer some alternative views for girls. It is particularly important that we also target our intervention efforts toward boys, as well as girls, because boys tend to hold more stereotypical attitudes and because negative male peer pressure can be a very powerful deterrent to adolescent girls' positive math attitudes. The strategies, activities, and resources described on the following pages are designed to change these stereotypical attitudes.

Strategies

1. Word problems in which women are depicted as technical career persons functioning at the center of problems can go a long way toward reducing stereotypes. Be sure your text contains many such examples. If it does not, you can supplement it with your own problems. For example, as an engineer, Mary needed to find out the total weight of vehicles crossing a bridge. If an average of 100 cars crossed the bridge every day and each weighed 700 pounds, how many tons of weight did the bridge handle per day?

2. Simple statistics that chronicle the low percentage of women graduates in math-related fields such as engineering, the lower pay for women, etc., can serve as springboards for discussions about the unfairness of stereotyping in language and social assumptions. These statistics are often available through local school counselors, as well as state and regional sex equity centers. Your state department of education probably has a sex equity coordinator who can provide resources or tell you where to get them.

3. The spoken language is an extremely powerful tool in building and reinforcing or tearing down stereotypes. Be sure that you are not using a generic "he" when referring to students, principals, doctors, etc. Also, don't use a generic "she" when referring to other teachers, nurses, or parents.

4. Just as math-related activities need to be specifically targeted toward a female audience, males must be encouraged to consider female-dominated areas. Whenever possible, display your lack of sex bias by discussing home economics and theater (or other subjects) with boys in mind.

5. For open house night or in a newsletter, prepare a brief pitch to parents to help them become more aware of how their incidental stereotypical remarks about math and women may be damaging their daughters' math potentials. For example, typical remarks might include "that's just like a woman," when someone is unable to solve a math problem or "women are just no good at math" or "she thinks like a boy," when describing a girl who is good at math. Remind parents that "Attitudes are contagious—is yours worth catching?"

Activity

Who Should?

Objective	To generate awareness among students of their own sex-role biases, and to provide them with a framework for becoming open to seeing both sexes in diverse roles
Grade Level	Grades K–6 (You may want to shorten or use only one or two sections of the questionnaire for K–2 students, and administer it orally for nonreaders.)
Time	20 minutes for the questionnaire, plus 10–20 minutes discussion time
Materials	"Who Should" worksheet on the following pages
Procedure	See instructions on the worksheet. Use this questionnaire or any of its sections, to determine the extent of sex-role stereotyping evidenced by your students. Also, use the form as a springboard for discussion in areas where the class openly disagrees.
Variation	You might ask other teachers to give the questionnaire to students in their classes, and complete a school survey of student attitudes. The list of careers used here can be expanded.

Worksheet

Who Should

Student Information

1. Please circle "Boy" if you are a boy or "Girl" if you are a girl.

   I am a: Boy   Girl

2. Please circle your grade: K   1   2   3   4   5   6

PART I.   For each of these jobs, circle "Woman" if you think only a woman *should* do the job; circle "Man" if you think only a man *should* do the job; circle "Both" if you think both a woman or a man *should* do the job. Be sure to circle only one answer for each job.

3. Airplane pilot	Woman	Man	Both
4. Artist	Woman	Man	Both
5. Astronaut	Woman	Man	Both
6. Carpenter	Woman	Man	Both
7. Cook	Woman	Man	Both
8. Doctor	Woman	Man	Both
9. Engineer	Woman	Man	Both
10. Forest ranger	Woman	Man	Both
11. Lawyer	Woman	Man	Both
12. Librarian	Woman	Man	Both
13. Nurse	Woman	Man	Both
14. President of the United States	Woman	Man	Both
15. Secretary	Woman	Man	Both
16. Scientist	Woman	Man	Both
17. Store clerk	Woman	Man	Both
18. Sixth grade teacher	Woman	Man	Both
19. Telephone operator	Woman	Man	Both
20. Truck driver	Woman	Man	Both
21. Nursery school teacher	Woman	Man	Both

PART II.   For the school work and activities listed below, circle "Boy" if you think only a boy *should* do this; circle "Girl" if you think only a girl *should* do this; circle "Both" if you think both a boy or a girl *should* do this. Be sure to circle only *one* answer for each activity.

22.	Solve difficult math problems	Boy	Girl	Both
23.	Learn to program computers	Boy	Girl	Both
24.	Take advanced math classes in high school	Boy	Girl	Both
25.	Play games on a computer	Boy	Girl	Both
26.	Read poems	Boy	Girl	Both
27.	Read lots of stories and books	Boy	Girl	Both
28.	Write funny stories	Boy	Girl	Both
29.	Learn a foreign language	Boy	Girl	Both

PART III.   When there are class jobs to be done, who do you think should do them? Circle "Girl" if you think only a girl *should* do them; circle "Boy" if you think only a boy *should* do them; circle "Both" if you think both a boy or a girl *should* do them. Be sure to circle only *one* answer for each activity.

30.	Messenger	Girl	Boy	Both
31.	Class president	Girl	Boy	Both
32.	Eraser cleaner	Girl	Boy	Both
33.	Check out game equipment	Girl	Boy	Both
34.	Class secretary	Girl	Boy	Both
35.	Water the plants	Girl	Boy	Both

PART IV.   Here is a list of spare time activities. Circle who *should* do them: a man, a woman, or both.

36.	Play football	Man	Woman	Both
37.	Swim	Man	Woman	Both
38.	Play the violin	Man	Woman	Both
39.	Go to sports events (like baseball)	Man	Woman	Both
40.	Gymnastics	Man	Woman	Both
41.	Help at a hospital every week	Man	Woman	Both

Activity

Textbook Awareness

Objective

To make students aware of potential and sometimes very subtle bias in math textbooks; to stress an appreciation for the changing and evolving roles of women in professional areas

Grade Level

Grades 4–6

Time

One class period

Materials

Large piece of butcher paper, math books from the 1960s and 1970s (Many schools store these in an old book room, but if yours does not, you may find old copies in the library.)

Procedure

Outline a tally sheet on the butcher paper as shown below.

Women or Girls in Traditional Roles	*Women or Girls in the World of Math*
_____	_____
_____	_____
_____	_____
_____	_____
_____	_____
_____	_____

Divide the class into random groups of three students. Half the groups will use their own books (from math class); the other half will use the vintage texts. You may want to examine some copies in advance to make sure there is actual variation between the books that students can identify. Try to have the groups using vintage texts use the same one.

Discuss the traditional roles of women versus today's roles. Each group will review their books to spot traditional or nontraditional (math-related) roles of females in the story problems, pictures, or in supplementary materials. Let the students know that examples of traditional roles for women can include traditional occupations—teacher, nurse, housewife, etc.—or they can include passive roles such as the assistant or the person who needs help. Students can write the names of the women and the page numbers for reference on their tally sheets.

Have students tally the numbers of women found in traditional and nontraditional roles. Compare the findings for current and older texts in a class discussion. Let students read some examples and comment on the positive changes. Let students discuss how the use of girls' and women's names in problems can subtly give the message that girls can or cannot do math problems. Let students suggest ways of rewriting problems to eliminate sex bias.

Note: If, in your survey, you and your students discover that your current arithmetic text reveals stereotyping and bias in the way females are depicted in problems, have students rewrite problems to correct this bias, and talk to your textbook committee.

Variation

Repeat the process above with a tally for men in the world of math. Compare and discuss the number of instances of men in math-related careers to the number of women in math-related careers.

Resources

American Institutes for Research. 1980. *Programs to combat stereotyping in career choice*. Palo Alto, CA: American Institutes for Research.

> In this book, sex stereotyping in career choice is discussed, and nine programs designed to expand students' career awareness and break stereotypical patterns are described. Many of the programs are suitable or modifiable for upper elementary students.

Council on Interracial Books for Children. 1984. *10 quick ways to analyze children's books for racism and sexism*. NY: Council on Interracial Books for Children.

> This one-page flier gives parents and educators ten guidelines for evaluating children's books for racist and sexist content.

Fraser, S., ed. 1982. *SPACES: Solving problems of access to careers in engineering and science*. Berkeley: University of California, Lawrence Hall of Science.

> A collection of thirty-two classroom activities designed to stimulate students' thinking about math-related careers, develop problem-solving skills, and promote positive attitudes toward math. Activities are designed for students in grades 4–10.

Glennon, L., ed. 1976. *The yellow, blue, and red book.* Seattle, WA: Highline School District, Project Equality.

> This colorful three ring binder is a collection of brief activities developed by and for K–6 teachers to help expand student awareness of traditional sex-role stereotyping. Activities are grouped into three time periods (10–20 minutes, 20–40 minutes, and 40 plus minutes) and have designated subject matter emphasis.

Gordon, R. 1981. *PEER report—ties that bind: The price of pursuing the male mystique.* Washington, DC: NOW Legal Defense and Education Fund.

> An excellent summary of the negative effects of sex-role stereotyping on men and boys. This material provides discussion topics about stereotyping for students in grades 5–6.

Kaseberg, A.; Kreinberg, N.; and Downie, D. 1980. *Use EQUALS to promote the participation of women in mathematics*. Berkeley: University of California, Math/Science Network.

> This handbook assists educators in conducting teacher training to increase awareness of the problem of female math avoidance, enhance female interest and competence in mathematics, and provide information about opportunities for women in nontraditional careers. Ultimately, the purpose of the program is to help teachers promote positive math attitudes and bring about changes in the occupational patterns of women. The book includes activities that increase girls' confidence in their math abilities and relate the usefulness of mathematics to future career choices. An excellent sampling of strategy games, spatial activities, and logic problems is also included, as well as bibliographies on problem solving in mathematics and sex-fair counseling and instruction. Materials are suitable for grades 4–12.

Skolnick, J.; Langbort, C.; and Day, L. 1982. *How to encourage girls in math and science: Strategies for parents and educators*. Palo Alto, CA: Dale Seymour Publications.

This excellent resource examines the effect of sex-role socialization on girls' math/science skills and confidence. It explains how attitudes, parenting and teaching practices, stereotypical play activities and books, peer pressure, and career and family expectations cause girls to question their abilities in math and science, and thus hinder their development in these areas.

In addition to the summary of the socialization process, the book contains a variety of compensatory educational strategies and activities that may be used to encourage females in mathematics. These particularly focus on increasing math confidence, spatial visualization skills, and problem solving. Both parents and educators can benefit from this book.

Encouraging Cooperative Learning

Our educational system was originally designed to meet the needs of male students. Many classroom activities and procedures emphasize competitive techniques, which have traditionally been thought to be more appropriate for the male learning style. However, recent studies have shown that most children learn more readily in cooperative situations. Competitiveness can interfere with learning because it (1) makes students anxious and interferes with their concentration, (2) doesn't permit them to share talents and learn from each other as easily, and (3) distracts them from what they are doing—they concentrate on the reward or on winning instead of on what they are learning. Evidence also indicates that girls learn more readily in cooperative situations that emphasize working with others and discussing how to solve problems.

- Of 109 studies conducted between 1924 and 1980 comparing competitive and cooperative learning, 60 percent found that students achieve higher levels when they work cooperatively as opposed to competitively. The reverse was true in only 7 percent, and no differences were found in one-third of the studies. The more complex the learning task, the worse children fared in a competitive environment. The superiority of cooperation was consistent for all academic subjects across all age groups. (Johnson, cited in Kohn 1986)

- In classes where boys and girls collaborate, sex stereotyping is reduced, girls display more positive self-esteem, and are more apt to assume leadership roles. (Campbell 1984)

- Girls are more likely to continue studying math when their math classes are interactive and instructive. (Stallings 1985)

- Many teachers have been taught to use competitive instructional strategies in the classroom. These can work to the disadvantage of female students who may feel more comfortable and perform at higher levels in cooperative situations. (*Concerns* 1985; Peterson and Fennema 1985)

- Academic work is rarely organized to encourage student collaboration, particularly cross-sex collaboration. In one study, only about 11 percent of instructional time was devoted to mixed-sex groups. (Lockheed, cited in Grayson and Martin 1988)

- The math achievement of fourth grade girls in both high- and low-level problem solving was found to be positively related to participation in cooperative mathematics activities, and negatively related to participation in competitive math activities. For boys, these relationships were reversed. (Peterson and Fennema 1985)

Although in testing situations, students still have to work on their own, the experience of cooperative learning has been shown in most cases to be a valuable one for both boys and girls. The ideas on the following pages include suggestions for incorporating cooperative activities in your classroom.

Strategies

The following are guidelines for creating a school environment that supports cooperative behavior:

1. Assign tasks on some basis other than gender. Every participant has resources useful to the group's problem-solving efforts.

2. Encourage females and males to sit next to each other.

3. Don't allow any single group member to dominate the group, activity, or the most desirable spaces in the group (e.g., head of the table).

4. Focus on the process of the cooperative activity. Recognize and share with students the results of cooperative efforts.

5. Expand the meaning of cooperation to include the whole school, families, neighborhoods, and workplaces.

6. Encourage students to study together—be aware of any learning group that forms naturally.

7. When leaders choose team members, make sure equal numbers of girls and boys are chosen for each team.

8. Possible barriers that sometimes hinder cooperative learning have been suggested by these research findings:

   a. Males are more likely to control discussion through introducing topics, interrupting, and talking more than females.

   b. Females talk less, often assume supportive rather than leadership roles in conversation, and receive less attention for their ideas from the group.

   c. Both males and females may expect group members to follow sex stereotypic roles that can limit each individual's contributions (e.g., males will be leaders, females will be secretaries).

   Take action to overcome these potential barriers by:

   a. Adhering to strict rules of class behavior, and using the same rules for boys and girls

   b. Placing girls in leadership roles and monitoring their performance

   c. Making students aware of stereotyping, expected roles, and how we are all free to choose and modify our roles

9. Some authors have suggested that if girls appear to have low confidence in their abilities, they need to be placed in single-sex groups to build confidence before joining mixed-sex groups.

10. Researchers have suggested several classroom cooperative techniques that can be adapted for teaching math. Some of these techniques are listed on the following pages. They include Math Teams Tournaments, Student Teams—Achievement Divisions, Jigsaw, and small group teaching. Use the resource list to find suggestions for additional ways to structure cooperative math learning activities.

Activity

Math Teams Tournaments (and other cooperative activities)

Objective   To allow students to experience cooperative learning; to learn and practice math skills

Grade Level   Grades 2–6

Time   Blocks of 30–45 minutes per session (may also be used as a free-time activity and/or a continuing activity)

Materials   Teacher-generated study materials, quiz questions for tournaments

Procedure   This technique combines elements of both cooperation (the teams) and competition (the tournaments). The primary function of the teams in this activity is to prepare members to do well in the tournament. First explain the procedure to students, letting them know how the teams will function, and that all teams have an equal chance of winning—it depends on how well they prepare.

Assign students to *heterogeneous* groups of four or five members. Each group should include females and males who vary in ability level and ethnic origin.

Instruct student teams to prepare for math tournaments that will be held once each week (or whenever you decide). Give students worksheets covering the academic material to be included in the tournament; teammates study together and quiz each other to make sure that all are prepared.

For the tournament, assign students to groups of three with *homogeneous ability* at each tournament table; assign the top three students in past performance to one table, the next three to another table, etc. For example, if your class had 30 students, you might have six 5-person teams and ten 3-person tournament tables. To avoid stigmatizing lower ability children, use various names for these tables, rather than numbers or letters. Also, don't automatically put the top students at the first table you assign and the lowest ability students at the last table you assign; assign a mid-level table first, then a lower ability table, etc.

Students at each table compete in simple math quizzes that cover content material that you have presented in class and on the study worksheets. You might, for example, quiz on math facts or problems and allow students to answer "Jeopardy" style, with the first right answer at a table earning a point, and an incorrect answer losing a point. Students compete as members of their teams, and the scores they earn at their tournament tables are added to make a total team score. Because students are assigned to ability-homogeneous competitive groups (the tournament tables), each student has an equal chance of contributing a maximum score to his or her team.

Following the competition, recognize successful teams and first place scorers at each tournament table. For future tournaments, members can remain on the same teams; however, you may have to change assignments to tournament tables to maintain equality of performance among each group of three students.

Variations

Student Teams—Achievement Divisions. In this variation, the same four to five member heterogeneous teams are used for studying math materials. Instead of competing in groups of three, however, have all students take a written 15-minute quiz. Compare scores of students within six student "achievement divisions," i.e., the top six students on past math performance would be in Division Yellow, the next six students would be in Division Green, and so on. Using this method, you will be comparing students within fairly homogeneous ability groups. Decide ahead of time how to assign points for division winners and runners-up. These scores contribute to an overall score for each original team. For example, the top scoring student in each achievement division might earn 10 points, the second student, 8 points, and so forth. For subsequent quizzes, change division assignments to maintain equality in the divisions, but leave students in their original teams. Again, recognize winning teams and individual winners and runners-ups in each division.

Jigsaw. In this variation, a student from each team focuses on learning and/or reviewing one particular skill or aspect of a problem solution. Members from different teams who are assigned to a particular topic or aspect study it together; then, each student teaches the material to their original teammates. All students take a quiz, and their scores are used individually or as contributions to team scores.

Small Group Teaching. In small group teaching, learning takes place through cooperative group inquiry, discussion, and data gathering by students. Students select subtopics within a general area selected by you—for example, salaries in different professions or occupations. Students then organize into small groups of two to six members and subdivide the topic into individual tasks to be performed by group members. Each group presents its findings to the class as a whole.

Activity

Cookie Store

Objective

To have students work cooperatively in teams to demonstrate their ability to make change, figure costs, reduce or increase amounts of ingredients in recipes; to allow students to practice computational skills and to give them a cooperative money-raising activity

Grade Level

Grades 3–6 (modify for grades 1–2 by omitting more complex aspects)

Time

1–2 class periods

Materials

Cookie ingredients and food preparation equipment, expense sheets

Procedure

Let students prepare cookies, figure costs, and be responsible for retailing the merchandise.

Divide students into the following *mixed-gender groups:* planners, bakers, accountants, and sellers. The planners will be responsible for deciding what kinds of cookies to make and for figuring the amount of ingredients to be purchased. The bakers will be responsible for preparing the cookies, including reducing or increasing recipes. The accountants (grades 3–6 only) will figure the amounts to charge for the cookies and the "net profit or loss," and the sellers will collect the money from the cookie sales and make change.

Implementing an Effort-Persistence-Mastery Approach to Problem Solving

The goal of this approach is for students to become interested in learning for its own sake and that they strive to understand and improve their own performance rather than judge themselves against other students. In other words, for students to become intrinsically motivated. For a variety of reasons, including the ways teachers assist them, girls are more likely than boys to exhibit "learned helplessness"; i.e., the feeling that one is incapable of learning without assistance. As discussed in the section on attribution patterns, girls typically attribute failure to lack of ability, which often results in lack of persistence and lack of motivation to master mathematical concepts.

- Teachers typically encourage boys to figure out the answers to a problem; they are more likely to help girls by giving them the answers. (Sadker and Sadker 1985)

- Math teachers have been found to give different feedback to boys and girls for wrong answers—telling boys to try harder, while praising girls for "just trying"; this finding is consistent with the "learned helplessness" syndrome. (Fox 1981)

- Teachers contribute to "learned helplessness" in girls by praising them for intellectually irrelevant aspects of their responses, even when the responses themselves are incorrect (Russo 1985). Some examples: teachers often praise girls for the neatness of their work, even though it may be of poor academic quality; teachers often say to girls who failed, "That's okay, as long as you tried."

- In one study, it was found that girls were seldom criticized for the neatness and form of their work; about 90 percent of teacher criticism directed at girls focused on intellectual inadequacy of their work; only about 50 percent of criticism directed at boys focused on intellectual inadequacy. (Dweck, cited in Grayson and Martin 1988)

- In another experiment, boys were found to be more likely than girls to persist in difficult tasks in which they had failed. (Hughes et al. 1986)

- Several investigators have found that females are not as involved in problem solving activities as males. However, the belief that females are not as intrinsically motivated in mathematics as males has been refuted by results from many other studies. (Fox et al. 1980)

- Teachers' encouraging comments are very important, particularly for girls. Some researchers have theorized that boys appear to have an intrinsic mastery motivation, whereas girls' motivation is related to extrinsic need for approval. (Story and Sullivan 1986)

The suggestions on the following pages are designed to help you increase an effort-persistence-mastery approach to problem solving for all students, and especially for girls.

Implementing an Effort-Persistence-Mastery Approach to Problem Solving   *159*

Strategies

1. In your mathematics instruction, include information for males and females about the importance and usefulness of math. Help students develop a desire to learn math for its own value, not because they "have to."

2. De-emphasize right and wrong answers. Give special attention to procedure, so students can identify their errors and focus on specific areas needing improvement to attain mastery. Students will be reassured of their ability to master mathematical skills if this approach is used.

3. Provide opportunities to increase problem-solving abilities. To expose students to problem solving, try the "two-problem approach." Each day at the beginning of the class, put two problems on the board for students to solve. Be sure to give students an opportunity to discuss their solutions and the merits to each approach in solving these problems.

4. Use guessing activities to help students develop estimating skills. Be careful not to reinforce wild guesses, but utilize wrong answers as a way of learning. Using probing questions can guide students to restructure their thought processes to appropriate responses. Such interaction will increase response opportunities and enhance self-concept—another area in which girls more than boys need assistance.

5. Ask students to state their problem-solving strategies, not just the answers. Focus on the use of appropriate methods and strategies rather than the "one right answer." Give students credit for using appropriate strategies. Explain that there is often more than one method to find the solution to a problem.

6. Use recreational and intuitive forms of mathematics. A technique to involve the whole class and to challenge students of all abilities is the use of "head" or oral problems. Head problems take a small amount of time and can be used to introduce new concepts and strengthen previous ones (see the first activity in this section for examples of head problems). If students' mathematical errors are due to lack of attention to detail, head problems should help remediate these types of errors. When possible, recreational mathematics should incorporate activities to strengthen spatial relationship skills for all students, especially the females.

7. According to Grayson and Martin (1988), the average time allowed by teachers for students to respond in class is 2.6 seconds. In mathematics classes, less response time is typically given to girls than to boys. Research suggests that this may be due to lower teacher expectations for girls in mathematics classes. Make sure you give all students an equal opportunity to answer. Don't answer for them or let girls "off the hook" too easily.

8. Encourage all students to figure out the answers to problems; don't give them the answers or do their work for them. Let girls know that their understanding is very important; it's not okay if they "just try."

9. When some students don't seem to understand, search for *alternative* ways to explain. Make sure that, if the class has to move on, those students who have not mastered the concepts aren't left in confusion. While they study the next topic, give them additional help and practice outside regular math classes until they catch up with the rest of the group.

10. Help students focus on what they've learned and understood, not just on grades. As much as possible, stress a concept mastery approach to math.

11. Make sure that all students understand that math is *not* a subject in which "either you catch on immediately" or "you don't catch on at all." Let them know that some topics in math are difficult, and that it takes persistence and practice to master math skills. Also let them know that even the top mathematicians in the world are working on problems they can't solve or understand yet. If you encountered *and overcame* difficulties in math, this is good information to pass along to your class.

12. Don't be so "kind" to girls that you let them get by without mastering basic mathematics. Often, girls are "good students," and teachers believe they are showing concern for them when they don't insist on concept mastery. Be really kind and concerned by being firm and insisting that no student get by without mastering all basic arithmetic skills. Students who have trouble with the basic concepts of mathematics in grade school will be at a tremendous disadvantage in future math classes.

13. Don't give in to girls' tears. Respond to the frustration, not to the tears. Try to help girls work through problems in a calm supportive way, without displaying undue solicitude or embarrassment.

Activity

Head Problems

Objective	To motivate students to concentrate and follow problem steps
Grade Level	Grades 2–6
Time	Each problem takes one or two minutes (This is a great activity for "spare minutes" before the bell rings.)
Materials	Suggested "head problems" below
Procedure	Instruct students that you will be reading a problem aloud, and that they need to pay attention to every part of it and work it out in their heads. Read each problem aloud slowly. Make sure students are following along and are not using paper and pencils.
Variation	You can make up many of these problems on the spur of the moment, or ask students to make up head problems. You can also use these types of problems to help students learn measurements, history facts, and so forth, while practicing computations. For students in higher grades, use fractions and decimals in problems. These problems and student-generated ones can also be put on cards and kept for future use. In this form the game is known as Crazy Cards.

Head Problems for Grades 2–3

1. Start with the number of days in a week; subtract the number of quarters in a dollar; add the number of toes on one foot. (Answer = 8)

2. Start with the number of cents in a dime; subtract the number of feet in a yard; subtract the number of eyes in your head. (Answer = 5)

3. Start with the number on the clock that comes after 12; add the number of cents in a nickel; subtract the number of legs on a dog. (Answer = 2)

4. Start with the number of sides on a square; add the number of cents in a penny; subtract the number of days you go to school every week. (Answer = 0)

5. If you had 3 tens and five ones, what number would you have? (Answer = 35)

6. If you had 2 hundreds, seven tens, and four ones, what number you would you have? (Answer = 274)

The idea for "head problems" was suggested by Tom Lester, San Juan Unified School District, Sacramento, California.

7. If you had nine ones and seven tens, what number would you have? (Answer = 79)

8. If you had four hundreds, three ones, and eight tens, what number would you have? (Answer = 483)

Head Problems for Grades 4–6

1. Start with the number of inches in a foot; divide by the number of legs on a horse; multiply by the number of days in a week; subtract the number of dimes in a dollar. (Answer = 11)

2. Start with the number of minutes in an hour; divide by three; add the number of sides on a triangle; add the number of years in a century; subtract the number of weeks in a year. (Answer = 71)

3. Start with the number of pounds in a ton; divide by 200; multiply by the number of sides in a rectangle; add the number of pints in a quart; subtract the number of weeks in half a year. (Answer = 16)

4. Start with the number of ounces in a pound; subtract the number that comes right before 14; multiply by the number of feet in a yard; divide by the number of tires on a bicycle. (Answer = 4 1/2)

5. Take the number 5; multiply by 2; add 4; subtract 7. What is the number? (Answer = 7)

6. Think of the number of months in a year; divide that number by 4; add 2; multiply by 5. (Answer = 25)

7. Think of the number of leaves on a "lucky" clover; add the number of years it took you to be 10 years old; divide by the number of days in a week. (Answer = 2)

8. Take the number of states in the United States; divide by 5; add the number of fingers on both hands; subtract 2. (Answer = 18)

9. Take the number that comes after 19; double it; divide by 8, and add 4. (Answer = 9)

10. Take the whole number that is greater than 7 and less than 9; add the digit that tells you how many hundreds there are in 1,582; subtract the number of things in a dozen. (Answer = 1)

11. Take the number that makes the sentence __ x 5 = 20 true; add the number of cents in a dime; subtract the first odd number that comes after 6. (Answer = 7)

12. Take the number 20; add 300; add 7; add 4,000. What is the number? (Answer = 4,327)

Activity

Goal Setting and Watching Your Progress

Objective

To help students learn to set academic goals and to monitor their progress in math; to teach them that they earn their grades

Grade Level

Grades 4–6

Time

5–10 minutes per day

Materials

"Math Progress Sheet" (shown on the next page), folders, felt tip markers or crayons

Procedure

Learning to set and obtain goals may be one of the most important life skills you can teach to a child. Looking at grades or scores is one way to monitor academic performance. Provide a folder for each student. Staple a "Math Progress Sheet" into each folder, and have students fill in their name and the date. Ask students what grade they plan to obtain this grading period. Have them place that grade on the blank titled "First Grading Period Goal." Ask students to answer the question "What will I do to accomplish this goal?"

For the first half of the grading period, give students weekly grades and test and/or quiz grades. Have students draw a line graph of their grades with felt tip pens or crayons. Have students keep important papers and tests or quizzes in the folder.

At the midterm point, give students an update on their grade, or have students complete their own averages. Have students place that grade on the blank titled "Midterm Grade." Discuss with each student how she or he can obtain her or his goal, and praise students who are accomplishing their goals.

These folders can also be used at parent conferences. It is helpful to parents to see their child's work and goals. At the beginning of each grading period, start fresh with a new "Math Progress Sheet."

Variation

Before a test or quiz begins, ask students to place the score they are trying for on the upper left-hand corner of their papers. When the graded papers are returned, students will see how their goal and actual score matched. If there is a huge difference, have a conference with the student to see if you can help. Students will begin to realize that goal setting, studying, and paying attention in class will enhance learning.

Worksheet

Math Progress Sheet

Name _____ Date _____

First Grading Period Goal _____

Midterm Grade _____

What will I do to accomplish this goal? _____

Weekly Grades

	9/15	9/22	9/29	10/6	10/13	10/20	10/27	11/3	11/10
A									
A-									
B									
B-									
C									
C-									
D									
D-									
F									

Tests or Quizzes

	1	2	3	4	5	6	7	8
A								
A-								
B								
B-								
C								
C-								
D								
D-								
F								

Activity

Problem Problems

Objective   To give students practice in solving problems

Grade Level   Grades 5–6

Time   Variable, depending on number of problems given

Materials   One problem at a time or several (see examples below)

Procedure   Use the problems on the following page or problems from books on resource list to challenge students. Other excellent sources of problems are your text or the math books of other publishers. Check your local university library for these.

1. A bottle and a cork cost one dollar and a dime. The bottle cost one dollar more than the cork. How much did the cork cost? (Answer = .05)

2. Sue bought a turtle for $5. She sold it to Ben for $7. Ben did not want to keep the turtle, so Sue bought the turtle from Ben for $8. By now the turtle was bigger so she sold it to Ann for $9. How much money did Sue make? (Answer = $3)

3. If there are four people in a room and everybody shakes hands once with each of the other people, how many hand shakes will there be? (Answer =6)

4. Ms. McDonald had a farm with ducks and donkeys. There were 24 animals in all. Counting webbed feet and hoofed feet, if these animals had a total of 62 feet, how many ducks and how many donkeys could Ms. McDonald have? (Answer = 17 ducks and 7 donkeys)

5. Susan had 5 cages and 10 guinea pigs. She wanted to put her guinea pigs into the cages so that each cage contained a different number of animals. How could this be done?

   Answer: Cage 1: 0
   Cage 2: 1
   Cage 3: 2
   Cage 4: 3
   Cage 5: 4
   $4 + 3 + 2 + 1 + 0 = 10$

Resources

Barrata-Lorton, M. 1976. *Mathematics their way.* Menlo Park, CA: Addison-Wesley.

> *Mathematics their way* features an activity-centered manipulative math program in which children learn through using all five senses. The curriculum consists of many activities that focus on counting, classifying, graphing, estimating, and measuring. All activities are appropriate for students in grades K–2. The program teaches math concepts so that all children understand them, regardless of their abilities.

Barrata-Lorton, R. 1977. *Mathematics, a way of thinking.* Menlo Park, CA: Addison-Wesley.

> This book describes an activity-centered manipulative mathematics program for grades 3–6. The program is similar to the one presented in *Mathematics their way.*

Casserly, P. L. 1983. "Encouraging young women to persist and achieve in mathematics." *Children Today* 12, no. 1: 8–12.

> Casserly's article summarizes the factors that enhance or inhibit young females in their study of mathematics. Many strategies for elementary classrooms, counselors, and parents are discussed.

DeRoche, E. F., and Bogenschild, E. G. 1977. *400 group games and activities for teaching math.* West Nyack, NY: Parker.

> This book includes 400 classroom-tested math strategies and activities suitable for use in cooperative mathematics learning for elementary and junior high students. The activities are enjoyable and focus on the practical implications of learning math in a cooperative classroom atmosphere.

Fennell, F., and Williams, D. 1986. *Ideas from the* Arithmetic Teacher: *Grades 4–6 intermediate school.* Reston, VA: National Council of Teachers of Mathematics.

> This is a collection of classroom-tested activities from the journal *Arithmetic Teacher*. Perforated pages allow easy duplication of activities, which can be used for supplementing, extending, or reinforcing daily mathematics lessons in numeration, whole number computation, rational numbers, geometry, measurement, and problem solving. Objectives, grade levels, directions, and answers are given for each activity.

Holden, L. 1987. "Math: Even middle graders can learn with manipulatives." *Learning 87* 16, no. 3: 52–55.

> Learning fractions can be frustrating; this article includes many ways that manipulatives can be used to help upper elementary students understand fractions and learn geometric concepts.

Jensen, R., and Spector, D. 1984. *Teaching mathematics to young children.* Englewood Cliffs, NJ: Prentice-Hall.

> This teacher and parent resource suggests many activities that can be used to explore math concepts with young children. The suggested vehicles for math instruction—including manipulative and creative movements, art activities, and games—allow children to tackle new ideas while developing problem-solving skills. Chapters and activities are organized from concrete to abstract, so that young children may master mathematical concepts and gain confidence in learning.

Overholt, J. L. 1978. *Dr. Jim's elementary math prescriptions.* Santa Monica, CA: Goodyear.

> *Dr. Jim's elementary math prescriptions* is a resource for educators in grades K–8 who are searching for effective methods of teaching mathematics. Each mathematical concept is presented with alternative methods to accommodate students with varied learning styles, abilities, and interests. Selected activities provide enjoyable mastery practice, so that students will develop mathematical competence and appreciation.

Stenmark, J. K.; Thompson, V.; and Cossey, R. 1986. *Family math.* Berkeley: University of California, Lawrence Hall of Science.

> If mathematics promotion is a goal of your teaching, *Family math* activities will help you introduce parents and children to ideas that improve their math skills and help them gain an appreciation for math. Hands-on mathematical experiences provide families an opportunity to develop problem-solving skills by using the following strategies: looking for patterns, drawing pictures, working backwards, working cooperatively with a partner, and eliminating possibilities. Spatial relationships (geometry), estimation, data interpretation (probability and statistics), and mathematical reasoning are the mathematical concepts learned from *Family math.* Materials suitable for ages 5–18.

Encouraging Independent Thinking, Intellectual Risk Taking, and Creative Problem Solving

Creativity in mathematics has been defined as "the ability to produce original or unusual applicable methods of problem solution; . . . combine ideas, things, techniques and approaches in a new way; [or] . . . analyze a given problem in many ways, observe patterns, see likenesses and differences, and on the basis of what has worked in similar situations, decide on a method of attack in an unfamiliar situation" (Aiken 1973). Creativity sometimes requires that a person take the risk of standing out or of being different from others. Creativity in mathematics requires that students feel independent and confident enough about themselves and their abilities to stand up for their own ideas.

Girls in our culture have traditionally been more dependent than boys; they are reluctant to take intellectual risks, and they are very concerned about "looking stupid" or being embarrassed. Girls are also less likely to ask questions or to experiment with different methods to solve a problem. Since creativity is extremely important in advanced mathematics, it is necessary that we help girls experience their own creativity at an early age.

- Girls are less willing than boys to be wrong and less likely to experiment with different ways to solve a problem. (Kreinberg and Stenmark 1984)

- Independence training facilitates math achievement, and the early socialization of girls typically includes less independence training than that of boys. (Stage et al. 1985)

- Humiliation by a teacher was a primary reason given by a sample of high school girls who decided not to continue in math. (Sherman 1982)

- In the classroom, boys are more often encouraged to be creative and to persevere, whereas girls are rewarded for being docile and conforming. (Harway and Astin 1977)

- Girls' creativity is decreased in competitive situations. (Amabile, cited in Kohn 1986)

- Although teachers' practice of not being too hard on girls in mathematics may be well intentioned, it results in girls not becoming independent problem solvers who do well in high-level cognitive tasks. (Fennema 1983)

- Teachers reported that they rewarded the creative activity of boys three times as often as that of girls. (Torrance, cited in Grayson and Martin 1988)

- In most traditional schools, boys become leaders in problem solving, whereas girls become followers. (Minuchin, cited in Fox et al. 1980)

- Children enter school with girls tending to be more dependent on others and boys tending to be more self-reliant; through classroom practices, schools reinforce and further develop these dependent/independent behaviors in each sex—particularly in mathematics. (Fennema 1983)

- Several authors have speculated that the greater attention, both positive and negative, that boys receive in the classroom makes them more exploratory, more autonomous, more independent, and more oriented toward achievement in mathematics. (Stockard et al. 1980)

The strategies and activities that follow are designed to help students experience creative problem solving and to help them learn to take risks without worrying about experiencing embarrassment or humiliation.

Strategies

1. To develop creativity in students, a teacher must provide opportunities for personal initiative and responsibility. Although girls seem to be more creative in group situations, they also need to learn to become independent.

2. Structure some activities in your math classes where guessing is encouraged, and there is no penalty for wrong answers. Make it "okay" and never embarrassing for girls to give a wrong answer. Use wrong answers to help the student think through the process and come up with the correct answer, without humiliation. Build a "safe" environment in your class in which everyone can take intellectual risks without fear of embarrassment.

3. Some teachers model making a mistake; they have the class help them think it through, and then find the correct solution to a problem. This can have a positive effect on your female students by letting them know that everyone makes mistakes, and that we can often learn more from our errors than from our successes.

4. Stress alternative approaches to problem solving and understanding mathematical concepts. Researchers have found that girls often use verbal strategies to solve problems, when spatial strategies such as diagramming, organizing the information into charts, or working backwards would be more helpful. Teach different approaches and strategies for problem solving, and encourage girls to use them—especially visual/spatial strategies and manipulatives—when they are unable to solve a problem using traditional methods.

5. Some authors have explained girls' relatively stronger math achievement in the elementary grades and lower achievement in high school by the following: girls are taught and encouraged to obey rather than challenge rules. This may prepare them very well for elementary math, but hinder them tremendously at advanced levels. Be sure you encourage both girls and boys to examine and challenge the "rules" of math; through this process comes understanding rather than dependence on rules.

6. The same types of guessing, checking, and estimating activities that are discussed in the section on building math confidence can also be appropriately used to increase students' intellectual risk-taking behaviors. Many teachers have noted that girls seem afraid to guess, and that they dislike estimating activities. Present these as "fun" activities, and be sure to specifically involve girls in these processes.

7. Some teachers have speculated that because of societal pressures, girls are expected to be well-behaved "perfect ladies," while boys are expected to "have more fun." This pressure can lead to girls' fear of taking risks and making mistakes. Try to ensure that girls enjoy and have fun with math activities.

8. Use "brainstorming" to encourage intellectual risk-taking. Encourage students to develop problems for the class to solve and to present their solutions for the class.

9. Insist that girls become independent, self-reliant problem solvers. Be sure to reward their creative efforts and intellectual risk-taking behaviors.

10. The books *How to encourage girls in math and science* by Skolnick et al., *Math for smarty pants* by Burns, *Solving problems kids care about* by Souviney, *Math for girls and other problem solvers* by Downie et al., *Mastermind* by Clements and Hawkes, and *Use EQUALS to promote the participation of women in mathematics* by Kaseberg et al. contain excellent examples of problems you can use to supplement texts. Additional information on these books can be found in the resource list following this subsection.

11. The ability to "break set," or see beyond the expected, is one element of independent thinking and creativity in mathematics. The book *Math for girls and other problem solvers* by Downie et al. contains a number of activities that help children learn to break set. These include a series of "mystery stories" in which students have to look beyond the obvious to solve a problem and toothpick puzzles, which involve breaking visual set to find new patterns.

12. Many of the books mentioned above also contain good examples of logic problems. These give students practice using deductive reasoning to solve problems. *Family math* by Stenmark et al. (see resource list) contains an excellent game called Rainbow Logic, which also teaches reasoning in a very interesting way.

13. Look at the "Mindwinders" columns in *Instructor* magazine for interesting logic problems for your students.

Activity

Strategy Games—Card Tricks and Mind Reader

Objectives	To allow students to discover and practice problem-solving strategies
Grade Level	Grades 1–6, depending on problem complexity
Time	10–20 minutes
Materials	A deck of playing cards with the jokers removed, scratch paper, and pencils

Procedure

Card Tricks. Demonstrate this trick first. Then play the game over, and let students guess the card. To play the game, you leave the room, and a student selects a card from the deck and places it face down on her or his desk. For grades 1–3, use only one suit with numbered cards 2–10; for grades 4–6 use the entire deck. Explain to the older group that the Ace is the lowest card and that the order of the cards above the 10 is Jack, Queen, King. Also define the suits. The object of the game is for the person doing the "card trick" to guess the turned down card in as few guesses as possible. For the younger group using only one suit with cards 2–10, the person doing the "trick" can ask the following types of questions: "Is it higher than 5?", "Is it lower than 3?", "Is it the 10?", and so forth. (Note that if the card was the 5 and the question was "Is it higher than 5," the answer would be "no." If the card in question was the 3 and the question was "Is it lower than 3," the answer would be "no." For the advanced trick with the entire deck of 52 cards, questions can *also* include the color and suit; i.e., "Is it a red card?" and "Is it a spade?" The answers can only be "yes" or "no."

Allow students to use paper and pencils to keep track of their guesses. For each game, designate one student as monitor, and let that student count and write the guesses on the board. Seeing how questioning has progressed should help students develop successful strategies. This game also teaches the value of wrong guesses, for example, when students ask, "Is it a red card?", and the answer is "no," they have now narrowed it down to a black card. In this game a "no" answer can sometimes be more valuable than a "yes."

After you have demonstrated the trick, select a card, place it face down, and let the class guess it. Help students focus on developing a strategy to narrow down their choices and rule out incorrect answers.

Mind Reader. This game is similar to Card Tricks. One person thinks of a number and writes it down. For grades 1–3, use single-digit or possibly two-digit numbers. For higher grade level students use 2–4 digit numbers. The person thinking of the number tells the group (the "mind readers") the number of digits. Students may ask questions similar to those in Card Tricks; i.e., "Is it higher than 5?", "Is it lower than 300?" Again, the object is to "read the person's mind" and guess the number in as few tries as possible.

Variation
The book *Use EQUALS to promote the participation of women in mathematics* by Kaseberg et al. (see resource list following this section) contains several strategy games, including Bagels, another interesting guessing game that involves strategy development and may be played on several levels of complexity.

Activity

Find the Missing Numbers

Objective	To allow students to practice adding and multiplying in a problem-solving format
Grade Level	Grades 2–3 (worksheet A), 4–6 (worksheets B and C)
Time	10–30 minutes
Materials	Worksheets A, B, and C on the following pages
Procedure	Duplicate one work sheet for each student. Ask students to complete the sheets using the hints provided on them. One set of answers is shown on the following page. Alternative solutions are possible for several of the squares. Suggest a trial-and-error approach to finding a set of missing numbers. To help students get started, you might supply one or more numbers for each square.

Answers to problems on worksheet A:

Row 1	2 4	2 3	5 4	2 5
	3 5	5 4	3 2	4 3
Row 2	1 5	3 7	5 1	7 9
	3 7	5 9	9 3	1 5
Row 3	1 2	5 5	4 1	4 6
	3 1	4 2	3 4	6 3
Row 4	1 4	4 5	7 2	9 2
	2 9	3 6	4 3	4 3

Answers to problems on worksheet B:

Row 1	1 3 6	3 1 2	5 1 5
	4 2 7	4 3 5	2 6 3
	2 5 2	7 6 3	5 4 7
Row 2	1 6 7	2 1 9	1 7 2
	5 3 4	4 5 7	4 5 3
	9 2 5	3 5 6	5 3 8

Answers to problems on worksheet C:

Row 1									
2	4	3		5	3	2		3	5	2
3	4	1		2	5	5		4	6	1
5	5	2		6	3	1		2	3	4

Row 2									
1	2	15		3	1	20		30	4	1
10	1	3		25	2	1		1	40	3
4	5	1		1	5	4		2	1	35

Variation

Encourage students to construct similar problems with or without hints for others to solve.

Worksheet

Find the Missing Numbers (A)

Directions: Fill in the squares so that the sum of each row and column of numbers is equal to the outside numbers. Hints about the missing numbers are given for each row.

Hint: The missing numbers for these squares can only be 2s, 3s, 4s, or 5s.

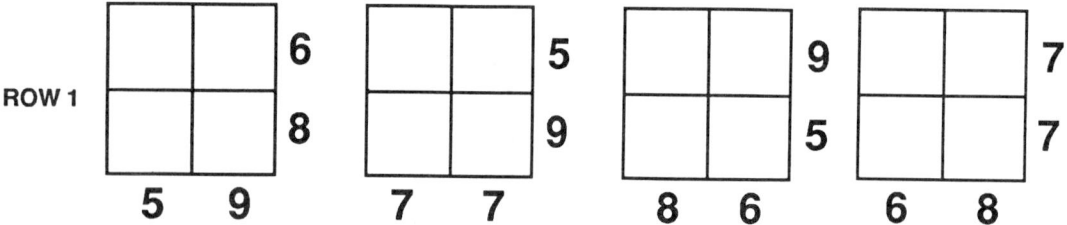

Hint: The missing numbers in these squares are odd numbers. Use numbers 1–9.

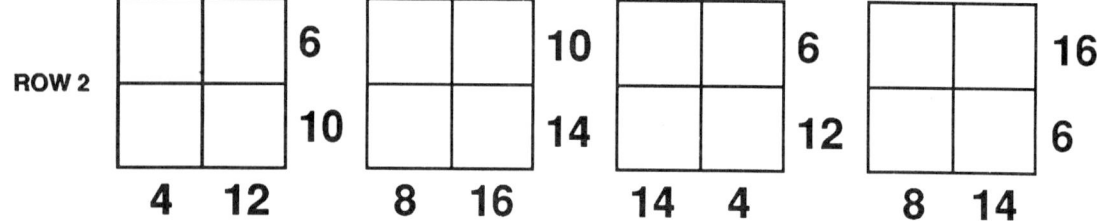

178   *Part 3   The Learning Environment*

Hint: In each of these squares, two missing numbers are the same. Use numbers 1–9.

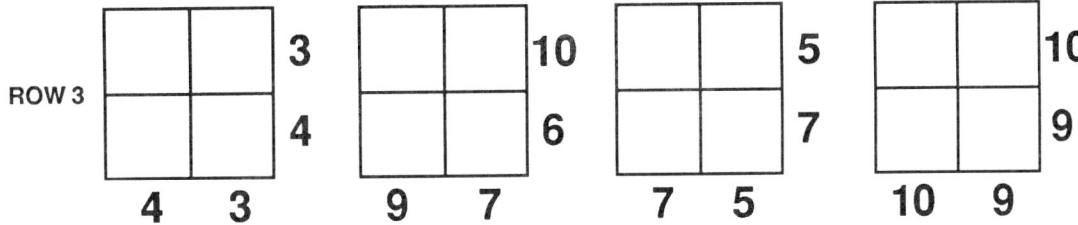

ROW 3

Hint: Half of the missing numbers below are odd, but you never add two odd numbers together. Use numbers 1–9.

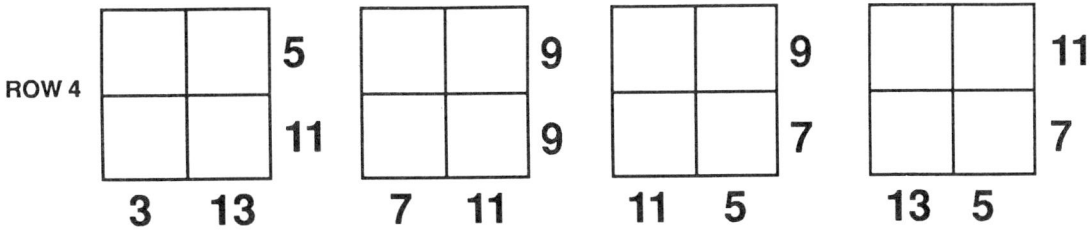

ROW 4

Worksheet

Find the Missing Numbers—Adding (B)

Directions: Fill in the squares so that the sum of each row and column of numbers is equal to the outside numbers. Hints about the missing numbers are given for each row.

Hint: In each square, one of the missing numbers is used twice; all answers use numbers 1–7.

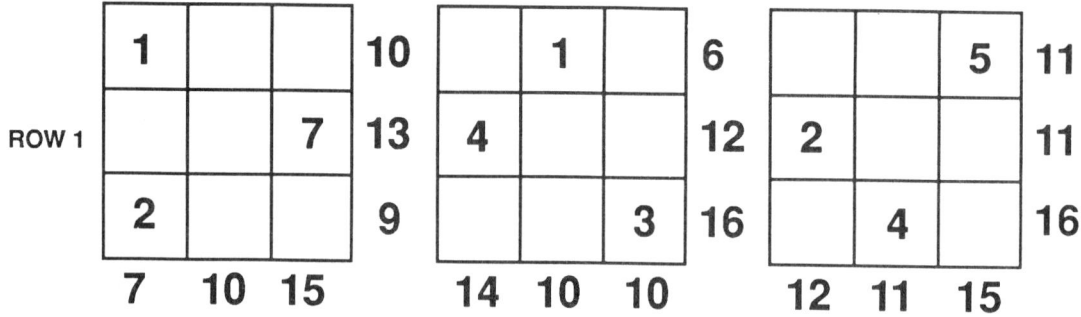

Hint: When you complete these squares, each row will contain two odd numbers; select from numbers 1–9.

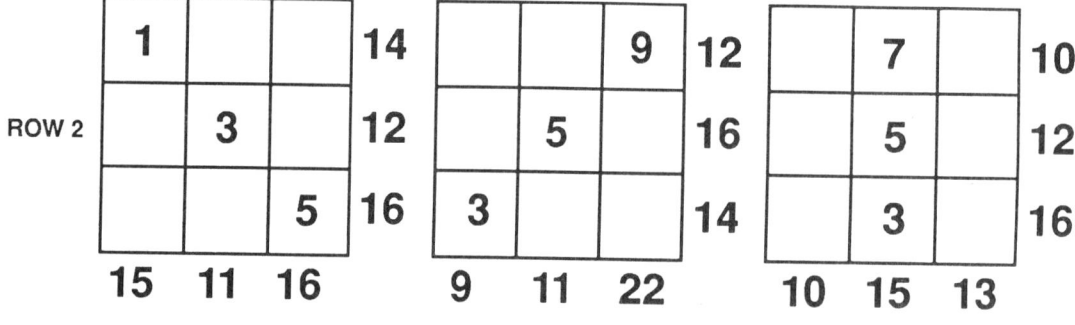

180   *Part 3   The Learning Environment*

Worksheet

Find the Missing Numbers—Multiplying (C)

Directions: Fill in the squares so that the outside numbers are the *product* of all the numbers in the row or column inside the grid.

Hint: In all three of these squares, two of the missing numbers are 2s. The other four numbers can be selected from numbers 1, 3, 4, 5, and 6. When it's finished, every square will contain two 3s.

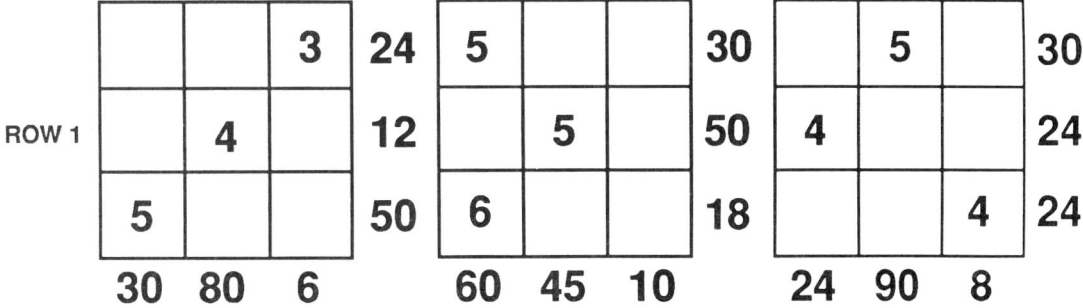

Hint: Each row in each square contains a 1. The other three numbers are 5s or multiples or 5, and they can go as high as 40.

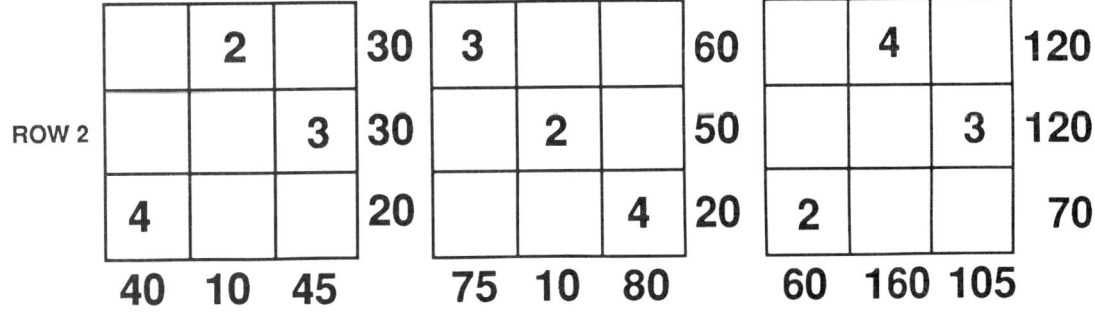

Activity

Parts Add Up

Objective	To allow students to experiment with mathematical strategies to discover how the numbers 1–9 can be arranged in groups of three to add up to the same total
Grade Level	Grades 3–6
Time	20 minutes
Materials	One copy of the triangle with "square" sides (see following page) for each student
Procedure	Hand out triangle with squares. Instruct students to arrange the numbers 1–9 in the squares so that each side has the same total. Each number can only be used once. Students can work individually or in groups. Encourage students to share the process they used to figure their answers.

Answer:

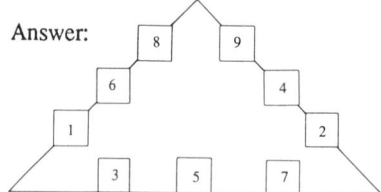

Possible strategies: Trial and error with the nine numbers. Add the numbers and divide by 3; that tells us that each side must equal 15. Now use trial and error to find sets of numbers that will add to 15.

Handout

Parts Add Up Triangle

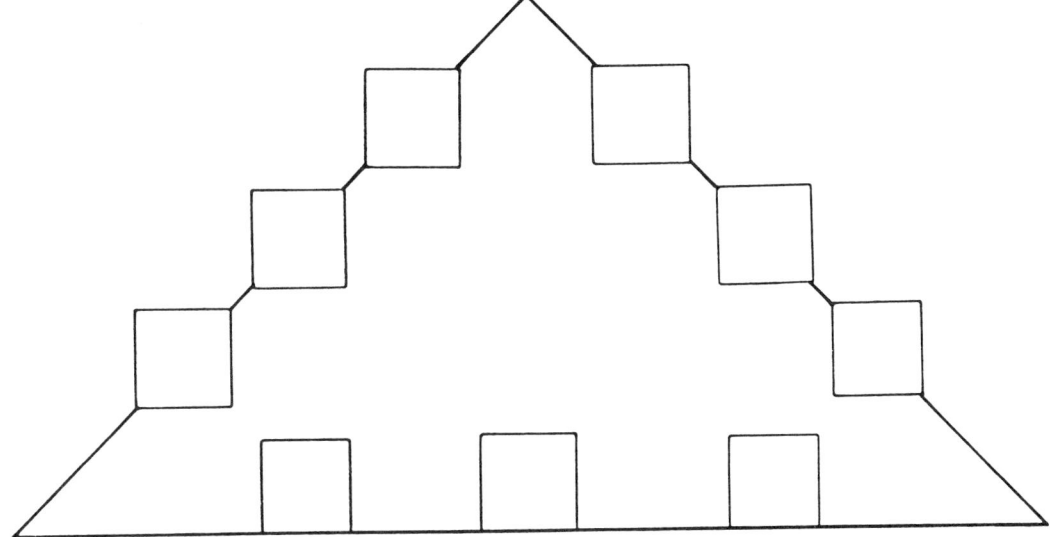

Activity

Creative Numbers

Objective	To help students become aware of the many ways numbers are used in everyday life
Grade Level	Grades 3–6
Time	20–30 minutes
Materials	One copy of "Creative Numbers" worksheet on the following page for each student (To use as an activity for the entire class, put on the board or use with an overhead projector.)
Procedure	Ask students to use their imaginations to determine what the letters stand for on the worksheet. Let the entire class brainstorm together or divide the class into small groups. Students can also be asked to make their own creative number statements or to take a copy of the sheet home and ask their parents for help. Note that students might come up with some answers that are different from the ones we show below. If the answer fits, it's correct!

Answers:

1. days of the week
2. months of the year
3. legs on a centipede
4. quarters in a dollar
5. cents in a dime
6. legs on a millipede
7. legs on a spider
8. things in a dozen
9. states in America
10. sides on a triangle
11. letters in the alphabet
12. planets in the solar system
13. stripes on the American flag
14. hours in a day
15. eyes on a face
16. legs on a human being
17. pins on a bowling lane
18. blind mice

Worksheet

Creative Numbers

Directions: We use numbers in many ways. Each statement below contains numbers and the initials of words that will make it correct. Fill in the correct words.

Example: 2 A on a B   Answer: 2 arms on a body

1. 7 D of the W   _____

2. 12 M of the Y   _____

3. 100 L on a C   _____

4. 4 Q in a D   _____

5. 10 C in a D   _____

6. 1000 L on a M   _____

7. 8 L on a S   _____

8. 12 T in a D   _____

9. 50 S in A   _____

10. 3 S on a T   _____

11. 26 L in the A   _____

12. 9 P in the S S   _____

13. 13 S on the A F   _____

14. 24 H in a D   _____

15. 2 E on a F   _____

16. 2 L on a H B   _____

17. 10 P on a B L   _____

18. 3 B M   _____

Activity

The Pizza Store

Objective	To allow students to discover the rules for determining combinations
Grade Level	Grades 4–6
Time	10 minutes per problem
Materials	"The Pizza Store" handout on the following page
Procedure	Divide students into small groups. Give each group a copy of the handout. Define "combinations," and then let students, working in groups, build their own pizzas using the problems outlined on the following page.

Answers:

a. 8 different pizzas

b. 4

c. 24

d. 48

e. The rule is to multiply the number of choices in each category together, i.e., A = 2 x 2 x 2 = 8, B = 1 x 2 x 2 = 4, C = 2 x 2 x 2 x 3 = 24, D = 2 x 2 x 2 x 2 x 3 = 48.

Handout

The Pizza Store

The pizza store started out with the following menu:

Crusts	Cheeses	Meats
Thick	Mozzarella	Sausage
Thin	Provolone	Pepperoni

a. If customers choose one kind of crust, one kind of cheese, and one kind of meat on every pizza, how many different kinds of pizza (or combinations) could the pizza store make? *Note:* cheese has to be in the middle with meat on top.

b. The pizza store owner decided that it was too much trouble to make two crusts, so she would make only thin crust. How many combinations could the store make if customers still choose only one kind of cheese in the middle and one kind of meat on top?

c. After a while, the pizza store owner decided that she would go back to making two kinds of crust, and she would also add three kinds of vegetables—onions, green peppers, and mushrooms. Now, if customers could choose one kind of crust, one kind of cheese, one kind of meat, and one kind of vegetable on each pizza, how many different kinds of pizza could she make?

d. What if the owner decided to have 2 sizes of pizza—small and large? Now how many different combinations would be possible?

e. Can you figure out a rule for solving these problems?

Activity

A Penny for Wages

Objective	To practice problem solving, learn the "time" value of money (Problem A), and practice finding different solutions (Problem B)
Grade Level	Grades 5–6
Time	10–15 minutes per problem (These could serve as problems of the day.)
Materials	"A Penny for Wages" handout on following page and calculators (if available) (For additional problems, use books from the resource list or math texts.)
Procedure	*Problem A:* Let students read the problems on the following page. Before they attempt a solution, let students estimate and vote on their answers. Then allow about 10–15 minutes for solution with calculators.
	Problem B: Working in pairs, have students read the word problem and find two different methods to solve it.
	Answers:
	Problem A: At one cent doubling every hour, Susan would earn $328 in two days.
	Problem B: Solution (1) Fill the 5-quart can. Then pour from it to fill the 3-quart can; you will have two quarts left in the 5-quart can. Pour the remaining 2-quarts into the mower. Repeat the process, filling the 5-quart, transferring to the 3-quart, and putting the remaining two quarts in the mower. Two quarts plus two quarts equals one gallon. Solution (2) Fill the 3-quart can; pour it into the 5-quart can. Fill the 3-quart can again, and use that gasoline to fill the 5-quart can. The 5-quart can will take two more quarts with one left over. Pour the remaining quart into the mower. Then transfer the gas from the 5-quart can back to the 3-quart can. Pour the full 3-quart can into the mower. One quart plus three quarts equals one gallon.
Variation	Using their calculators, students can develop and find solutions for many problems about savings and interest that involve compounding. For example, if you put $10,000 in a savings account that earns 8% interest per year, and you left all the money in the account, how long would it take before your money had doubled (to $20,000)?

Handout

A Penny for Wages

Problem A

Susan was offered a job that would last only two days. Her boss said she could choose either to earn $100 a day or to start at one penny per hour and then her salary would double every hour. If Susan were going to work two days for eight hours per day, which would be the best deal—$100 per day or starting at one cent per hour with her wage doubling every hour? Which would you choose?

Problem B

Monica and Pat were making money by mowing lawns during the summer. The mower used a mixture of gasoline and oil. They had two gas cans—one held three quarts and the other held five quarts. The cans were not marked in any way. The lawn mower required exactly one gallon of gasoline to be mixed with one quart of oil. Using the 3-quart and 5-quart containers, how did the girls measure exactly one gallon of gasoline? Write two different solutions.

Resources

Burns, M. 1982. *Math for smarty pants.* Boston: Little, Brown.

This book contains a wide range of accessible activities presented in an entertaining format. It would be particularly useful for expanding upper elementary students' perceptions of mathematics.

Burns, M. 1975. *The I hate mathematics book.* Boston: Little, Brown.

For those students who "seemingly" hate mathematics, this book provides many relevant activities to boost confidence and aspirations. Positive attitudes toward mathematics develop as students experiment with and investigate the uses of mathematics in solving everyday problems. Suitable only for upper elementary students.

Clements, Z. J., and Hawkes, R. R. 1985. *Mastermind: Exercises in critical thinking, grades 4–6.* Palo Alto, CA: Scott, Foresman.

Mastermind activities are designed to foster the mathematical learning strategies of talented and gifted students in elementary school. The activities have been grouped according to Bloom's Taxonomy of Educational Objectives. The hierarchical organization of the book's contents enables educators to use the activities as supplemental material to develop new critical thinking skills and to reinforce previously learned mathematical skills.

Cook, M. *Math materials.* Catalog. Balboa Island, California.

These materials include tiling sets, task cards, and books designed to add variety to math. The materials emphasize problem solving and focus on active student involvement. Also included are several books on cooperative learning. A catalog is available from Marcy Cook, P.O. Box 5840. Balboa Island, CA 92662.

Coombs, B.; Harcort, L.; Travis, J.; and Wannamaker, N. 1987. *Explorations.* Menlo Park, CA: Addison-Wesley.

A mathematics program based on learning through the use of manipulatives and interacting with the environment. Programs for K–2 and 3–6 integrate math and language arts.

Downie, D.; Slesnick, T.; and Stenmark, J. K. 1981. *Math for girls and other problem solvers.* Berkeley: University of California, Math/Science Network.

The activities in this book encourage independent thinking and creativity in mathematics. Students and teachers are encouraged to think about problem solving in versatile ways and forms. Although this book was originally designed for females, the activities are appropriate and interesting to both boys and girls, ages 7–14.

Fennell, F., and Williams, D. 1986. *Ideas from the arithmetic teacher: Grades 4–6 intermediate school.* Reston, VA: National Council of Teachers of Mathematics.

This is a collection of classroom-tested activities from the journal *Arithmetic Teacher.* Perforated pages allow easy duplication of activities, which can be used for supplementing, extending, or reinforcing daily mathematics lessons in numeration, whole number computation, rational numbers, geometry, measurement, and problem solving. Objectives, grade levels, directions, and answers are given for each activity.

Fraser, S., ed. 1982. *SPACES: Solving problems of access to careers in engineering and science.* Berkeley: University of California, Lawrence Hall of Science.

A collection of thirty-two classroom activities designed to stimulate students' thinking about math-related careers, develop problem-solving skills, and promote positive attitudes toward math. Activities are designed for students in grades 4–10.

Holden, L. 1987. "Math: Even middle graders can learn with manipulatives." *Learning 87* 16, no. 3: 52–55.

Learning fractions can be frustrating; this article includes many ways that manipulatives can be used to help upper elementary students understand fractions and learn geometric concepts.

Jensen, R., and Spector, D. 1984. *Teaching mathematics to young children.* Englewood Cliffs, NJ: Prentice-Hall.

This teacher and parent resource suggests many activities that can be used to explore math concepts with young children. The suggested vehicles for math instruction—including manipulative and creative movements, art activities, and games—allow children to tackle new ideas while developing problem-solving skills. Chapters and activities are organized from concrete to abstract, so that young children may master mathematical concepts and gain confidence in learning.

Kaseberg, A.; Kreinberg, N.; and Downie, D. 1980. *Use EQUALS to promote the participation of women in mathematics.* Berkeley: University of California, Math/Science Network.

This handbook assists educators in conducting teacher training to increase awareness of the problem of female math avoidance, enhance female interest and competence in mathematics, and provide information about opportunities for women in nontraditional careers. Ultimately, the purpose of the program is to help teachers promote positive math attitudes and bring about changes in the occupational patterns of women. The book includes activities that increase girls' confidence in their math abilities and relate the usefulness of mathematics to future career choices. An excellent sampling of strategy games, spatial activities, and logic problems is also included, as well as bibliographies on problem solving in mathematics and sex-fair counseling and instruction. Materials are suitable for grades 4–12.

Overholt, J. L. 1978. *Dr. Jim's elementary math prescriptions.* Santa Monica, CA: Goodyear.

Dr. Jim's elementary math prescriptions is a resource for educators in grades K–8 who are searching for effective methods of teaching mathematics. Each mathematical concept is presented with alternative methods to accommodate students with varied learning styles, abilities, and interests. Selected activities provide enjoyable mastery practice, so that students will develop mathematical competence and appreciation.

Silvey, L., and Smart, J. R., eds. 1982. *Mathematics for the middle grades (5–9).* Reston, VA: National Council of Teachers of Mathematics.

This book was developed to aid teachers in promoting the mathematical development of students in grades 5–9. The three sections of the book cover critical issues in mathematics education, unique learning activities, and strategies for teaching problem solving.

Skolnick, J.; Langbort, C.; and Day, L. 1982. *How to encourage girls in math and science: Strategies for parents and educators.* Palo Alto, CA: Dale Seymour Publications.

> This excellent resource examines the effect of sex-role socialization on girls' math/science skills and confidence. It explains how attitudes, parenting and teaching practices, stereotypical play activities and books, peer pressure, and career and family expectations cause girls to question their abilities in math and science, and thus hinder their development in these areas.
>
> In addition to the summary of the socialization process, the book contains a variety of compensatory educational strategies and activities that may be used to encourage females in mathematics. These particularly focus on increasing math confidence, spatial visualization skills, and problem solving. Both parents and educators can benefit from this book.

Souviney, R. J. 1981. *Solving problems kids care about.* Palo Alto, CA: Scott, Foresman.

> *Solving problems kids care about* is divided into two parts. The first section includes notes and strategies for teaching mathematical problem solving. The second section contains thirty real-world problems that encourage divergent and logical thinking. Many of the problems have a range of acceptable solutions and multiple solution strategies, so students have the opportunity to be creative and independent thinkers. Activities are designed for elementary through junior high school students; teachers will enjoy them also.

Stenmark, J. K.; Thompson, V.; and Cossey, R. 1986. *Family math.* Berkeley: University of California, Lawrence Hall of Science.

> If mathematics promotion is a goal of your teaching, *Family math* activities will help you introduce parents and children to ideas that improve their skills and help them gain an appreciation for math. Hands-on mathematical experiences provide families an opportunity to develop problem-solving skills by using the following strategies: looking for patterns, drawing pictures, working backwards, working cooperatively with a partner, and eliminating possibilities. Spatial relationships (geometry), estimation, data interpretation (probability and statistics), and mathematical reasoning are the mathematical concepts learned from *Family math.* Materials suitable for ages 5–18.

Wiebe, A., and Hillen, J., eds. 1986. *AIMS Newsletter.* Fresno, CA: AIMS Education Foundation.

> This newsletter describes the AIMS (Activities that Integrate Math and Science) program. The program includes a wide range of science and math activities and focuses on the integration of learning experiences, problem-solving activities, and cooperative learning.

Improving Spatial Visualization Skills

Spatial visualization involves the visual imagery of objects as they are rotated, reflected, and/or translated; in other words, the mental manipulation of objects and their properties. Although the results are not entirely consistent, many investigators have found that junior high and high school boys perform better than girls on visual-spatial measures. The magnitude of this difference varies according to many factors, including students' personality characteristics, previous experience with spatial activities, and the particular test given. Evidence on how spatial visualization skills affect mathematics achievement is mixed, but many investigators believe that increased training and early experience with spatial visualization can help girls, especially in geometry.

- Fennema and Ayer (1984) concluded that if spatial visualization skills do affect the learning of mathematics, the influence must be extremely subtle; however, other researchers have found strong relationships between spatial skills and mathematics achievement test scores. (Stage et al. 1985)

- Relatively large sex differences have been found on a test measuring the rotation of objects in three-dimensional space. (Sanders et al., cited in Chipman and Wilson 1985)

- Exposure to different toys and recreational activities has been linked to sex differences in spatial skills (Stage et al. 1985). Math readiness is linked to preschool games and toys (blocks, construction sets, cars, tools, etc.) that lead to an understanding of shapes and how things work. Males tend to have had more experience with these types of toys than females have. (Grayson and Martin 1988)

- There is much evidence that spatial visualization skills can be trained (Stage et al. 1985); yet this type of training is not usually made a part of the mathematics curriculum. (Fox 1981)

- Spatial visualization skills that require students to select three-dimensional shapes that would be formed by folding two-dimensional shapes have been found to be strongly related to mathematics achievement. (Chipman and Wilson 1985)

The activities on the following pages are designed to give all students, but especially female students, some of the practice and experience they need to sharpen their spatial visualization skills. It is important to begin training girls at an early age, while they are still performing on a par with boys.

Strategies

The components of spatial visualization skills include memory of shapes, figure completion, mental rotation of objects, finding hidden shapes, and the creation of three-dimensional objects from two-dimensional patterns. The first six strategies below, which will allow students to practice each of these skills, were suggested in Blackwell's *Spatial encounters* workbook (see resource list following this section).

1. To help with memory of shapes, let students try to draw familiar objects from memory or look at a picture and then try to remember the exact details of its shapes.

2. To help with figure completion skills, encourage students to work jigsaw puzzles or visualize shapes in clouds.

3. To help with mental rotation of objects, students can make stick figures of geometric shapes with Tinkertoys or toothpicks and observe how figures look when turned around.

4. To help with spatial memory and rotation, let students practice visualizing what is behind them. Have them make a sketch and then check. Also, studying a single picture and then trying to draw it as if it had been rotated 90 to 180 degrees is a good exercise.

5. To help with the ability to locate hidden shapes, students can try to distinguish geometric shapes in simple pictures or even the clouds. They can also look for common shapes in everyday objects and in stylized artwork.

6. To help with the ability to go from two-dimensional to three-dimensional space, students can assemble models of toys.

7. Use a copy of Blackwell's *Spatial encounters* to allow students to practice the various component skills that constitute spatial visualization.

8. Especially in the early elementary years, make sure that the girls in your classes play with blocks, legos, and other manipulative building toys as much as the boys do. To encourage girls to play with these toys, model that behavior.

9. In high school geometry, students organize and structure their spatial experiences. Few elementary math curricula stress informal geometry. Research findings indicate that both elementary and pre-geometry high school students exhibit many misconceptions about shapes. The article by Burger (see resource list) provides several suggestions for activities that will help elementary students become familiar with informal geometry concepts and prepare them for geometry.

Activity

Closed Curve Art

Objective	To draw and understand the definition of a closed curve; to become aware of symetrical patterns
Grade Level	Grades 3–6
Time	15 minutes
Materials	One sheet of 8 1/2" x 11" white or colored construction paper, scissors for each student
Procedure	Have each student fold his or her paper in half from the 11" side to the 11" side:

Keeping the paper folded and the fold at the bottom,

each student should write his or her name (first or last) in very large cursive letters.

Then have the student make a cut from the left *through the fold* and continue cutting around the outside of the letters, being careful *not to cut through the fold* until he or she reaches the last letter. Then, cut through the fold.

These can be decorated with markers or crayons.

Variation	Students can also draw other types of designs on their folded papers. For example:

or   or

individual letters

Cut around the outside of the designs and unfold to create symmetrical multisided straight line designs.

Activity

Dice Patterns

Objective

To practice spatial visualization skills by observing and constructing a three-dimensional object

Grade Level

Grades 3–6

Time

Variable—up to 25 minutes

Materials

Heavy paper or cardstock, transparent tape or glue, scissors, several dice, copies of the dice patterns on the following pages

Procedure

This activity may be done individually or in two- or three-person groups. Give each group of students one die and copies of the dice patterns. Ask students to observe the dice carefully. Do all of the dice have the same pattern of dots on their sides? (that is, the one opposite the six, the two opposite the five, and three opposite the four). Is there a pattern in the patterns? (opposite sides always add to seven).

After students have observed the dice, ask them to decide which of the patterns on the following pages could be cut out and folded to make a dice. Have students verify their decisions by cutting each of the patterns out of heavy paper or cardstock. The patterns should be cut on the solid outside lines, creased on all dotted lines, and shaped with the edges glued or taped at the tabs. ("E" is the only pattern that can be shaped into a cube.)

Variation

Let advanced students design their own patterns. How many different patterns can be folded into a dice with the correct number of dots on each side.

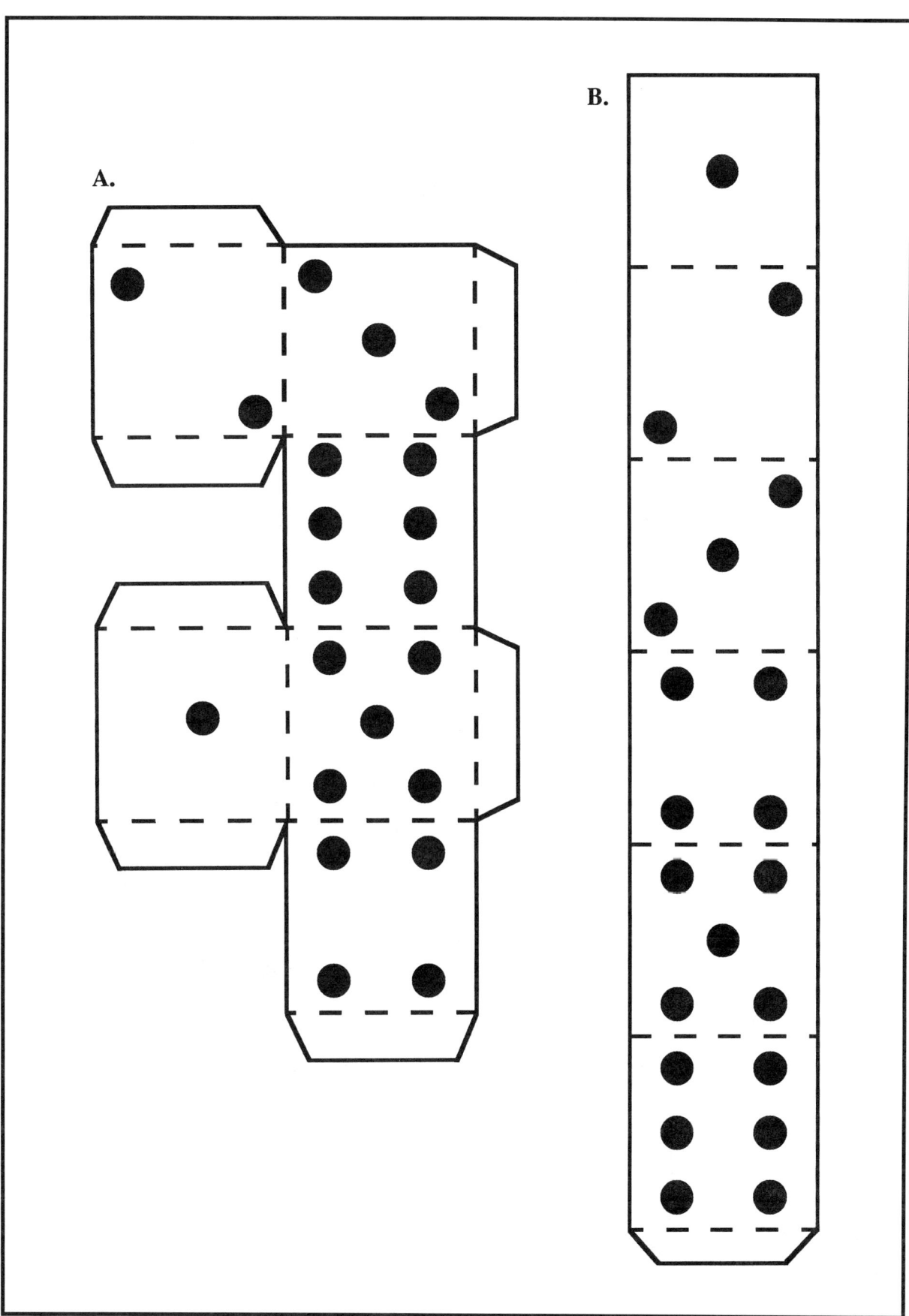

A.

B.

C.

D.

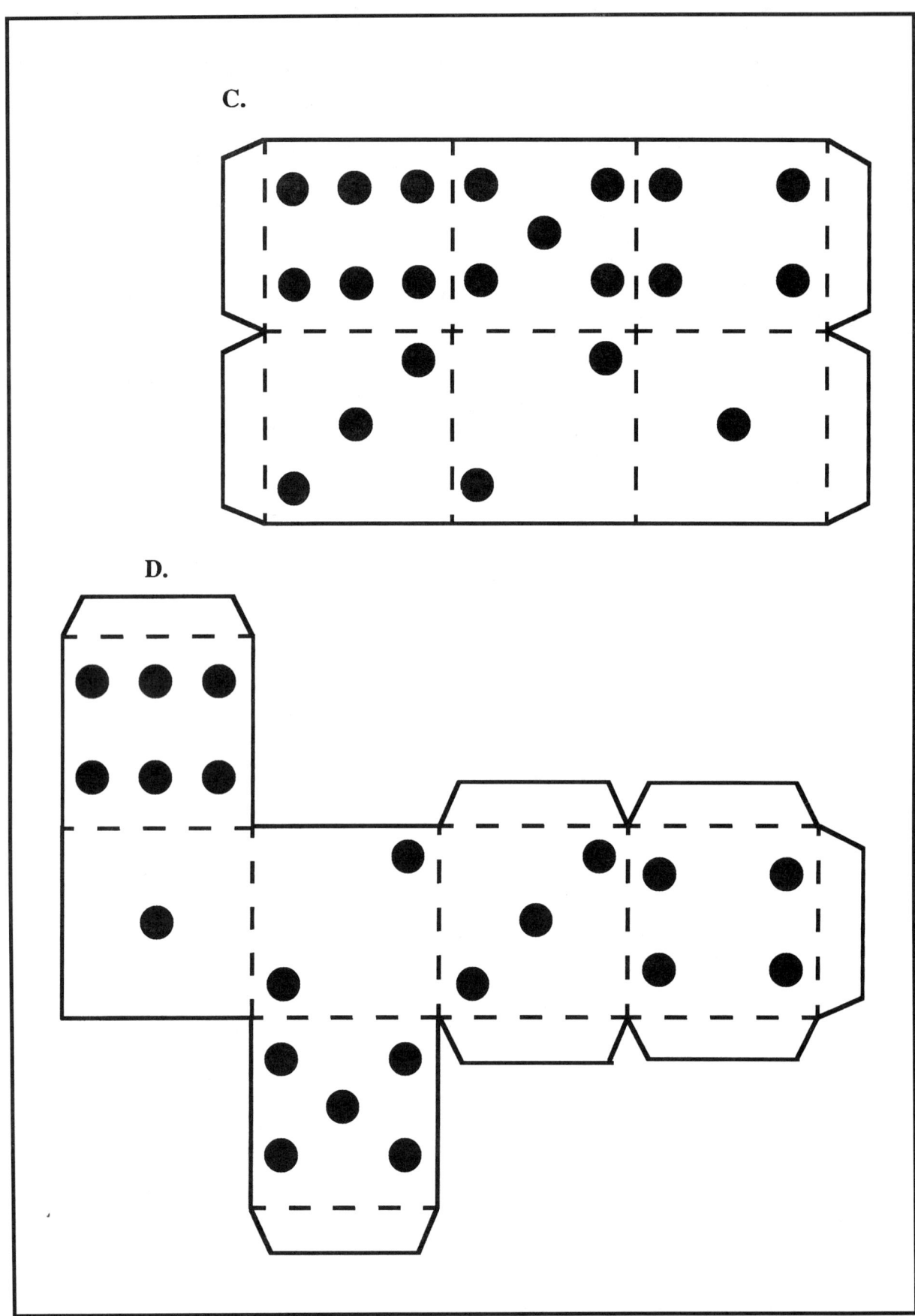

E.

Activity

Count the Shapes

Objective	To help students discover the characteristics of single shapes
Grade Level	Grades 4–6 (See variation for grades 2–3.)
Time	20 minutes
Materials	Shapes drawing on the following page
Procedure	The figure on the following page contains a number of common shapes. Make a copy for each student. Define and discuss the properties of various shapes. Then ask students to count the numbers of triangles, quadrilaterals, and pentagons.
Variations	1. Let students draw their own puzzles and demonstrate the properties of these or other shapes.
	2. *Variation for grades 2–3.* Let students color all of the triangles one color, all of the quadrilaterals a second color, and all of the pentagons a third color.

Resources

Blackwell, P. J. 1982. *Spatial encounters: Exercises in spatial awareness.* Newton, MA: Women's Educational Equity Act Publishing Center/EDC.
  A fun book for students of all ages who need to develop their spatial visualization skills. *Spatial encounters* contains a variety of games and exercises that involve figure completion, memory of shapes, and rotation. The book is particularly helpful for girls whose skills have been hampered by lack of practice.

Burger, W. F. 1985. "Geometry." *Arithmetic Teacher* 32, no. 6: 52–56.
  This article describes common misconceptions students hold about shapes and suggest a number of activities to introduce them to informal geometry and strengthen their skills in analyzing and classifying shapes by their properties. The article contains an excellent bibliography of materials and activities useful in teaching informal geometry.

Crowley, M. L. 1987. "The van Hiele model for the development of geometric thought." In *Learning and teaching geometry, K–12,* edited by M. M. Lindquist, 8–12. Reston, VA: National Council of Teachers of Mathematics.
  Manipulatives are used to help children understand geometric concepts in the many activities suggested in this article.

Downie, D.; Slesnick, T.; and Stenmark, J. K. 1981. *Math for girls and other problem solvers.* Berkeley: University of California, Math/Science Network.
  The activities in this book encourage independent thinking and creativity in mathematics. Students and teachers are encouraged to think about problem solving in versatile ways and forms. Although this book was originally designed for females, the activities are appropriate and interesting to both boys and girls, ages 7–14.

Fraser, S., ed. 1982. *SPACES: Solving problems of access to careers in engineering and science.* Berkeley: University of California, Lawrence Hall of Science.
  A collection of thirty-two classroom activities designed to stimulate students' thinking about math-related careers, develop problem-solving skills, and promote positive attitudes toward math. Activities are designed for students in grades 4–10.

Hill, J. M., ed. 1987. *Geometry for grades K–6: Readings from* Arithmetic Teacher. Reston, VA: National Council of Teachers of Mathematics.
  This collection of articles and activities from the *Arithmetic Teacher* features a hands-on approach to teaching the foundation principles of geometry to students in grades K–6.

Kaseberg, A.; Kreinberg, N.; and Downie, D. 1980. *Use EQUALS to promote the participation of women in mathematics.* Berkeley: University of California, Math/Science Network.
  This handbook assists educators in conducting teacher training to increase awareness of the problem of female math avoidance, enhance female interest and competence in mathematics and provide information about opportunities for women in nontraditional careers. Ultimately, the purpose of the program is to help teachers promote positive math attitudes and bring about changes in the occupational patterns of women. The book includes activities that increase girls' confidence in their math abilities and relate the usefulness of mathematics to future career choices. An excellent

sampling of strategy games, spatial activities, and logic problems is also included, as well as bibliographies on problem solving in mathematics and sex-fair counseling and instruction. Materials are suitable for grades 4–12.

Silvey, L., and Smart, J. R., eds. 1982. *Mathematics for the middle grades (5–9)*. Reston, VA: National Council of Teachers of Mathematics.

This book was developed to aid teachers in promoting the mathematical development of students in grades 5–9. The three sections of the book cover critical issues in mathematics education, unique learning activities, and strategies for teaching problem solving.

Skolnick, J.; Langbort, C.; and Day, L. 1982. *How to encourage girls in math and science: Strategies for parents and educators*. Palo Alto, CA: Dale Seymour Publications.

This excellent resource examines the effect of sex-role socialization on girls' math/science skills and confidence. It explains how attitudes, parenting and teaching practices, stereotypical play activities and books, peer pressure, and career and family expectations cause girls to question their abilities in math and science, and thus hinder their development in these areas.

In addition to the summary of the socialization process, the book contains a variety of compensatory educational strategies and activities that may be used to encourage females in mathematics. These particularly focus on increasing math confidence, spatial visualization skills, and problem solving. Both parents and educators can benefit from this book.

Stenmark, J. K.; Thompson, V.; and Cossey, R. 1986. *Family math*. Berkeley: University of California, Lawrence Hall of Science.

If mathematics promotion is a goal of your teaching, *Family math* activities will help you introduce parents and children to ideas that improve their math skills and help them gain an appreciation for math. Hands-on mathematical experiences provide families an opportunity to develop problem-solving skills by using the following strategies: looking for patterns, drawing pictures, working backwards, working cooperatively with a partner, and eliminating possibilities. Spatial relationships (geometry), estimation, data interpretation (probability and statistics), and mathematical reasoning are the mathematical concepts learned from *Family math*. Materials suitable for ages 5–18.

Improving Test-Taking Skills

Although girls generally do as well as boys on tests that cover materials learned in the classroom, their performance tends to be poorer than that of boys on national math tests, such as the Scholastic Aptitude Test (SAT) or the American College Testing Program (ACT). Many educators prefer to deemphasize the use of standardized tests; however, in today's educational climate, these tests can have a great impact on students' futures. In some school systems, elementary students are tracked based on standardized test scores such as the California Achievement Test (CAT). Test scores also impact students' placement in advanced or enriched programs, entrance to particular colleges, and awards of scholarships or other forms of financial aid for postsecondary school. Thus, even as we work toward better ways of assessing student achievement, we still need to ensure that students know how to do as well as they can on standardized tests. There are some indications from research that girls may be attending to different cues and taking in too much distracting information during testing. Others hypothesize that boys may do better because they are more competitive and may be less anxious about performing under pressure.

- In results of recent national testing of high school students, boys out-scored girls on the mathematics portion of the SAT by 47 points, which represents an average score that was 10.4 percent higher for boys than for girls. On the ACT mathematics subtest, boys' average scores were 15.5 percent higher than the average scores for girls. ("Gaps persist between sexes" 1987)

- Even when the SAT is given to seventh graders, there is a large difference (35 points) favoring boys; this difference cannot be explained by differences in course-taking at school. (Fox 1981)

- Some researchers have suggested that sex differences in math scores may be due to content bias in the tests; others suggest that girls use a poorer test-taking strategy than boys do. (Dwyer 1987)

- Testing often represents a pressure-filled situation for students. Increased evaluative pressure has been found to enhance the performance of boys, but to impair the performance of girls. (Dweck and Gilliard, cited in Russo 1985)

- Slightly speeded tests which feature easy-to-hard item order favor male students more so than any other item arrangement. In one study, girls performed best when items were arranged in many small clusters of easy-to-hard order. (Plake et al. 1982)

- Female students generally experience significantly more test anxiety, as measured by the Mathematics Anxiety Rating Scale, than do boys. (Plake et al. 1982)

The following suggestions should help all students improve their test-taking skills and should provide practice that will build a foundation to assist girls in scoring better on national tests.

Strategies

1. *Vary* the format of tests; give objective math tests.

   a. Design a test supplying all the answers. Students need to work the problems to eliminate the incorrect answers.

   b. Design a matching test. This works well, especially when teaching properties in math.

   c. Design a multiple choice test. Include choices of "not here" or "none of the above" if you want to prevent guessing.

   d. Design an oral math test. This can be dictated onto a tape. In the primary grades, this can be useful for basic facts.

2. Let parents know from the beginning of the year how their child can get the most out of your math class (see the parental involvement section). Send home information about your grading procedures and how or when you schedule tests.

3. Provide frequent, mixed, cumulative reviews. Have an ongoing maintenance program for previously learned topics.

4. Cover key topics before test-taking time.

5. There is usually more than one right way to do a math problem, so accept a students' alternative procedure if it and the solution are correct.

6. Encourage students to take time to work problems out with paper and pencil.

7. Don't always push for speed in solving math problems. Quickness doesn't ensure correctness, and most good, interesting math problems require considerable time for solution.

8. Determine how many problems you *really* need on a test to test what you want to know.

9. Teach students the skills involved in test taking—examining problems for key words, using a process of elimination, etc.

10. To help students learn to solve word problems, (a) have them rewrite problems and (b) let them write their own problems for other students to solve.

11. Let students make up their own tests along with the answers. Constructing test items helps students think through the steps involved in solving problems.

Activity

Practice Test

Objective	To familiarize students with the testing format and practice skills required for a specific test
Grade Level	Grades 1–6
Time	30–45 minutes for preparation, 30 minutes (variable) for activity
Materials	Practice test
Procedure	This activity requires extra time from teachers, but definitely helps students study for a standardized test. Using the test format and type of problem(s) to be tested, select two sample problems (similar, not actual test questions) from each area. Design a practice test following the exact format of the standardized test. After the practice test, it is crucial that the answers (for skill understanding) and test format (for test-taking skills) be thoroughly reviewed with students. With very young children it is especially important to go over the format and instructions many times so that their anxiety will be reduced.

Activity

Math Bowl

Objective   To review skills to be tested

Grade Level   Grades 1–6

Time   20 minutes for preparation, plus 20–30 minutes for the activity

Materials   3" x 5" cards

Procedure   Depending on the type of test—cumulative, standardized, criterion-referenced, teacher-prepared, chapter test, and so forth—select sample problems. Put each problem on a 3" x 5" card with the solution. Divide the class into teams with an equal number of students with similar abilities. Problems can be given individually (as in a spelling bee) until all members of a team have been eliminated or cooperatively whereby the team decides on the solution. This activity is easily adaptable to all levels of math, requires little work of teachers, and encourages students to review on school time, not their time!

Activity

"Test-Wiseness" Training

Objective

To teach students how to score well on standardized tests

Grade Level

Grades 4–6

Time

20 minutes per session

Materials

Copies of "Hints to Make You 'Test-Wise'" on the following page

Procedure

Test-wiseness has been defined as the capacity to use the characteristics and formats of the test and/or test-taking situation to obtain high scores—independent of knowledge about the subject matter of the test. In other words, a test-wise person can increase his or her test scores by use of a number of strategies and attention to cues that have little to do with the content of the test. The results of many studies have shown that test-wiseness can be learned, that training is not effective until students reach grade 4, and that 9–14 hours of training are best for maximum results.

Four skills have been found to produce significant improvement in the test performance of elementary school children who have had little or no previous test-taking experience with standardized achievement tests. These skills are following directions, using time wisely, guessing strategies, and answer changing. In addition, multiple choice questions allow the test-wise student to eliminate distractors by attending to cues unrelated to test content.

On the following page, some hints for students are listed that are partially based on material from a 1986 article by Benson, Urman, and Hocevar (see resource list).

Before using these hints, make sure that they are appropriate for the standardized tests you give; modify them if necessary. For example, the second hint under "Following Directions" is only appropriate if questions are allowed throughout the testing period. Some standardized tests require that no student questions be answered once the test begins. The hint, "Do not leave any question blank" (under "Using Time Wisely"), as well as all of the guessing hints are only good strategies if there is no penalty or correction for guessing.

Go through the hints with students, helping them with examples from a previous math test. You may want to have younger students practice only one strategy at a time.

For a copy of a test-wiseness curriculum manual covering the four strategies, write to Jeri Benson, Assistant Professor, Department of Measurement, Statistics, and Evaluation, University of Maryland, College Park, Maryland, 20742.

Handout

Hints to Make You "Test-Wise"

Following Directions

Read and listen carefully to directions before starting the test.

If you don't understand something about the test or how to fill it out, ask your teacher during the test.

Fill in only one answer for each question on the answer sheet.

Fill in the answer space completely.

Using Time Wisely

Know how much time you have to take the test.

First, answer all the questions you know.

Go back to the harder questions later.

Do not leave any question blank.

Guessing

Make your best guess instead of leaving a question blank.

When guessing, reread the questions carefully.

When guessing, read every option carefully.

Guess only after going through these steps.

Changing Answers

Change an answer when you know you marked it wrong.

Change an answer when another seems better.

Activity

Math Words

Objective

To write steps and solutions to problems for test review

Grade Level

Grades 3–6

Time

Variable—20–30 minutes

Materials

Problems from textbook

Procedure

Give each student a different set of problems from a specific page in the text. Instruct the student to fold a piece of paper in half (lengthwise). On the left side, the student works each problem; on the right side, she or he explains in words, how to solve the problem.

Use this activity with individual students as a check for understanding or in a partner situation to help students who are having difficulty. This technique is especially effective in solving word problems, but is also helpful in breaking down understanding of the multiple steps required in math problems.

Resources

Benson, J.; Urman, H.; and Hocevar, D. 1986. "Effects of test-wiseness training and ethnicity on achievement of third- and fifth-grade students." *Measurement and Evaluation in Counseling and Development* 18, no. 4: 154–62.

This article summarizes findings from a study on math achievement after test-wiseness training. The study includes information from a test-wiseness training manual and lists a number of strategies for students.

Grassick, P. 1983. *Making the grade: How to score higher on all scholastic tests.* New York: Arco.

Includes a number of tips and strategies for test-taking.

Kroen, W. 1987. "Make your students test savvy." *Instructor* 96, no. 7: 66.

A brief article that provides several suggestions to help students become test-wise.

Margenau, J., and Sentlowitz, M. 1977. *How to study mathematics.* Reston, VA: National Council of Teachers of Mathematics.

A study guide that gives math students suggestions for developing study and test-taking skills. The guide is illustrated with cartoons to appeal to the junior high or high school student. Elementary teachers, especially those of grades 4–6, may find some tips to pass along to their students.

INDEX